EXPANDING ISSUES IN DESALINATION

Edited by **Robert Y. Ning**

Expanding Issues in Desalination
Edited by Robert Y. Ning

Published by InTech
Janeza Trdine 9, 51000 Rijeka, Croatia

Publishing Process Manager Sandra Bakicc

Technical Editor Teodora Smiljanic

Cover Designer InTech Design Team

Additional hard copies can be obtained from orders@intechopen.com

Expanding Issues in Desalination, Edited by Robert Y. Ning
p. cm.
ISBN 978-953-307-624-9

For EU product safety concerns:
IN TECH d.o.o., Prolaz Marije Krucifikse Kozulić 3, 51000 Rijeka, Croatia,
info@intechopen.com or visit our website at intechopen.com.

Meet the editor

Dr. Robert Y. Ning is Vice President, Science and Business Development of King Lee Technologies based in San Diego, CA, USA. This company was founded in 1977, and is specialized in chemistry and chemicals used in reverse osmosis (RO) membrane plants. Since 1995, Bob is responsible for science and product development, as well as establishment of a network of process design engineers and service providers around the world to design and service RO plant processes.
Bob has a B.S. degree in Chemistry from Rochester Instutitute of Technology, Ph.D. in Organic Chemistry from University of Illinois, MBA from Fairleigh Dickinson University, and was a post-doctoral fellow at California Institute of Technology. He has 25 years of process chemistry experience in the pharmaceutical and biotech industries, prior to his current specialization in water and wastewater treatments.

Contents

Preface

For this book, the term desalination is used in the broadest sense of the removal of dissolved, suspended, visible and invisible impurities in seawater, brackish water and wastewater, to make them drinkable, or pure enough for industrial applications like in the processes for the production of steam, power, pharmaceuticals and microelectronics, or simply for discharge back into the environment.

From space, the earth is a blue planet, covered three quarters of surface by oceans and one quarter by land. Seawater contains 35 to 40 grams per liter of dissolved salts, too salty to drink or use. At any moment in time, about 0.04% of the water is in the process of being recycled through the atmosphere by the heat from the sun. Distilled water from clouds, condense as rain or snow, is falling mostly in the ocean. Only 2.5% of the Earth's total water is fresh water, but two-thirds of this water is locked away from man's use in ice caps and glaciers. The estimated one third (0.8%) of fresh water sink into the deep aquifers, or flows into lakes and rivers then to the sea.

Water, being a most powerful solvent, leaches soluble salts and erodes rocks from the ground, and becomes increasingly saline and turbid before reaching the sea, or equilibrates as brackish water in the lakes and aquifers. Being the essential medium for microbial and aquatic life, it is filled with living and non-living biotic organic matter. Much of the complex suspended organic and inorganic matter in water are colloidal particles, smaller than 1 micron (1 nanometer) in size, hence invisible to naked eyes.

Human needs drive the technology of desalination of water. The need is exacerbated by the mismatch of population densities with the natural distribution of available water on land. Like other basic technologies, the mountain of published information on this subject can hardly be captured in one book. Major desalination processes involve thermal evaporation known as Multi-Stage Flash Distillation (MSF), Multiple-Effect Distillation (MED) and Vapor Compression Distillation (VC). Since 1970s, technologies using semi-permeable membranes grew rapidly to compete with distillation to minimize energy consumption. Reverse Osmosis (RO) now is becoming favored over MSF even in oil-rich countries of the Persian/Arabian Gulf. Ion-exchange resins are used to soften or deionize water. Electrodialysis (ED) commercially introduced in the early 1960s also uses semi-permeable membranes. It finds greater application in industrial rather than municipal applications. Minor desalination processes include freezing, membrane distillation and solar humidification.

For students and workers in the field of desalination, this book provides a summary of key concepts and keywords with which detailed information may be gathered through internet search engines. Papers and reviews collected in this volume covers the spectrum of topics in desalination of water, too broad to delve into in depth. The literature citations in these papers serve to fill in gaps in the coverage of this book. Contributions to the knowledge-base of desalination is expected to continue to grow exponentially in the coming years.

Robert Y. Ning, Ph.D.
Vice President, Science
King Lee Technologies
USA

Part 1

Reverse Osmosis

1

A Large Review of the Pre Treatment

Kader Gaid
Technical Department, Veolia water
France

1. Introduction

Desalination using seawater reverse osmosis (SWRO) technology is an important option available to water-scarce coastal regions. Worldwide sea water desalination is a very effective and economical way of producing potable water for drinking and industries. Reverse osmosis plants to convert sea water to potable drinking water and for other usages have been prevalent throughout the world for more than 4 decades. Design and operation of seawater reverse osmosis plants strongly depend on the raw seawater quality to be treated. The performance of desalination reverse osmosis (RO) systems relies upon the production of high quality pre treated water, and the selection of the best pre treatment technology depends on the raw seawater quality and its variations. Number of full-scale experiences has shown that pre treatment is the key for this application of reverse osmosis technology. It is why during these last years, an import effort has been done to identify and to characterise the diverse organic and mineral components present in the seawater in a view to optimise the seawater pre-treatment and to develop advanced analytical methods for feed water characterization, appropriate fouling indicators and prediction tools.

This Chapter describes firstly a comprehensive approach to characterize raw seawater samples through analytical tools which allow the knowledge of the characterization of seawater from many aspects:

(a) inorganic content, (b) natural organic matter, (c) enumeration of micro-organisms and phytoplankton.

Secondly, this Chapter describes the effect of each of these parameters on the fouling of the reverse osmosis membrane. Finally, this chapter describes the different possible pre-treatments available to reduce or remove the elements or substances up-stream reverse osmosis stage.

2. Sea water characterization

Seawater is a mixture of various salts, organic substances, algae, bacteria and micro particles present in the water. Advanced analytical tools have been developed to allow thorough characterization of seawater samples from many aspects: (a) inorganic content,(b) natural organic matter,(c) enumeration of micro-organisms and phytoplankton.

The types of foulants (figure 1,table 1) most commonly encountered in RO systems include:

- Inorganic & particle fouling: Accumulation of particles on the membrane surface not removed from the raw water during the filtration process in the pre-treatment. The indicators of sufficient reduction of suspended solids and particles are turbidity values of less than 0.5 Nephelometric Turbidity Unit (NTU) and silt density index (SDI) values of less than 4.

- Colloidal fouling: Deposition of metal oxides, proteins, silicates, organic matter, and clay creating a colloidal slime on the membrane surface. Colloidal fouling is due to the presence of suspended solids in water, such as mud and silt, and tends to cause gross plugging of the device rather than fouling of the membrane surface.
- Biological fouling: Build-up of a microbial community on the membrane surface including microbes and their by-products, resulting in a slime layer. Bio-fouling is a special case of particulate fouling that involves living organisms and can be a serious problem. Biological material growing on membrane surfaces not only causes loss of flux but may physically degrade certain types of membranes.
- Organic fouling: Adsorption of organic matter, particularly humic and fulvic acids, on the membrane surface. Organic fouling is most complex in nature and can cause hydrocarbon oils (naturally occurring or as a result of pollution) and have been known to cause performance deterioration.
- Scaling of RO membrane surfaces is caused by the precipitation of sparingly soluble salts from concentrated brine.

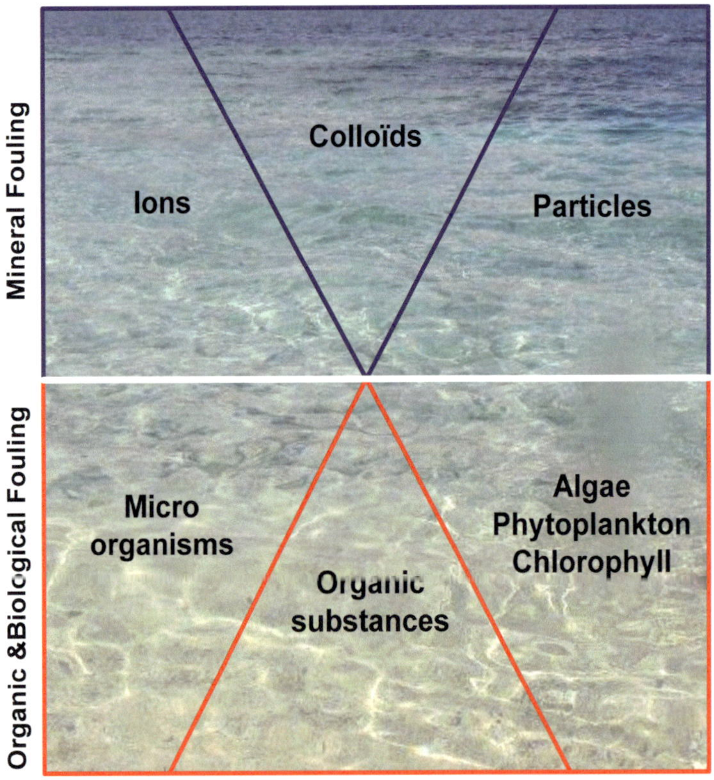

Fig. 1. Fouling potential

The quality will be determined by analysis of physical, chemical, and bacteriological contents to determine the level of treatment to supply the necessary water quality for the reverse osmosis membranes

Inorganic	Organic	Biological
Silica Quartz Silt Carbonates/sulphates	Lipids Proteins Polysaccharides	Algae Plankton Unicellular organisms

Table 1. Seawater constituents and potential membrane foulants

2.1 Physical characteristics

2.1.1 Density

At zero degrees Celsius liquid water turns into ice and has a density of approximately 917 kilograms per cubic meter. Liquid water at the same temperature has a density of nearly 1,000 kilograms per cubic meter. The density of seawater generally increases with decreasing temperature, increasing salinity, and increasing depth in the ocean. The density of seawater at the surface of the ocean varies from 1,020 to 1,029 kilograms per cubic meter. Highest densities are achieved with depth because of the overlying weight of water. In the deepest parts of the oceans, seawater densities can be as high as 1,050 kilograms per cubic meter.

The other physical characteristics of the sea water that must be evaluated are total suspended solids (TSS) and temperature, Turbidity and silt density index (SDI).

2.1.2 Total suspended solids

The total suspended solids level must be evaluated to determine the level of pre-treatment processes required. Sea water having low total suspended solids levels generally requires less pre-treatment.

2.1.3 Temperature

The temperature of the sea water source must be matched to the specific desalination process because this parameter may control the desalination process design. The evolution of the temperature during the year must be made prior to determine the seasonal maximum and minimum water temperatures of the sea water.

2.1.4 Turbidity

This parameter provides the amount of fine particulate matter in the water. Turbidity is measured in Nephelometric Turbidity Units (NTU).

2.1.5 Fouling index

Fouling is the major issue when using membranes for water treatment. Several parameters have been proposed for measuring a fouling potential and using it as a predictive tool for assessing the adequacy of pre-treatment. The Silt Density Index (SDI) and the Modified Fouling Index (MFI) are presently the only standard methods, even if they do not reflect the real potential of fouling because particles smaller than 0.45 μm responsible for fouling are not taken into consideration. Their limitations have been evidenced by several studies [Khirani et al., 2005; Junga & Son, 2009.Hong et al., 2009].

2.1.5.1 Silt density index (SDI)

Silt density index (SDI) is a parameter characterising the fouling potential of water. Particulate, colloidal matter and micro-organisms (figure 2) have a natural tendency to deposit themselves on the membrane, thus impairing its effectiveness. It is one of the most important parameter for the design and operation of reverse osmosis membrane process. SDI analytical protocol is standardized in the ASTM D 4189-95, re-approved 2002, and it evaluates the amount of 0.45-micron filter plugging caused by passing a sample of water through the filter for 15 minutes. SDI is recognised as the standard test to estimate membrane fouling potential (Iwahori et al., 2003; Kim et al., 2006, Kremen & Tanner, 1998; Mosset et al., 2008). It is strongly depending on the amount of particles but also representative of other fouling compounds

Fig. 2. Origin of the fouling compounds according to SDI membrane appearance

The protocol for this measurement is standardised (see Standard ASTM-D4189-07). The SDI must be as low as possible to limit the fouling of the filtration membranes. The principle of this protocol is to measure the time required to filter a clearly defined volume of water (500 ml) with a new test membrane and then compare this with the time required to filter the same volume after 15 minutes of filtration. The increase in the time required for filtration of the 500 ml is used to calculate an index (figure 3). The minimum SDI_{15} value is 0 and the maximum value, corresponding to an infinite filtration time, is 6.67. In practice, it is never possible to obtain SDI_{15} = 0. The test is carried out at a pressure of 2.05 bars through a membrane with a cut-off threshold of 0.45µm. Conventionally, the measurement is made over a period of 15 minutes (SDI_{15}) on the pre treated water. When the water has very high

fouling properties, it may be made over a period of 10, 5 or 3 minutes. Note that the ranges of values are not at all the same for different measurement periods (SDI $_3$: 0→33.3; SDI_5: 0→20; SDI_{10}: 0→10; SDI_{15}: 0→6.67). It is therefore expressed in %/min.
Because the filter is more or less plugging versus time, the rate of plugging (SDI) is more or less important. SDI is then calculated according to the following formula:

$$SDI_{Td} = \frac{100}{Td} \times \left(1 - \frac{T_i}{T_f}\right)$$

Where:
Td is the overall filtration time (3, 5, 10, 15 minutes)
T_i is the initial time (in s) to filter 500 ml of water on a 0.45 µm membrane at 2.05 bar
T_f is the final time (in s) to filter 500 ml after 15 min.
Standard ASTM D 4189 does not stipulate the material of the test membrane or its supplier. The nature of the test membrane is a critical parameter because it has been demonstrated that the choice of type of membranes used for the test is primordial. An SDI value given without specifying the type of membrane used for the measurement is meaningless.
The Factors interfering with SDI measures are:
The influence **of pH** shows an increase of SDI values from 4 to 6 when pH is increased from 7 to 8 and is mainly explained by the presence of dissolved substances (Ca, Mg...), which precipitate with the increasing of the pH.
The type of **Membrane** is determinant for SDI values. A comparative study between hydrophobic and hydrophilic membrane (figure 3) shows that higher results are obtained for hydrophobic membrane compared to hydrophilic membrane.

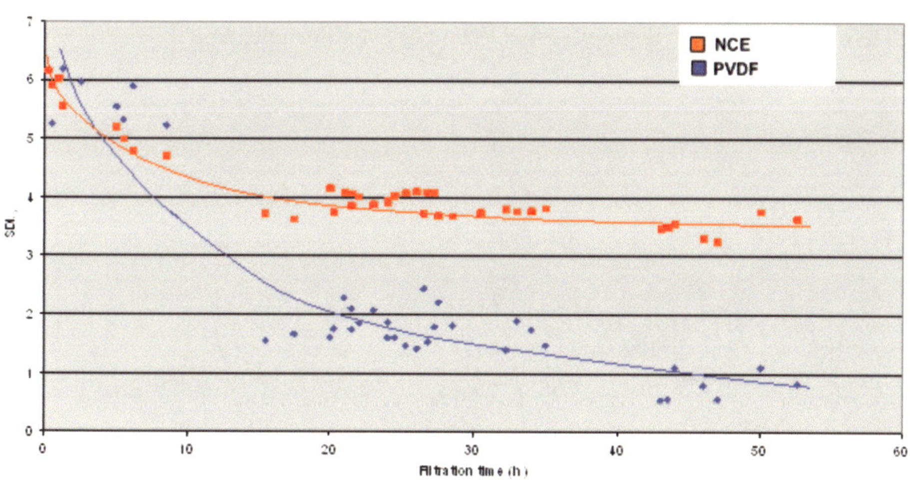

Fig. 3. SDI 15 Comparison Nitrocellulose membrane vs PVDF membrane

SDI is the essential parameter to control the fouling potential of water. Compared to others parameters like turbidity or suspended solids, it is more sensitive. The figures 4 & 5 show the evolution of the SDI during spring and autumn on a site located on the Mediterranean sea. This evolution could influence the pre treated water quality if the selected technology and the design are not based on the worst and the most disadvantageous values. The operation of the plant concluded that raw water quality seems to be the most obvious answer. The SDI results for the 3-minute test are significant only if inferior to 33. Therefore, all measured values superior to 33 for seawater were discarded. When looking at the evolution over the seasons, it can be observed that, during the autumn and winter months, the average SDI is low with little evolution. On the contrary, during the spring the SDI becomes higher and varies greatly.

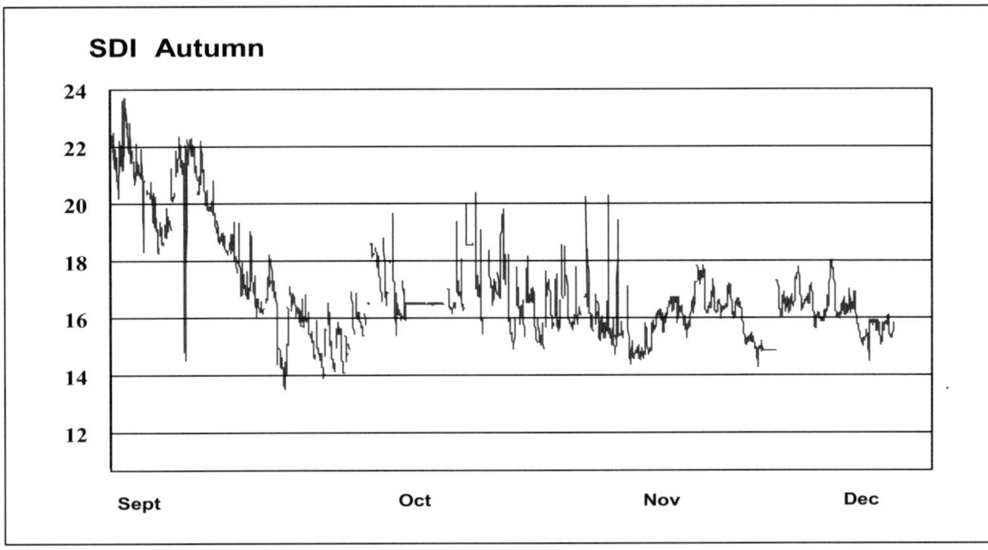

Fig. 4. SDI_3 of seawater from September to December (Mediterranean Site)

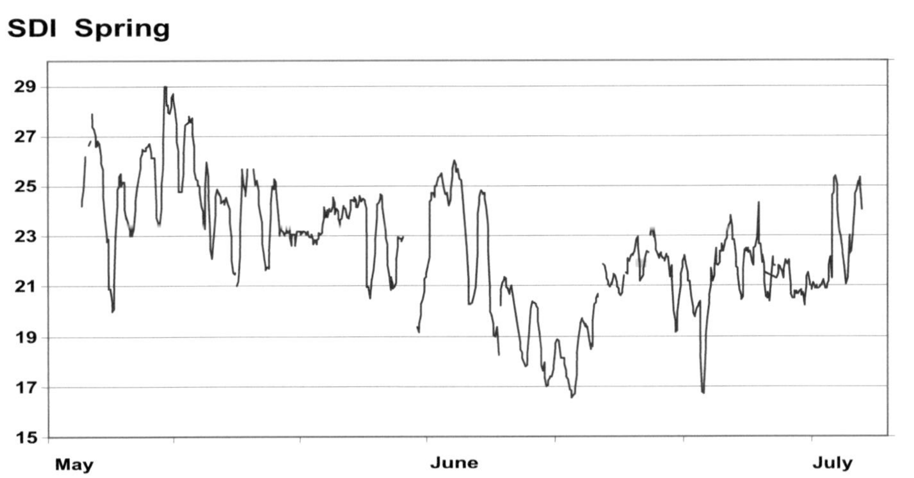

Fig. 5. SDI_3 of seawater from May to July (Mediterranean Site)

2.1.5.2 Modified Fouling Index

Schippers & Verdouw have proposed a fouling index called "Modified Fouling Index" (MFI) which takes into account fouling mechanisms (Schippers & Verdouw, 1980). They considered that the fouling of a flat-sheet membrane in dead-end filtration at constant transmembrane pressure takes place in three steps: (1) pore blocking, (2) formation of an incompressible cake and (3) formation of a compressible cake. This mechanism is based on the laws of dead-end filtration at constant transmembrane pressure or constant flux which give explicit relationships between filtration time and permeate flow rate (Boerlage et al., 1997; 2002a ; 2002b; 2004). This is illustrated by figure 6 which represents the evolution of the ratio t/V as a function of V, where t is the filtration time and V the cumulated permeate volume.

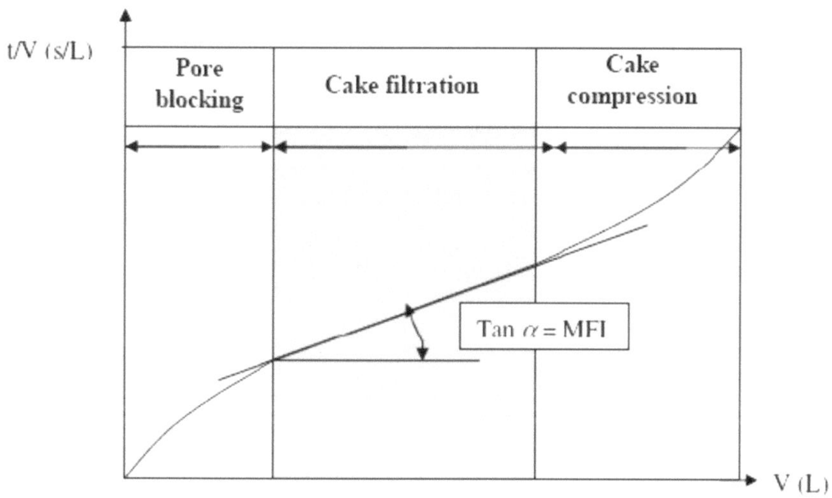

Fig. 6. Evolution of the t/V ratio vs. Volume

Theoretical background making the hypothesis that the only mechanism that increases the apparent resistance during the filtration test is the formation of a cake on the membrane surface. The global relation is given by :

$$\frac{t}{V/A} = \frac{\eta\, Rm}{\Delta P} + MFI\frac{V}{A} \quad \text{with} \quad MFI = \frac{\eta\, \alpha\, Cp}{2\Delta P}$$

where t is filtration time (s), V/A is the permeate volume produced per membrane area (m^2), ΔP is the TMP (Pa), A the membrane area (m^2), Rm the resistance of the membrane (m^{-1}), Rc the resistance of the cake (m^{-1}), α is the specific cake resistance, Cp is the concentration of particles in the feed water, and η is the dynamic viscosity of the water (N $s.m^{-2}$).
The MFI could be represented by the value of the specific resistance of the cake formed by the fouling components of the water deposited on a membrane during a standard filtration test. The main advantage of the MFI over SDI thus lies in the fact that MFI is a dynamic index which takes into account the evolution of membrane fouling all along a filtration test whereas SDI is only based on an initial and a final measurement.

2.2 Chemical constituents

The chemical constituents of the raw water must be determined to provide information for the pre treatment selection.

2.2.1 Ions content

2.2.1.1 Total dissolved solids (TDS)

Most of the dissolved chemical constituents or salts found in seawater have a continental origin. Only six **elements** and **compounds** comprise about 99% of sea salts: chlorine (Cl^-), sodium (Na^+), sulphates ($SO4^{-2}$), magnesium (Mg^{+2}), calcium (Ca^{+2}), and potassium (K^+)
Because salinity is directly proportional to the amount of chlorine in sea water, and because chlorine can be measured accurately by a simple chemical analysis, salinity *S* was redefined using chlorine content. The following relation is often used:

$$S(g/L) = 1.80655 \text{ Cl } (g/L)$$

The TDS of sea water (usually 35 g/L) is made up by all the dissolved salts present in the water. Landlocked seas like the Black Sea and the Baltic Sea have differing concentrations. This world map shows how the TDS of the oceans changes slightly from around 32 g/L (3.2%) to 40 g/L (4.0%). Low TDS is found in cold seas, particularly during the summer season when ice melts.
High salinity is found in the ocean coinciding with the continental deserts. Due to cool dry air descending and warming up, these desert zones have very little rainfall, and high evaporation. The Red Sea located in the desert region but almost completely closed shows the highest salinity of all (42 g/L) but the Mediterranean Sea follows as a close second (38 g/L). Lowest salinity is found in the upper reaches of the Baltic Sea (5 g/L). The Dead Sea is 240 g/L saline, containing mainly magnesium chloride $MgCl_2$. Shallow coastal areas are 2.6-3.0 g/L saline and estuaries 1-3g/L. The overall ion content of the Arabian Gulf is higher as compared to the oceans and the Mediterranean Sea, the sites located on the Pacific Ocean and the Atlantic Ocean show a slightly lower salt content than the Mediterranean Sea, and this could impact in some cases the design of the RO systems, notably with respect to the boron removal. Overall, these differences of salt content will not impact the selection of the pre treatment strategy, but will impact the sizing of the reverse osmosis systems (Blute et al., 2008).
However, if the major part of ions analysed in the sea water will not impact the pre treatment design, iron and manganese have to be removed before the water feeds the reverse osmosis membrane. The explanation is presented above.

2.2.1.2 Specific ions: Iron & manganese

Iron oxides as well as manganese play an important role in the removal of trace elements from seawater. In the sediments, iron and manganese oxides transported with settling particles are reduced to ferrous and manganous ions during oxidation of organic matter.
Ferrous and manganous ions diffuse upward through interstitial water and are transformed again to iron and manganese oxides at the sediment-water interface. Iron and manganese oxides take up dissolved trace elements released from settling particles during diagenesis.

2.2.1.2.1 Iron (Fe)

The behavior of iron is greatly different from that of manganese. Iron chemistry, such as inorganic speciation and organic complexes, in seawater is very complex and not yet fully understood. The hypothesis that dissolved iron concentration is a key variable that controls phytoplankton processes in ocean surface waters is proved today. Iron is an essential micronutrient for phytoplankton growth, as an important component of such biochemical processes as photosynthetic and respiratory electron transport, nitrate and nitrite reduction, chlorophyll synthesis, and a number of other biosynthetic or degradative reactions (Weinberg, 1989; Kuma, 1996; Geider & Roche, 1994).

The oxidation rate constant of Fe^{2+} tends to increase with increasing pH and temperature, and decrease with increasing salinity (ionic strength). Millero (Millero, 1980) proposes the relationship given by:

$$\text{Log } k = 21.56 - 1.545/T - 3.29\, I^{1/2} + 1.52\, I$$

where k is the oxidation rate constant, T is the absolute temperature and I is the ionic strength.

The inorganic speciation of Fe^{3+} in seawater is dominated by its hydrolysis behavior and ready tendency to nucleate into particulate Fe^{3+} hydroxides. In general, iron in oxic seawater around pH 8 is present predominantly in the particulate iron oxyhydroxide (FeOOH), which has an extremely low solubility, (Millero,1987, 1988) and thermodynamically stable. Numerous studies of both the solubility of iron in seawater and of the detailed hydrolysis behavior of Fe^{3+} as a function of pH have been undertaken over the last 25 years. Number of studies of the Fe^{3+} hydroxide solubility in seawater suggest that the Fe^{3+} solubility is controlled by organic complexation (Kuma et al., 1996; Millero, 1998; Liu & Millero, 2002; Tani et al., 2003), which, subsequently, regulates dissolved iron concentrations in seawater (Kuma et al., 1998, 2003; Johnson et al., 1997; Archer & Johnson, 2000; Nakabayashi et al., 2001). In general, the dissolved Fe concentrations in the surface mixed layer were lower than those in mid- depth and deep waters and the values of Fe^{3+} solubility in the surface water, resulting from the active biological removal of dissolved Fe and excess concentration of Fe-binding organic ligands (Rue & Bruland, 1995 & 1997; Kuma et al., 1998).The dissolved Fe profiles generally show low concentrations at the surface (0.2 - 50 µg/L), abroad maximum from 500 m to 1000 m (10 -150 µg/L). The vertical profiles are similar to those of Fe^{3+}solubility, suggesting that dissolved Fe concentrations in deep ocean waters are controlled primarily by the Fe^{3+} complexation with natural organic ligands, which were released through the oxidative decomposition and transformation of biogenic organic matter in mid-depth and deep waters. In oxic seawater, iron is present predominantly in the insoluble (extremely low solubility). Therefore, phytoplankton growth is controlled by the Fe^{3+} solubilities and the iron dissolution rates of colloidal Fe^{3+} phases (Wells et al., 1983, Stumm & Lee, 1961; Gabelich et al., 2005)).

In previous studies (Kuma et al. 1999 & 2000), it has been suggested that the natural organic-Fe^{3+} complexes and acidic Fe^{3+} supplied by river inputs play an important role in supplying supersaturated bioavailable Fe^{3+} , above the equilibrium concentration of Fe^{3+}, in estuarine mixing systems and coastal waters through its dissociation and hydrolytic precipitation at high pH of seawater and high levels of seawater cations (Stumm & Morgan 1962). The exchange reaction between organic-Fe^{3+} complex and major alkaline earth metals (such as Ca^{2+} and Mg^{2+}) in seawater is one of the most important processes resulting in dissociation of organic-Fe^{3+} complexes and subsequent Fe^{3+} hydrolytic precipitation. The high concentration of alkaline earth cations in seawater probably caused the dissolution of organic Fe^{3+} complexes through the metal exchange reaction. In estuarine and coastal waters, the natural dissolved organic-Fe^{3+} complexes supplied by river input such as fulvic-Fe^{3+} may play an important role in the supply of biological available iron by heightening the dissolved inorganic Fe^{3+} concentration, through the dissociation of organic-Fe^{3+} complexes during mixing with seawater.

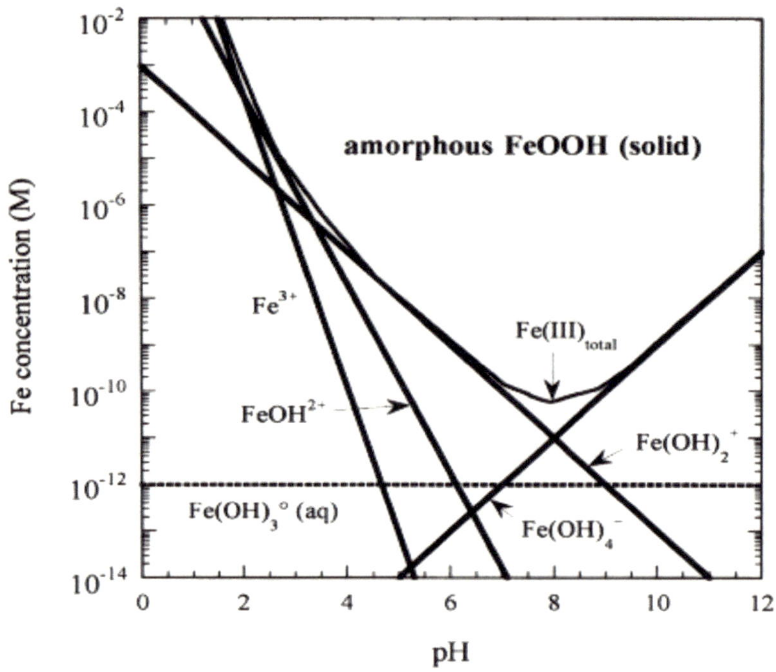

Fig. 7. Iron solubility of solid amorphous FeOOH

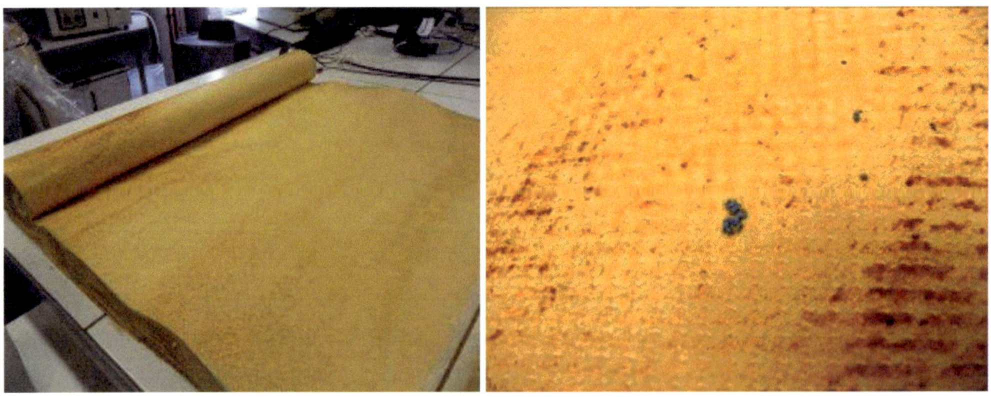

Fig. 8. Iron deposit on RO membrane

The limitation recommended by the membrane suppliers for iron is low than 50 µg/l due to its possible oxidation on the membranes which damages irreversibly the membrane surface.

2.2.1.2.2 Manganese (Mn)

The chemistry of manganese in seawater is complex and is largely governed by pH and redox conditions. Mn^{2+} dominates at lower pH and redox potential, with an increasing proportion of colloidal manganese oxyhydroxides above pH 5.5 in non-dystrophic waters (Lazerte & Burling, 1990). Oxidation rates of manganese increase with increasing pH The Mn^{2+} ion is more soluble than Mn^{4+}, therefore, manganese will tend to become more bio-available with decreasing pH and redox potential. The presence of chlorides and sulphates increases manganese solubility (Schaanning et al., 1988).
Manganese exists in the seawater in two main forms: Mn^{2+} and Mn^{4+}. Transition between these two forms occurs via oxidation and reduction reactions.
Based on laboratory experiments, the oxidation of Mn^{2+} to Mn^{4+} occurs as a two-step process in which solid phase Mn-bearing oxides (e.g., Mn_3O_4) or oxyhydroxides (e.g., ß- MnOOH)

$$3Mn^{2+} + 1/2\,O_2 + 3H_2O \rightarrow Mn_3O_4 + 6H^+$$

$$Mn^{2+} + 1/4\,O_2 + 3/2H_2O \rightarrow MnOOH + 2H^+$$

are initially formed and then undergo slower disproportionation or protonation reactions, ultimately forming Mn^{4+} oxides (MnO2) (Tebo et al.,2004).
Then, the stoichiometry of Mn^{2+} oxidation based on measurements of O_2 consumption and H^+ production follows the chemical reaction typically written for Mn^{2+} oxidation (de Vrind et al. 1986, Adams & Ghiorse 1988):

$$Mn^{2+} + 1/2\,O_2 + H_2O \rightarrow MnO_2 + 2H^+$$

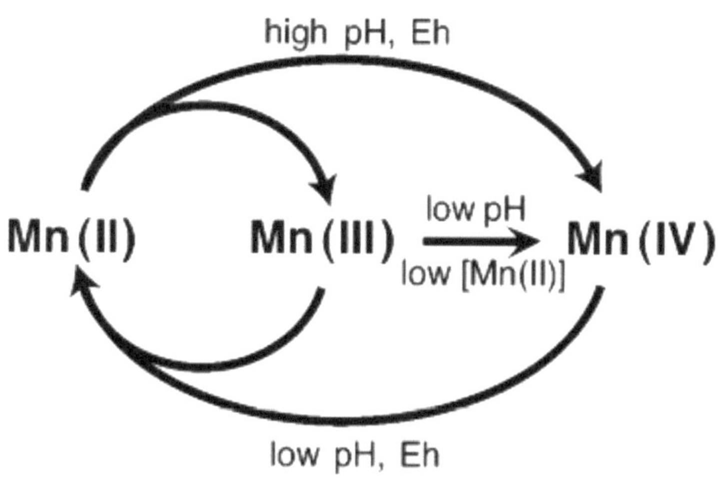

Fig. 9. The Mn cycle of oxidation states

In oxygenated waters, Mn^{2+} is thermodynamically unstable with respect to the oxidation to insoluble manganese oxides. However, owing to the relatively slow kinetics of oxidation of Mn^{2+} in seawater, the low equilibrium concentrations are rarely attained. The ocean distribution of the metal appears to be dominated by external input sources which lead to maxima in the surface waters.

Concentrations of manganese in open seawater range from 0.4 to 10 µg/litre. In the North Sea, the north-east Atlantic Ocean, the English Channel, and the Indian Ocean, manganese content was reported to range from 2 to 230 µg/litre. Levels found in coastal waters of the Irish Sea and in the North Sea off the coast of the United Kingdom ranged from 2 to 25.5 µg/litre (Alessio et al.,2007). Hypoxic concentrations below 16% saturation can increase the concentration of dissolved manganese above that normally found in seawater to concentrations approaching 1500 µg/litre (Mucci,2004). The concentration of dissolved Mn^{2+} in the anoxic waters is probably limited by its solubility with respect to MnCO3.

Seawater	Manganese concentration (µg/L)	Iron Concentration (µg/L)
North Sea	< 5 to 4	5 to 4
Indian Ocean,	2 to 180	2 to 220
north-east Atlantic Ocean,	5 to 80	25 to 230
English Channel	3 to 4	3 to 4
Irish Sea	2 to 25	2 to 25
Arabian sea	< 5	< 10
Red sea	5-10	5 - 100
Mediterranean sea	5 - 10	< 15
Oman sea	< 5	< 10

Table 2. Evolution of the Fe & Mn concentrations in different seawaters

Neutral streams with elevated levels of iron and manganese can develop blooms of ferromanganese-depositing bacteria with oxide deposition zones. The limitation recommended by the membrane suppliers for manganese is 20 µg/L, due to its possible oxidation on the membrane which will damage irreversibly the membrane surface. The following figures (figures 10 & 11) show the damage observed on the reverse osmosis membrane due to the oxidation of Mn^{2+} into Mn^{4+}. Scanning Electron Microscopy - Energy Dispersive X-ray Analysis (SEM-EDXA) are used also to study membrane surface and identify the elemental composition of the foulant.

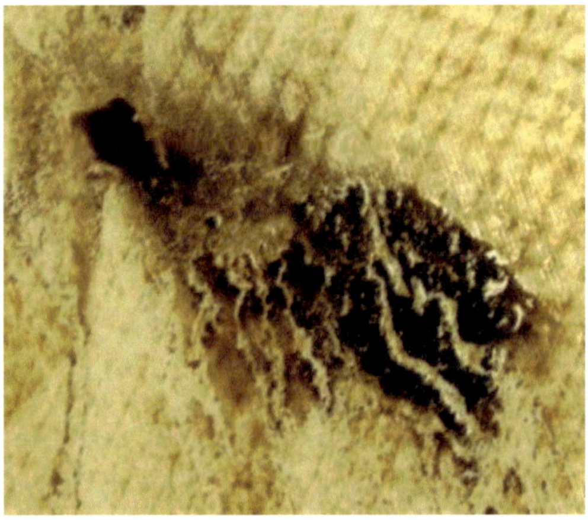

Fig. 10. Deposit of MnO2 on RO membrane

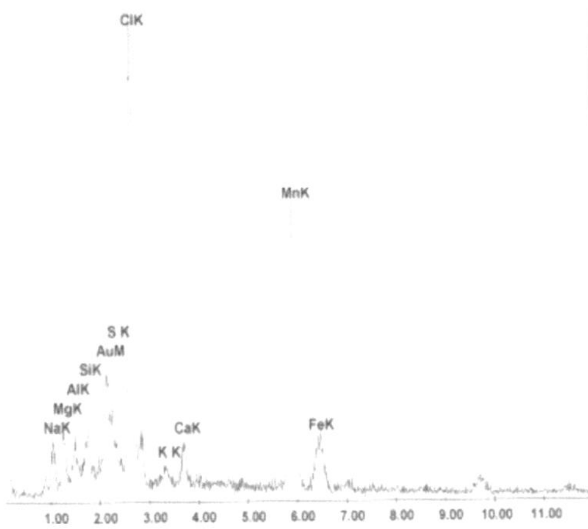

Fig. 11. Abnormal presence of Mn on RO membrane

Elemental determination with the SEM-EDXA system is based on analysis of X-rays produced via electron beam excitation of a sample area. This technique allows analysis of a sample in selective areas. The limited depth of analysis (typically a few microns), and the possibility to select the area of interest under the electron beam, allows for local analysis to reveal differences in composition.

The identification and measurement of individual peak intensities in the X-ray spectrum is done with a computerized multi-channel analyzer. Samples are covered by gold (Au) for analysis.

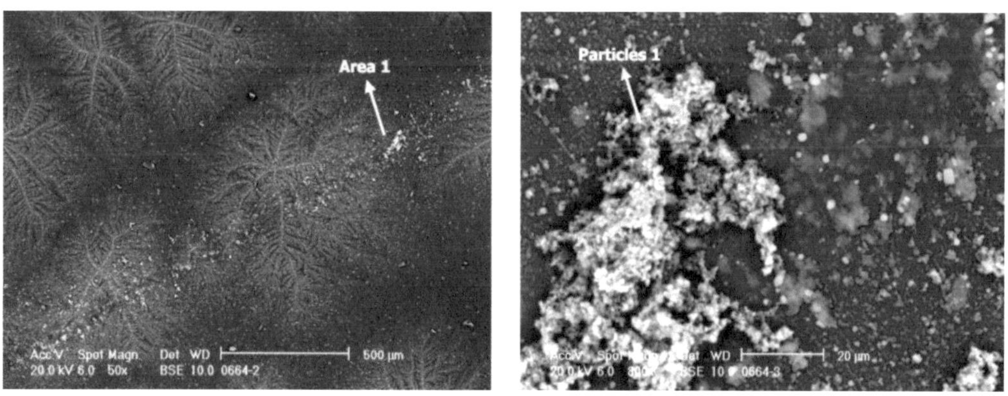

Fig. 12. Microphotographs 1 & 2 General view of membrane and its foulant.

The microphotographs show membrane surface which is completely covered by a deposit, composed of granulated particles (figure 12). EDX analysis on the particles show the presence of chlorine (Cl), sodium (Na), manganese (Mn), sulphur (S), magnesium (Mg), iron (Fe), calcium (Ca) and potassium (K). A survey of the Mn concentration in the feed water before the RO membrane is recommended.

2.3 Organic substances

It is well known that fouling in Reverse osmosis membranes causes serious problems including a gradual decline of membrane flux thereby decrease in permeate production, an increase in AP thereby increasing requirement of high pressure pump rating and a degradation of membrane itself. All these factors reflect on the cost of water production. Hence, now-a-days attempts are being made to deplete the concentration of organic from the feed to RO to overcome these problems by various pre-treatment methods. Various studies have been carried out to find the factors affecting organic fouling. In order to understand organic fouling, it has been necessary to thoroughly characterize organic matters.

The dissolved organic matter is not a single substance but a mixture of many aliphatic and aromatic compounds. However, among the total dissolved organic substance in seawater. 90% of them are represented by the humic materials or substances. Number of studies (Amy,2008) have demonstrated that the two main types of bulk organic matter (OM) of interest in seawater desalination plant are:

Allochthonous natural organic matter (NOM) dominated by humic substances and Autochthonous or algal organic matter mainly consisting of extra cellular macromolecular and cellular debris.

Moreover, organic matters usually have functional groups such as carboxyl (-COOH) and phenolic groups (-OH). It has been known that these functional groups play a key role in organic fouling since the functionality could change depending on water chemistries. Humic substances (HS) are generated from the degradation of organic matter and represent a significant fraction of the total organic matter in water. The HS are mostly constituted of humic acids (HA) and fulvic acids (FA) in natural water. Humic and fulvic acids possess a significant negative charge density and a bulky macromolecular shape. Subsequently, humic and fulvic acids are not as easily adsorbed onto such a membrane, even if it is intrinsically hydrophobic. Natural Organic matters exhibit relatively high specific UV absorbance values and contain relatively large amounts of aromatic carbon. It has been known that the rate and extent of organic fouling tends to be accelerated with decreasing pH, increasing ionic strength and increasing divalent cations . The electrostatic repulsion was increased at low pH condition, almost completely deprotonating carboxylic and phenolic groups. The electrostatic repulsion was reduced at higher ionic strengths and higher divalent cation concentrations, due to electric double layer around charged organic matters is compressed (Kim et al.,2009b, Krasner et al.,1996)).

There are several measurement procedures for the OM used for the characterization of the organic substances present in the seawater, including:

- Total organic Carbon (TOC) representing the total amount of OM including the particles content
- Dissolved organic Carbon (DOC) representing the amount of OM dissolved in the raw water UVA absorbance @254 nm, reflecting the aromatic character of OM
- SUVA (ratio UVA_{254}/DOC) representing the part of the humic substances versus the non-humic substances.

The LC-OCD stands for "Liquid Chromatography-Organic Carbon Detection". It consists of a size exclusion chromatography column, which separates hydrophilic organic molecules according to their molecular size. Its values refer to"mass of organic bound carbon" (OC), not to total mass of compounds. The underlying principle is the diffusion of molecules into the resin pores (Her et al.,2002; Serkis & Purdue,1990). This means that larger molecules elute first as they cannot penetrate the pores very deeply, while smaller molecules take more time to diffuse into the pores and out again. The separated compounds are then detected by two different detectors: a UV detector (absorption at 254 nm) and a DOC detector (after inorganic carbon purging). Depending on the size of the molecules, the composition of the organic matter can be obtained (figure 13). With a bespoke algorithm program, the different peaks can be integrated to evaluate the proportion of each organic fraction. The Dissolved Organic Carbon measurement can be carried out using a by-pass mode. In this case, the samples go straight through the TOC reactor and analyzer. Figure 15 depicts a typical chromatograph with the different peaks and their associated organic fractions.

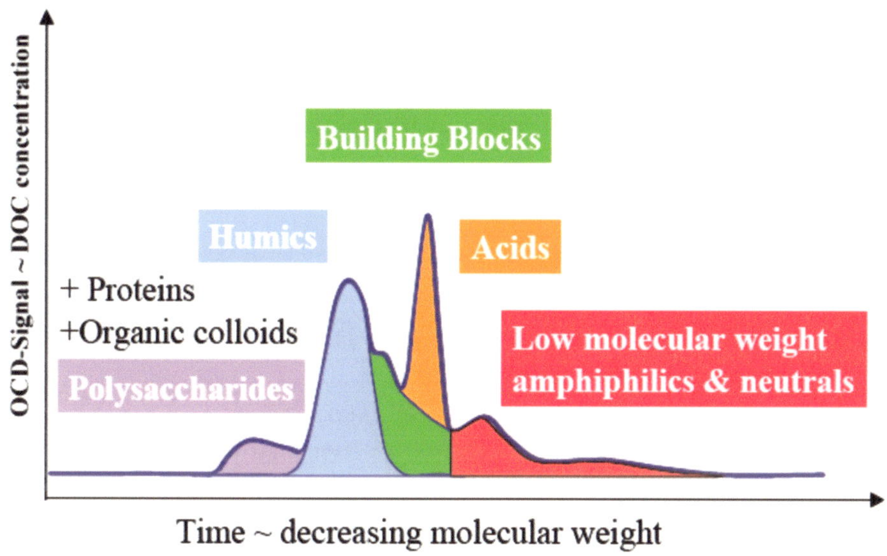

Fig. 13. Chromatograph obtained with LC-OCD chromatography

Biopolymers (polysaccharides amino sugars, polypeptides, proteins; "EPS"): This fraction is very high in molecular weight (100.000 - 2 Mio. g/mol), hydrophilic, not UV-absorbing. Polysaccharides exist only in surface waters.
Humics (HS): There is a tight definition for HS based on retention time.
Building Blocks (HS-Hydrolysates): The HS-fraction is overlain by broad shoulders which. are sub-units ("building blocks") of HS with molecular weights between 300-450 g/mol. Building Blocks are perhaps weathering and oxidation products of HS.
LMW (Low Molecular Weight) & Organic Acids: In this fraction, all aliphatic low-molecular-mass organic acids co-elute due to an ion chromatographic effect.
LMW Neutrals: Only low-molecular weight weakly charged hydrophilic or slightly hydrophobic compounds appear in this fraction, like alcohols, aldehydes, ketones, amino acids. The hydrophobic character increases with retention time, e. g.pentanol at 120 min.
A number of membrane related studies have demonstrated the use of LC-OCD in characterising dissolved organic matter (DOM) in surface waters (LeParc et al.,2007; Hong & Elimeleh,1997; Kim et al.,2009) and to identify the constituents that cause organic fouling.

DOC fractions	Size range (Da)	Composition
Biopolymers	> 20 000	Polysaccharides (e.g;TEP) & proteins
Humic substances	~ 1000	Humic and fulvic acids
Building Blocks	300 - 500	Oxidation products of humics
LMW organic acids	< 350	All aliphatic low molar weight organic acids
LMW neutrals	< 350	Alcohols, aldehydes, ketones and amino acids

Table 3. Typical sizes of DOC fractions detected by LC-OCD

Each LC-OCD system has an online organic carbon detector (OCD) that can measure carbon down to a low ppb-range. An online organic nitrogen detector is also connected to the system to measure levels of organic nitrogen. (Passow,2000,2002 ; Wotton,2004,LeParc et al.,2007):

1. Total exopolysaccharides (TEP) monitoring showed that colloidal TEP (82-93%) was more abundant than particulate TEP (7-18%) in the coastal seawater source.
2. The observed increase of total TEP in the raw water in spring coincided with an increase in chlorophyll-a and TOC.
3. LC-OCD analysis results show that biopolymers in the raw water, which were dominated by polysaccharides, doubled during spring and summer periods.

Seawaters collected from open intakes at various sites had a fairly low and stable TOC levels, ranging from 0.8 to 1.5 mg/L (Leparc et al., 2007). Figures 15 & 16 demonstrate the advantage of using beach well as seawater feed as compared to open intake. Firstly, the TOC levels of beach well seawater are slightly lower, but most importantly, the polysaccharides are almost completely removed through the slow filtration occurring when beach wells are used. Beach wells are therefore an excellent line of defense against organic and biological fouling on the RO membranes as polysaccharides are easily absorbed onto spiral-wound membranes. Then, polysaccharides foster the microbial attachment onto the reverse osmosis membranes, and these high molecular weight compounds can also be used as nutrients by the bacteria, thus facilitating the development of a biofilm onto the membranes.

The NOM characterization through LC-OCD chromatography allows to demonstrate that the NOM content (Her et al.,2002; Mitra et al.,2009) may vary depending on the seasons. Figure 5 notably shows that samples collected during fall have lower polysaccharides levels than samples during summer (LeParc et al.,2007). These lower polysaccharides levels during colder seasons correspond also with lower $SDI_{3\ min}$ values, and lower bacterial counts. It already appears that the bacterial and algal activities, enhanced with warmer water temperatures and higher sun exposure, are major water quality factors impacting the fouling potential of open intake seawaters, and therefore, will impact the selection of the pre treatment strategy.

NOM can be fractionated into hydrophobic, transphilic and hydrophilic acid fractions according to the XAD-8/4 resin method (Krasner et al.1996). Conventional methods such as coagulation or filtration through activated carbon are efficient to remove a part of the organic load from the feed of RO.

2.4 Algae

Algae, dinoflagellates and cyanobacteria are a large and varied group of photosynthetic organisms that are found in oceans. Algal and cyanobacterial cells contain chlorophyll and other photosynthetic pigments. They exist in a wide variety of forms; from single cells and strings of cells, through to complex multicellular seaweeds. The most familiar algae are red, brown and green seaweeds, which are part of a group of large multicellular algae known as macroalgae. However, the majority of algae and cyanobacteria are single-celled species that float freely in the water column; they are invisible to the naked eye and collectively form a group known as phytoplankton .Excessive growth of phytoplankton can occur in coastal seawater and estuaries causing the seawater to appear coloured typically red, or brown close to the surface of the seawater (figure 14) due to the density and numbers of algae.

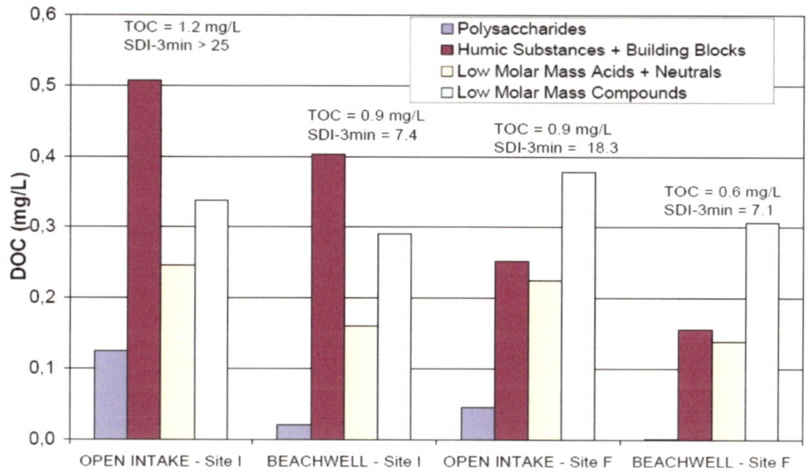

Fig. 14. Impact of the seawater intake type on the NOM content

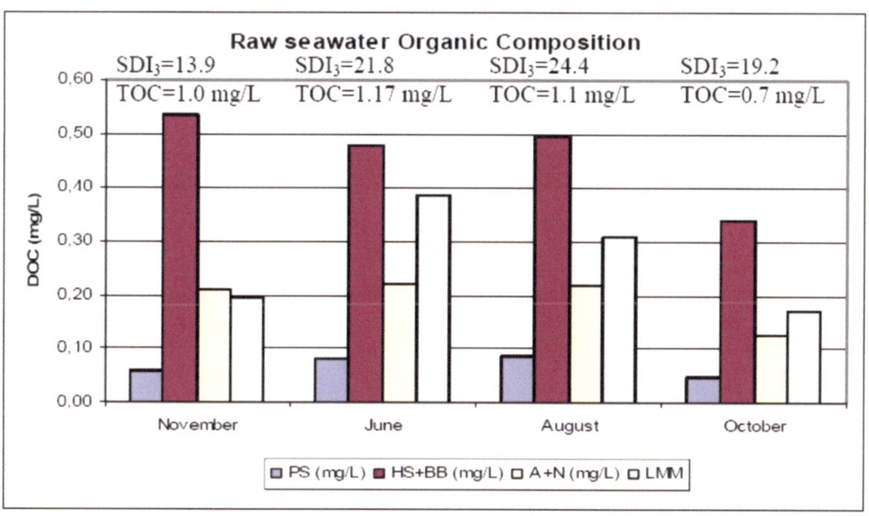

Fig. 15. Seasonal variations of the NOM content of the raw seawater- Mediterranean Sea – Open Intake

This is commonly referred to as an algal bloom. Strictly speaking the most accurate term is "phytoplankton bloom". Algal blooms can pose problems to the operation of a desalination plant. Extremely high algal numbers result in a high suspended solids load and organics. Most marine algal blooms are harmless, resulting only in a discolouration of the water. Algae exist in natural waters in a variety of sizes, geometric structures and cell wall materials. (figures 16 & 17). Although most algae are microscopic (ranging from 2 µm to 100 µm), a number of forms are macroscopic, with some species growing to lengths over 100 ft (Brock & Clyne, 1984). Plankton organisms are classified by size from femtoplankton (smaller than 0.2µm), picoplankton (0.2-2µm) to megaplankton (0.2-2mm). Phytoplankton consists of organisms from bacteria to diatoms and large dinoflagellates (like sea spark, *Noctiluca scintillans*). Their biomass can be estimated by measuring their chlorophyll (green pigment) from light measurements. However, other pigments (brown, red) are also common and the amount of chlorophyll is only a small part of biomass. So, even quantifying the amount of phytoplankton is almost impossible.

Fig. 16. Algae bloom in different sea waters

Advanced analytical tool was developed to allow thorough characterization of seawater samples the enumeration of phytoplankton and bacteria. Results (Leparc et al.,2007) obtained on raw seawater samples showed that the bacteria and phytoplankton counts appear to be positively correlated with (a) the concentration of polysaccharides, organic compounds highly fouling for reverse osmosis, and with (b) the SDI values of both the raw and pre-treated seawaters. The other conventional water quality parameters such as turbidity and TOC does not show any correlation with the fouling potential of both the raw and pre-treated seawaters. Indeed, biofouling due to bacteria attachment and growth on the membranes is one of a major threat for seawater reverse osmosis plant and the presence of polysaccharides in the pre-treated water increase that threat as these organic compounds are very prone to absorb onto the RO membranes and then be used as nutrients by bacteria.

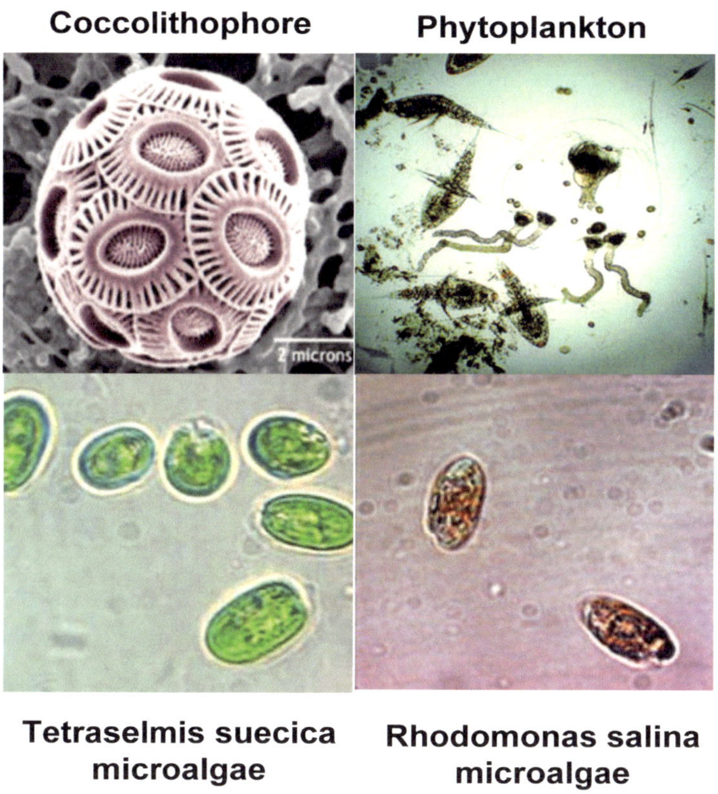

Fig. 17. Types of algae found in the seawater

Overall, the use of these complementary water quality parameters should provide engineers with valuable information to design, build, and operate more efficient and sustainable seawater reverse osmosis plants as future design and operation engineering practices will take into account more detailed information on site-specific water quality challenges.

Picophytoplankton species corresponds to the smaller size species of phytoplankton. The concentrations of picophytoplankton species appeared interesting to be monitored in both raw and pretreated seawaters because phytoplankton species with a size greater than 100 µm are very likely to be removed through the pretreatment processes and therefore, smaller size algal organisms, such as picophytoplankton, are the most likely to pose a threat to the RO membranes.

Figure 18 shows the concentrations of phytoplankton species (Le Parc et al.,2007) at various seawater desalination sites. The following observations can be made:

- the Arabian Gulf seawater has a significantly higher algal activity as compared to the Mediterranean Sea (and other oceans – data not shown),
- the positive impact of the beachwell is again demonstrated, as concentrations of phytoplankton species in beachwell seawater is more than one level of magnitude lower than that of surface seawaters.

Red Tide Events - algae bloom.

Red tide is a complex phenomenon involving many different types of creatures with different characteristics covering large areas. "Red Tide" is a common name for such a phenomenon where certain phytoplankton species contain reddish pigments and the water appears to be coloured red. They disrupt the ecosystems causing large scale environmental damage. Most of the red tides cause large scale fish kill and the killed fish will be washed to the shores resulting with a bad smell on the beaches. Red tide events may occur in the spring-summer period of the year and may result in increase of algae content in the source water (intake turbidity increases to up to 10 NTU); creased organics (TOC concentration increases to 4 -.5 mg/L) and apparent colour and odour. One or more sequential red tide events may occur per year and each event may last 6 to 8 weeks. Recent red tide in 2006 and 2008 in the Arabian Gulf caused considerable environmental damage and economic losses in the Gulf countries (Bauman et al;2010, Choules et al.,2007) and also in Iran, Iraq and Pakistan.

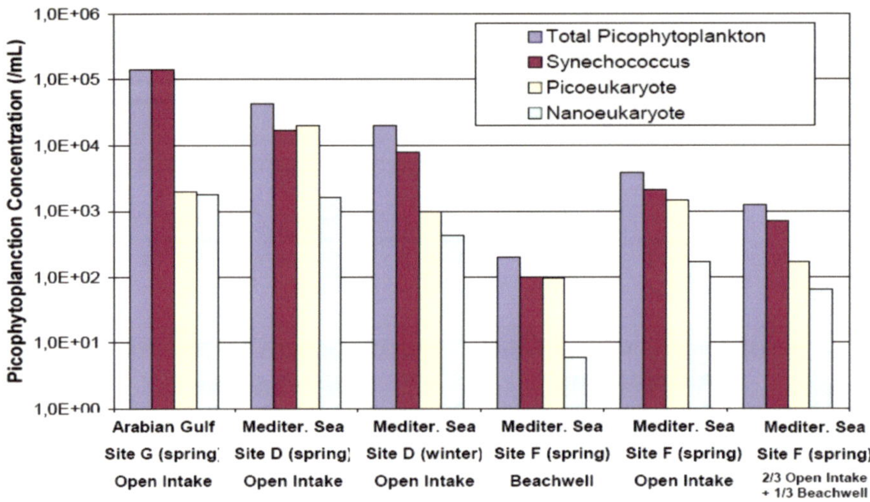

Fig. 18. Phytoplankton concentrations in various raw seawaters (log-scale)

Red tide is not caused by any single organism, although some are more common than others. Many of these species are regional and are quite adoptive. The two main types of toxic red tide creatures are certain phytoplanktons which produce mostly chemical toxins harmful to fisheries and the environment and a group of dinoflagellates that produce mostly neurotoxins harmful to humans and marine mammals. Most of the harmful algal blooms from 1988 to 2008 in Oman were caused by some type of dinoflagellates

Seawater desalination plants, power plants and other plants that use seawater for cooling purposes were forced to close during the last red tide in the Arabian Gulf region to avoid the fouling and blockage problems.

Algae blooms may occur in freshwater as well as marine environments. Typically only one or a few phytoplankton species are involved and some blooms may be recognized by discoloration of the water resulting from the high density of pigmented cells. Although there is no officially recognized threshold level, algae can be considered to be blooming at concentrations may reach millions of cells per mL, depending on the causative species (figure 20). Colours observed are green, yellowish-brown, or red. As more algae and plants grow, others die. This dead organic matter becomes food for bacteria that decompose it. Algal blooms may also be of concern as some species of algae produce neurotoxins .At the high cell concentrations reached during some blooms, these toxins may have severe biological impacts on wildlife. Algal blooms (Hallegraeff, G.M.,1993) known to naturally produce biotoxins are often called Harmful Algal Blooms (HABs).

2.5 Oil and chemical spills

Sometimes, oil and chemical spills have been detected in the seawater. The range of the concentration is 0 – 10 mg/L. These can affect the desalination plant. Emulsified oil and grease are the principle sources of immiscible liquid fouling in desalination facilities. Flotation appears as the most efficient treatment for this contaminant. A polishing on granular activated carbon is sometimes used to maintain acceptable levels upstream the reverse osmosis membranes.

Fig. 19. Algal bloom in Fujairah coast

3. Selection of the pre treatment

The characterization of the seawater through the main parameters which could be removed along the pre treatment process such as, Fe, Mn, natural organic matter, SDI, bacteria and algal are very useful from many aspects: - better understanding of site-specific seawater quality and its seasonal variation; - improved assessment tools to evaluate and predict the impact of raw seawater quality on the performance of a conventional pre treatment process, - additional and complementary indicators to the conventional water quality parameters (SDI, turbidity) for quantifying the risks of fouling on the RO units.

Before raw water is desalinated, the undesirable materials will be removed or reduced to acceptable levels. Without adequate pre treatment, desalination facilities are destined for reduced lifetimes, shortened periods of operation, and high maintenance.

After the completion of physical, chemical, and bacteriological analysis of the selected feed water, the type of pre treatment can be examined and is used to bring a saline feed water within limits so that a desalination process can be used. One of the most significant factors in successfully (and cost-effectively) operating a reverse osmosis (RO) desalination plant is the ability of the pre treatment system to consistently produce well-filtered and relatively particle- and microbe-free water for feed to the RO system Pre treatment is critical in RO applications because it directly impacts fouling of the RO membranes. Fouling of the RO membranes results in increased operating cost from increased cleaning demands, increased feed pressures, and reduced membrane life. Additionally, fouling can result in reduced permeate water quality and permeate quantity, thereby impacting production from the RO facility.

Parameters to be removed or reduced during the pre-treatment	Limit recommended up stream reverse osmosis membrane
Iron	< 50 μg/L
Manganese	< 20 μg/L
Turbidity	< 0.5 NTU
Suspended solids	< 1 mg/L
SDI	< 4 (95 % of time)
Algae	< 100 u/L
Chlorophyll	< 2 μg/L
Organic substances (Dissolved Organic Carbon)	< 2 mg/L
Hydocarbons	< 0.1 mg/L

Table 4. Limits recommended up stream RO membranes

Both processes may be implemented in series with other typical water treatment processes such as clarification and flotation. Seawater is usually chemically conditioned as part of the pre-treatment process. This may include pH adjustment, coagulation and flocculant dosing. Physico-chemical selection (figure 21) would depend on process choice, feed water quality and other environmental and design parameters (Gaid & Treal,2007; Choi et al.,2009).
The pre-treatment process would:

- remove Fe, Mn, turbidity , suspended solids & SDI
- manage risks from human activities such as oil leaks from shipping
- manage risks from naturally occurring events such as algal blooms & red tides
- reduce dissolved organic carbon

3.1 Unit operation and process of the pre treatment

To achieve these goals, a variety of treatment operation and processes (figure 20) are utilized, which exploit various physical and chemical phenomena to remove or reduce the undesirable constituents from the water. Each unit operation / process used plays an important role at the various stages of the pre treatment. The predominant role and responsibility of the design engineer is the selection and the design of the appropriate pre treatment operation/ process. The type of pre-treatment required depends on the characteristics of the raw water. The characteristics of the sea water is assessed by taking sample of water from the source during different seasons of the year and analyzing for physical, chemical and bacteriological quality parameters. Initial screening equipment will remove the (mobile) larval stages of these types of organisms from the raw water supply.

3.1.1 Prechlorination

The addition of chemical oxidants, such as chlorine, bromine, iodine, or ozone, can provide biological disinfection before membrane processes. Because, the first stage of fouling formation is an uncontrolled growth of microbial organisms on surfaces, with a preliminary formation of slime, which gives a biofilm, produced by the living cells and their metabolic by-products. The term biofouling refers to the final deposit, resulting from the mixture of bio-film (microbial and their extra-cellular polymeric substances (EPS), suspended solids, corrosion products and macro-organisms finally adhering and growing on the surface. The fouling layer reaches the maximum development with the adhesion of marine animals (figure 21) such as Crustacea & Molluscs,. Mussels are considered the most characteristic macro-fouling species and are the main species responsible for clogging of industrial pipes. It is very difficult to destroy and detach mussel shells from pipe walls due to their strong adhesion.

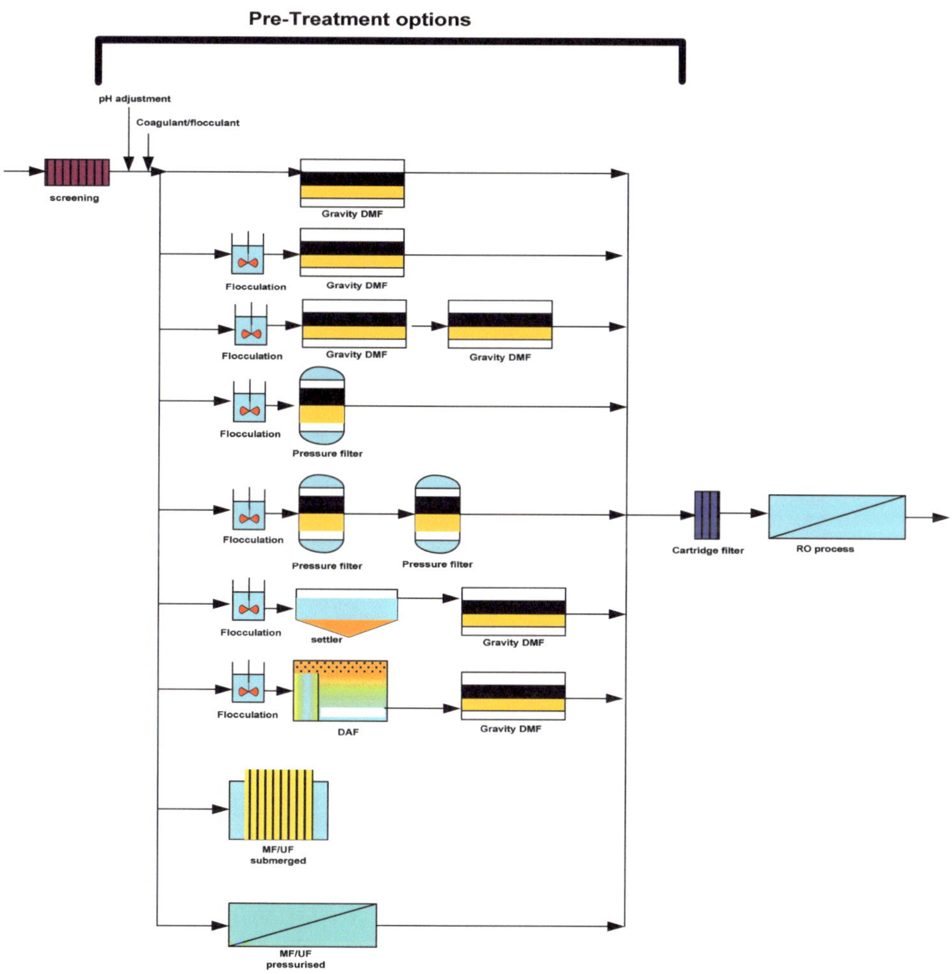

Fig. 20. Pre-Treatment options

Due to the anaerobic conditions, the activity of sulphate reducing bacteria (SRBs) is favourised and allows the corrosion phenomena on metallic surfaces of the pipes. It is why that a critical planning consideration for the full-scale seawater desalination facility is the risk of bio-fouling of intake and membrane equipment caused by marine organisms. The bio fouling risk is dynamic, changing with seasonal variances in source water quality parameters, such as nutrient loading, freshwater inflow, contamination, oil spills, and algae blooms.To minimize the problems related to micro and macro-fouling in desalination plant, continuous or intermittent injection of oxidant is added into the seawater at the intake. The pre chlorination is the most common method for bio-fouling control in seawater applications, especially where large water quantities are needed for desalination plants.The use of chlorine must be monitored carefully to keep the chlorine below 0.1 milligrams per liter of free chlorine residual that would even damage most of RO membrane used by the constructors. This dechlorination is accomplished chemically through sulfite compound addition .When organic substances are chlorinated, the resulting chlorine oxidation generates halogenated carbon compounds, such as the trihalomethane class of compounds. Complete dechlorination and destruction of the chlorine residual by reducing compounds will ensure that chemicals do not attack these sensitive membrane systems.

Chlorination chemistry

Chlorine is most commonly available as chlorine gas and sodium and calcium hypochlorites. In water, they hydrolyze instantaneously to hypochlorous acid:

$$Cl_2 + H_2O \rightarrow HOCl + HCl$$

$$NaOCl + H_2O \rightarrow HOCl + NaOH$$

$$Ca(OCl)_2 + 2H_2O \rightarrow 2HOCl + Ca(OH)_2$$

Fig. 21. Mussels development on intake (left) and biofilm on pipes (right)

Hypochlorous acid dissociates in water to hydrogen ions and hypochlorite ions:

$$HOCl \leftrightarrow H^+ + OCl^-$$

The sum of Cl_2, NaOCl, $Ca(OCl)_2$, HOCl, and OCl^- is referred to as free available chlorine or free residual chlorine, expressed as mg/L Cl_2. Sodium metabisulfite (SMBS) is commonly used for removal of free chlorine. Other chemical reducing agents exist (e.g., sulfur dioxide), but they are not as cost-effective as SMBS. When dissolved in water, sodium bisulfite (SBS) is formed from SMBS:

$$NaS_2O_3 + H_2O \rightarrow 2NaHSO_3$$

SBS then reduces hypochlorous acid according to:

$$2NaHSO_3 + 2HOCl \rightarrow H_2SO_4 + 2HCl + NaSO_4$$

In theory, 1.34 mg of sodium metabisulfite will remove 1.0 mg of free chlorine. In practice, however, 3.0 mg of sodium metabisulfite is normally used to remove 1.0 mg of chlorine.
Efficient bio-fouling control is achieved at concentrations in the range of 1– 3 mg/l when it is used under a continuous procedure, and in the range of 5-10 mg/l when it is used intermittently for some hours per day. Dechlorination upstream of the membranes is required, however, to protect the membranes from oxidation.
To day, the main question is: Continuous chlorination or intermittent chlorination?
According to seasonal and/or daily parameters (temperature, organisms population, light), to operational parameters, chlorine can be dosed through continuous or intermittent (higher dosages for shorter time at fixed intervals of time) way providing always a good mix with the feed water.

Continuous chlorination

The continuous dosing of chlorine had been confirmed as effective in the long history of RO systems operation.
However, this requires sensitive carefulness in operation. Chlorine is added continuously at the intake, and a reaction time of 20–30 min should be allowed. A free residual chlorine concentration around 0.5 mg/L should be maintained through the whole pretreatment line.

- But, it is well known that very close attention is required to minimize deterioration by oxidization of the membranes if a small residual of free chlorine is still present on the RO feed water.
- There is always a risk that membrane deterioration takes place rapidly under the presence of heavy metals such as Fe, Mn,Cu, Co, and others in the system.
- Bio-fouling problem downstream of the point of dechlorination is still common because the chlorine reacts with the organic matter in the water and breaks it down to more biodegradable fragments. Since there is no chlorine present on the membranes, microorganisms can grow with an enhanced nutrient offering, unless the system is sanitized very frequently. It is why, it is admitted that bio-fouling refers to the undesirable accumulation of a biotic deposit on a surface.

Therefore, the continuous chlorination/dechlorination method is becoming less popular.

Intermittent or shock chlorination

Instead of continuous chlorination, chlorine is more and more applied preferably periodically. Chlorine is added intermittently for some hours per day at the intake at concentrations in the range of 5 - 10 mg/l. Shock dosages can be extremely effective and provide a high inactivation rate of the organisms. Before the system goes into operation again, all chlorine containing feed water has to be rinsed out carefully, and the absence of chlorine must be verified (e.g., by monitoring of the oxidation-redox potential (ORP)). In the shock dosage, the chlorine dose must satisfy the "chlorine feed water demand" at the forecast contact time and a chlorine residual of about 0.1 mg/L should be present. The shock dosing is carried out for 10 minutes every 12 hours with only 3 ppm dosage. No algae or mussels growth was noticed in the seawater intake therefore the process appears to be very effective (Sommariya et al.,,2009)
In order to achieve the long membrane life that is desired for seawater desalination RO modules, optimization of the chlorine injection method becomes indispensable. Therefore, in order to reduce chlorine load to the RO module, the intermittent or shock chlorination method is more and more recommended instead continuous chlorination method.

Chlorine dioxide

Chlorine dioxide (ClO_2) is a greenish-yellow gas, highly soluble in water. It is generated "on site", mainly according to the following process with sodium chlorite as reagent:

$$2\,NaClO_2 + Cl_2 \rightarrow 2\,NaCl + 2\,ClO_2$$

$$5\,NaClO_2 + 4\,HCl \rightarrow 4\,ClO_2 + 5\,NaCl + 2\,H_2O$$

$$5\,NaClO_2 + 4\,HCl \rightarrow 4\,ClO_2 + 5\,NaCl + 2\,H_2O$$

In a pH range of 6–8.5, chlorine dioxide remains in solution as dissolved gas. The Jumeirah Palm project is the first desalination project in the Gulf to adopt chlorine dioxide for both seawater and potable water sterilization (Petrucci & Rosellini, 2005).The limitations of the Chlorine dioxide observed on site are the chlorites (ClO_2^-) production which can be 30 % of the ClO_2 concentrations. Due to the fact that the chlorites are not removed during the pre treatment, their impact on the reverse osmosis membranes through an eventual oxidation is possible but nor clearly proved due to the small desalination plants using this oxidant. The second limitation is often due to the operating cost because ClO_2 is more expensive than the sodium hypochlorite.

3.1.2 pH adjustment

The pH adjustment step of pre-treatment must result in the optimal pH level for the desalination system. After coagulants have been added, the pH is often changed significantly. In most cases, the pH must be returned to a neutral or a slightly acid level. Adjustment chemicals to lower the pH include sulfuric acid and hydrochloric acid.

3.2 Filtration

The type and choice of pretreatment depend on the extremes of raw water characteristics. Different source waters require varying levels of pre-treatment to ensure maximum RO membrane. With multiple technologies available for the pre-treatment, desalination engineers can look forward to satisfactory fouling index, efficient down stream RO plant and equipment operation

The most common pre treatment for open seawater is multimedia filters. It is possible to use a single stage filtration if the feed water is constantly of high quality. Double stage filtration is required if the seawater is degraded.

Regarding applications of filtration, it is noted that the extent and complexity of the pre-treatment systems for removing or reducing colloidal and organic fouling depend on site conditions. In case of open seawater intake, reverse osmosis membranes should be protected against a variety of foulants, necessitating an extensive pre-treatment process. For example, the use of coagulants and sedimentation or flotation equipment maybe necessary, followed by media filtration. Alternatively, granular media filtration can be replaced by low pressure membrane systems such as ultrafiltration or microfiltration.

In all water purification processes, filtration will be an integral step if not the main step. Filtration is an essentially mechanical operation and its goal is to trap particles larger than 10 microns (100,000 angstroms). In granular filtration, interception, gravitational sedimentation, and Brownian diffusion are the key mechanisms of colloidal particle transport from the pore fluid to the surface of a filter grain (Yao & Habibian,1971)

Granular media filters have two different design configurations:

- single media filter or dual media filter
- gravity filter or pressure filter

These two configurations can be also used as single stage filtration or double stage filtration

3.2.1 Single media filter and dual media filter

Single media filter - Gravity filter

Direct filtration is proposed when the seawater presents a low level of particles and low SDI. Single-media filtration is used when the SDI is lower than 10. Dual-media filtration is used when the SDI is lower than 25. The conditions of the use of the direct filtration is summarised on the table 5.

Seawater parameters	Single media filter	Dual media filter
Turbidity, NTU	1 - 2	< 10
Suspended solids, mg/L	< 3	< 15
SDI	10 - 15	30
Algae, u/L	-	2 000
Chlorophyll µg/L	-	< 5
DOC , mg/L	< 1.0	< 2.0

Table 5. Limit recommended of the feed water

Single-media filtration consists of one media (figure 22). This media is often small-grained silica sand based on 0.8 m for the effective size and 1.3 for the uniformity coefficient. The height of the media ranges between 1.0 - 1.5 m. This type of filter is mainly proposed when the SDI and the suspended solids are very low. A critical factor in designing pre-treatment is the possible use of an intake well, in particular a beach-well. If such an intake well exists, it is essentially part of the pre-treatment process because of the capacity of the sand (usually present at the sea bed) to act effectively as a first filter medium for the suspended solids in the seawater. Then, a simple form of pre-treatment by granular filter media, even without addition of coagulants, maybe adequate. Very few references exist for open intake and most of references concern groundwater or beach well.

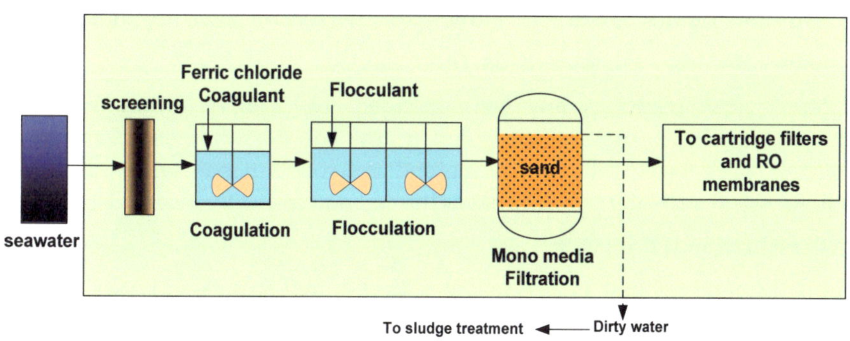

Fig. 22. Principle of coagulation-flocculation-single media filtration

Granular filter media must satisfy various specifications before they can be considered for applications. These include grain size, grain surface condition, density, particle porosity, solubility, durability, settling rate . The void fraction of the granular bed formed by the grains is also important. The shape of the grains used in filtration media mainly depends on the origin of the material. Grains collected from river beds are usually rounded and smooth. Grains resulting from the crushing of larger pieces are jagged and angular. Although inadequately studied so far, improved performance of crushed particles over rounded grains has been demonstrated for water filtration. For instance, the shape of the grains affects the bulk porosity of the bed, which is strongly related to the increase in head loss that results from deposits in the filter. However, systematic studies on the effect of grain shape on filtration are rather limited. Regarding grain surface roughness effects, although intuition and recent evidence suggest that they should play a significant role in filtration..

Single-media filtration – Continuous backwash up-flow sand filter

The filter is the Continuous Backwash Up-flow Sand Filter, commercially available for over 25 years, and over 8,000 units have been installed worldwide. Treatment begins when influent feed water enters at the top of the filter and flows downward through an annular space between the feed pipe and airlift housing. The feed is then introduced into the media bed through distribution radials, which are open at the bottom. As the influent flows upward through the moving granular bed, the solids are captured in and on the media while clean water continues rising into the filtrate pool above the bed. The filtrate then exits at the top of the filter over the effluent weir. Simultaneously, the granular media is being cleaned and recycled throughout the filter via an airlift pipe and media washer mechanism. The solids-laden filter media is drawn downward towards the intake of the airlift pipe located in the center of the filter bottom. A small, steady stream of compressed air is introduced into the airlift bottom, which draws the granular media and solids into the airlift. The solids-laden media is scoured as it rises in the airlift. Upon reaching the top of the airlift, the solids and granular media are released into the central reject compartment. The heavy filter media grains are returned to the bed after falling through the washer. As the filter media falls through the washer, which consists of several concentric stages, a small amount of filtered water passes upward, hydraulically lifting the solids or "dirt", while allowing the heavier and coarser granular media to fall. This counter-current flow of filtrate quality water is created by a difference in filtrate and reject weir heights. The cleansed media is then deposited at the top of the filter bed. This method of cleaning provides continuous, uninterrupted flows of filtrate and reject water (Dynasand). The application of continuous backwash up-flow for desalination is combined with a second stage filtration.

Single-media filtration – Diatomeceous filter

Diatomaceous earth media is low recommended for primary filtration because of its characteristic high head loss and short fun times. The diatomaceous earth precoat filter technology is long-term, established method in conventional water and wastewater treatment. Its use in seawater applications is limited and few references exist worldwide.

Dual media filtration (DMF)

Dual-media filtration consists of two media with different specific gravities (figure 23). The difference creates a two-layer separation effect. Use silica sand for one layer; use anthracite or pumice or equivalent media for the other layer. Anthracite is a black coal which allows for longer run times than can be achieved by sand alone. The use of dual media will allow larger quantities of material to be filtered and will reduce head loss during operation. The first layer is anthracite (or pumice) with a large diameter size which gives to the media bed, larger void spaces with greater solids holding capacity. It acts as a very robust roughing layer capable of handling heavy solids loading conditions associated with certain seasonal conditions such as suspended solids, algae development. The second layer is sand with a smaller diameter which acts as a final barrier for the fine particles responsible of the fouling. Considering an open intake, the seawater would in all cases be chemically conditioned to coagulate and flocculate the suspended matter (colloids, particles, algae) for removal in the pre-treatment filters. Coagulation is a process where a coagulant (Al or Fe salt) is added to the water to destabilise small particles suspended in the water. Coagulation is a rapid process requiring rapid mixing to disperse the coagulant in the feed water. The coagulation is done through a static mixer (most of cases) or through a coagulation tank with an adapted mixer. Since the pH of coagulation is critical (>8.5), an acid (mainly H_2SO_4) is added prior to coagulation to maintain optimum pH conditions. The coagulated water would discharge to flocculation tanks with mixers to provide gentle mixing of the coagulated water for the destabilised particles to form flocs. At the entry to the flocculation tanks, flocculant (polyelectrolyte) would be added to aid the process. Flocculated water would be removed from the water by passing through the gravity filters filled with sand granular medium for the single media filter and with anthracite (coal) / sand for the dual media filter. Inline coagulation is used and the dosage of the coagulants depends of the seawater quality. The most relevant process conditions for inline coagulation with metal salts are pH, dose, velocity gradient (G), shear rate (Gt) and temperature .pH affects the surface charge of colloids and determines the predominant coagulant species. Therefore, floc size and structure (porosity, density) may differ as a function of pH. Flocs formed at low pH and low dosage are reported to be denser and less porous than those formed at high pH and high dosage (Shin & O'Melia,2006). The physical properties of flocs are sensitive to flocculation conditions such as G and Gt. Increased shear reduces the average steady-state size of flocs. Higher G values lead to larger fractal dimension, which is related. The design of the flocculation tank depends of the seawater parameters. An important consideration in inline coagulation applications is the fate of coagulated & flocculated particles in the pipe network feeding the pre- treatment plant. It is important to consider if the flow regimes in the pipes favour floc growth or break the flocs. Studies on turbulent pipe flow for particle destabilization and aggregation show that for Reynolds number between 8 000 and 16000, the reaction rate for particle aggregation increased. Beyond 16000, the reaction rate decreased and may be attributed to a reduced collision efficiency of the primary particles and/or disruption of microflocs if the turbulence intensity in the pipe reactor exceeds a certain critical value. Reports on flocculation experiments in pipes of various diameters (8–600 mm) show that under steady-state conditions, a decreased floc size is observed with an increasing flow velocity (Johir et al.,2009; Mitrouli et al.,2008)).

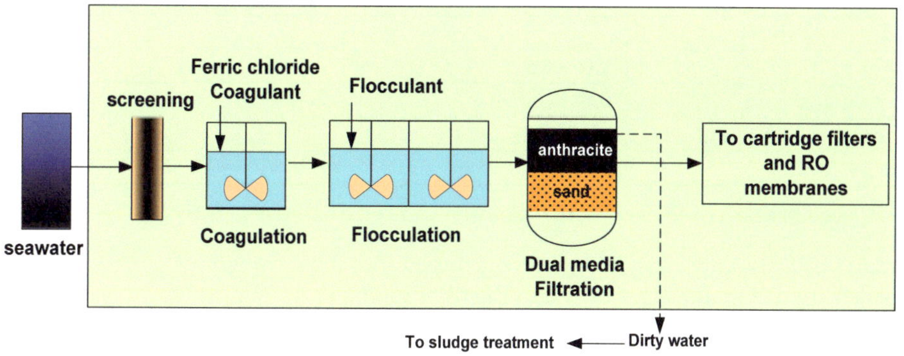

Fig. 23. Principle of coagulation-flocculation-dual media filtration

Anthracite or Pumice can be replaced by Filtralite which has excellent properties for use in pre-treatment filters in desalination plants. Filtralite media is made by burning of clay at about 1200° C, followed by crushing and sieving. The material has a porous structure and when crushed, a large surface area is exposed. The aggregates do not release harmful substances, and the acid solubility is minimal. With Filtralite, time between backwashes can be increased by about 25%, then reducing use of backwash water (Mitrouli et al.,2009).

The use of two media types introduced in a DMF will provide a good coarse-of-fine filtration process for desalination facilities. The total height of the two media ranges between 1.4 - 1.6 m based on a half layer part for each media. Pre-treatment processes are similar to the processes utilised for treating fresh water (in surface water drinking supplies). It is expected that sand and anthracite would be replaced about every 10 years.

A new development of DMF for seawater pre treatment have been proposed based on a total height of the media of 3 m -4 m and related with a higher filtration velocity (15 m/h). A linear relation ship has been demonstrated between the total height of the media , the filtration velocity and the final performance of the DMF.

The pre-treatment filters will also be provided with a backwash system. All chemical feed systems have been designed using prudent engineering practices and providing at least one standby chemical feed pump per system and adequate chemical mixing upstream of the pre-treatment filters.. Filters will be equipped with distribution boxes, crosswalks, wash headers, filtered water weirs, automatic backwash controls, backwash waste piping and valves and cell isolation gates.

Algal cells proved difficult to remove by direct filtration. The filter clogging (headloss) as a function of algal content has been evaluated in number studies and proved that this process is limited by this parameter. Petruchevski (Petrusevski et al., 1995) has demonstrated that algal removal efficiency by dual media filters was shown to vary strongly with algal species, suggesting that properties other than size and shape (algal motility, presence of outer mucilaginous layer, algal cell form) may have had a significant impact on filterability.

Parameters	Single media filter	Dual media filter
anthracite		
Height m		0.7 - 0.8
Effective size mm		1.0 - 1.5
Uniformity coefficient		≤ 1.4
sand		
Height m	1.0 - 1.5	0.7 - 0.8
Effective size mm	0.8	0.6
Uniformity coefficient	≤ 1.3	≤ 1.3
Velocity m/h	6 - 8	7 - 10

Table 8. Specification of the single media filter and the dual media filter

Filtration problems caused by algae were grouped into three categories:

1. penetration of stable algal cells into the filter media;
2. interferences with coagulation/flocculation caused by extracellular algal materials; and
3. short filter runs (increased headloss due to filter clogging) with increased need for backwash.

Fig. 24. Single media Pressure filters (Oman Sur desalination Plant)

Fig. 25. Gravity dual media filters (Australian desalination plant)

3.2.2 Double stage filtration

The pre-treatment system can include a double stage filtration (figure 26). The 1st stage filter can be a dual media with 1.2-1.6 m of total effective media depth. The specifications of the two media are summarised on the table 9. The first layer is anthracite (or pumice) with a large diameter size which gives to the media bed, larger void spaces with greater solids holding capacity. The second layer is sand with a smaller diameter which acts as a final barrier for the fine particles responsible of the fouling. The 1st Stage filters operate as roughing filters and are sized for a filtration velocity around 13 m/h. The 2nd Stage filter is a standard-bed sand filter with 1.0 m of effective sand media depth. The 2nd Stage filter uses a smaller diameter sand size (0.4-0.5 mm ES) which maximizes the sand media surface area and subsequently the possibility for attachment of finer particles still in the water. The 2nd Stage acts as a polishing filter to remove the majority of the remaining particles producing an effluent with a low Silt Density Index. The 2nd Stage filters operate as polishing filters and are sized at a greater hydraulic loading rate (16 - 18 m/h). The combination of increased sand media surface area and a higher hydraulic loading rate enables to meet the expected performances. The experiences shows that the 2nd stage filter can remove 0.3 – 0.4 SDI. The seawater would in all cases be chemically conditioned to coagulate colloids and micro particles. The filtration system is sized to optimize solids removal performance.

Parameters	1st stage filtration	2nd stage filtration
anthracite		
Height m	0.5 – 0.7	
Effective size mm	1.0 - 1.2	
Uniformity coefficient	≤ 1.4	
sand		
Height m	0.5 – 0.7	1.0
Effective size mm	0.6	0.4 - 0.5
Uniformity coefficient	≤ 1.3	≤ 1.3
Velocity m/h	10 - 13	14 - 18

Table 9. Specifications of the double stage filtration

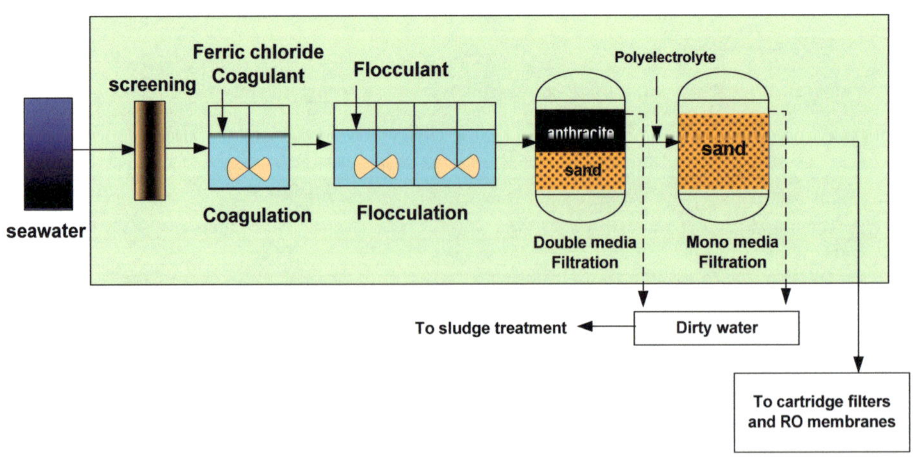

Fig. 26. Double stage filtration system

The figure 27 shows the performance of the double stage filtration for the removal of the SDI 15 (Millipore membrane). The SDI 15 after the double stage filtration is better than that recorded after a single dual media filter. A difference between 0.4 - 0.5 is observed between the two configurations. The SDI_3 of the raw water was in the range 16 - 29 %.min.

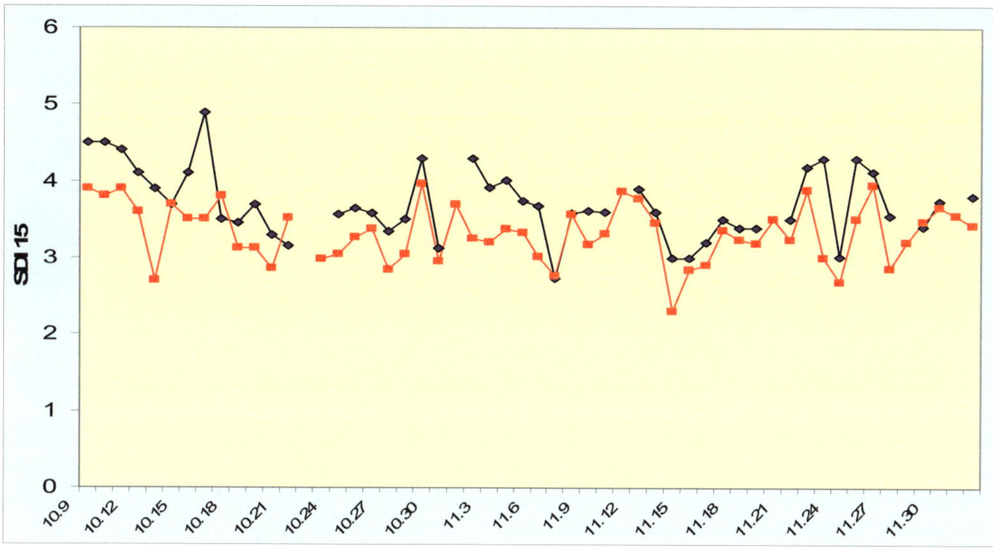

Fig. 27. Comparison of the SDI reduction between single DMF and double stage filtration (red line corresponds to double stage filtration, black line corresponds to DMF)

An in-line SDI monitoring meter is recommended to be installed downstream of each second stage filter cell. If the in-line monitors indicate that the SDI reading is above the design criterion for this parameter (SDI < 3-4), an alarm will give to operators to manually isolate the filter train . An in-line turbidimeter is also recommended to be installed in the first and second stage effluent channel.

Leparc (Leparc & al., 2008) has developed a new concept for the double filtration based on the following advantages points:

- Full-scale application with battery of filters: improved process stability and robustness against upsets of 1st stage
- Improved SDI mainly during first few hours of filtration
- Reduce the size of the 2nd stage (30 - 70% reduction based on raw seawater quality)
- Avoid useless cost for over-quality as compared to a full double stage pretreatment
- Provide better stability of water quality as compared to single-stage filtration

3.2.3 Three media filtration

When three media are used in filters, a better coarse-to-fine filtration pattern can be constructed. High-density silica sand, garnet, and anthracite are commonly used to provide the filter bed. This gradual change in media size provides a gradient from coarse to fine and creates a media flow pattern necessary to achieve a very low silt density index. However, it is observed that after a backwash (air + water), the different media do not stratify completely. Some desalination pre-treatment systems use an alternate media such as greensand to remove iron and manganese compounds.

3.2.4 Gravity filters or pressure filters

The open filter cells will be covered to minimize algae growth. Protective coating suitable for seawater applications will be used in the filter cells as the structural integrity of the concrete structure will be adequate. for example, pressure granular media filters (figure 30) are used for Oman Sur desalination plant (Oman). Pressure filters are also widely used in small plants worldwide because they are cost-competitive, space efficient and easier and faster to install and operate when compared to granular media gravity filters. Often when the source seawater is collected via open intake, two-stage dual media (sand and anthracite) pressure filters are applied.

4. Enhanced coagulation

For a degraded seawater containing high suspended solids, high concentration of algae, oil , organic matter, a pre treatment including only a direct filtration is not enough. It is necessary to add a solid -liquid separation systems to remove these physico-chemical parameters.

4.1 Flotation

As opposed to settling, flotation is a solid–liquid separation technique that is applied to particles whose density is lower or has been made lower than the liquid they are in.. The Dissolved Air Flotation (DAF) process is proving to be a very efficient and cost effective pre-treatment option. Several suppliers have selected the DAF as the preferred pre-treatment provider for a large RO desalination plant.

4.1.1 Principle of flotation

The DAF process is an efficient process for the separation of suspended matter (turbidity, algae etc.) and SDI from seawater following the addition of a coagulant chemical and flocculation(figure 28). Dissolved air flotation (DAF) utilises the property of micro-bubble adherence to suspended solids, increasing the tendency of the particles to float. The flocculated water meets a water flow with supersaturated air (85% to 95%) which is supplied through nozzles. Due to the pressure drop of the supersaturated water at the nozzles, small air bubbles are formed. These micro-bubbles attach themselves to discreet floc particles created in the flocculation process. The rising velocity of the air bubbles is higher than the water velocity and the air bubbles will thus collide with the flocs in water. The density of the aggregates decreases until values below the water density. As a consequence, the aggregates will float on the water surface (Shawwa & Smith, 2000; Edzwald et al., 1999; Haarhoff, 2008; Peleka & Matis,2008). As they rise to the surface, the buoyant flocs form a stable sludge layer above the water surface. Mechanical scrapers skim the solids from the surface into a collecting bin. When surface scrapers are used a sludge with a dry solids content in excess 2 -3% may be produced.
The size of the bubbles greatly affects the efficiency of the flotation process, with bubbles smaller than 100 µm considered the most effective (Edzward,1995; 2007a ; 2007b). Air bubbles of 20 to 50 µm are considered the best for the recovery of fats. The air to solids ratio has a major effect on the performance of a DAF unit. The proportion of the TS present as suspended solids is also critical in determining efficiency (Arnold et al,1995; Edzwald & Wingler, 1990), Depending on the raw water quality and the efficiency of mixing of the recycle stream with the flocculated water, the amount of recycle required typically lies somewhere between 8% and 12% of the influent flow (Peleka & Matis,2008).

The amount of air that can be dissolved in a certain volume of water depends on the pressure and the water temperature and can be calculated wit Henry's law:

$$Cs \;\; = k_D \, C_l = k_D \, \frac{MW \, p}{R \, T}$$

Cs= saturation concentration of gas in water (g/m3)..k_D = distribution coefficient ,.C_l= specific density of air at the prevailing temperature and pressure (g/m3). MW= molecular weight of gas (g/mol), p = total air pressure (Pa),.R is the universal gas constant = 8,3142 (J/(°K mol), T = temperature (K)
Assuming a laminar flow and spherical aggregates, the rising velocity can be calculated with Stokes' law

$$V_{st} \;\; = \frac{1}{18} \frac{g}{\nu} \frac{(\rho w - \rho a)}{\rho w} \quad d^2 a$$

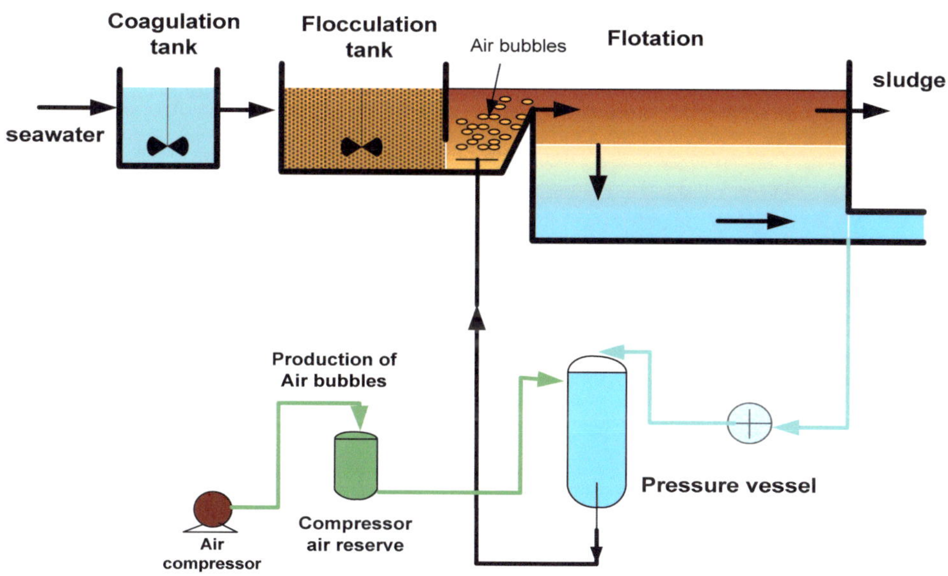

Fig. 28. Schematic of a traditional dissolved air flotation system

Vst= rising velocity of the air bubble floc-aggregate (m/s); ρa = density of the aggregate (kg/m3); ρw= density of the water (kg/m3); da = diameter of the aggregate (m)
For seawater, high rate DAF processes have been developed at loadings of 15 - 50 m/h. For example, Spidflow™ which works at 30 -40 m/h for seawater application, comprises a coagulation stage, followed by a flocculation step and a clarification phase through fast flotation (figure 29). The flocculation stage may also use a Turbomix™ when dealing with cold water. The fine air bubbles, formed by pressurising air in water (at pressures of 5 to 6 bar) when producing white water, are injected through specific nozzles into the Spidflow™ flotation units through a dedicated distribution system. This ensures the separation of Suspended Solids (SS), algae, oil, and hydrocarbons, which are trapped in hydroxide flocs formed by the addition of coagulant (figures 30 & 31).

The hydraulic sequencing of the various compartments of the Spidflow™ process has been designed in accordance with specific Computerized Fluid Dynamics (CFD) type studies. Spidflow™ has a floor for the distribution of flocculated water, which is located before the mixing step with white water. It also includes anti-spiral flow plates that break down any short circuits and collection lines which uniformly distribute water flow. This unparalleled process optimisation ensures that Spidflow™ achieves levels of treatment efficiency which allow it to operate at clarification rates between30 and 50 m/hour.
Spidflow™ fits specifically well seawater desalination pre treatment, as an upstream step of a reverse osmosis membrane treatment chain. Spidflow™ is especially efficient during red tide algal bloom periods. This process significantly maximises filtration cycles duration following pre treatment steps and protects reverse osmosis membranes against ill-timed clogging. As a result, Spidflow™ guarantees very low SDI (Silt Density Index) figures that remain stable over time.

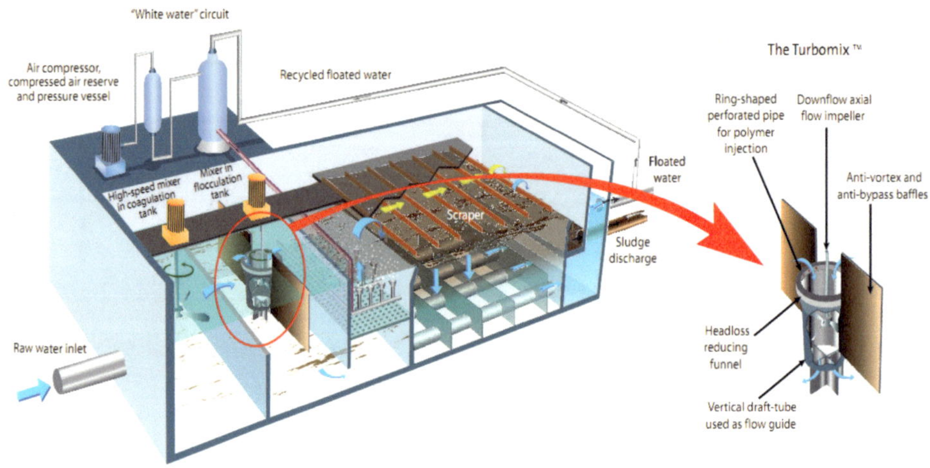

Fig. 29. Spidflow system

Fig. 30. White water formation (Spidflow™)

Fig. 31. Sludge blanket at the top of the flotator (Spidflow™)

4.1.2 Performances expected with the flotation

4.1.2.1 Oil removal

The limit accepted by the RO membranes is 0.1 mg/L. In lot of cases, desalination plants facilities are built in or near industrial zones and draw their raw water from sometimes very busy shipping zones. A carefully designed intake structure is essential and oil has to be removed to ensure the systems do not get damaged. The DAF and is one of the best suited to remove oil from sweater. If oil were to get through the flocculation zones onto the flotation zone as free oil it will almost certainly be trapped by the micro bubbles and be floated of into the sludge blanket for removal along with the floc. The percent removal is around 90 %.

The filtration stage will not removed oil content, it is why the DAF has to be well designed to ensure the system against oil presence.

4.1.2.2 SDI removal

Silt density index (SDI, according to ASTM) is an easy and useful tool for particle evaluation and has been widely applied to determine the fouling characteristics of membranes. In principle, the removal efficiency for SDI is largely dependant on the chemical dosing adopted for the plant. If the chemistry is not right, the clarification and filtration plant will not be able to achieve the required treated water standards no matter how high the efficiency of the process or the quality of the equipment. It is now common knowledge that in order to achieve the required SDI levels it is not sufficient to just remove solids and algae from the water. It is important to look at reduction of the dissolved organic matter as well.

4.1.2.3 Dissolved Organic Carbon (DOC)

Dissolved Organic Carbon (DOC) can be reduced using enhanced flocculation (figures 32 & 33). Enhanced flocculation is performed at a lower pH, typically requires a more coagulant ($FeCl_3$ most of time) and possibly some acid to depress the pH to the right level. Ferric salts are generally considered to be more effective for the removal of organics when compared to Aluminium salts. A coagulation aid polymer can be included in the design. The DAF process combined with the use of enhanced coagulation, the small foot print and short retention times have also been demonstrated to provide good TOC reduction (resulting in very low SDI's) and minimising the potential for issues such as bio-fouling.

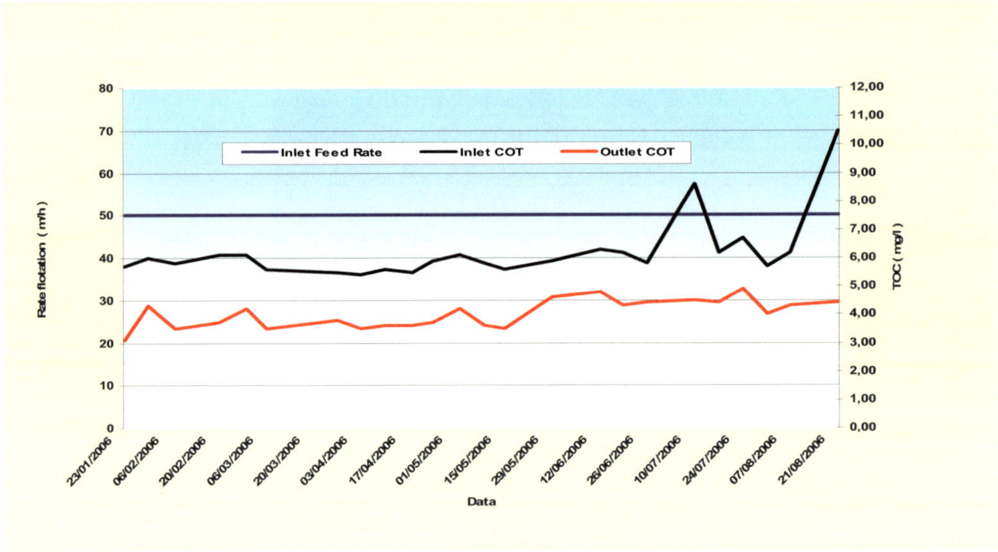

Fig. 32. Performances of the removal of TOC with DAF

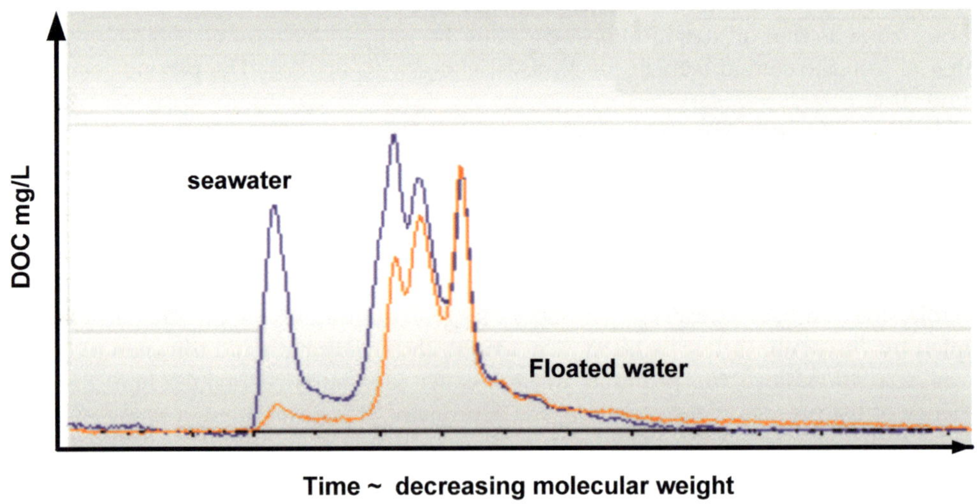

Fig. 33. LC-OCD of the raw water and floated water

4.1.2.4 Algae

One of main applications of DAF is for the removal of algae. Algae are difficult to remove by conventional treatment such as sedimentation, as they are naturally less dense than water so doesn't settle well (Al-Leyla & Middlebrooks, 1974; Vlaski et al., 1997).Algal species

typically are characterized by a negative surface charge in natural water environments. In the work done by Bernhardt and Clasen (Bernhardt & Clasen, 1991), algal destabilization by charge neutralization was achieved through chemical coagulation and flocculation. Maximum filterability of suspended algae was observed when algal cells were destabilized and in aggregate form.
Poor removal of algae (diatoms, green algae, flagellates, blue green algae) can lead to clogging of granular media filters and short filter runs. 99 - 99.9 % removal can be obtained with DAF (figure 34) compared with 90 - 99 % by sedimentation. The removal of algae with flotation is more effective than sedimentation (60 - 90%) depending of the inlet concentration of algae (Henderson et al., 2008, Edwards , 2007a).

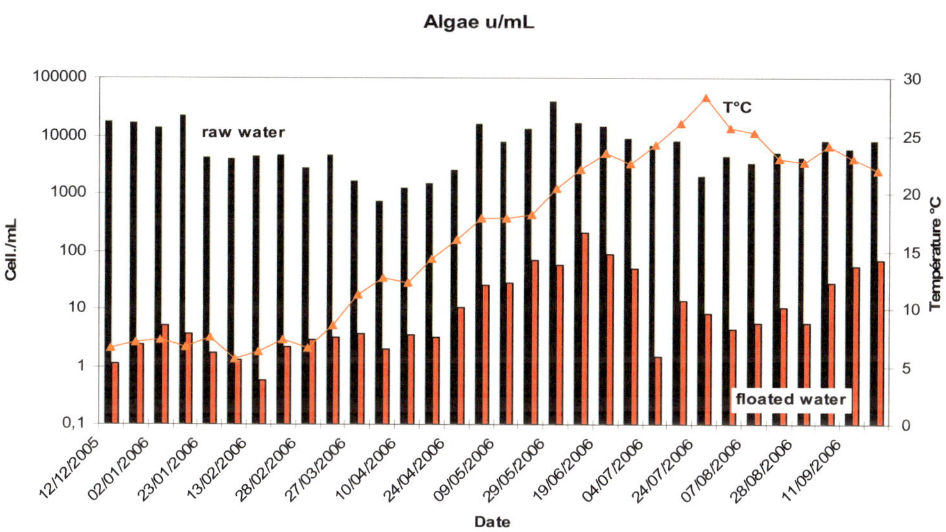

Fig. 34. performances of the removal of algae with DAF

4.1.2.5 Conclusion for flotation process

If the chemistry is right, the chances of achieving the treated water quality is much better with Dissolved Air Flotation followed by Filtration than with any other process and that could possibly include membrane filtration. The reason for this - we believe - is that the very small particles that make up the material that is collected on the filter paper when one analyses for SDI is the very small material that for some reason is not captured in the floc or did break away from the floc and has a density equal to or even a little lighter than water.
In sedimentation processes (in particular the very high rate versions) this material finds its way onto the filters and inevitably a percentage will break through. The chances of capturing these particles in a DAF plant are much better as the particles are small and light.
Low SDI's are further assured by spending some further attention to detail when designing the DAF plant such as: the correct flotation and filtration rates; the method of mixing the saturated water with the flocculated water; the method of removing the collected solids

without disturbing the sludge blanket; distribution of the clarified water over the filter media; washing methods etc.

4.2 Sedimentation

The principles of the coagulation & flocculation are simulars of those described in the flotation chapter. The Coagulation process is the dosing of coagulant in water, resulting into the destablisation of negatively charged particles. a. Coagulation. Non settleable solids and some suspended materials do not precipitate because of electrical charges on the surface of the particles. If the charges on the particles can be reduced, the particles may precipitate. Chemicals that lower surface charges are lime, alum, ferric salts, and polyelectrolytes (Smoluchowki,1916).

During adsorptive coagulation, micro-particles present in the seawater are adsorbed to the positively charged hydrolysis products. The optimal pH-range for adsorptive coagulation with iron salts is between 6 and 8, the optimal pH-range with aluminium salts is more narrow and is about 7.

Flocs are formed from a combination of suspended materials in the raw water together with adsorbed and precipitate solids gained via coagulation. For seawater pre treatment, frequently iron chloride ($FeCl_3$) is used as coagulant. Alternatively, aluminium sulphate ($Al_2(SO_4)_3$) can be applied.

4.2.1 Different types of clarifier used as pre treatment

4.2.1.1 Floc blanket clarifier

The character and behaviour of the suspended bed within a floc blanket clarifier (FBC) lies at the heat of the process. The blanket can be envisaged as being held in place as a result on a balance between the flux of material associated with the upflow velocity and a downward flux associated with sedimentation. It consists of assemblage of flocs whose sizes are relatively large compared with the incoming feed floc and serves as a filter of the incoming material. As collection progresses, there is an accumulation of solids held within the suspension, this manifesting as an increase in the blanket depth. Periodically, excess solids are extracted. The impact of the flocs in the FBC can be examined in term of blanket dynamics and clarification capacity.

When using hydrolysing metal coagulants, the greater the dose, the grater is the propensity to trap water. One of the advantages of optimising the dose and pH is that the floc density appears to be maximised at the optimum conditions. The advantage of using polymer in solid liquid separation the dose of metal coagulant (by charge neutralisation ,is reduced

4.2.1.2 Actiflo

Ballasting refers to the use of materials added during flocculation, that increase the density of flocs resulting in faster settlement. The ACTIFLO® process is a high rate settling process that combines the advantages of ballasted flocculation and lamella clarification (figure 35). The aim of Actiflo is to remove suspended solids, reduce turbidity and remove dissolved organic water. The use of microsand gives the Actiflo® process the advantage of treating a wide range of raw water. Density and shape of microsand particles increase the flocculation and settling efficiencies. As a consequence it is easier and faster to create a strong floc, even with lighter solids coming in. Once the floc is formed and attached to the microsand, the floc settling velocity is high enough to allow for high rise rates.

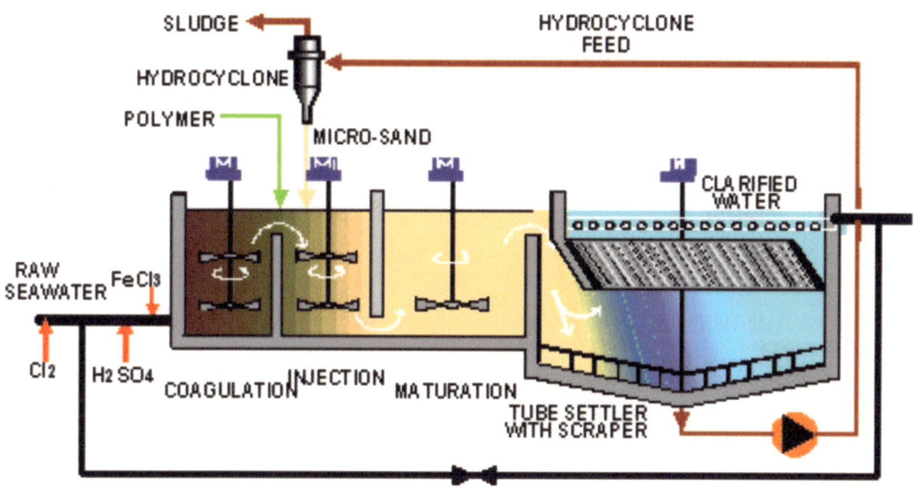

Fig. 35. Actiflo settler

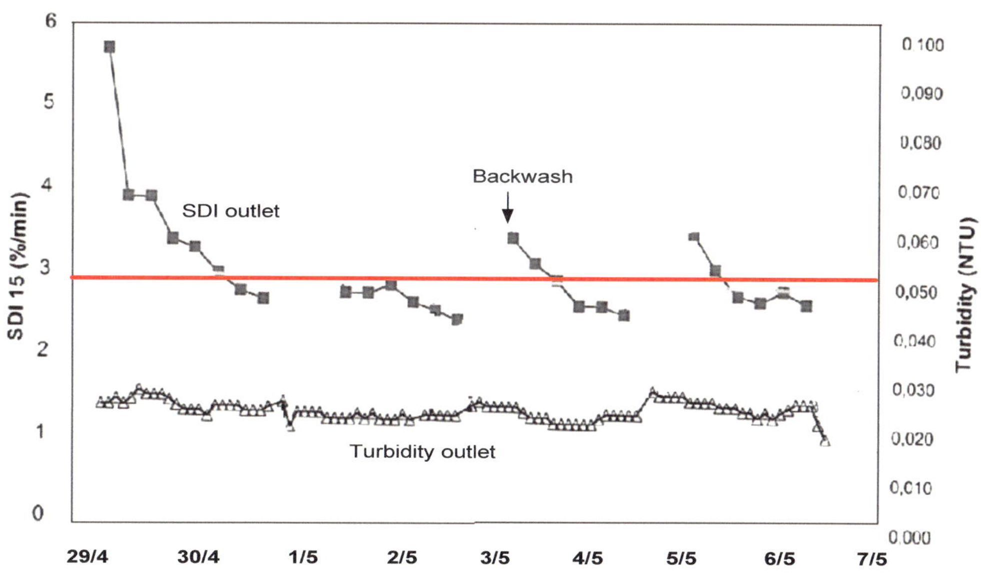

Fig. 36. Performances of Actiflo on SDI and turbidity for seawater pre treatment application

Ferric chloride is injected into the coagulation chamber at a dosing rate of 10 ppm. Rapid mixing in this basin starts the coagulation process. Coagulated raw water then enters the injection tank where micro-sand and polymer are added. Polymer is injected at a dosing rate of 0.2 ppm. Dynamic mixing induces a high probability of contact between coagulated solids, polymer and microsand. Flocculated water enters the maturation chamber where a slow mixing process enables floc maturation and the increase of floc size. The micro-sand then becomes the nucleus of the newly formed floc. Rapid settling of ballasted floc is achieved in the lamella clarifier section. Micro-sand allows for high rise rate design of the lamellar clarifier. Clarified water is then collected at the surface. Sludge is extracted from the bottom of the clarifier by means of an abrasion-resistant centrifugal pump. Using the centrifugal vortex principle, a hydrocyclone then separates the sludge from the reusable microsand. These microsand particles are discharged through the underflow of the hydrocyclone and recycled in the injection tank. The fine and lighter sludge particles move upwards with the major fraction of the water out through the vortex overflow. The application of the Actiflo proicess for seawater pre treatment gives a good efficiency for the removal of suspended solids, turbidity, SDI, dissolved organic carbon and algae (figure 36).

5. Membranes (ultrafiltration and microfiltration)

Most of the desalination plants use conventional pre treatment processes (i.e. dual media filtration preceded by coagulation and sometimes by sedimentation or air flotation for more challenging seawaters). Despite these conventional processes are quite efficient in decreasing the fouling ability of the raw seawater, they could present some difficulties to maintain at anytime a SDI below 3 when this low value is required and when high variations of the seawater quality is observed. However, they could easily meet a value of SDI ≤ 4 which is value largely accepted by the membrane suppliers. The main advantage of the conventional pre treatment processes is the use of adequate and high dosage of the coagulant (such Fe) which implies a reduction of high organic matter which control the bio-film on the RO membranes (Laine et al.,2003).

In drinking water plants fed from surface or ground water, low-pressure membrane processes such as microfiltration (MF) or ultrafiltration (UF) are used to produce high quality water. In the last ten years, ultrafiltration (UF) or microfiltration (MF) pre-treatment has gained widespread attention as potential pre-treatment to seawater desalination by seawater reverse osmosis (SWRO). While in the period until 2002, mostly pilot studies were undertaken, in recent years there have been about 10-15 seawater reverse osmosis (SWRO) plants implemented using ultrafiltration pre-treatment (Bonnélye et al.,2008; Jerowska et al.,2009; Kim & Yoon, 2005).

Membrane filtration pre-treatment involves forcing seawater through a membrane with very fine pores. Particles that are larger than the pores are filtered out. Current large plants utilise low Coagulant dosage since dissolved organics can still pass through the membrane. There has been increasing utilisation of membrane filtration for reverse osmosis systems over the last few years. It is why, low pressure membranes such as ultrafiltration (UF) and microfiltration (MF) are choose for pre-treatment in RO systems (figure 37), primarily due to their effectiveness in removing potential foulants in sea waters. The table 12 summarizes the different specifications of some membranes used for the seawater pre treatment. In treating bad seawater quality, MF and UF are susceptible to fouling for which backwashing without chemicals is sometimes no longer effective. This may result to an increase in chemical consumption for membrane cleaning as well as for in-line coagulation. Several studies reported that polysaccharides are the main cause of fouling in MF/UF membranes (Yoon et al;2004 ; Yang & Kim,2009;Kruithof et al.,1998).

The origin of the flux decline can be accounted for by using different theoretical kinetics models commonly employed for systems showing flux decline (McCabe et al., 1985). Hermia (Hermia.J, 1982) and Van Hoof (Van Hoof et al., 2001) introduced some filtration models: (a) complete blocking, (b) intermediate blocking, (c) standard blocking, and (d) cake filtration which aim to describe fouling mechanism. The complete blocking model occurs when particles arrive at the membrane and block some pore or pores with no superposition of particles. An intermediate blocking model is when particles can settle on other particles previously arrived that already block some pores. A standard blocking model is when particles arrive at the membrane and are deposited onto the internal pore wall. The cake filtration model is when particles are located on other already arrived and are already blocking some pores. During the filtration mode, all the solids in suspension retain on the membrane surface. These retained solids are generally referred to as fouling and the filtration process cannot be maintained indefinitely, due to the fact that the driving force across the membrane has to be increased constantly to keep the flow through the membrane constant. Hence, the system is backwashed (BW) by reversing the flow direction through the filter at regular intervals (Xia et al.,2004; Teuler et al.,1999). The solids are washed away to drain and the whole process repeats itself. While backwashing can remove most of the solids from the system, chemical cleaning methods have to be applied to completely clean the membrane. Some substances tend to adhere to the membrane surface so they cannot be removed by mechanical force alone. (Boerlage et al.,1997) These substances, often of organic and microbial origin, tend to slowly but surely block the membrane. This blocking of the membrane is what ought to be called fouling since the removal of these particular substances is most often not the main purpose of the process. They are sometimes in solution (small organic compounds) and would pass through the membrane, if not for their strong tendency to adhere to the surface. They could also constitute micro-organisms that are removed by the membrane, but start producing extra cellular substances once they have settled onto the membrane surface. The primary goal of in-line coagulation (Qin et al.,2006) is stabilization of the dead-end UF process between 2 chemical cleanings. Dosing of some ppm's coagulant is stabilizing the dead-end filtration process resulting in a slowly rising resistance-profile in time (Maa et al.,2006; Liang et al.,2008;Chen et al/,2007; Knox-Holmes et al.,1994). A recent study by Villacorte (Villacorte et al.,2009) reported significant amounts of TEP in the raw water and after MF/UF pre treatment. It was also found that acidic polysaccharides smaller were up to 5 times more abundant than those larger than 0.40 μm. The presence of colloidal TEPs is also essential considering that MF/UF pre-treatment may not completely remove this fraction from the RO feed water. Systems composed of ultrafiltration (UF) pre-treatment for seawater reverse osmosis (SWRO) desalination are often termed "integrated membrane system" or "dual membrane system".

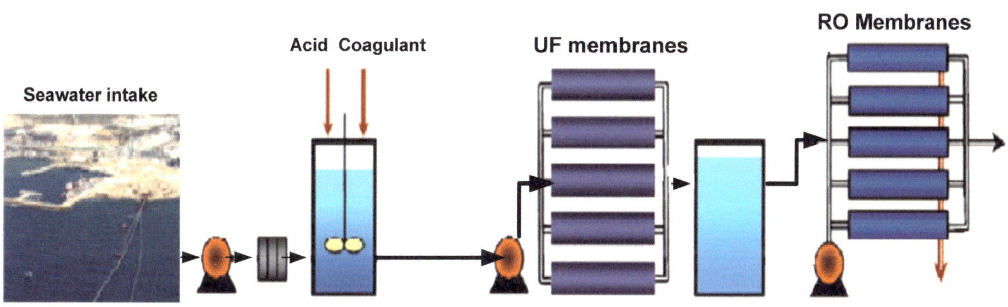

Fig. 37. Systems composed of ultrafiltration (UF) pre-treatment for seawater reverse osmosis (SWRO)

There have been few studies about seawater RO pre-treatment by membrane processes in the past. Kumar (Kumar et al.,2007) compared MF and UF membranes in pre-treatment to determine differences in filtrate quality: 0.1 µm MF and 100 kDa UF membranes showed no difference in term of flux decrease in the RO element, suggesting equal fouling potential of the filtrate. On the contrary, a 20 kDa UF membrane resulted in a reduced flux decline in the RO element, suggesting less membrane fouling. In 2003, Vial & Doussau (Vial & Doussau, 2002) tested 0.1 µm hollow-fibre membranes for the pre treatment of Mediterranean seawater. They observed no influence of turbidity and SDI peaks on permeate turbidity and SDI. In 2004, Pearce (Pearce,2004) used an UF membrane pre treatment at Port Jeddah, Saudi Arabia, as an alternative to its conventional pre treatment facility, which could not meet targeted feed water quality during algal blooms and storms. The implementation of membrane pre treatment with daily air-enhanced backwashes achieved an average filtrate SDI of 2.2, which corresponded to an SDI improvement of two units compared to the previous conventional pre treatment. Higher RO feed water quality hence resulted in reduced fouling of the RO element by 75%.

Most of these studies about seawater RO pre treatment by membrane processes are based on an evaluation of pre treatment performance (Choules et al., 2009); Gaid & Craig,2009) through conventional and limited analytical tools such as SDI, turbidity or particle counts. Moreover, few of these studies presented a side-by-side comparison of conventional and membrane pre treatment fed at the same time by the same seawater.

Membrane element characteristic	ZeeWeed 1000 Ultrafiltration	Norit Ultrafiltration	DOW Ultrafiltration	Hydranautics Ultrafiltration	Pall Microfiltration
Active Membrane Area (m^2)	55.8	40	51	46	50
Flow Path (In-Out; Out-In)	Outside-In	Inside-Out	Outside-In	Inside-Out	Outside-In
Molecular Weight Cutoff (Daltons)	100 000	150 000	150 000	150 000	-
Nominal Membrane Pore Size (microns)	0.02	0.05	0.05	0.05	0.10
Absolute Membrane Pore Size (microns)	0.1	0.075	0.075	0.075	-
Membrane Material	PVdF	PES	PVdF	PES	PVdF
Membrane Hydrophobicity	Hydrophilic	Hydrophilic	Hydrophilic	Hydrophilic	Hydrophobic
Membrane Charge	Slightly negative	Neutral	Neutral	Neutral	Slightly negative
Acceptable Range Operating pH Values	5 - 10	2 -12	2 -12	2 -12	1 - 10
Maximum TMP for System (bar)	0.9	2.5	2.1	2.5	3
Chlorine/Oxidant resistance	Resistant to NaOCl, ClO2, KMnO4	200 ppm continuous	>5 000 000 ppm.hrs	200 ppm	Chlorine 10 000 ppm, oxidant resistance
Required Pre screening (microns)	100	100	100	100	200

Table 12. Main characteristics of some membranes used for pre treatment of seawater

5.1 Performances of the ultrafiltration and microfiltration membranes

5.1.1 Biofilm control on Reverse osmosis membranes

Controlling the persistent problem of organic and biological fouling, is still a major challenge in many reverse osmosis (RO) plants (Shon et al.,2009, Wittmann ,2011). A membrane bio-film is known to offer significant advantages for micro-organisms embedded within its matrix, including better stability against mechanical cleaning and higher tolerance to biocides. Bio-film studies cited the importance of extra-cellular polymeric substances (EPS) in the cohesion of microbial cells and other particles, as well as adhesion to surfaces. Whether the accumulation of EPS in RO membranes is mainly due to local production by bio-film micro-organisms or by gradual deposition of EPS from the RO feed water, is still not clearly known (Aleem et al.,1998; Fujiwara & Matsuyama,2008; Tekeuchi et al.,2008)).

Over the last decade, the discovery of a formerly undetectable but abundant type of EPS called transparent exo-polymer particles (TEP), has led to a better understanding of its role in the carbon cycle and biological life in aquatic systems. TEP is distinct among EPSs in many ways. Unlike most EPS, TEP exist as individual particles rather than as cell coatings or dissolved slimes. They have been characterised as transparent, sticky, gel-like substances which are comprised mainly of acidic polysaccharides. TEPs are hydrophilic and known to exist in different shapes (blobs, clouds, sheets, fibers or clumps) and sizes (~0.4 to 200 μm). As a planktonic type of EPS, they are ubiquitous in most fresh (terrestrial/surface) and marine waters and have also been found in wastewater.

Most TEP originates from polysaccharides released by phytoplankton and bacterio-plankton, which subsequently coagulates to form TEP. However, they are also exuded or lysed out from macro algae and some higher marine organisms . The majority of TEP are formed abiotically from colloidal polysaccharides, 1-3 nm in diameter by hundreds of nanometers long, which are flexible enough to pass through 8 kDa pore size membranes. Thus, these colloidal polysaccharides or the so called "colloidal TEPs" are capable of passing through MF/UF pre-treatment and may compromise the operation of the reverse osmosis (RO) system downstream.

5.1.2 Algae & membranes

Algal species are larger than the nominal pore size of both microfiltration and ultrafiltration membranes; therefore, rejection of algae occurs to a large extent on the basis of size. However, the trapped cells may release extra-cellular matter that can block the module outlet and lead to severe flux decline or TMP increase. Membrane fouling often occurs as a result of accumulation and/or adsorption of rejected materials at the membrane surface. However, in the presence of algae, the layer of rejected material at the membrane surface should not be considered as a non-adhesive cake layer (Adham S.,1997). In the absence of any pre-treatment, algae cells and organic matter can easily enter the UF module. Organic matter can penetrate within the membrane pore and cause evident flux decline. Live algae cells can also deposit on the surface of membrane and release extra-cellular material during filtration. The released polysaccharides can bond with other organic species and increase the resistance for filtration. For this reason, a direct UF system is not suitable for algae-rich water treatment (Passow, 2000 and 2002; Kwon et al.,2005). Instead, the extra-cellular algal slimes (mucilaginous or gelatinous materials) are likely to serve as "cement" for the particulate material rejected by the membrane surface. The presence of extra-cellular algal slimes is likely to contribute to a greater superimposed resistance to filtration (Thornton, 2004) than would be observed for a non-adhesive cake layer.

Coagulation pre-treatment can improve UF by flocculation of the material depositing on the membrane surface, which can then form a cake layer. The low molecular organic matters can be trapped within this layer and the fouling is reduced. During the algae-rich water treatment, coagulation - flocculation can only adsorb and flocculate the algae cells, while it can not make them inactive. Coagulation pre-treatment alone can not satisfy the needs of UF application in algae-rich water treatment.

5.1.3 Feed back of pilot plants and existing plants

a. The seawater of the Palm Jumeirah plant (figure 38) is characterized by a very high seawater SDI measured as SDI 2.5 min. Furthermore, as is common in the seawaters of the Gulf, the contribution to the SDI is given by particles of very small size and this accounts for the relatively poor turbidity values observed upstream of the Norit UF membranes. On the other hand excellent permeate quality was obtained during plant operation with excellent permeate SDI regardless of the very high feed seawater quality (around 30 NTU). Measured values of permeate SDI are in range from 0,5 to 3 with the average about 1,6. SDI values obtained during operations (Sommariva et al.,2009). Experiments were performed on site using 10μ, 5 μ, 1 μ and 0.47 μ test filter papers. It was found that the 10μ and 5μ filter papers removed almost no particles at all; the 1 micron filter paper was able to remove around 10% of the particles; a further 45% of particles were removed by the 0.47μ filter and the remaining 50% had a particle size of less than 0.47 μ. It is possible that some of the particles would be small enough to pass even the UF membrane (or migrate through faulty seals). For this reason a small amount of Ferric Chloride is dosed as a coagulant filtration aid (~0.25 mg/l as total iron was found to be optimal). The high volume of very small particle size silt and the consequent necessity to dose ferric chloride led to deposits forming on the membrane surface and to escalating operating pressure of the UF units as backwashing and CEB could not fully restore the trans-membrane pressure differential. A 'clean-in-place' (CIP) regime was implemented to restore the membranes to their original performance levels. The selected cleaning chemical was a mixture of oxalic acid ($H_2C_2O_4$) and ascorbic acid ($C_6H_8O_6$). The cleaning solution consisted of RO permeate containing 1% oxalic acid and ¼% ascorbic acid heated to 36°C and the procedure was a series of soak and recirculation stages followed by rinsing with fresh, warm RO permeate (Ingham et al. 2009).

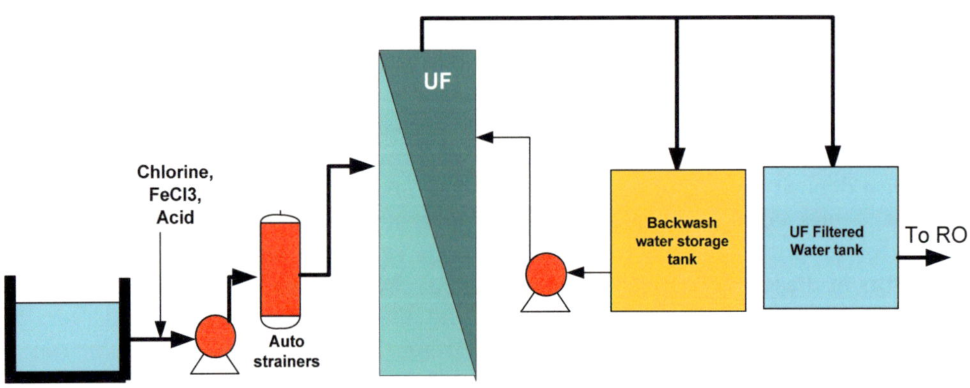

Fig. 38. Palm Jumeirah Pre treatment including Norit X-Flow UF

b. Codemo (Codemo et al.,2009) said that since UF membranes provide a barrier to particulates, they provide significant benefits to the RO and overall system design. Everybody must be aware that if UF is not handled together with respect RO as one compact system - significant problems in RO performance may occur. RO performance could suffer due to heavy bio-fouling despite the fact that UF filtrate will provide very good membrane feed water with no particulates. But despite this - one important parameter may be out of control - bio-fouling. Bio-fouling is much bigger and more dangerous threat for UF-RO system than i.e. fiber breakage.

c. Busch (Busch et al., 2009) described the cleaning occurred on Wang Tan power plant where the modules been autopsied and showed a significant amount of fouling on the module. The fouling appeared to be of brown to red colour. After analysis, it was concluded that the colour of the fouling can be explained by the presence of high organic and iron levels. The most appropriate cleaning condition was as follows: oxalic acid 2%, temperature 35°C, circulation time 2 h, then soaking 3 h, then backwashing. It can be seen that after this protocol, the module can be very well cleaned again and the full permeability can be restored.

d. Knops (Knops & Lintelo, 2009) describe the tests done in 2007 on the Colakoglu steel mill plant where they have observed a loose of the permeability on the seaguard UF Membrane. One membrane element was removed from the UF system and returned to the factory for autopsy. The immediate investigation revealed a dark red discoloration of the membrane element . EDX analysis of the membrane confirmed that the colour was caused by iron, with trace amounts of silica and aluminium being present. This indicated that the most probable cause of the drop in performance was fouling with the inorganic coagulant ($FeCl_3$). Laboratory scale testing of the performance revealed a sharp drop in permeability when compared to the original value for the new membrane.. A plant cleaning in place (CIP) was performed to remove the bulk of the fouling (Junga & Son, 2009).

5.1.4 Lessons learned from the UF membrane SWRO pre-treatment

If there is any doubt about the long term quality of the feedwater then the following should be considered when designing a desalination plant:

- UF will provide effective treatment of the raw water; therefore the RO stages can therefore be designed quite aggressively.
- Turbidity alone is not a good predictor of fouling potential.
- The UF system should be designed using conservative flux and net recovery assumptions. Physical space should be left for additional UF plant in case anticipated flux values and net recovery are difficult to achieve. Additional capacity should be designed into the screening and straining stage.
- Ultrafiltration provides more stable water quality than a multi media filtration

But an irreversible membrane fouling has to be considered and could be a limitation in the application of this technology depending of the seawater quality. The accumulation of the retained matter on the membrane surface leads to an increase in operating costs, due to progressively increasing energy consumption and the necessity of periodic cleaning. To reduce these operating costs, it is necessary to control the fouling behaviour. This can be achieved by in-line coagulation. In addition to performance enhancement, in-line coagulation promotes the retention of organic macro-molecules and phosphates, which helps to reduce bio-fouling in downstream processes, such as SWRO. However, the application of in-line coagulation does have drawbacks. Firstly, it forms a large portion of the operating costs, due to chemicals consumption and the increased disposal costs of the concentrate stream. Secondly, coagulant residuals in the permeate, caused by over dosing, reduce the product quality and can lead to issues in downstream processes, for example SWRO (Futelaar et al., 2009; Yacubowicz, 2010).

6. Acknowledgments

Thanks to the Veolia Environment Research and Innovation (VERI) Division for the support and the results of the pilot tests (Philippe Breant, Jerome Leparc and Jean Cantet)

7. References

Adams, L. F., & W. C. Ghiorse, W.C., (1987), Characterization of extracellular Mn2+-oxidizing activity and isolation of a Mn^{2+}oxidizing protein from Leptothrix discophora SS-1. *J. Bacteriol.*, vol.169, pp.1279-1285

Adham S., (1997), Control of membrane fouling due to algae, Montgomery Watson Technology Transfer Note

Applied Research Department No. 12, January 1997, pp.1-4

Aleem, F.A.A.E., Al-Sugair, K.A. & Alahmad., M.I., (1998), Biofouling problems in membrane processes for water desalination and reuse in Saudi Arabia, *Intern.Biodeter. Biodegrad.*, vol.41, pp. 19–23.

Alessio, L., Campagna, M., & Lucchini, R., (2007), From lead to manganese through mercury: mythology, science, and lessons for prevention, *American journal of industrial medicine,* vol. 50, n°11, pp-779–787

Al-Leyla,M.A. & Middlebrooks,E.J., (1974), Algae removal by chemical coagulant, Water *and Sewage Works*, pp.76-80.

Amy, G., (2008), Fundamentals understanding of organic matter fouling of membranes, *Desalination*, vol.231, pp.44-51

Archer, D. E. & Johnson K. S., (2000), A model of the iron cycle in the ocean, *Global Biogeochem. Cycles*, vol.14, pp.269–279.

Arnold,S.R.,Grubb,T.P. & Harvey,P.J.,(1995), Recent applications of dissolved air flotation pilot studies and full scale design, *Water Science and Technology*,vol.3,N°14,pp.327-340

Bauman, A.G., Burt, J.A., Feary, D.A., Marquis, E., & Usseglio,P.,(2010),Tropical harmful algal blooms: An emerging threat to coral reef communities?, *Marine pollution bulletin*,vol.60,n°11,pp.2117-2122

Bernhardt, H. and Clasen, J., (1991), Flocculation of micro-organisms. *J Water SRT-Aqua,* vol. 40, pp.76 - 87

Blute, N.K., McGuire, M.J., West, N., Voutchkov, N., Maclaggan, P. & Reich K., (2008), Integration of desalinated seawater into a distribution system: a corrosion pilot study, *Journal of the American Water Works Association*, September, vol.100:9, pp.117-131

Boerlage, S.F.E., Kennedy, M., Dickson, M.R.,El-Hodali, D.E.Y. & Schippers, J.C., (2002), The modified fouling index using ultrafiltration membrane (MFI-UF): characterization, filtration mechanisms and proposed reference membrane,*J.Membr.Sci.*, 197, pp.1-21

Boerlage, S.F.E., Kennedy, M., Tarawneh, Z., Faber, R.D. & Schippers, J.C., (2004), Development of MFI-UF in constant flux filtration, *Desalination, vol.161, pp.103*-113

Boerlage, S.F.E., Kennedy, M., Bonne, P.A.C., Gilbert, G.& Schippers, J.C.,(1997), Prediction of flux decline in membrane systems due to particulate fouling ,*Desalination*,vol.113,pp.231-238

Boerlage, S.F.E., Kennedy, M., Tarawneh, Z., Aniye, M.P., Abrogrean, E.M., El-Hodali, D.E.Y. & Schippers, J.C., (2002), The modified fouling index using ultrafiltration membranes (MFI-UF): characterization, filtration mechanisms and proposed reference membrane, *J.Membr.Sci.*, vol197, pp.1-9

Bonnélye, V., Guey, L. & Del Castillo, J., (2008), UF/MF as RO pre-treatment: the real benefit, Desalination, vol.222, pp.59-65

Brock, T.D. & Clyne, J., (1984), Significance of Algal Excretory Products for Growth of Epilimnetic Bacteria, *Appl. Environ. Microbiol.*, Vol. 47, n°4, pp.731-734

Busch,M., Chub,R., Kolbec,U., Mengd, Q. & Li,S.,(2009), Ultrafiltration pre-treatment to reverse osmosis for seawater desalination three years field experience in the Wangtan Datang power plant, *Desalination Publications*, vol.10, pp.1-20

Chen,Y.,Dong,B.Z.,Gao,N.Y.,Fan,J.C.,(2007), Effects of coagulation pre-treatment on fouling of an ultrafiltration membrane,*Desalination*,vol.204,pp.181-187

Choi, Y.H., Kweon, J.H., Kim, D.I. & Lee, S., (2009), Evaluation of various pre-treatment for particle and inorganic fouling control on performance of SWRO, *Desalination, vol.247, pp.137*-147

Choules,P., Schrotter, J.C., Leparc, J., Gaid,K.& Lafon D.,(2007), Reverse osmosis: improved operation through experience with SWRO plants, presented at the American Membrane Technology Association's Conference, Las Vegas, Nev., July 23-26

Choules,P., Ben Boudinar,M.& Mack,B.,(2009), Membrane (MF&UF) pre-treatment design and operational experience from three seawter RO plants, presented at IDA World Congress - Atlantis, The Palm - Dubai, UAE November 7-12

Codemo, G., Awadalla, A.R., Parker,M.J., Banham,J. & Rybar,S.,(2009), Hamriyah SWRO Desalination Plant - Largest Sea Water IMS Plant, presented at the IDA World Congress - Atlantis, The Palm - Dubai, UAE November 7-12

de Vrind, J.P., Boogerd,F.C. & de Vrind-de Jong, E.W., *(1986)*, Manganese reduction by a marine Bacillus species.*J Bacteriol,* vol. 167, n°1, pp. 30-34

Edzwald, J.K., (1995), Principles and applications of dissolved air flotation, *Water Science and Technology*, vol.31, n°3-4, pp.1-23

Edzwald, J.K., (2007a), Fundamentals of dissolved air flotation, *Journal of the New England Water Works Association*, vol.121, pp.89-112

Edzwald, J.K., (2007b), Development of high rate dissolved air flotation for drinking water treatment, *Journal of Water Supply: Research and Technology - Aqua*, vol.56, n°6-7, pp.399-409

Edzwald,J.K.,Tobiason,J.E.,Amato,T.& Maggi,L.J.,(1999), Integrating high rate dissolved air flotation technology into plant design ,*Journal of the American Water Works Association*, vol.84,n°3,pp.92-100

Edzwald, J.K. & Wingler, B.J., (1990), Chemical and physical aspects of dissolved air flotation for the removal of algae, *Journal of the Water Supply: Research and Technology –Aqua*, vol.39, n°2, and pp.24-35

Fujiwara, N. & Matsuyama, H., (2008), Elimination of biological fouling in seawater reverse osmosis desalination plants, *Desalination*, vol.227, pp.295-305

Futselaar,H., Blankert, B., Spenkelink, F.& Rosberg R., (2009), Ultrafiltration used as Pre-treatment for SWRO Desalination: Dynamic Coagulant Control and Optimization, presented at the IDA World Congress - Atlantis, The Palm - Dubai, UAE, November 7-12

Gabelich, C.J., Frankin, J.C., Gerringer, F.W., Ishida, K.P. & Suffet. I.H., (2005), Enhanced oxidation of polyamide membranes using monochloramine and ferrous iron, *Journal of Membrane Science*, vol.258, pp.64-70.

Gaid, A. & Treal, Y., (2007), le dessalement des eaux de mer par osmose inverse, *Desalination*, vol.203, pp1–14.

Gaid, A. & Craig, K., (2009), Success story for reverse osmosis technology: two new plants, presented at IDA World Congress – Atlantis, The Palm – Dubai, UAE November 7-12

Geider, R. J.& Roche, J. L.,(1994), The role of iron in phytoplankton photosynthesis, and the potential for iron-limitation of primary productivity in the sea, *Photosynthesis Res.*, vol.39,pp.275–301

Haarhoff,J.,(2008),Dissolved air flotation and prospects for drinking water treatment, *Journal of Water Supply: Research and Technology-Aqua*,vol.57,n°8,pp.555-567

Hallegraeff, G.M.,(1993), A review of harmful algal blooms and their apparent global increase, *Phycologia*, vol.32,pp- 79-99.

Henderson, R.K., Parsons, S.A. & Jefferson, B., (2008), Surfactants as bubble surface modifiers in the flotation of algae: Dissolved air flotation that utilises a chemically modified bubble surface. *Environmental Science and Technology*, vol.42, pp.4883-4888

Her, N., Amy, G., Fos, D., Cho, J., Yoon, Y. & Kosenka, P., (2002), Optimization of method for detecting and characterizing NOM by HPLC-size exclusion chromatography with UV and on-line DOC detection, *Environ. Sci. Technol.*, vol.36, pp.1069-1076

Hermia J., (1982), Constant pressure blocking filtration laws: application to power-law non-Newtonian fluids, *Trans. Inst. Chem. Eng.*, vol.60, p.183

Hong, K. ,Lee, S., Choi, S., Yu, Y., Hong, S., Moon, J., Sohn, J. & Yang J.,(2009), Assessment of various membrane fouling indexes under seawater conditions, *Desalination*,vol.247,pp.247-259

Hong, S.& Elimeleh, M.,(1997), Chemical and physical aspects of natural organic matter (NOM) fouling nanofiltration membranes, *J.Membr Sci.*,(1997),vol.132, pp.159 - 181

Ingham,R.A., Mansour,L., Qadan,T., Al Hindi, M. & Attawneh, O.,(2009), Pre-treatment Design and Performance under Challenging Conditions in the Arabian Gulf, presented at the IDA World Congress - Atlantis, The Palm - Dubai, UAE November 7-12

Iwahori, H., (2003), Peculiar or unexpected behavior of Silt Density Index of pre-treated seawater for RO desalination, presented at the IDA Paradise Island Bahamas, November

Jerowska, A., Bottino, A., Capanelli, G., Fabbri, C. & Migliorini, G., (2009), Ultrafiltration as direct pre-treatment of seawater- a case study, *Desalination*, vol.245,pp.723-729

Johir, A.H., Khorshed, C., Vigneswaran, S. & Shon, H.K., (2009), In-line flocculation as pre-treatment to reverse osmosis desalination, *Desalination*, vol.247, pp.85-93

Johnson, K. S., Gordon, R. M. & Coale, K. H., (1997), What controls dissolved iron concentrations in the world ocean?, *Mar. Chem.*, vol.57, pp.181–186.

Junga, C.W. & Son H.J., (2009), Filtration model: Evaluation of membrane fouling mechanism in various membrane pre-treatment processes, *Desalination and Water Treatment*, vol. 2, pp.195–20

Khirani, S., Ben aim, R., Manero, M.H., (2007), Improving the measurement of the modified fouling index using nanofiltration membrane (NF-MFI), *Desalination*, vol.191, pp.1-7

Kim, P.H., Hong, S. & Choi, S., (2006), Variation and prediction of membrane fouling index under various feed water characteristics,*J.Membr.Sci.*, vol.284, pp.248-254

Kim, S., Chon, K., Kim, S.J., Lee, S., Lee, E., Cho, J., (2009), Uncertainty in organic matter analysis for seawater reverse osmosis 5SWRO) desalination, *Desalination*, vol.238, pp.30-36

Kim, S.H. & Yoon, J.S., (2005), Optimization of microfiltration for seawater suffering from red tide contamination, *Desalination, vol.182, pp.315*-321

Knox-Holmes, B., Hassan, A.M., Williams, E.E. &. Al-Tisan, I., (1994), Fouling prevention in desalination plants, presented in the Second Gulf Water Conference, Bahrain, Vol.1, pp.66-75

Knops, F. & Lintelo R., (2009), Long-term operating experience of Seaguard UF as pre-treatment to SWRO, in the Mediterranean region, *Desalination and Water Treatment*, vol.5, pp. 74–79

Krasner, S.W., Croué, J.P., Buffle, J. & Perdue, E.M.,(1996), Three Approaches for Characterizing NOM, *Jour. AWWA*, vol.88, pp.66-79

Kremen, S.S. & Tanner, M., (1998), Silt density index (SDI), percent plugging factor (%PF) their relation to actual foulant deposition, *Desalination*, vol.119, pp.360-372

Kruithof, J.C., Schippers, J.C., Kamp, P.C., Folmer H.C. & Hofman, J., (1998), Integrated multi-objective membrane systems for surface water treatment: pre-treatment of reverse osmosis by conventional treatment and ultrafiltration. *Desalination*, vol.117, pp.37-48

Kuma, K., Nishioka, J. & Matsunaga, K., (1996), Controls on iron (III) hydroxide solubility in seawater: The influence of pH and natural organic chelators, *Limnol. Oceanogr.*, vol. 41, pp.396–407.

Kumar, M., Adham, S., Decarolis, J. & Pearce, W., (2007), Comparative seawater RO pre-treatment evaluation using bench- and pilot-scale testing, *Journal of the American Water Works Association*, September, vol.99:9, pp.168-178

Kwon, B., N. Park, N. & Cho,J.,(2005), Effect of algae on fouling and efficiency of UF membranes, *Desalination*, vol.179 ,pp.203–214.

Laîné, J.M., Campos, C., Baudin, I. & Janex, M.L., (2003), Understanding membrane fouling: a review of over a decade of research, *Water Sci.Technol.: Water Supply*, vol.3, n°5-6, pp.155-164

LaZerte, B.D. & Burling, K., (1990), Manganese speciation in dilute waters in the Precambrian shield, Canada, *Water Res.*, vol.24, pp.1097-1101

Leparc, J., Rapenne, S., Courties, C., Lebaron, P., Croue, J.P., Jacquemet, V. & Turner, G.,(2007), Water quality and performance evaluation at seawater reverse osmosis plants through the use of advanced analytical tools,*Desalination*,vol.203,pp.243-255

Leparc J., Breant, Ph., (2008), procédé de traitement d'eau à filtration séquencée, France ,11 february, n°2.927.325

Liang,H., Gong,W.,& Li.G.,(2008), Performance evaluation of water treatment ultrafiltration pilot plants treating algae-rich reservoir water ,*Desalination*,vol.221,pp.345-350Liu, X. & Millero, F. J. (2002): The solubility of iron in seawater, *Mar. Chem.*, vol. 77,pp. 43–54

Maa,W., Zhao.Y.& Wanga,L.,(2006),The pre-treatment with enhanced coagulation and a UF membrane for seawater desalination with reverse osmosis,*Desalination*,vol.203,(2006),pp.256-259

McCabe, W.L., Smith, J.C. & Harriott, P., (1985), Unit Operation of Chemical Engineering, McGraw-Hill, New York,

Millero, F. J., Sotolongo, S. & Izaguirre, M., (1987), The oxidation of Fe (II) in seawater, *Geochim. Cosmochim Acta*, vol.51, pp.793–801

Millero, F. J. (1998), Solubility of Fe (III) in seawater, *Earth Planet Sci. Lett.*, vol.154, pp.3232–329

Mitra, S.S., Thomas, A.R. & Gang, G.T.., (2009), Evaluation and characterization of seawater RO membrane fouling, *Desalination, vol.247, pp.94*-107

Mitrouli, S.T., Yiantsios, S.G., Karabelas, A.J., Mitrakas, M., FØ llesdal, M. F. & Kjolseth, P.A., (2008), Pre-treatment for desalination of seawater from an open intake by dual-media filtration: Pilot testing and comparison of two different media, *Desalination*, vol. 222, pp.24-37

Mitrouli, S.T., Yiantsios, S.G., Karabelas, A.J. & Kjolseth, P.A., (2009), New granular for dual-media filtration of seawater: pilot testing, separation and purification technology, vol.65, pp.147-155

Mosset, A., Bonnélye V., Petrya, M. & Sanz, M.A., (2008), The sensitivity of SDI analysis: from RO feed water to raw water, *Desalination*, vol.222, pp.17-23

Mucci, A.,(2004),The Behavior of Mixed Ca–Mn Carbonates in Water and Seawater: Controls of Manganese Concentrations in Marine Porewaters, *Aquatic Geochemistry* ,Vol.10, N°1-2, pp.139-169

Nakabayashi, S., Kusakabe, M., Kuma, K. & Kudo, I., (2001), Vertical distributions of iron(III) hydroxide solubility and dissolved iron in the north-western North Pacific Ocean, *Geophys. Res. Lett.*, vol.28, pp.4611–4614

Passow, U., (2002), Transparent exopolymer particles (TEP) in aquatic environments, *Progress in Oceanography*, vol.55, pp. 287-333

Passow, U., (2000), Formation of transparent exopolymer particles, TEP, from dissolved precursor material. *Mar. Ecol. Prog. Ser.*, vol.192, pp.1-11

Pearce, G.K., (2007), The case of UF/MF pre-treatment to RO in seawater applications, *Desalination*, vol.203, pp.286-295

Peleka, E.N., Matis, K.A., (2008), Application of flotation as a pre-treatment process during desalination, *Desalination*, vol.222, pp.1-8

Petrucci, G. & Rosellini, M., (2005), Chlorine dioxide in seawater for fouling control and post-disinfection in potable waterworks, *Desalination*, vol.182, pp.283-291

Petrusevski, B., Vlaski, A., van Breemen, A.N. & Alaerts, G.J., (1995), GJ. Optimisation of Coagulation Conditions for Direct Filtration of Impounded Surface Water. Journal of Water Supply: Research and Technology-Aqua, Vol.44, No.2, pp.92-102

Petrusevski, B., van Breemen, A.N., Alaerts, G.J., (1996), Effect of permanganate treatment and coagulation with dual coagulants on algae removal in direct filtration. *Journal of Water Supply: Research and Technology-Aqua*, Vol.45, No.5, pp. 316-326

Rue, E.L. & Bruland, K. W., (1995), Complexation of iron (III) by natural organic ligands in the Central North Pacific as determined by a new competitive ligand equilibration /adsorptive cathodic stripping voltammetric method, *Mar. Chem.*, vol.50, pp.117–138

Rue E.L. and Bruland K. W. (1997): The role of organic complexation on ambient iron, chemistry in the equatorial Pacific Ocean and the response of a mesoscale iron addition, experiment. Limnol. Oceanogr., 42, 901–910.

Qin, J.J., Oo, M.H., Kekre, K.A., Knops, F. & Miller, P., (2006), Reservoir water treatment using hybrid coagulation-ultrafiltration, *Desalination*, vol.193, pp.344–349

Schaanning, M., Naes, K., Egeberg, P.K., (1988), Cycling of manganese in the permanently anoxic Drammens fjord, *Marine Chemistry*, vol. 23, pp.365-382

Serkiz S.M.& Perdue, E.M.,(1990), Isolation of Dissolved Organic Matter from the Suwannee River Using Reverse Osmosis, *Water Res.*, vol.24, pp.911-916

Schaanning M, Naes K, Egeberg PK, et al. 1988. Cycling of manganese in the permanently anoxic Drammens fjord. Marine Chemistry 23:365-382.

*Schippers, J.C.& Verdow, J.,(1980), The modified fouling index. A method of determining the fouling characteristics of water, *Desalination*,vol.32, pp.137-145

Shawwa,A.R. & Smith,D.W.,(2000), Dissolved air flotation model for drinking water treatment, *Canadian Journal of Civil Engineering*,vol.27,pp.373-382

Shin, J.Y. & O'Melia, C.R., (2006), Pre-treatment chemistry for dual media filtration: model simulations and experimental studies, *Water Sci.Technol.* vol.53,n°7,pp.167-175

Shon, H.K., Kim, S.H.,Vigneswaran, S., Ben Aim R. ,Lee S. & Cho J.,(2009), Physicochemical pre-treatment of seawater fouling reduction and membrane characterization, *Desalination*,vol.238,pp.10-21

Smoluchowski,M.V.,(1916), Drei Vorträge über Diffusion, Brownsche Bewegung und Koagulation von Kolloidteilchen, *Physik. Z.*, vol.17, pp. 557–585.

Sommariva, C., Al Hindi,M.& Fabbri,C., (2009), Palm Jumeirah: The first large SWRO with Ultra-filtration Plant in the Gulf, presented at the IDA World Congress - Atlantis, The Palm - Dubai, UAE November 7-12

Stumm, W., & Lee, G F., (1961), Oxygenation of ferrous iron, *Intl. Eng.Them.*, vol. 53, pp.143-146

Stumm, W., & Morgan, J J., (1962), Chemical aspects of coagulation, *J. American Water Work. Assn. vol.*54, pp.971-94

Tani, H., Nishioka, J., Kuma, K., Takata, H., Yamashita, Y., Tanoue, E. & Midorikawa, T.,(2003), Iron(III) hydroxide solubility and humic-type fluorescent organic matter in the deep water column of the Okhotsk Sea and the northwestern North Pacific Ocean, *.Deep-Sea Res. Part I*, vol.50, pp.1063–1078

Tebo,B.M., Bargar,J.R., Clement,B.G., Dick,G.J., Murray,K.J., Parker,D., Verity,R., & Webb, S.M., (2004), Biogenic Manganese oxides: Properties and Mechanisms of Formation, *Annual Review of Earth and Planetary Sciences*, Vol. 32,pp. 287-328

Tekeuchi,K., Iwahashi, H., Kishi,M., Goto,T., Hirai,M.& Al Mohannadi,F.H.,(2008), Who plays an Important Role for Bio-fouling in SWRO Plant , presented at the IDA World Congress - Atlantis, The Palm - Dubai, UAE November 7-12, pp.1-6

Teuler, A., Glucina, K. & Laine, J.M., (1999), Assessment of UF pre-treatment prior RO membranes for seawater desalination, *Desalination, vol.125, pp89-96*

Thornton, D.C.O., (2004), Formation of transparent exopolymeric particles (TEP) from macroalgal detritus, *Marine Ecology Progress,* Series 282, pp.1-12.

Van Hoof, S.C.J.M., Minnery, J.G. & Mack, B.,(2001), Dead-end ultrafiltration as alternative pre-treatment to reverse osmosis in seawater desalination: a case study,*Desalination*,vol.139,pp.161-168

Vial, D. & Doussau, G., (2002), The use of microfiltration membranes for seawater pre-treatment prior to reverse osmosis membranes, *Desalination*, vol.153, pp.141-147

Villacorte,L.O., Schurer,R., Kennedy, M.D., Amy, G.& Schippers, J.C.,(2009), Removal and deposition of transparent exopolymer particles (TEP) in Seawater UF-RO System, presented at IDA World Congress - Atlantis, The Palm - Dubai, UAE November 7-12

Vlaski, A., van Breemen A.N. & Alaerts, G.J., (1996), The algae problem in the Netherlands from a water treatment perspective, *Journal Water SRT-Aqua*, vol.45, n°4, pp.184-194

Weinberg E. D. (1989), Cellular regulation of iron assimilation, *Q. Rev. Biol.*, vol.64, pp.261–290

Wells, M. L., Zorkin, N. G. & Lewis A. G. (1983), The role of colloid chemistry in providing a source of iron to phytoplankton,. *J. Mar. Res.*, vol.41, pp.731–746

Wittmann,E., Belkhodja, M., Lee Y., Remize,O.J., Brouwer,H. & Schrotter J.C.,(2011), RO membrane biofouling: Problems and solutions, presented at the World Congress Perth Convention and Exhibition Centre (PCEC), Perth, Western Australia September 4-9

Wotton, R., (2004), The essential role of exopolymers (EPS) in aquatic systems *Oceanography and Marine Biology: An Annual Review*, vol.42, pp.57–94

Xia,S., Nan,J., Liu, R. &. Li, G.,(2004), Study of drinking water treatment by ultrafiltration of surface water and its application to China, *Desalination*, vol.170,pp.41–47.

Yang, H.J. & Kim, H.S., (2009), Effects of coagulation on MF/UF for removal of particles as a pre-treatment in seawater desalination, *Desalination, vol.247, pp.45*-52

Yacubowicz H., (2010), Large-scale SWRO system supplies Chilean copper gold mine, *World Water*, july-august, pp.41-42

Yao, K.-M. & Habibian, M.T., (1971), Water and Waste Water Filtration: Concepts and Applications, E*nvironmental Science and Technology*, vol. 5, n°11, pp.1105-1115.

Yoon,J.S., Kim,S.H., Yoon,C.H. & Kim G.T.,(2004), Microfiltration as pre-treatment of reverse osmosis for seawater desalination, *J.Korean Soc.Environ.Eng.*,vol.26,N°4,pp.475-480.

2

Pretreatment for Reverse Osmosis Systems

Robert Y. Ning
King Lee Technologies, San Diego, California
United States of America

1. Introduction

Reverse osmosis (RO) membranes are very sensitive to fouling by seawater or brackish water on which desalination is conducted. In this chapter, the nature of foulants, the mechanism of fouling and the methods of monitoring and diagnosis of RO fouling are discussed. Based on this, methods of pretreatments are presented, and recommended specifications for the pretreatment of RO feedwater are described.

2. Fouling of RO membranes

2.1 Commonly observed foulants on reverse osmosis (RO) membranes

(Ning and Shen, 1998; Ning, 2004) can be divided into hard scales and soft amorphous complexes. The most common scales from brackish waters, seldom seen on seawater membranes, are calcium carbonate, calcium sulfate, strontium sulfate, barium sulfate, calcium fluoride and calcium phosphate. Amorphous gels are composed of natural exocellular polysacharrides from microorganisms, hydrated silica, ferric and aluminum hydroxides, colloidal iron and manganese compounds and pretreatment polymers. Fine silt, clay and media from filters such as activated carbon and manganese greensand that escape guard filters also deposit on membranes.

2.2 Mechanisms of fouling

Scales result from super-saturation of the RO brine with respect to salts of low solubility as those mentioned above. RO systems are typically operated with recoveries in the range of 50-90%. This means feed waters entering the RO are concentrated 2-10 fold. Under suitable conditions of concentrations of total dissolved solids, intrinsic crystallization rates of specific compounds and presence of seed crystals, scales can form, build up and move forward from the back-end of the RO system where concentrations are the highest towards the front where the concentrations are lower. With the advent of effective modern antiscalants (Ning, 2003 and Ning et al. 2009) scaling today is a much smaller problem than colloidal fouling by coagulation of particles invisible to the naked eyes. Natural waters are full of suspended particles that are extremely small (<0.1 micron, defined as colloidal). At such sizes, surface to mass ratio are so much larger than visible particles that large surface to surface interactions cause them to self-assemble in discernable patterns (termed Nanotechnology when synthetic and controlled) or coagulate. The coming together of particles and sticking to membrane surfaces results in what we observe in membrane autopsies as amorphous gels. Such foulants are complex mixtures and are difficult, sometimes impossible to clean. Larger, visible particles if not removed from RO feed water will naturally plug the feed flow channels in the membrane elements. To prevent such fouling, RO feed water needs to have turbidities of less than 1 NTU, and Silt Density Index (SDI), a flowrate over time through a 0.45 micron filter, of less than 4.0. Turbidity and SDI do not detect colloidal fouling potentials.

To control scaling, antiscalants are used to bind to nascent seed crystals preventing them from growing into scales and safely discharged with the reject water. For this reason it is commonly called threshold inhibition mechanism. Antifoulants for controlling colloidal fouling works on the principle of keeping the colloidal particles from coagulating or even dispersing the coagulate once formed on the membrane (Ning et al. 2005; Ning and Troyer, 2007; Ning, 2009) . Colloidal iron and manganese compounds, due to their positively charged characteristics are particularly sticky on the negatively charged membranes (Ning, 2009). Special mention should be made on the mechanism of fouling by colloidal silica and silicates due to the spontaneous polymerization of monomeric silicic acid in all natural waters (Ning, 2002; Ning, 2010). Silicic acid [$Si(OH)_4$] is ¡the reactive silica species commonly measured by the molybdate colorimetric assay. It spontaneously polymerizes by elimination of water during RO concentration, generating in the RO concentrate a reaction mixture of oligomeric silica and silicates. The representation of silica in the ultimate dehydrated state is SiO_2 (eg. sand, quartz). When hydroxides of iron, aluminum, magnesium and calcium are involved in copolymerization with silicic acid, complex silicate oligomers are formed in the RO concentrate, some of which may be large enough and sticky enough to end up on the membrane.

2.3 Monitoring and diagnosis of fouling

Due to the myriad possibilities of irreversible fouling of RO membranes, careful monitoring of the RO performance is necessary for operation and maintenance and solution of fouling problems as soon as they appear. Monitoring is accomplished ideally with normalized values of permeate flowrate, differential pressure and salt passage plotted as trend charts (Ning, 2004; Troyer et al., 2006). For two stage RO systems, it is best to install an inter-stage pressure gauge so that differential pressures for the first and the second stages can be monitored separately.

To diagnose fouling, scaling occurs in and grows from the back end forward, and shows prominent increases in differential pressure. Colloidal fouling begins with severe normalized permeate flowrate, long before significant differential pressures appear. Fouling by visible particles is more severe in the lead elements, often limited to the element in the first position, and accompanied by rapid fouling of the 1 to 20 micron cartridge guard filters.

3. Seawater pretreatment

3.1 Intake systems

Seawater can be drawn into the plant from open sea, or through beach well or seabed. Popular these days is to co-locate the seawater RO next to an existing power plant that discharges seawater used in cooling residual heat. Intake system design has considerable impact on the feed water quality, hence the entire pretreatment design. Open intake suffers from fluctuating, often high turbidities and colloidal contaminants such as algal matter, transparent exopolymer particles (about 0.4-200 microns), colloidal extracellular polymeric substances (<0.4 microns), humic substances, oils, clay and other organic and inorganic matter from surf action and nearby discharges from shore. Beach well and under-seabed intakes act as filters to reduce suspended particles as well as colloidal dispersions to some extent. Depending on the sites, beach wells can introduce foulants more commonly seen on land. The advantage of using power plant cooling seawater is the reduced cost of building a new intake and discharge structure, while using warm water that increases membrane flux and reduces pump power consumption..

3.2 Recommended RO feedwater quality

The quality of seawater is defined by turbidity, Silt Density Index, Total Organic Carbon, Critical Flux (the maximum membrane flux above which colloidal fouling will occur for a particular pretreated water), Red Tide algal bloom events, colloidal destabilization caused by blending with river discharge or chemical dosing such as with chlorine. While these parameter provide useful data for designing and operating pretreatment unit operations, no model currently exists that can predict with certainty, the maximum RO fouling rate of a pretreated RO seawater. For this reason, careful monitoring is essential for the operation and maintenance of the system.

3.3 Shock vs continuous chlorination

Chlorination (sodium hypochlorite or chlorine dioxide) is needed to minimize the population of growing sea organisms such as algae, mussels, clams and microorganisms that inhabit the surfaces of pipes and tanks in the intake system. Long term experiences in many seawater RO plants have shown that shock chlorination is better than continuous chlorination. Continuous chlorination chemically reacts with and adds its weight to planktonic (moving) colloidal organic matter and causes destabilization and more coagulation of the natural colloidal polymers. It is possible also that continuous chlorination only irritates the larger anchored sea organisms in the intake system rather than killing them, allowing them to multiply and continue to shed foulants.. Biofilms and microbial control in plant equipment contacting water remains a major challenge to RO pretreatment.

3.4 Clarification

Coagulants are added to seawater before sand or multi-media filters to enhance removal of suspended particles. Dissolve air floatation by deliberate introduction of a dispersion of air bubbles while extending contact time with coagulants improves the removal of algae and natural organic materials (NOM) sometimes called transparent exopolymeric particles (TEP), or extracellular polymeric substances (EPS). These coagulated sticky transparent gels are composed of primarily complex polysaccharides.

3.5 Microfiltration/ultrafiltration

Membranes having tortuous pores, with diameters in the approximate range of 0.05 to 1.5 microns (MF) and smaller pores in the range of 0.002 to 0.10 microns (UF) are used as alternatives to traditional clarification with media filters. Timed backwashing cycles are necessary to recover trans-membrane pressures, along with periodic chemical cleanings to recover productivity. MF and UF membranes, usually called low-pressure membranes in contrast to RO membranes suffer from fouling by cake-layers of foulants as well as plugging of pores. Pores on the membranes once plugged are difficult if impossible to recover. For this reason, low flux and judicious use of coagulants to form cake-layers of protective foulants are strategies to extend service lives of the membranes.

3.6 Antiscalant

Due to the high dissolved total solid concentrations of seawater, and the resulting increased solubilities of scales, along with low water recoveries of about 50% to minimize energy consumption, the need for antiscalants in seawater ROs are minimal. Almost all fouling seen are due to colloidal organic matter on which antiscalants exert little effect. Carryovers of iron and aluminum coagulants, or colloidal inorganic particles picked up from beach wells however can be controlled by appropriate antiscalants.

3.7 Antifoulant

The application of antifoulants as anti-coagulants and dispersants of colloidal matter in seawater (Ning, 2003; Ning et al, 2005) show promise, but not yet adopted on large scale.

4. Brackish water pretreatment

4.1 Well vs surface water

Due to long equilibration times and adsorption and filtration actions of soils in the aquifer, well waters have low turbidities and Silt Density Indeces compared with surface waters from rivers, lakes and ponds. Surface waters in contrast have high turbidities and SDI, and natural organic matter derived from abundance of biotic life. Reactive silica can be very high from wells, especially in the volcanic regions of the world, but low in surface waters. Silica and silicates in surface waters tend to be in colloidal non-reactive forms. Ferrous ions associated with iron-reducing bacteria survive air oxidation in the anaerobic environments of wells, so does sulfide ions associated with sulfate-reducing bacteria in some wells. Upon emerging to the atmosphere on the surface, ferrous ions oxidize to the red ferric hydroxide species in surface water and sulfide waters form colloidal sulfur and hydrogen sulfide gas.

4.2 Clarification

Deep-well waters can typically be pumped directly to the RO system through cartridge guard filter after dosing with antiscalant (Ning, 2003). Surface waters need clarification to turbidity of <1 NTU and SDI of <3-4. Coagulant, sand or multimedia filters are used. Filtration through activated carbon should be avoided if possible due to propensity for microbial growth. MF and UF are sometimes used, especially for brackish municipal wastewater. This expensive pretreatment is primarily for the purpose of gaining virus removal credits as process steps for certifiable process designs for treating municipal wastewater.

4.3 Chlorination

Surface waters and shallow well waters are sometimes chlorinated followed by dechlorination with bisulfite, metabisulfite or activated carbon before entering RO. In successful resolution of membrane biofouling problems, we have found that this pretreatment is not necessary. Short contact time of low dosages of chlorine affect planktonic bacteria to a degree but is ineffective in controlling biofilms. Membrane damage from iron-catalyzed oxidation by residual chlorine or even chloramines can be rapid and severe. Activated carbon filters always grow bacteria and shed microbial slime, and are difficult to maintain. Effective control of biofilm on RO membrane is best performed during periodic maintenance cleaning and sanitization of the system.

4.4 Antiscalant

Antiscalants as threshold inhibitors in pretreatment greatly simplifies the design of RO pretreatment process (Ning, 2003). Scaling, dissolved iron, aluminum and manganese, and silica polymerization chemistry are almost always controlled in the typical range of 75-85% RO water recovery. For extreme recoveries of 90-98% of brackish water, antiscalants play essential roles. In inland regions where RO concentrates are expensive or impossible to dispose of, ROs are being tested to the limiting opposing osmotic pressure of 1000 psi (68 atmospheres) to reduce the volume of RO concentrate requiring disposal (Ning et al, 2010).

4.5 Antifoulant

Antifoulants acting as anti-coagulants for colloidal foulants and dispersants for colloidal coagulates are specifically developed for colloidal sulfur, silica and microbial slime in brackish water (Ning et al, 2005). Simultaneous use of an antiscalant along with an antifoulant has long been used in medium-sized RO systems. Application in major municipal RO plant is thus far rare due to the high cost of required dosages that are larger than those of antiscalants.

5. Summary comments

5.1 Limits to scaling and fouling projections

Although proprietary software programs that predict scaling potentials are available from membrane and RO chemical suppliers, no standard formulae exist today on predicting scaling potentials for RO. Textbook Solubility Product Constant values are adopted, but proprietary formulae used are varied and unpublished. It must be understood that solubility products are reported for single pure substances at equilibrium in distilled water, and very difficult to measure experimentally with any precision. Further, the confounding factors when applied to RO concentrates are disequilibrium conditions, variable interferences from other ions on crystallization rates, intrinsic crystallization rates, total dissolved salt concentrations, effects of seeding by particles and surfaces, just to mention a few. Although it is necessary to use arbitrary calculations of scaling potentials to determine antiscalant dosages needed, large body of field data correlating RO performance with water chemistry, antiscalant selection and antiscalant dosages provide some confidence in the selection of antiscalants and dosages. Likewise, the complexity of the surface properties and variety of uncharacterizable colloidal particles in raw water precludes possibility of projecting fouling potential and interactions with antifoulants. Antifoulants and dosages are developed entirely empirically.

5.2 Importance of trend-charting

Due to the sensitivity of RO systems to fouling, good pretreatments need to be coupled to continuous monitoring to avoid irreversible fouling. It is the general practice that whenever productivity, differential pressure and salt passage change by 10-15%, cleaning is called for. The longer the system is left uncleaned, the less recoverable the lost performance is. Since system feed pressure and water temperature can change day-to-day that affects the three parameters mentioned, normalization to the same temperature and feed-pressure is necessary to compare performance data from day to day and from year to year. This is accomplished by using software for normalization and trend-charting (Troyer et al, 2006). Beyond pretreatment process upsets, since feedwater quality and operating conditions can change from time to time, trend-charting is also necessary to detect ineffectiveness and insufficiencies of maintenance cleaning procedures. When this occurs, optimized cleaning methods need to be validated on-line with trend charts. Likewise, optimization of pretreatment steps are accomplished with trend charts of normalized data.

6. References

Ning, R.Y. (2010). Reactive silica in natural waters -A review. *Desalination and Water Treatment*, Vol. 21, 79-86.

Ning, R.Y., Tarquin, A.J., and Balliew, J.E. (2010). Seawater RO treatment of RO concentrate to extreme silica concentrations, *Desalination and Water Treatment*, Vol. 22, 286-291.

Ning, R.Y. (2009). Colloidal iron and manganese in water affecting RO operation. *Desalination and Water Treatment*, Vol. 12, 162-168.

Ning, R.Y., Troyer, T.L. and Tominello, R.S. (2009). Antiscalants for near complete recovery of water with tandem RO process. *Desalination and Water Treatment,* Vol. 9, 92-95.

Ning, R.Y. and Troyer, T.L. (2007). Colloidal fouling of RO membranes following MF/UF in the reclamation of mucicipal wastewater. *Desalination*, Vol. 208, 232-237.

Troyer, T.L., Tominello, R.S. and Ning, R.Y. (2006). A method to automate normalization and trending for RO plant operators. *Ultrapure Water,* September, 37-43.

Ning, R.Y., Troyer, T.L. and Tominello, R.S. (2005). Chemical control of colloidal fouling of reverse osmosis systems. *Desalination*, Vol. 172, 1-6.

Ning, R.Y. (2004). Sustaining the productivity of reverse osmosis plants, *The Analyst*, Fall, 1-5.

Ning, R.Y. (2003). Process simplification through the use of antiscalants and antifoulants. *Ultrapure Water*, September, 17-20.

Ning, R.Y. (2002). Discussion of silica speciation, fouling, control and maximum reduction. *Desalination* Vol. 151, 67-73.

Ning, R.Y. & Shen, P.T.L. (1998). Observations from analysis of reverse osmosis membrane foulants, *Ultrapure Water,* April, 37-40.

3

Membrane Cleaning

José Miguel Arnal, Beatriz García-Fayos and María Sancho
Instituto de Seguridad Industrial,
Radiofísica y Medioambiente (ISIRYM) / Universitat Politècnica de València
Spain

1. Introduction

In last years, desalination by reverse osmosis (RO) process has experienced a significant development, and it has become one of the major technologies for producing potable water throughout the world. Despite this relevant growth, reverse osmosis, and membrane processes in general, has several drawbacks to overcome. Specifically, fouling of membranes is the most important problem in reverse osmosis desalination since economics of the process is still highly influenced by membrane fouling rate and effectiveness of fouling control.

1.1 Fouling agents

Membrane lifetime and permeate productivity are primarily affected by concentration polarization and fouling at the membrane surface. IUPAC defines membrane fouling as "a process resulting in loss of performance of a membrane due to the deposition of suspended or dissolved substances on its external surfaces, at its pore openings or within pores". The main mechanisms of membrane fouling are the following (Goosen et al., 2004):

- *Adsorption,* due to chemical affinity or interaction between solutes and membrane material. This can happen at membrane surface or inside the pores.
- *Pore blockage,* when solutes go inside membrane pores.
- *Gel formation,* as a consequence of molecule accumulation at the film layer of the membrane. This is very typical in solutions containing proteins.
- *Biofouling,* cause by bacterial adhesion and growth at membrane surface, besides the production of extracellular polysaccharides (EPS) by some genera of bacteria, which in fact are the substance responsible for the biofilm (Baker & Dudley, 1998).

In desalination by RO membranes, the most common fouling types are: organic fouling due to natural organic material (NOM) such as humic and fulvic acids, protein and carbohydrate; inorganic fouling due to depositions of inorganic scales (mainly $BaSO_4$, $CaSO_4$, $CaCO_3$); and biofouling due to microbial attachment to membrane surface followed by growth and multiplication (Chesters, 2009; Al-Amoudi & Lovitt, 2007). Biofouling is one of the most serious forms of membrane fouling because once it forms; biofilm is very difficult to remove.

1.2 Consequences of fouling in membrane performance

As a result of membrane fouling, membrane resistance increases with time due to accumulation of foulants on membrane surface and/or inside the membrane (Tay & Song,

2005). The main consequences of fouling are: flux decline, permeate quality deterioration and energy consumption increase. Since operating costs of reverse osmosis highly depend on membrane useful life (membrane replacement involves 20-30% of the operating costs), fouling control is essential for increasing membrane operational life and thus reducing economics of the process (Butt et al., 1997).

1.3 Preventing and minimizing fouling

Possible ways for fouling control are feed pre-treatment, operation conditions and membrane cleaning.

1.3.1 Pretreatment

It is essential to establish a good pre-treatment to avoid or minimize fouling, so productivity loss would be lower (Figure 1). Nevertheless, in spite of a good pre-treatment, membranes have to be periodically cleaned to remove reversible fouling. In case pre-treatment was inadequate, higher frequency of cleanings will be necessary, and the restoration of membrane performance will be worse (Sadhwani & Veza, 2001).

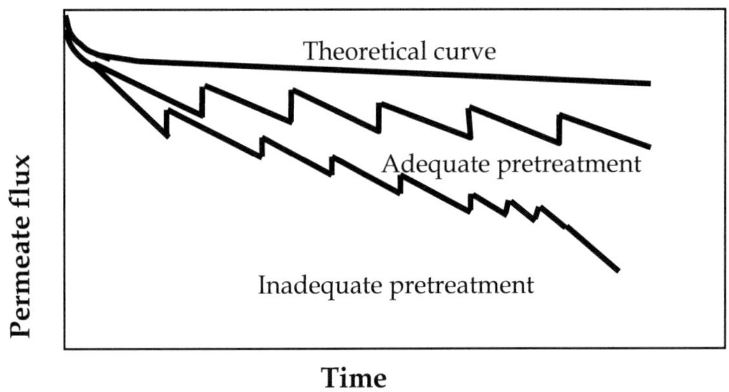

Fig. 1. Influence of pre-treatment on membrane productivity

1.3.2 Operation conditions

In spite of the pre-treatment of feed water, operation conditions such as temperature, pressure and crossflow velocity can be considered to reduce fouling (Mohammadi et al., 2002). Furthermore, it has been showed that there is a critical permeate flux, operating above which productivity can be maximized at the same time that membrane cleaning frequency is reduced (Goosen et al., 2004).

1.4 Modes of fouling detection and characterization

It is vey important to know the characteristics of the foulant deposited on the membrane in order to select the most effective cleaning procedure. The best method of foulant identification is membrane autopsy after fouling, which is considered as an off-line method for fouling characterization. However, this is a destructive and expensive method that may be considered to enhance system performance when fouling is complex, when cleaning fails to restore membrane performance (Pontié et al., 2005; Al-Amoudi & Lovitt, 2007), and when membranes result damaged after cleaning.

Membrane autopsies are characterized for using techniques that are able to identify the inorganic and organic material found at the membrane surface:

- *Scanning Electron Microscope* (SEM) and *Scanning Electron Microscope coupled with Energy Dispersed X-ray* (SEM-EDX) to determinate inorganic foulant elemental composition.
- *Fourier Transform Infrared Spectroscopy* (FTIR) to identify organic compounds.
- *Targeted Energy Dispersive X-Ray Analysis* (T-EDXA) as an indication of inorganic compounds.

For identifying biofouling, several techniques can be achieved during autopsies, such as:

- *Total Direct Count*, which determinates in the membrane surface the number of bacteria dead or alive, growth in a medium and counted with microscope per cm^2.
- *Heterotrophic Plate Count*, which scrapes a sample of the biomass attached on the membrane surface to determinate colony forming units (CFU) growth in a special medium.
- *Adenosine Triphosphate* (ATP), which is considered an essential indicator of living bacteria and gives a measure of the amount of active biomass over the membrane surface.

Even though, recent trends about fouling detection and characterization are pointed to develop and use non-destructive and visual on-line methods that help to study the fundamentals of membrane fouling, and also work as indicators and early warnings fouling detectors in real-time. Some examples of this recently developed technology are the following:

- *Membrane Fouling Simulator* (MFS): uses the same materials as spiral-wound RO and NF membranes. This tool is very useful to study and monitor fouling and biofouling over spiral wound membranes, monitoring pressure drop, in situ-real time observation of membrane surface and multiple analysis of coupons sampled form the membrane. The small size and low water and chemical use of the MFS facilitate to perform systematic parallel studies. With the MFS, fouling of membranes applied in water treatment can be characterized (Vrouwenvelder et al., 2006; Vrouwenvelder et al., 2007a; Vrouwenvelder et al., 2007b).
- *Visual Observation methods*: recently, a new high-pressure optical membrane module has been developed to enable direct visual observation of bacterial deposition onto reverse osmosis membranes under practical seawater desalination operating conditions. The new optical membrane module withstands pressures up to 8.3MPa while allowing high transmission of UV and visible light. Computational fluid dynamic analyses of the optical module suggest practically relevant cross-flow hydrodynamic conditions exist through the viewing area of the optical window (Huang et al., 2010).
- *In-line biofouling monitor*: this recent development is based on accurate monitoring of normalized pressure drop, flow and temperature across first stage. It is very useful to monitorize biofouling and scaling.
- *Ultrasonic Time-Domain Reflectometry* (UTDR): as an in situ, non-invasive real-time technique has been successfully used to quantify membrane fouling and cleaning. This ultrasonic technique is useful for quantitative study of the effect of magnetic fields on $CaCO_3$ scale deposition on the membrane surface during cross-flow nanofiltration (NF) and also in ultrafiltration (UF) membranes (Li et al., 2002). These results have been corroborated by comparing this technique with SEM and flux decline data also (Li et al., 2007).

- *Magnetic Resonance Imaging* (MRI): this has been recently applied to biofouling studies of RO membranes, identifying both biofilm accumulation within the module and the resultant effects on water flow. MRI thus offers an in situ and non-destructive method to quantify biofilm directly. It is also a tool for direct studies of chemical cleaning of biofouled RO membranes, specially for studying cleaning effectiveness (Creber et al., 2010).

There are also alternative methods that help to identify fouling and the convenience of cleaning, such as chemical and biological analysis of the current feedwater, and the analysis of acid, caustic and distilled water extracts of the prefilter (Graham et al., 1989).

In fact, there are some parameters that can help to determinate the fouling potential capacity of feed water. The most usual ones are the following:

- *Silt Density Index* (SDI): is a common parameter applied for the fouling potential of feed water for Reverse Osmosis and Nanofiltration membranes. It is based on the measurement of plugging a membrane filter having 0.45 μm at a pressure of 210 kPa. Even though it has been reported several deficiencies about the use of this parameter, it is still used in operation plants as fouling indicator.
- *Modified Fouling Index* (MFI): developed to overcome the main deficiencies of SDI, it is performed in the same equipment that SDI and it is based on the occurrence of cake filtration during a filtration test in RO and NF membranes.
- *MFI-UF*: recently developed, this parameter has been used to detect fouling potential capacity due to particles smaller than 0.45 μm, in RO and NF membranes. It requires more sophisticated equipment and it is not yet standardized.
- *Langelier Saturation Index* (LSI): this parameter correlates the presence of calcium carbonate with its potential precipitation. It is commonly used for low salinity waters.
- *Stiff & Davies Index* (S & DI): this is similar to LSI, but more accurate for high salinity waters, since it adds the effect of ionic strength and temperature.

In general, fouling removal must be done when some of the following situations are reached:

- 10-15% of permeate flux decline
- 10-15% of permeate solute concentration increasing
- 15-20% of pressure drop increase in a pressure vessel

When some of these situations occurred, membrane must be cleaned to restore system performance. The cleaning procedure depends mainly on the type of fouling.

2. Conventional methods of cleaning

Membrane cleaning methods can be divided into physical, chemical and physio-chemical. In practice, physical cleaning methods followed by chemical cleaning methods are widely used in membrane applications. However, only the chemical cleaning methods are widely applied for RO desalination.

2.1 Physical cleaning methods

Physical cleaning methods use mechanical forces to dislodge and remove foulants from the membrane surface. Physical methods include sponge ball cleaning, forward and reverse flushing, backwashing, air flushing (also called air sparging, air scouring or air bubbling) and CO_2 back permeation (Ebrahim, 1994; Al-Amoudi & Lovitt, 2007). Ultrasonic, electrical fields and magnetic fields are other physical cleaning methods that are described in detail in non-conventional cleaning methods point of this work.

2.1.1 Sponge ball cleaning

In this method, sponge balls made of polyurethane or other materials are inserted into the membrane modules for a few seconds for scrubbing the foulant from membrane surface (Ebrahim, 1994). Sponge balls are normally used for cleaning large diameter tubular membranes when treating heavily polluted solution such as wastewater and industrial process water (Psoch & Schiewer, 2006).

2.1.2 Forward and reverse flushing

Forward flushing consists in pumping permeate water at high cross-flow velocity through the feed side in order to remove foulants from the membrane surface (Ebrahim, 1994). Because of the more rapid flow and the resulting turbulence, particles absorbed to the membrane are released and discharged. In the reverse flushing method the direction of the permeate flush is alternated for a few seconds in the forward (feed to brine) and for a few seconds in the reverse direction (brine to feed) (Figure 2). Forward flush techniques are particularly useful in removing colloidal matter.

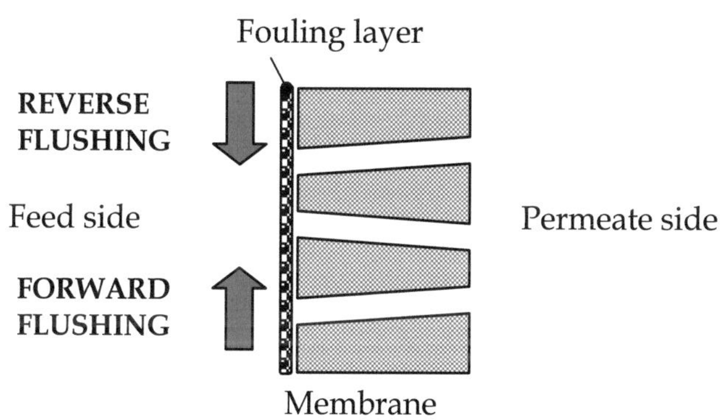

Fig. 2. Cleaning flow direction in forward and reverse flushing

2.1.3 Backwashing

This is a reversed filtration process in which permeate is flushed through the membrane to the concentrate side. In porous membranes, when backward flush is applied, the pores are flushed inside out. The pressure on the permeate side of the membrane is higher than the pressure within the membranes, causing the pores to be cleaned (Figure 3).

In reverse osmosis membranes, backwash is based on flow induced by osmotic pressure as direct osmotic cleaning. This cleaning process is based on negative driving pressure between the operating pressure and the osmotic pressure of the water solution in the feed side. This can be done either by reducing operation pressure below the osmotic pressure of the feed solution or by increasing the permeate pressure (Sagiv & Semiat, 2005). Backflow from the permeate to the feed side of the membrane expands the thickness of the fouling layer and fluidizes it. After this, a forward flush is usually used to wash out the detached layer or dilute the fouling layer. The best cleaning performance is generally reached optimizing the two flows (forward flow and backflow). Some significant factors affecting physical cleaning when combining forward and backward flushing are production interval between cleans, duration of backwash and pressure during forward flush (Chen et al., 2003).

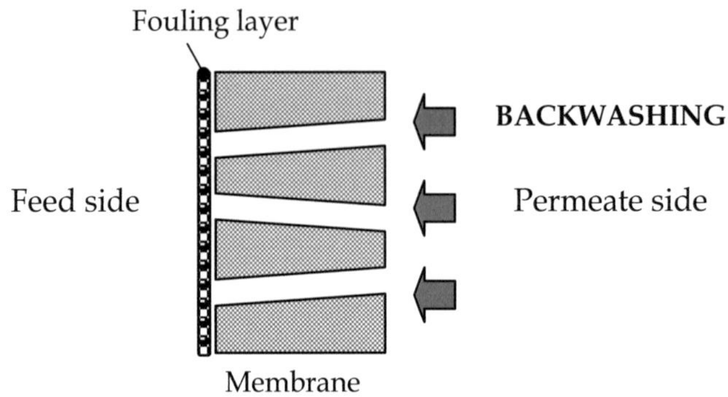

Fig. 3. Cleaning flow direction in backwashing

A main factor of the backwash cleaning is the driving force, which results from a combination of concentrations and applied pressures from both sides of RO membrane. From the point of view of operating conditions, backwash cleaning may be divided into two cases: (1) with zero feed velocity and applied pressure, and (2) with feed velocity and pressure. Some works have shown that methods without pressure seem to have some advantages on methods with pressure (Sagiv & Semiat, 2010).

From the point of view of salt concentration difference between both sides of membranes, there is recent research on the use of hypersaline solutions in the feed side, which is described in more detail in section 3.1 of this work.

2.1.4 Air flushing

The air flushing or air sparging method generates a two phase flow to remove external fouling and thus reduces the cake layer deposited on the membrane surface. There are several flow patterns possible depending on the superficial liquid and air velocities. It has been showed that slug flow is the most effective pattern to enhance mass flow (Psoch & Schiewer, 2006; Mercier et al., 1997). The type of gas used for the sparging also seems to have an influence on the cleaning efficiency. In a recent research, it has been shown that water/CO_2 mixture performes better cleaning results in comparison to water/N_2 mixture (Ngene et al., 2010).

Air sparging can be applied either during the course of filtration to reduce fouling deposition (Cui & Taha, 2003; Mercier et al., 1997; Cabassud et al., 2001) or periodically to remove already formed deposits. Air sparging is typically applied in MF and UF membranes, and it seems to work best for tubular and flat sheet membranes and to a lesser extent in hollow fiber and spiral wound modules (Cui & Taha, 2003). Anyway, it is clear that the use of air leads to an enhancement of flux in MF and UF. This positive effect is due to the presence of air bubbles which increase turbulence in the feed side of the membrane, thus increasing permeate flux as well as solute separation efficiency (Cabassud et al., 2001; Ducom & Cabassud, 2003).

There have been little attempts to apply air sparging in NF membranes. In some studies, air sparging was applied to prevent fouling during NF, and it was prooved that poor benefits can be expected from air sparging in case of solutions containing only salt in water (Ducom & Cabassud, 2003). (Cornelissen et al., 2007) used air/water cleaning for biofouling control showing that the best results in spiral wound NF modules were obtained when combining daily copper sulphate dosing with sporadic air/water cleaning.

More recently, (Qaisrani & Samhaber, 2011) have evaluated the potential of air sparging and backflushing for MF membrane regeneration. They have found that the combination of backflushing and air bubbling is the best method for improving membrane performance both in terms of fouling control and membrane cleaning time.

2.1.5 CO_2 back permeation

This is a method traditionally used for hollow fiber configuration in which CO_2 gas is forced from the permeate side through the internal fiber and out through them (Ebrahim, 1994). More recently, (Fritsch & Morau, 2008) applied a CO_2 backpulsing system for cleaning a tubular MF membrane used in a dairy industry. They state that this is a very promising technique for maintaining a greater and more stable permeate fluxes.

2.2 Chemical cleaning methods

Membrane fouling can be classified as physically reversible fouling which can be totally eliminated by physical cleaning or certain pretreatment, and physically irreversible fouling which cannot completely removed by physical cleaning or pretreatment (Hiroshi et al., 2007, as cited in Gao et al., 2011). Such irreversible fouling can only be overcome by chemical cleaning, which has to be limited to a minimum frequency since repeated chemical cleaning may affect membrane life (Kimura et al., 2004).

Chemical cleaning is the most common membrane cleaning method, especially in reverse osmosis membranes. In this type of cleaning, the choice of the cleaning agent is critical. The optimal selection of the cleaning agent depends mainly on membrane material and type of foulant. These agents must be able to dissolved most of the deposited materials on the surface and removed them from the surface but not damaging membrane surface, thus maintaining membrane properties. Commercial cleaning products are often, Most chemical cleaning agents are commercially available, they are often mixtures of compounds, and many of them are recommended by membrane manufacturers according to the type of foulant, although in most cases the actual composition is not clearly specified (Ang et al., 2006). Anyway, in general acid (nitric, phosphoric, hydrochloric, sulphuric and citric) are often used to remove precipitated salts or scalants, while alkaline cleaning is suitable for organic fouling removal. Other categories of chemical cleaning agents are: metal chelating agents, surfactants and enzymes (Mohammadi et al., 2002). In addition, disinfectants (O_3), oxidants (H_2O_2, $KMnO_4$) or sequestration agents (EDTA) are often used for chemical cleaning of membranes (Lin et al., 2010).

Table 1 summarizes the most suitable and usual cleaning agents according to the type of fouling (Fritzmann et al., 2007).

Type of fouling	Chemical agent
Colloidal	NaOH solutions, chelating agents and surfactants
Organic	NaOH solutions, chelating agents and surfactants
Metal oxides	Citri acid with low pH or $Na_2S_2O_4$
Silica	NaOH solutions with high pH
Carbonate scales ($CaCO_3$)	Citric acid or HCl with low pH
Sulphate scales ($CaSO_4$, $BaSO_4$)	HCl solutions or sequestration agents (EDTA)
Biofilms	NaOH solutions, chelating or sequestration agents, surfactants and disinfectants

Table 1. Cleaning solutions according the type of foulant

A cleaning cycle generally includes several stages: product removal, rinsing with water, cleaning in one or more steps, and rinsing with water. In order to obtain a good cleaning effect, cross-flow velocity should be higher and the pressure lower than those used during normal operation (Tragardh, 1989). A typical cleaning procedure consists of the following steps (Fritzmann et al., 2007):

- Low flow pumping of the pre-heated cleaning solution at minimum pressure.
- Recycling of cleaning solution until a stable temperature is achieved. In this stage, pH can be adjusted when necessary.
- Soaking the RO membranes with the cleaning solution for 1 to 15 hours, depending on the type and grade of fouling.
- High flow operation to flushed foulants and chemical solution out of the system.

When the cleaning agents come into contact with the fouled layer, some physical transformations and chemical reactions take place. Physical transformations include melting, mechanical and thermal stress, wetting, soaking and swelling, within others. The chemical reactions involved are hydrolisis, peptization, solubilization, dispersion, chelation, sequestering and suspending (Tragärdh, 1989).

Other important aspects concerning chemical cleaning are temperature, chemical concentration, pH, pressure and flow, and time (Liikanen et al., 2002; Chen et al., 2003). Some studies suggest that there is an optimal temperature for chemical cleaning (Barlett et al., 1995). Usually, increasing temperature (always below membrane maximum temperature) increases cleaning efficiency, cross-flow velocity seems to have no effect on cleaning results, whereas increasing trans-membrane pressure may even decrease cleaning efficiency. Zero transmembrane pressure is recommended for maximum efficiency in deposit removal (Barlett et al., 1995). With regard to the time required for cleaning it varies according to the foulant and the cleaning process.

2.3 Physico-chemical cleaning methods

The physico-chemical cleaning methods use physical cleaning methods with the addition of chemical agents to enhance cleaning effectiveness. Physico-chemical cleaning methods are not implemented widely in the RO industry. The applications usually consist in forward flushing with permeate between cleanings when more than one chemical cleaning is used (Ebrahim, 1994), but not in a simultaneous use of physical and chemical cleaning actions.

Very few works have researched on simultaneous combination of physical and chemical methods for membrane cleaning in spiral wound membranes. (Arnal et al., 2009b) studied the cleaning of UF membranes by the application of air in combination with different chemical solutions, with the objective of testing both hydraulic and chemical actions simultaneously in spiral wound membranes. The chemical solutions used in these experiments were sodium hypochlorite, hydrogen peroxide, sodium hydroxide and deionised water. The experimental results were compared to those obtained with chemical solutions without air bubbles (Arnal et al., 2009a), and showed a significant improvement in the cleaning effectiveness when using air bubbles with chemical solutions, also when comparing the results with the ones obtained with air sparging without chemical agents.

3. Non conventional cleaning methods

Many techniques have been developed to overcome fouling. Backwashing and backflushing are typically used to clean membranes, while more persistent foulants are removed by the use of chemical reagents, as itwas described previously in section 2. However, these chemicals sometimes damage the membrane materials reducing the lifetime and efficiency of the membrane and even cause a secondary pollution over cleaned membrane. Moreover, these methods have proven to be inefficient for some kind of fouling and also create environmental issues related to the waste chemical disposal. In other cases they add costs of cleanup, handling, and transporting dangerous chemicals, waste energy by decreasing and then increasing pressures needed for the membrane system to work, and waste cleaned water (Lu et al., 2009). It has been reported that, in general, 5-20% of the operating costs of a large plant are associated to cleaning procedures (Madaeni et al., 2001). Therefore, an intense research work is being done to develop new cleaning methods, called non-conventional or emerging techniques, more efficient and more environmental safe. Some of them are being applied in RO membranes and others just started for MF and UF membranes.

These novel cleaning techniques are described now and specifically for RO membranes used in desalination.

3.1 Osmotic backwashing with hypersaline solution

Pressure driven backwashing is common practice in filtration processes, including microfiltration and ultrafiltration, offering an effective means of fouling control. However, it was not extensively employed for membranes used in desalting, due to high back-pressure required for a hydraulically driven backwash since it may rupture the composite membrane used (Ramon et al., 2010).

Osmotic backwash may be induced when the feed-side osmotic pressure exceeds the applied hydraulic pressure across the membrane. It is based on water back flow from the permeate side to the feed side driven by osmotic forces or salt (mainly NaCl) concentration differences between both sides of the RO membranes. When water enters the feed channel, it dilutes the concentration polarization (CP) layer and helps in cleaning the membrane surface to resume its original flow rate.

Based on this technology, osmotic backwash induced by the use of hypersaline solution has been developed in the last years. It is considered an innovative, effective and potentially chemical-free cleaning technique. It is performed by injecting into the feed channel a pulse of high concentration solution called hypersaline solution (HS) that can supply more than 100 bar of the cleaning force and which is used to remove the foulants from the membrane surface, specially colloidal matter and biofilm material. Hypersaline solution injection promotes direct osmosis (DO) across the RO membrane and the flow of the water from the permeate side to the feed. The reversible flow helps to dislodge any foulants and scaling on the membrane surface and promotes lifting, sweeping and removing of the CP layer.

In fact, hypersaline solution injection promotes different effects such as (Liberman & Liberman, 2005):

- Lifting of an existing fouling layer due to backflow of permeate through the membrane. Even the osmotic pressure of the high-salinity solution decreases on its way, but it remains strong enough to create back permeate flow and lift foulants until it reaches the end of the pressure vessel.
- Sweeping and stripping of the lifted fouling layer. Increasing velocities inside the feed channel mechanically removes foulants from the feed and reduces the pressure drop along the pressure vessel.

- Separation of biomass from the membrane surface due to bio-osmotic shock. Water is sucked out from cytoplasm such as bacteria, fungi and algae, cell membranes shrink and detach from the membrane surface.
- Dissolving of micro-crystals due to the ionic strength of the hypersaline solution.

Membrane cleaning by hypersaline solutions ensured stable operation with significantly improved membrane performance, consistently reduced low pressure drop and increased membrane salt rejection. The main advantages of this cleaning method are the following:

- On-line technique without interruption of the operation.
- High effectiveness.
- Low cost.
- Ease of implementation at existing facilities and in the design of new plants without great changes to plant equipment.
- Simplified disposal due to a reduction of cleaning chemicals.
- Low membrane replacement rate.

It is considered a novel backwash approach for on-line membrane cleaning in RO operation without stopping the RO pump. There are short studies about this promising technique but all of them provide a preliminary indication regarding the potential for employing an osmotic backwash useful for membrane cleaning. First studies were published by Liberman in 2005 (Liberman & Liberman, 2005). In that work, a short injection of feed water with increased salt concentrations (25% NaCl solution) with an associated osmotic pressure of 200 bar overcomes feed pump gauge pressure and reverse osmosis shifts to direct osmosis, leading to a permeate backwash stream through the reverse osmosis membrane, as it is shown in following figure (Figure 4).

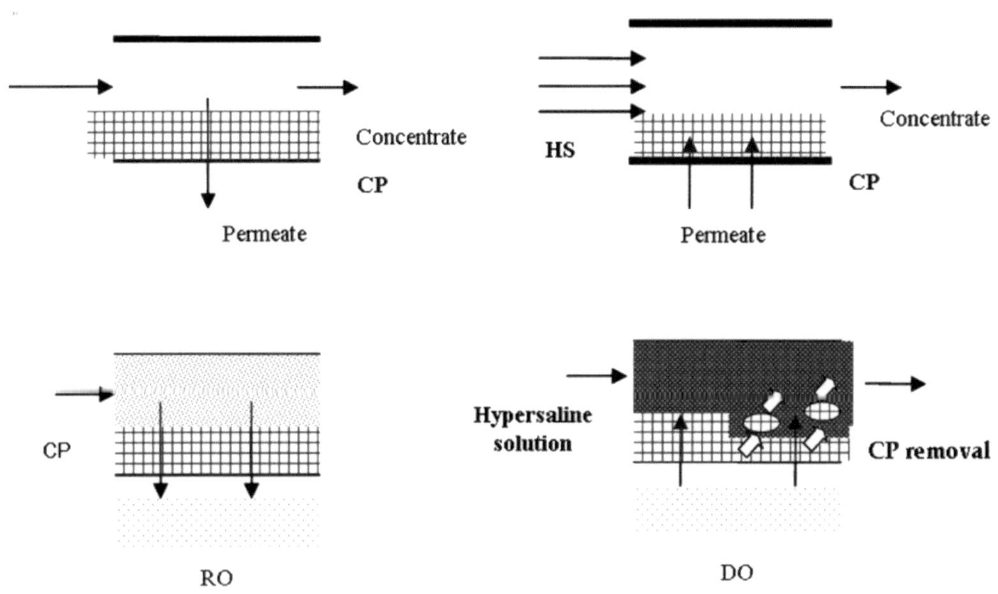

Fig. 4. Hypersaline solution backwash cleaning

In other studies, resistance of the RO membrane to water passage was considered a factor affecting the uniformity of the cleaning process. It was suggested to increase the gauge pressure at the permeate side of the membrane to a level slightly below the feed pressure of the raw saline solution, which is done by back pumping permeate for a few seconds. Results from that study, stated that permeate back pressure was to be at least 4 bar in order to make sure the DO backwash cleaning would be efficient. This remark increases the performance of the cleaning process but increases its cost due to investment needed to equipment for the permeate side working at high pressure (Liberman, 2004).

It was also reported the use of a super-saline solution made of 17% NaCl with osmotic pressure of 170 bar could be injected into the feed stream over a few seconds in order to create a very strong net driving force of 55 bar for the DO backwash process (Liberman, 2005). Recent studies (Qin et al., 2010) employed this technique to remove fouling from a RO system fouled with a secondary-treated waste-water. Research observed a rise of the turbidity of the brine stream (from 0.6-5 NTU), indicating the foulant removal from the membrane.

Main parameters that should be studied for proper implementation of this novel technique are saline concentration, duration of the pulse injection and flow rate of hypersaline solution in compliance with feed water chemistry, configuration of the existing RO train and pumps' curves. Improper implementation of the technique can damage the membrane element (Qin et al., 2009).

For a proper choice of some major parameters, it is generally recommended that the concentration of HS solution injection is 5-10% NaCl and the flow rate of HS solution is close to the normal RO operation as a starting point.

To date, most recent studies about the use of hypersaline solutions for osmotic backwashing show the importance of the duration of pulse duration. Pulse duration should be longer than the residence time for a maximum achievable cycle-averaged permeation rate. A shorter pulse is ineffective in inducing osmotic permeation particularly towards the end of the channel. In fact, short pulse is significantly diluted on the membrane surface to the point where its concentration may drop below that required for inducing osmotic flow. The pulse concentration and duration must be optimized if efficient osmotic cleaning is to be achieved throughout the full length of a membrane train (Ramon et al., 2010).

3.2 Ultrasonic fields

Application of ultrasound for enhancing permeation in membrane processes has been extensively studied, specially for enhancing microfiltration (Kyllönen et al., 2005) and ultrafiltration (Chen et al., 2006; Juang & Lin, 2004; Saxena et al., 2009) membrane separation, and it has also been considered as a pretreatment to minimize fouling specially due to particulate or organic matter.

Ultrasound is defined at frequencies ranging from 16 kHz to 1019 kHz produced by electromechanical transducers based on piezoelectric effect. The basic physical phenomenon behind the effect of ultrasound is cavitation, that starts between this range of frequency, and it is promoted by the pass of the ultrasound waves through the liquid medium in a series of alternate compression and expansion cycles. Cavitation mainly promotes formation, growth and implosive collapse of bubbles in the liquid that has significant mechanical and chemical effects. In fact, it is reported that each cavitation bubble generates temperatures of 4000-6000 K and pressures of 100-200 MPa, acting as active "hot spots" (Feng et al., 2006). The acoustic streaming and shear forces imposed by cavitation bubbles reduce the fouling on membrane surface. Further, when a cavitating bubble is oscillating near a solid surface, it does so asymmetrically, resulting in the generation of microjets (microstreams) of high velocity and shock waves (Li et al., 2002) as it is shown in next figure (Figure 5). Fluid flowing at these high velocities can decrease the thickness of boundary layer and diffusional resistance and therefore enhance the rates of mass transfer.

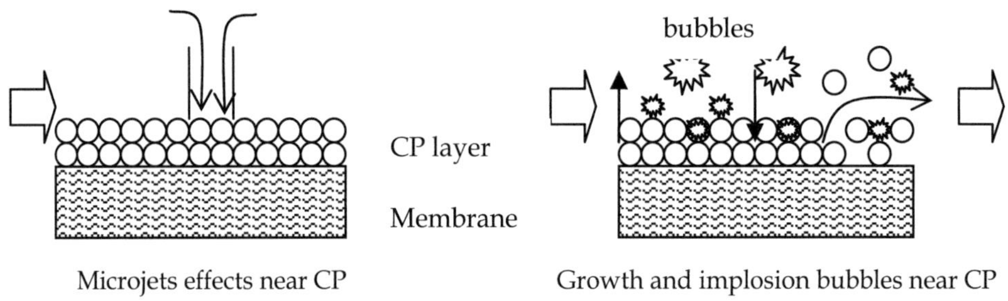

Fig. 5. Ultrasound effect on membrane cleaning

Main advantages of ultrasound cleaning are the following (Lu et al., 2009):

- Membranes can be cleaned while they are in use.
- There are no secondary pollutants and problems of transporting and handling of dangerous chemicals as in the chemical cleaning.
- Hydrogen peroxide (H_2O_2) and hydroxyl free radical (.OH) produced by ultrasound can be used for disinfection of the distribution systems of drinking water, reducing the use of chlorine that produces carcinogenic by-products and thus is harmful to humans.

Variables needed to be investigated include the intensity of the ultrasound field, the solute concentration and the irradiation time to increase effectivity of the process without damaging microstructure of the membrane surface due to sonication process.

Nowadays, there are short studies that reported the performance of applying this technology to reduce membrane fouling in reverse osmosis membranes specifically. The patent of Harvey (Harvey, 1965, as cited in Feng et al., 2006) proposed the use of an acoustic liquid whistle or ultrasound transducers to produce cavitation to prevent clogging of membranes and to remove concentration–polarisation in a water desalination process with semi-permeable reverse osmosis membranes.

Just Feng (Feng et al., 2006) reported that on-line ultrasonic defouling of the polyamide-based reverse osmosis membranes resulted in a significant increase in the permeate flux. In fact inorganic foulants such as $CaSO_4$ and $Fe(OH)_3$ and organic foulants such as carboxymethyl cellulose were effectively removed from the membrane surface.

In spite of this, it has been demonstrated in various laboratory scale studies that ultrasound can be used very effectively to clean membranes; so far, there are no ultrasound technologies that are used in a large-scale drinking water treatment. Main reasons for this situation are:

- The cost of energy needed by ultrasound cleaner would be high.
- Ultrasound transducers that could handle a high power to produce cavitations would be very costly, bulky, and brittle.

More research needs to be done in the ultrasound field application to improve ultrasound transducers and to reduce energy cost of the cleaning process.

3.3 Electric fields

Electric field is considered as another way to reduce membrane fouling but also a way to favour separation process. Electric field is based on a phenomenon generally referred as electrokinetics. Electrokinetics is defined as the "liquid flow that occurs along a solid/liquid interface as a result of an applied potential gradient or conversely to the potential developed when a liquid is made to flow along an interface" (Sennett & Oliver, 1965). This phenomenon has a beneficial effect in the filtration process which contributes as an additional driving force without increasing shear stress.

As a result of the electrokinetics, electrophoresis and electroosmosis effects are observed:

- Electrophoresis is related to the movement of the solids or charged species (such as proteins).
- Electrosmosis phenomenon is about the movement of the fluid, such as permeation of the fluid through a porous medium (membrane).

Application of an electric field on the membrane surface would potentially attract the particles from the membrane surface, lifting them and carrying them over. The electric field imposes an electrophoretic effect on the charged molecules dragging them away from the membrane surface, reducing the concentration polarization layer and increasing the flux. Besides the occurrence of electrophoresis and electroosmosis, other effects such as electrolysis, Joule's heating or ion migration also happen due to the application of an electric field, which can affect filtration performance and fouling process.

The applied electric field strength depends on the conductivity of the feed, electrode placement and material. The electrochemical properties of the membrane surface and the dispersed materials or solutes can also have a significant influence on the nature and magnitude of the interactions between the membrane and the substances being used, and their separation characteristics. Finally, zeta potential and charge of the species present in the feed should be studied with the aim of optimizing the application of this technology as it is represented in next figure (Figure 6).

Different kinds of cell and configurations are used, being most of them flat sheet or spiral wound membranes with platinum electrodes. Process used to be enhanced by the presence of dynamic turbulence promoters as static metal deployed sheet, oxygen bubbles or activated alumina in UF membranes (Mameri et al., 1999). In other cases, gases generated during the resulting water electrolysis have a "sweeping action" on the membrane surface, that keeps the membrane surface clean.

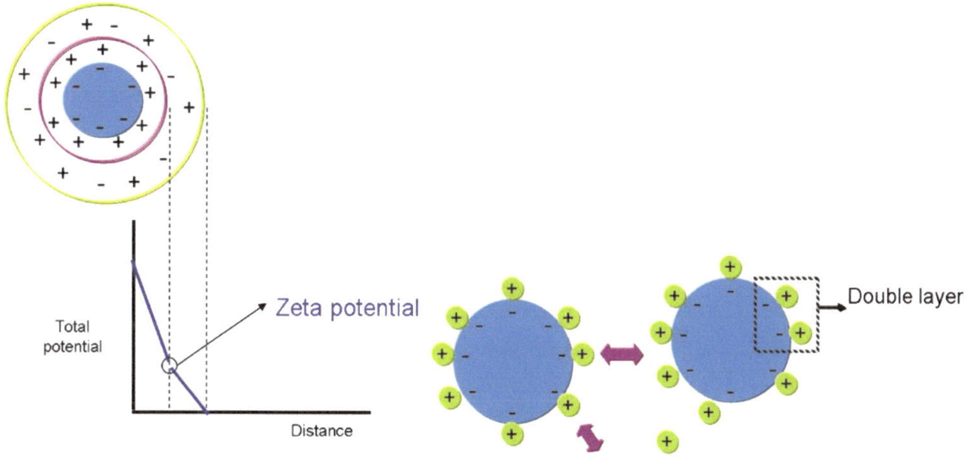

Fig. 6. Zeta potential of a colloidal particle

Main advantage of this cleaning technique is that electric field can be applied in the work cycle without interruption, in a similar way of hypersaline and ultrasound cleaning techniques described before. Electric field is applied either continuously (Iritani et al., 2000) or intermittently, being this second option more suitable because it reduces energy consumption, reduces the heat production and also reduces the induction time to obtain undesirable chemical changes in the process due to electrochemical reactions (Oussedik et al., 2000; Weigert et al., 1999, Webster et al., 2000). Best results have been obtained with electric fields of high voltage, intermittently applied and long time applied (Ahmad et al., 2002).
Therefore, it is considered an emerging technology for its application in RO due to some disadvantages that needed more research such as:

- The use of uneconomic power levels to decouple the solutes from the permeate or bulk stream which can limit this approach in water treatment using RO.
- The cost of the electrodes. The electrode materials chosen should be compatible with the membrane material as well as with the feed stock being processed. Furthermore, the electrode areas should not interfere with the feed and permeate flows (Jagannadh et al., 1996).

Main research about the use of this technique for membrane cleaning is related with UF membrane. Among all the research reviewed, Tarazaga has important results, obtained with their studies with bovine plasma over UF membranes. These studies demonstrate the huge potential of applying an electric field under cross-flow filtration, to reduce membrane fouling of the membrane (Tarazaga et al., 2006a). Application of a rectified medium wave pulse made an increase of the permeate flux till values almost initial, being proportional the increase of the flux to the increase in the electric field. Polarization concentration and fouling of the membranes became a reversible phenomenon due to application of electric fields in different cycles, allowing to recover permeate initial flux in all the experiments. Even more, some studies state that the use of electric fields has demonstrated that the initial permeate flux of protein UF fouled membranes can be restored almost instantaneously by means of an exponential curve (Tarazaga et al., 2006b).
Application of an electric field used to be very effective for protein separation, because they present an electric charge that helps protein separation by ultrafiltration (Saxena et al., 2009). This technique is called electro-ultrafiltration and is represented in next figure (Figure 7). It is also applied in membranes previously fouled with colloidal matter or oils.

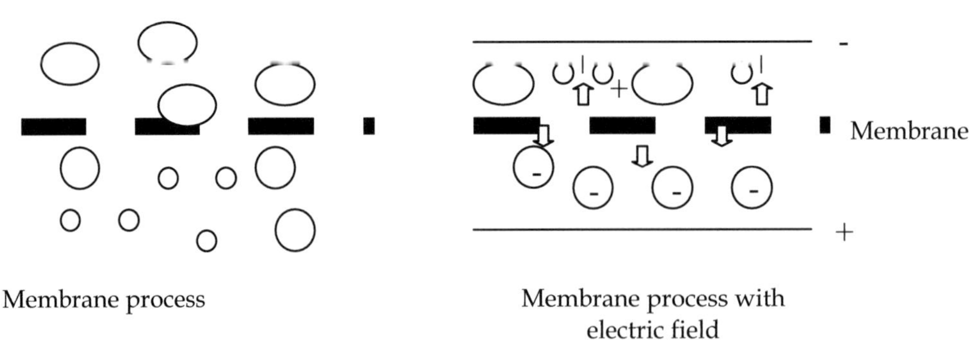

Fig. 7. Effect of the electric field in separation process

Some specific studies related with RO membranes are published. Romo and Pitts (Romo & Pitts, 1998), stated the application of electric field was effective in dispersing colloidal particles contained in water from a soft drink bottling plant. The effect of the high voltage electric field (30 kV DC power supply) created by a ceramic electrode on RO membranes kept them free from biological deposits during 9 months operation. After 11 months chemical cleaning was performed as a preventive maintenance measure. Performance of the membranes was not enhanced by chemical cleaning, proving that membranes had remained clean throughout the testing period with the electric field.

3.4 Magnetic fields

Calcium carbonate precipitation has been the focus of numerous investigations because of its importance in several industrial processes. For instance, it is involved in many industrial applications as pigment, brightener filler, adsorbent, and in different biomedical uses. However, calcium carbonate scale deposition induces important damage in the domestic, agricultural or industrial installations using natural waters (pipe blocking, membrane clogging, efficiency decay of heaters or heat exchangers, etc.) (Alimi et al., 2009). Various methods were used to prevent scaling, for example by water decarbonation using electrochemical processes, seeding or acid addition. Other way consists in blocking the nucleation and crystal growth by adding chemical inhibitors called antiscalants. However, when these strategies do not give enough good results, alternative non-chemical processes have therefore been performed. They are preferable due to the fact that they do not change the properties of water. In particular, magnetic water treatments may be considered, due to some advantages such as low cost, simple operation and no harm to the environment.

It is believed that magnetic water treatment decreases the hardness of water and either reduce and remove existing scale or produce a softer and less tenacious scale. This effect is because magnetic treatment alters the morphology and adhesion of calcium carbonate scale, changing calcite, into aragonite. These are the two common natural forms of $CaCO_3$, with rhombohedra and orthorhombic crystal structures, respectively. Aragonite has higher density, is more soluble and does not precipitate, avoiding and reducing scale formation. Even though, there is some dispute as to whether the deposits of calcium carbonate from magnetically treated water are predominantly: calcite or aragonite (Coey & Cass, 2000).

Scaling or precipitation fouling of reverse osmosis and nanofiltration membranes is also a common phenomenon occurring wherever the solubility of any sparingly soluble salt in the feed water is exceeded. Inorganic scale formation on the membrane shortens the life of the membrane system and increases operating and maintenance costs. Sometimes may also lead to physical damage of membranes due to the difficulty of scale removal and to irreversible membrane pore plugging. The most common constituents of scale are $CaCO_3$, $CaSO_4.2H_2O$ and silica, while other potential scaling species are $BaSO_4$, $SrSO_4$,$Ca_3(PO_4)_2$ and ferric and aluminium hydroxides. A certain degree of fouling control can usually be achieved by optimisation operating conditions with specific reference to factors having an influence on concentration polarisation (CP). In other cases, chemical reagents are needed to reduce scaling by lowering pH, adding antiscalants or a mix of both. However, when these strategies do not give enough good results, alternative non-chemical processes have therefore been performance, by inducing turbulence in the module to reduce boundary layer (Baker et al., 1997).

Magnetic fields are applied normally in the feed flow, in the storage tank or even over pipe that takes the flow into the membrane module. Devices for generating magnetic fields is not normalized, and some of the researchers use electromagnets and others permanent magnets. They should accomplish a dynamic magnetic treatment, since the solution has to move fast enough through the magnetic field. Better results are obtained for systems with continuous recirculation where water is exposed to magnetic fields several times (Baker et al., 1996).
Al-Qahtani (Al-Qahtani, 1996) concluded that pretreatment of the solution before entering to the RO membrane module, decreases salt concentration of the permeate when comparing results to those obtained in the same process but with untreated feed flows. Application of a magnetic field improved conductivity and pH of the treated solution and concentrate of RO membrane increased when magnetic treatment was introduced. Other researchers such as Vedavyasan (Vedavyasan, 2001) stated that fouling improves when magnetic field is applied, decreasing energy cost in a rate of 18 % and cleaning cost in a 33 % percentage.
However, Baker (Baker et al., 1996), stated that magnetic field treatment for cleaning membranes was not useful to prevent scaling when RO membranes are working at a high recovery rate in systems with saturated solutions of calcium carbonate in one step systems and with recirculation. Opposite results were observed in NF membranes by Li (Li et al., 2007).
Since some results are contradictory, more research is needed about the mechanism that magnetic field has over formation and growing of precipitates over the RO membrane surface, and about analytical methods to study in a better way the performance of this treatment. In this way, UDTR has recently developed an existing ultrasonic technique for quantitative study of the effect of magnetic fields on $CaCO_3$ scale deposition over NF membranes with encouraging results (Li et al., 2007).

3.5 Electrochemical methods

Electrochemical methods allow cleaning membranes at low cross-flow velocity due to the use of potential application between electrodes. Studies done by using this cleaning method are not very extensive and nowadays are not being implemented in any large plant. Some examples of this methodology are found in metallic microfilters cleaning when fouled with albumin and phosphate. Results showed best cleaning performance and recovery when negative current was applied over the microfilter at a low pH cleaning medium. Even more, changes in pH near interface due to water reduction and hydrogen gas bubbles also helped membrane cleaning (Webster et al., 2000).

4. Evaluation of cleaning performance

Generally, the effectiveness of the cleaning process is checked by measuring the water flux after cleaning at defined operating conditions. However, this is not a reliable measure because a good water flux does not guarantee a good operational flux. In fact, many plant operators have noted that even when the permeate flux is fully restored, it cannot be maintained for as long as when membranes were initially used (Tay & Song, 2005). Anyway, a low water flux after cleaning may indicate that the cleaning is not sufficient (Tragärdh, 1989).
There are different parameters in the literature for evaluating cleaning efficiency. In some cases, cleaning efficiency is evaluated by the *flux recovery* and the *fouling ratio* (Liikanen et al., 2002; Chen et al., 2003):

$$Flux\ recovery = J_c / J_0 \quad (1)$$

$$Fouling\ ratio = J_f / J_0 \quad (2)$$

Where J_c is the flux after cleaning, J_0 is the flux of the virgin membrane, and J_f is the flux for the fouled membrane.
The *flux recovery* has been also defined as (Madaeni et al., 2001):

$$FR(\%) = [(J_c - J_w) / (J_0 - J_w)] \cdot 100 \quad (3)$$

Other works evaluate cleaning efficiency in terms of resistance, such as the *resistance removal* (*RR*) calculated as (Madaeni et al., 2001):

$$RR(\%) = [(R_f - R_c) / R_f] \cdot 100 \quad (4)$$

Where R_f is the resistance after fouling, and R_c is the resistance which remains after cleaning. Both resistances can be calculated from the water flux after washing with water, and the water flux after chemical cleaning, respectively.
(Chen et al., 2003) have also considered the enhancement in membrane selectivity by the improvement of TDS rejection calculated as the TDS rejection % after cleaning minus the TDS rejection % before cleaning.

5. Conclusions

Nowadays, fouling is the most critical phenomenon in membrane performance. Since the economical incidence of fouling can be very significant, minimizing this phenomenon is a must in membrane plants. In spite of methods for preventing fouling, such as good pretreatment and suitable operating conditions, membrane cleaning is always necessary to improve membrane performance over its useful life. Research on membrane cleaning must continue in order to enhance a process that represents an important percentage of the operating costs. Within physical cleaning methods, the use of hypersaline solutions for backwashing and the application of ultrasonic or some kinds of magnetic and electrical fields seem to be very promising techniques in membrane cleaning. With regard to chemical cleaning, research should be focused on the combination with physical methods, such as air flushing, since greater efficiencies can be achieved when combining physical and chemical actions simultaneously.

6. References

Ahmad, A.L.; Ibrahim N. & Bowen W.R. (2002). Automated electrophoretic membrane cleaning for dead-end microfiltration and ultrafiltration. *Separation and Purification Technology*, Vol. 29, No. 2, (November 2002), pp. 105-112, ISSN 1383-5866

Al-Amoudi, A. & Lovitt, W.L. (2007). Fouling strategies and the cleaning system of NF membranes and factors affecting cleaning efficiency. *Journal of Membrane Science*, Vol. 303, Issues 1-2, (October 2007), pp. 4-28, ISSN 0376-7388

Alimi, F.; Tlili, M.M.; Ben Amor, M.; Maurin, G. & Gabrielli C. (2009) Effect of magnetic water treatment on calcium carbonate precipitation: Influence of the pipe material. *Chemical Engineering and Processing*, Vol. 48, Issue 8 (August 2009), pp. 1327-1332, ISSN 0255-2701

Al-Qahtani H. (1996). Effect of magnetic treatment on Gulf seawater. *Desalination*, Vol. 107, Issue 1, (September 1996), pp. 75-81, ISSN 0011-9164

Ang, W.S.; Lee, S.; Elimelech, M. (2006). Chemical and physical aspects of cleaning of organic-fouled reverse osmosis membranes *Journal of Membrane Science*, Vol. 272, Issues 1-2, (March 2006), pp. 198-210, ISSN 0376-7388.

Arnal, J.M.; Garcia-Fayos, B.; Sancho, M. & Verdu, G. (2009a). Ultrafiltration membrane cleaning with different chemical solutions after treating surface water. *Desalination and Water Treatment*, Vol. 7, Issues 1-3, (July 2009), pp. 198-205, ISSN 1944-3994

Arnal, J.M.; Garcia-Fayos, B.; Sancho, M. & Verdu, G. (2009b). Cleaning ultrafiltration membranes by different chemical solutions with air bubbles. *Desalination and Water Treatment*, Vol. 10, Issues 1-3, (October 2009), pp. 198-205, ISSN 1944-3994

Baker, J.S. & Judd, S.J. (1996). Magnetic amelioration of scale formation. *Water Research* Vol. 30, Issue 2, (February 1996), pp. 247-260, ISSN 0043-1354

Baker, J.S.; Judd, S.J. & Parsons A.S. (1997). Antiscale magnetic pretreatment of reverse osmosis feedwater. *Desalination* Vol. 110, Issues 1-2, (August 1997), pp. 151-166, ISSN 0376-7388

Baker, J.S. & Dudley, L.Y. (1998). Biofouling in membrane systems_A review. *Desalination*, Vol. 118, Issues 1-3 (September 1998), pp. 81-90, ISSN 0376-7388

Bartlett, M.; Bird, M.R. & Howell, J.A. (1995). An experimental study for the development of a qualitative membrane cleaning model. *Journal of Membrane Science*, Vol. 105, Issues 1-2, (September 1995), pp. 147-157, ISSN 0376-7388

Butt, F.H.; Rahman, F. & Baduruthamal, U. (1997). Characterization of foulants by autopsy of RO desalination membranes. *Desalination*, Vol. 114, Issue 1, (December 1997), pp. 51-64, ISSN 0011-9164

Cabassud, C.; Laborie, S.; Durand-Bourlier, L. & Lainé, J.M. (2001). Air sparging in ultrafiltration hollow fibers: relationship between flux enhancement, cake characteristics and hydrodinamic parameters. *Journal of Membrane Science*, Vol. 181, Issue 1, (January 2001), pp. 57-69, ISSN 0376-7388

Chen, J.P.; Kim, S.L. & Ting, Y.P. (2003). Optimization of membrane physical and chemical cleaning by a statistically designed approach. *Journal of Membrane Science*, Vol. 219, Issues 1-2, (July 2003), pp. 27-45, ISSN 0376-7388

Chen, D.; Weavers, L.K. & Walker, H.W. (2006). Ultrasonic control of ceramic membrane fouling: effect of particle characteristics. *Water Research*, Vol. 40, No. 4, (February 2006), pp. 840-850, ISSN 0043-1354

Chesters, S.P. (2009). Innovations in the inhibition and cleaning of reverse osmosis membrane scaling and fouling. *Desalination*, Vol. 238, Issues 1-3, (March 2009), pp. 22-29, ISSN 0011-9164

Coey, J.M.D. & Cass S. (2000). Magnetic water treatment. *Journal of Magnetism and Magnetic Materials* Vol. 209, Issues 1-3, (February 2000), pp 71-74, ISSN 0304-8853

Cornelissen, E.R.; Vrouwenvelder, J.S.; Heijman, S.G.J.; Viallefont, X.D.; Van Der Kooij, D. & Wessels, L.P. (2007). Periodic air/water cleaning for control of biofouling in spiral wound membrane elements. *Journal of Membrane Science*, Vol. 287, Issue 1, (January 2007), pp. 94-101, ISSN 0376-7388

Creber, S.A.; Vrouwenvelder,J.S.; van Loosdrecht, M.C.M. & M.L. Johns. (2010). Chemical cleaning of biofouling in reverse osmosis membranes evaluated using magnetic resonance imaging. *Journal of Membrane Science*, Vol. 362, Issues 1-2, (October 2010), pp. 202–210, ISSN 0376-7388

Cui, Z. & Taha, T. (2003). Enhancement of ultrafiltration using gas sparging: a comparison of different membrane modules. *Journal of Chemical Technology and Biotechnology*, Vol. 78, Issues 2-3, (February-March 2003), pp. 249-253, ISSN 1097-4660

Ducom, G. & Cabassud, C. (2003). Possible effects of air sparging for nanofiltration of salted solutions. *Desalination*, Vol. 156, Issues 1-3, (August 2003), pp. 267-274, ISSN 0011-9164

Ebrahim, S. (1994). Cleaning and regeneration of membranes in desalination and wastewater applications: state-of-the-art. *Desalination*, Vol. 96, Issues 1-3, (June 1994), pp. 225-238, ISSN 0011-9164

Feng, D.; van Deventer, J.S.J. & Aldrich C. (2006). Ultrasonic defouling of reverse osmosis membranes used to treat wastewater effluents. *Separation and Purification Technology*, Vol. 50, No. 3, (July 2006), pp. 318–323, ISSN 1383-5866

Fritsch, J. & Morau, C.I. (2008). Development and optimization of a carbon dioxide-aided cold microfiltration process for the physical removal of microorganisms and somatic cells from skim milk. *Journal of Dairy Science*, Vol. 91, Issue 10 (October 2008), pp. 3744-3760, ISSN: 0022-0302

Gao, W.; Liang, H.; Ma, J.; Chen, Z.; Han, Z. & Li, G. (2011). Membrane fouling control in ultrafiltration technology for drinking water production: a review. *Desalination*, Vol. 272, Issues 1-3, (May 2011), pp. 1-8, ISSN 0011-9164

Goosen, M.F.A.; Sablani, S.S.; Al-Hinai, H.; Al-Obeidani, S.; Al-Belushi, R. & Jackson, D. (2004). Fouling of reverse osmosis and ultrafiltration membranes: a critical review. *Separation Science and Technology*, Vol. 39, No. 10, pp. 2261-2297, ISSN 0149-6395

Graham, S.I., Reitz, R.L. & Hickman, C.E. (1989). Improving reverse osmosis performance through periodic cleaning. *Desalination*, Vol. 74, pp. 113-124, ISSN 0011-9164

Huang, X.; Guillen, G.R. & Hoek E.M.V. (2010). A new high-pressure optical membrane module for direct observation of seawater RO membrane fouling and cleaning. *Journal of Membrane Science*, Vol. 364, Issues 1-2, (November 2010), pp. 149–156, ISSN 0376-7388

Iritani, E.; Mukai, Y. & Kiyotomo Y. (2000). Effects of electric field on dynamic behaviours of dead-end inclined and downward ultrafiltration of protein solutions. *Journal of Membrane Science*, Vol. 164, Issues 1-2, (January 2000), pp. 51-57, ISSN 0376-7388

Jagannadh, S.N. & Muralidhara H.S. (2006) Electrokinetics methods to control membrane fouling. *Industrial Engineering & Chemistry Research*, Vol. 35, Issue 4, (April 2006), pp. 1133-1140, ISSN 0888-5885

Juang, R.S. & Lin K.H. (2004). Flux recovery in the ultrafiltration of suspended solutions with ultrasound. *Journal of Membrane Science*, Vol. 243, Issues 1-2, (November 2004), pp. 115-124, ISSN 0376-7388

Kimura, K.; Hane, Y.; Watanabe, Y.; Amy, G. & Ohkuma, N. (2004). Irreversible membrane fouling during ultrafiltration of surface water. *Water Research*, Vol. 38, Issues 14-15, (August-September 2004), pp. 3431-3441, ISSN 0043-1354

Kyllönen, H.M.; Pirkonen, P. & Nyström M. (2005). Membrane filtration enhanced by ultrasound: a review. *Desalination*, Vol. 181, Issues 1-3, (September 2005), pp. 319-335, ISSN 0011-9164

Li, J.; Sanderson, R.D. & Jacobs, E.P. (2002). Non-invasive visualization of the fouling of microfiltration membranes by ultrasonic time-domain reflectometry. *Journal of Membrane Science*, Vol. 201, Issues 1-2, (May 2002), pp. 17-29, ISSN 0376-7388

Li, J; Liu J.; Yang, T. & Xiao, C. (2010) Quantitative study of the effect of electromagnetic field on scale deposition on nanofiltration membranes via UTDR. *Water Research*, Vol. 41, Issue 20, (December 2007), pp. 4595–4610, ISSN 0043-1354

Liberman I. RO membrane cleaning method. PCT, WO 2005/123232 A2 0181497 (2005)

Liberman, B. & Liberman, I. (2005). Replacing membrane CIP by Direct Osmosis cleaning. *Desalination Water Reuse*, Vol. 15, No. 1, (August/September 2005), pp. 28–34, ISSN 1022-5404

Liberman, B. Direct osmosis cleaning. Patent application WO 2004/062774 (2004) and U.S. patent application 20070246425 (2007)

Liikanen, R.; Yli-Kuivila, J. & Laukkanen, R. (2002) Efficiency of various chemical cleanings for nanofiltration membrane fouled by conventionally-treated surface water. *Journal of Membrane Science*, Vol. 195, Issue 2, (January 2002), pp. 265-276, ISSN 0376-7388.

Lin, J.C-T.; Lee, D-J. & Huang, C. (2010). Membrane fouling mitigation: Membrane cleaning. *Separation Science and Technology*, Vol. 45, Issue 7, pp. 858-872, ISSN 0149-6395.

Lu, J.Y.; Du, X. & Lipscomb G. (2009). Cleaning Membranes with Focused Ultrasound Beams for Drinking Water Treatment, *Proceedings of IEEE International Ultrasonics Symposium Proceedings*, pp. 1195-1198, ISBN 978-1-4244-4390-1/09, Rome, September, 2009

Madaeni, S.S.; Mohamamdi, T. & Moghadam, M.K. (2001). Chemical cleaning of reverse osmosis membranes, *Desalination*, Vol. 134, Issues 1-3, (April 2001), pp. 77-82, ISSN 0011-9164

Mameri, N.; Oussedik, S.; Yeddou, R.; Piron, D.L.; Belhocine, D.; Lounici, H. & Grib, H. (1999). Enhancement of ultrafiltration flux by coupling static turbulence promoter and electric field. *Separation and Purification Technology*, Vol. 17, No 3, (December 1999), pp. 203-211, ISSN 1383-5866

Mercier, M.; Fonade, C. & Lafforgue-Delorme, C. (1997). How slug flow can enhance the ultrafiltration flux in mineral tubular membranes. *Journal of Membrane Science*, Vol. 128, Issue 1, (May 1997), pp. 103-113, ISSN 0376-7388

Mohammadi, T.; Moghadam, M.K. & Madaeni, S.S. (2002). Hydrodinamic factors affecting flux and fouling during reverse osmosis of seawater. *Desalination*, Vol. 151, Issue 3, (January 2003), pp. 239-245, ISSN 0011-9164

Mohammadi, T.; Madaeni, S.S. & Moghadam, M.K. (2002) Investigation of membrane fouling. *Desalination*, Vol. 153, Issues 1-3, (February 2003), pp. 155-160, ISSN 0011-9164

Ngene, I.S.; Lammertink, R.G.H.; Kemperman, A.J.B.; Van de Ven, W.J.C.; Wessels, L.P.; Wessling, M. & Van der Meer, W.G.J. (2010). CO_2 nucleation in membrane spacer channels remove biofilms and fouling deposits. *Industrial and Engineering Chemistry Research*, Vol. 49, No 20, (September 2010), pp. 10034-10039, ISSN 0888-5885

Oussedik, S.; Belhocine, D.; Grib, H.; Lounici, H.; Piron, D.L. & Mamer N. (2000). Enhanced ultrafiltration of bovine serum albumin with pulsed electric field and fluidized activated alumina. *Desalination*, Vol. 127, Issue 1, (January 2000), pp. 59-68, ISSN 0011-9164

Pontié, M.; Rapenne, S.; Thekkedath, A.; Duchesne, J.; Jacquemet, V.; Leparc, J. & Suty, H. (2005). Tools for membrane autopsies and antifouling strategies in seawater feeds: a review. *Desalination*, Vol. 181, Issues 1-3, (September 2005), pp. 75-90, ISSN 0011-9164

Psoch, C. & Schiewer, S. (2006). Direct filtration of natural and simulated river water with air sparging and sponge ball for fouling control. *Desalination*, Vol. 197, Issues 1-3, (October 2006), pp. 190-204, ISSN 0011-9164

Qaisrani, T.M. & Samhaber, W.M. (2011). Impact of gas bubbling and backflushing on fouling control and membrane cleaning. *Desalination*, Vol. 266, Issues 1-3, (January 2011), pp. 154-161, ISSN 0011-9164

Qin, J.J.; Liberman, B.; Kekre, K.A. (2009). Direct Osmosis for Reverse Osmosis Fouling Control: Principles, Applications and Recent Developments. *The Open Chemical Engineering Journal*, Vol. 3, No. 1, pp. 8-16, ISSN 1874-1231

Qin, J.J.; Oo, M. H.; Kekre, K.A. & Liberman, B. (2010). Development of novel backwash cleaning technique for reverse osmosis in reclamation of secondary effluent. *Journal of Membrane Science*, Vol. 346, Issue 1, (January 2010), pp. 8–14, ISSN 0376-7388

Ramon, G.; Agnon, Y. & Dosoretz, C. (2010). Dynamics of an osmotic backwash cycle. *Journal of Membrane Science*, Vol. 364, Issues 1-2, (November 2010), pp. 157-166, ISSN 0376-7388

Romo, R.F.V., & Pitts, M.M. Application of electrotehcnology for removal and prevention of reverse osmosis biofouling. *Proceedings of AIChE Spring meeting*, pp. 1-11, New Orleans (USA), March, 1998

Sadhwani, J.J. & Veza, J.M. (2001). Cleaning tests for seawater reverse osmosis membranes. *Desalination*, Vol. 139, Issues 1-3, (September 2001), pp. 177-182, ISSN 0011-9164

Sagiv, A. & Semiat, R. (2005). Backwash of spiral wound membranes. *Desalination*, Vol. 179, Issues 1-3, (June 2005), pp. 1-9, ISSN 0011-9164

Sagiv, A. & Semiat, R. (2010). Parameters affecting backwash variables of RO membranes. *Desalination*, Vol. 261, Issue 3, (October 2010), pp. 347-353, ISSN 0011-9164

Saxena, A.; Tripathi, B.P.; Kumar M. & Shahi V. K. (2009) Membrane-based techniques for the separation and purification of proteins: An overview. *Advances in Colloid and Interface Science*, Vol. 145, Issues 1-2, (January, 2009), pp. 1–22, ISSN 0001-8686

Sennett, P. & Oliver, J.P. (1965). Colloidal dispersions electrokinetic effects and the concept of zeta potential. *Industrial Engineering Chemical*, Vol. 57, Issue 8, (August 1965), pp. 32-49, ISSN 0019-7866

Tarazaga, C.C.; Campderros M.E. & Padilla, A.P. (2006a). Physical cleaning by means of electric field in the ultrafiltration of a biological solution. *Journal of Membrane Science*, Vol. 278, Issues 1-2, (July 2006), pp. 219-224, ISSN0376-7388

Tarazaga, C.C.; Campderros M.E. & Padilla, A.P. (2006b). Characterization of exponential permeate flux by technical parameters during fouling and membrane cleaning by electric field. *Journal of Membrane Science*, Vol. 283, No. 1-2, (October 2006), pp. 339-345, ISSN 0376-7388

Tay, K.G. & Song, L. (2005). A more effective method for fouling characterization in a full-scale reverse osmosis process. *Desalination*, Vol. 177, Issues 1-3, (June 2005), pp. 95-107, ISSN 0011-9164

Tragardh, G. (1989). Membrane cleaning. *Desalination*, Vol. 71, Issue 3, pp. 325-335, ISSN 0011-9164

Vedavyasan, C.V. (2001). Potential use of magnetic field a perspective. *Desalination* Vol. 134, Issues 1-3, (April, 2001), pp. 105-108, ISSN 0011-9164

Vrouwenvelder, J.S.; vanPaassen, J.A.M.; Wessels, L.P.; van Dam, A.F. & Bakker, S.M. (2006). The Membrane Fouling Simulator: A practical tool for fouling prediction and control. *Journal of Membrane Science*, Vol. 281, Issues 1-2, (September 2006), pp. 316-324, ISSN 0376-7388

Vrouwenvelder, J.S; Bakker, S.M; Wessels, L.P & vanPaassen, J. A.M. (2007a). The membrane fouling simulator as a new tool for biofouling control of spiral wound membranes. *Desalination*, Vol. 204, Issues 1-3, (February 2007), pp. 170-174, ISSN 0011-9164

Vrouwenvelder, J.S; Bakker, S.M.; Cauchard, M.; Le Grand, R.; Apacandie, M.; Idrissi, M.; Lagrave, S., Wessels, L.P.; vanPaassen, J.A.M.; Kruithof J.C. & Van Loosdrecht M.C.M. (2007b). The membrane fouling simulator: a suitable tool for prediction and characterisation of membrane fouling. *Water Science and Technology*, Vol. 55, Issues 8-9, (September, 2007), pp. 197-205, ISSN 0273-1223

Webster, R.D.; Chilukuri, S.V.V.; Levesley J.A. & Webster, B.J. (2000). Electrochemical cleaning of microporous metallic filters fouled with bovine serum albumin and phosphate under cross-flow velocities. *Journal of Applied Electrochemistry*, Vol. 30, Issue 8, (August 2000), pp. 915-924, ISSN 0021-891X

Weigert, T.; Altmann J. & Ripperger S. (1999). Cross-flow electrofiltration in pilot scale. *Journal of Membrane Science*, Vol. 159, Issues 1-2, (July 1999), pp. 253-262, ISSN 0376-7388

4

Reject Brines from Desalination as Possible Sources for Environmental Technologies

Caterina De Vito[1], Silvano Mignardi[1], Vincenzo Ferrini[1] and Robert F. Martin[2]
[1]*Dipartimento di Scienze della Terra, Sapienza Università di Roma,*
[2]*Department of Earth and Planetary Sciences, McGill University, Montreal,*
[1]*Italy*
[2]*Canada*

1. Introduction

Desalination of seawater represents the way to increase the supply of water for domestic, agricultural and industrial purposes. The potential environmental impact of desalination needs to be thoroughly evaluated, and any environmentally adverse consequences must be promptly addressed in order to assure a sustainable development of this technology.
The main concern in desalination is the management of the brines whose uncontrolled discharge has significant negative impacts on the environment. Indeed, even if the production costs have decreased in the last decade, those related to the disposal of the brine have shown only limited reduction of the relative costs. New production strategies benefit from modern and efficient freshwater-generation plants, but the management strategies are based only on the few traditional options for the disposal of wastes.
Sustainability is currently a main focus of high-level political discussions in the world, as it directly involves environmental pollution and global warming. Economic geologists are very well placed to understand and contribute to issues of sustainability, and to evaluate alternative resources for the future development of humanity. We, as economic geologists, are trained to unravel the complex series of events that led to the formation of natural resources. Consequently, we can use this expertise to help develop manmade processes to convert waste into resources.
Today, the challenge is to consider the brine, instead of a harmful by-product, as a valuable source for several compounds to be involved in diverse industrial processes. For example, the brines could provide significant amounts of magnesium, resulting in a very promising source of this element as an alternative to traditional mining. Indeed, today there are stringent regulations concerning mining in all most all countries to prevent or at least minimize damage to exploited properties. As traditional approaches in mining are costly in time or inapplicable in abandoned and remote regions, research needs to be focused on non-traditional sources such as brines coming from desalination plants. These brines certainly have several advantages, and in time can replace, at least in part, the conventional sources with an alternative, economically attractive process. Dual benefits can be achieved in recovering and recycling Mg from reject brines: reduction of large volumes of brines to be

discharged, as well as decrease of environmental concerns due to industrial extraction from silicate minerals.

In this chapter, we focus on novel processes making use of magnesium chloride of the reject brines as a reactant for the synthesis of minerals to convert the saline waste into useful and reusable solid products.

2. Background

2.1 The environmental concerns of reject brines

The increasing global population and industrial and agricultural production are driving a constant growth in the demand for fresh water worldwide. To meet the growing demand, non-typical sources of water have to be found. Water desalination is the main key option to provide the suitable supply of potable water. Recent estimates suggest that up to 25 million m^3 of desalinated water is produced daily worldwide (Lattemann & Höpner, 2008) of which almost two thirds is produced in countries in the Middle East and North Africa. In this area are located 77% of all desalination plants, whereas in the Asia-Pacific region, in the Americas and in Europe, there are 6, 7 and 10%, respectively (Lattemann & Höpner, 2008).

Desalination technologies can be divided in two main categories: thermal distillation (multi-stage flash, MSF and multiple effect distillation, MED) and membrane separation (reverse osmosis, RO). Hybrid plants integrating thermal and membrane processes also exist (Hamed, 2005). Other technologies that are of minor application include vapour compression (VC) and electrodialysis (ED). New engineering technologies, still under research and development, include capacitance deionization (CDI), membrane distillation (MD), forward osmosis (FO), gas hydrates (GH) and freezing, humidification dehumidification (HDH) (Mezher et al., 2010; Joint Water Reuse, Desalination Task Force [JWR & DTF], 2005).

The worldwide desalination capacity by process is shown in Fig. 1. Desalination plants use mostly seawater, followed by brackish, river and waste water (Fig. 1).

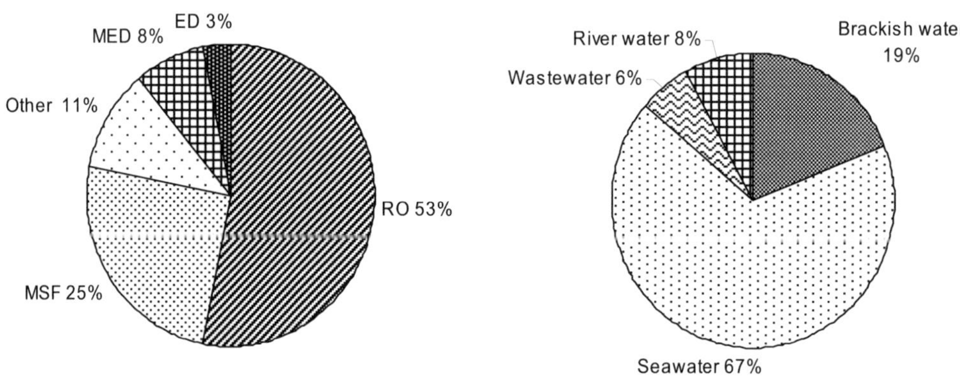

Fig. 1. Global desalination capacity by process (left) and worldwide feed water quality used in desalination (right) (ESCWA, 2009)

The importance and role of desalination do not deserve further comments. However, concerns are rising over strong potential negative impacts of desalination processes on the environment (Lattemann & Höpner, 2008; Maugin & Corsin, 2005; Miri & Chouikhi, 2005; Winters et al., 1979). The potential damages to the environment include: a) the uncontrolled discharge of concentrated brines which can impair aquatic ecosystems and water quality of aquifers; b) air pollution due to carbon dioxide emissions related to the energy used in the desalination processes. Desalination plants produce fresh water and discharge high salinity water (reject water or brine). The brine flow rates are large and generally vary with the type of technology; they are up to 40% for RO and up to 90% for MSF of the intake flow rate (Mickley, 2010). The reject water is a concentrate stream having a salinity higher than source feed water, up to 65,000-85,000 mg L^{-1} (Abdul-Wahab & Al-Weshahi, 2009), double or close to double that of natural seawater. The disposal of such a huge amount of high salinity water is one of the main environmental concerns. The brine may also have increased in temperature as values ranging from 5 to 15°C above intake waters may be reached (Lattemann & Höpner, 2008). Besides the high concentrations of salts, the brine contains residues of chemicals used for pre-treatment, by-products formed during the treatment, and heavy metals due to corrosion. The chemicals are used against scaling, biofouling, foaming, suspended solids and corrosion. The additives increase the detrimental impact of the brine as they are discharged along with the concentrate in the environment (Abdul-Wahab & Al-Weshahi, 2009; Lattemann & Höpner, 2008; Mezher et al., 2010).

In general, metal concentrations are low. However, copper in reject streams of thermal plants where Cu-Ni heat exchangers are used, can adversely affect the environment.

Several alternative approaches for handling the high salinity water exist (GHD, 2003; Lattemann & Höpner, 2008; Mezher et al., 2010; Mickley et al., 1993). The ways of disposing of the brine (Table 1) include deep well injection, land application, delivery to evaporation ponds, discharge in a sewer system, direct discharge to both surface water and submerged, aquaculture and beneficial re-use of the brine through salt recovery and harvesting systems.

Country	Disposal method
Qatar	Land application and evaporation ponds
Jordan	Land application and evaporation ponds
Oman	Land application and evaporation ponds
Kingdom of Saudi Arabia	Evaporation ponds
Australia	Evaporation ponds
Kuwait	Evaporation ponds
China	Land application
United Arab Emirates	Surface water
Spain	Surface water
Japan	Surface water
Algeria	Surface water
United Kingdom	Sewer system blending and land application
United States of America	Surface water, sewer system blending, land application, evaporation ponds, and deep well injection (in increasing order of % use)

Table 1. Disposal methods used by some countries
(Mezher et al., 2010 and references therein)

The selection of a disposal method is influenced by different factors such as volume of brine, constituents, geographical location of the discharge point, availability of receiving site, permissibility of the approach, public acceptance, capital and operating costs, ability of the facility to be expanded, and economics of the recovered products (Abdul-Wahab & Al-Weshahi, 2009; Ahmed et al., 2001; Mickley et al., 1993).

The cost of the brine disposal plays an important role in the selection of the method (Economic and Social Commission for Western Asia [ESCWA], 1993) as it could range from 5 to 33% of the total cost of desalination (Mohamed et al., 2005).

In the last decade, there has been an increased awareness that the main limitation to the further development of desalination is represented by brine management, a term that allows for beneficial use and volume reduction of brine in addition to its direct disposal. The challenge is to change the concept from saline waste to saline resource opening several opportunities to provide a reduction of brine disposal costs and volumes involved. The beneficial uses reported in the literature include aquaculture, oil and gas well field injection to maintain pressures, feedstock for hypochlorite production, dust control and de-icing, soil remediation, zero liquid discharge, wetlands treatment, salt recovery, and solar ponds (e.g., Abdul-Wahab & Al-Weshahi, 2009; GHD, 2003; Mickley, 2010).

An innovative and non-traditional use of the brine is its involvement in several industrial processes with the dual beneficial objective to reduce its volume to be discharged and mitigate the negative effects of some contaminants such as CO_2 and animal waste.

2.2 The problem of carbon dioxide

Daily we are all bombarded by alarming news about the rising levels of CO_2 in the atmosphere, caused largely by the combustion of fossil fuels, and about the deleterious impact of such loadings on our climate. Currently, many groups are actively seeking practical ways to sequester CO_2 being produced industrially, and to mitigate the environmental impact due to the uncontrolled discharge of concentrated brines from desalination plants. According to the Third Assessment Report of the Intergovernmental Panel on Climate Change (IPCC) (Morita et al., 2001), with its proposal to stabilize the CO_2 concentration between 450 and 750 ppmv, cumulative emissions ranging from 220 to 2,200 $GtCO_2$ (60-600 GtC) would need to be attained during this century. With the continued growth of emerging densely populated countries like China and India, there is the prospect of tripling the loading of CO_2 to the atmosphere by the end of the 21st century.

Clearly, the scientific community must implement quickly the research on effective methods of sequestering CO_2. All the approaches to reduce the global CO_2 emission must be examined carefully, even if most of them are not very encouraging as far as their quantitative potential to compensate for the huge amounts of fossil-fuel-derived CO_2 already in the atmosphere.

Numerous CO_2 sequestration approaches, including ocean, terrestrial, geological, biological and chemical options are currently being studied (Conrad & Klose, 1999; Freund & Ormerod, 1997; Friedmann, 2007; Gunter et al., 1997; Holloway, 1997; Lackner et al., 1995; Lackner et al., 1998; Lackner, 2003; Metz et al., 2005; O'Connor et al., 2002; Parson & Keith, 1998; Xu et al., 2004). As a substantial share of CO_2 emissions must be stored, the retention or sequestration of CO_2 in geological reservoirs is currently the applied option (e.g., Weyburn, Canada; Sleipner, North Sea; Holloway et al., 2007). Deep aquifers and depleted reservoirs in oil and gas fields have stored natural gas and crude oil for million of years. They thus offer possible sites for long-term storage of CO_2 (Gunter et al., 1997; Holloway, 1997). A variant of the depleted reservoir solution involves injecting

CO_2 in ultrabasic and basic rocks, such as might be encountered in ophiolitic complexes (Lackner et al., 1995; Robertson, 2004). These rocks are pervasively fractured owing to the tectonic forces active during their emplacement. However, the behaviour of CO_2 in such media, and the physical and chemical changes that can be expected in rocks if CO_2 is injected, are still a concern in the scientific community (Holloway et al., 2007; Huijgen et al., 2006).

The sequestration capacity of deep oceans is about 104 GtC (e.g., Freund & Ormerod, 1997; Lackner, 2003). The depth of injection would have to be ~1500 m, which raises serious technological challenges. This option could reduce the level of CO_2 in the atmosphere significantly. However, about 20% of injected gas can be expected to return to the atmosphere in ~300 years (Freund & Ormerod, 1997). Furthermore, the biological impact of such massive injection of CO_2 is of serious concern, as the resulting decrease in pH of seawater would most likely not be tolerated by the marine ecosystem (Freund & Ormerod, 1997; Lackner, 2003). In addition, there are thorny legal issues concerning disposal in the deep ocean, and possible catastrophic events during the injection process.

Existing photosynthetic systems naturally reduce the level of CO_2 in the atmosphere. They do not require pure CO_2, and incur no costs for preparation of the reactants. However, unrealistic areas of land and ocean would have to be used and could hardly be expected to compensate for CO_2 emissions of a single power plant (Parson & Keith, 1998). Efforts are under way to develop non-photosynthetic microbiological processes capable of converting CO_2 into methane and acetates (Conrad & Klose, 1999). It is appropriate to recall the small-scale success of the Fischer-Tropsch reaction in converting CO and H_2 into an array of hydrocarbons by reaction with iron-rich minerals, most recently to supply the domestic needs of South Africa during the apartheid-related oil embargo (Marsh et al., 2002). Again, unrealistic amounts of solid reactants would be needed to have any impact whatsoever.

Complementary CO_2 retention technologies, such as mineral sequestration *via* reaction of CO_2 with Mg-Ca-silicate rocks and/or neoformation of carbonate minerals in aqueous solutions, are attractive for their permanent and safe CO_2 storage in a solid form (Soong et al., 2006).

The methods based on neoformation of carbonates or bicarbonates, requires cations to neutralize CO_2 (Lackner et al., 1995; O'Connor et al., 2002). If a supply of the appropriate cations can be assured, such carbonation reactions offer virtually unlimited capacity and the promise of safe, permanent storage of CO_2. This idea was first proposed by Seifritz in 1990. There is little risk of unexpected release of CO_2 back to the atmosphere, because the resulting carbonates are thermodynamically stable (Ballirano et al., 2010; Ferrini et al., 2009). The process occurs naturally on a small scale during the weathering of rocks (Robertson, 2004), and has been shown to be important locally in ultrabasic and ophiolitic complexes (Hansen et al., 2005). The source of the neutralizing ions could be magnesium and calcium silicates such as olivine, serpentine-group minerals and clinopyroxene. Unfortunately, the industrial extraction of Ca and Mg from silicate minerals requires expensive pre-processing, which contributes to the problem rather than to the solution (Lackner et al., 1997). Furthermore, this option is not at all practical in many countries owing the paucity of exposed basic and ultrabasic rocks.

An attractive alternative would involve the interaction of ions in aqueous solution with CO_2. In this case, the precipitation of carbonates of Ca and Mg proceeds much faster than if the cations are locked in a silicate structure. The source of such a process could be a brine

(Ferrini et al., 2009) or even seawater. There is an advantage in focusing on Mg, because a greater weight proportion of CO_2 is found in Mg carbonates than in Ca carbonates. The process represents a variant of the Solvay process (Lackner, 2002) and is based on carbonating alkaline brines, which would produce carbonates from a chloride-rich reactant and hydrochloric acid as a by-product (Lackner, 2002).

2.3 The problem of animal waste

Large quantities of animal waste are produced in millions of tonnes each year worldwide. The intensive scale of animal farming, indeed, produces a large amount of waste that must be discharged. For this reason their management is of great concern, as the improper management of manure can adversely affect soil, water and air quality due to leaching and run-off of organic constituents, nutrients and heavy metals (Hsu & Lo, 2001; Wadman et al., 1987). In traditional agriculture, as the animal wastes represented the main source of crop nutrient, they were re-used for this purpose, generating equilibrium between waste production and their re-use and discharge. Today, the development of modern agriculture and intensive breeding has broken this equilibrium.

The amount of animal waste produced commonly is greater than that disposable locally on agricultural soil with no risks to the groundwater. In recent years the demand for livestock and poultry production grew in both developed and developing countries (Yetilmezsoy & Sapci-Zengin, 2009). For example, the total amount of organic dry substance produced in Germany by about 16 million cattle, 26 million pigs, 114 million poultry and 4 million sheep and horses accounts for 57,500 tonnes organic dry substance in 2001, whereas in Japan the total amount of animal waste is about 90 million tonnes per year (Haga, 1998; Köttner, 2001). Manure production in Spain was about 190 million tonnes per year (Moral et al., 2008). The yearly production of poultry and livestock waste in Russia exceeds 700 million m^3 (Kalyuzhnyi et al., 1998). Moreover, in most developing countries the poultry production is rapidly growing; for example, in Nigeria the amounts of waste produced represents millions of tonnes.

The disposal of such huge amounts of animal waste while minimizing the impact on the environment is one of the most serious challenge facing modern agriculture. The discharge of nutrient-rich wastewater to the environment increases the risk of eutrophication of natural waters owing to high levels of phosphorus and nitrogen (Yetilmezsoy & Sapci-Zengin, 2009). The removal of P and N in water is now regulated in many countries, and minimal concentrations of P in effluents are imposed. In particular, the European Community Urban Waste Water Treatment Directive [EC UWWTD] 97/271/EC (1991) has dealt with this pollution problem (Doyle & Parsons, 2002). The processes for P removal are traditionally based on P fixation in activated sludge by a biological or chemical method. Anaerobic lagoons are widely used in the treatment of liquid manure (Vanotti et al., 2007). In general, these processes can reduce the concentration of P in wastewater effluents, but they cause the accumulation of this element in sludge as well as a net increase in the volumes of sludge (Le Corre et al., 2009). Moreover, environmental problems connected with this method of treatment include emission of ammonia, odours, pathogens, and the potential deterioration of water quality (Aneja et al., 2000; Loughrin et al., 2006; Mallin, 2000; Szogi et al., 2006; Vanotti et al., 2005). New options for nitrogen removal are based on the so-called anammox, the combination of partial nitrification and anaerobic ammonium oxidation (Hwang et al., 2005), but they are still under development.

Besides P and N, animal waste pose a further risk of contamination owing to their contents in heavy metals. High concentrations of Cu, Zn and Mn occur in manure as these elements are provided in feed because their antimicrobial and growth-stimulation effects (Kessler et al., 1994). Most of this metal load (in particular Cu and Zn) is largely added at levels exceeding physiological requirements (Jondreville et al., 2003) and therefore, they are excreted in animal slurry. The application of the manure to agricultural soil results in a significant loading in soils. Some authors reported contents of Cu and Zn up to 3.3 kg/ha and 2.2 kg/ha, respectively. In addition to these elements, Pb, Ni, Cr and Cd are also present in the animal diet in Western Europe; these elements have phytotoxic effects even at low concentrations.

The animal wastes, containing mainly phosphorus, ammonium-bound nitrogen and heavy metals, can usefully be involved in a process *via* mineralization aimed at the discharge of reject brines from desalination plants. The process offers a dual-purpose approach for the management of reject brines and the removal of P and NH_4-N from wastewater produced by biological industrial activities. Moreover, the recovery of P through the synthesis of P-bearing minerals could represent a highly competitive source of this element, alternative to the limited reserves of natural phosphate rocks.

3. Novel technological strategies

3.1 Challenge 1: The role of reject brines in the sequestration of carbon dioxide *via* mineralization

Carbonation reactions are common in nature but the idea of developing this process for the disposal of CO_2 based on the chemical fixation of the carbon dioxide in the form of carbonate minerals was only proposed in 90s using calcium in brines, and then by carbonation of magnesium-bearing minerals from ultramafic igneous rocks. There are serious problems with that approach: the reaction kinetics for magnesium silicates tends to be too slow, and environmental concerns are associated with mining and processing very large amounts of magnesium-bearing rocks.

Our group has recently described a carbonation process involving the reaction of gaseous CO_2 with a magnesium chloride solution producing a hydrated carbonate stable at Earth surface conditions (Ferrini et al., 2009). Additional experiments intended to simulate more closely the reject brines that are industrially available have been performed to explore the influence of the salinity on the carbonation process. The results of the experimental work suggest that the goal of CO_2 sequestration and reject brines discharge seems to be realistically obtained also using the more common saline wastewater. Moreover, the solid products and the by-products of the process can be applied in a large number of industrial processes, resulting in an attractive way safely manage hazardous wastewater and greenhouse gases.

3.1.1 Reject brines disposal and CO_2 sequestration within carbonate minerals

Our initial work on the carbonation process involved a relatively dilute solution of magnesium chloride, roughly 7 g L^{-1} of Mg, at room temperature.

In fifty experiments we synthesized carbonates of Mg using $MgCl_2 \cdot 6H_2O$ as a source of magnesium and sparging CO_2 through the Mg chloride solution. The time of interaction of the flux of CO_2 with the solution was prolonged until the point of saturation of the solution in CO_2, as suggested by the stabilization of pH values.

The reaction rate was found to be rapid, the deposition of nesquehonite, $MgCO_3 \cdot 3H_2O$, being virtually complete in about ten minutes (Fig. 2) according to the following reaction.

$$CO_2 + MgCl_2 + 4H_2O \rightarrow MgCO_3 \cdot 3H_2O + 2HCl \quad (1)$$

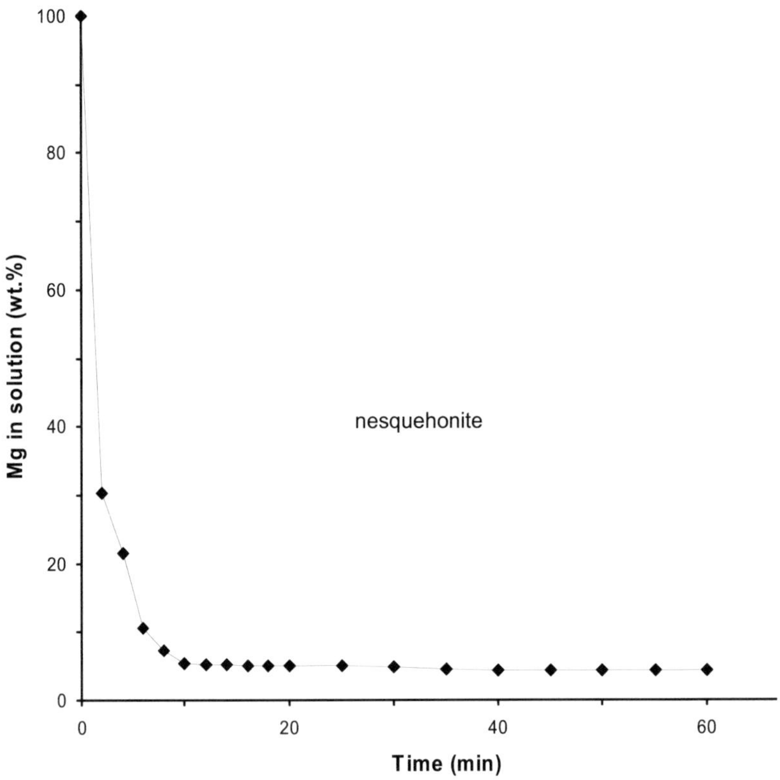

Fig. 2. Kinetics of the carbonation reactions

Nesquehonite removed about 80% of the carbon dioxide in solution, and the pH of the solution, adjusted by adding ammonia solution (25%), was made to range at an optimum value between 7.8 and 8.2.

To test the efficiency of the carbonation process in more concentrated solutions, having salinity more close to that of the reject brines, further experiments have been carried out using solutions with different degree of salinity (16 and 32 g L^{-1} of Mg).

A comparison of the efficiency of the CO_2 mineralization process among sets of experiments performed at low and high salinity shows that in more concentrated solutions, the efficiency was reduced to about 65%.

It is known that the degree to which CO_2 dissolves in water is determined by the solubility constant $K_0 = [CO_2^*]/[CO_{2(g)}]$, where CO_2^* is the sum of the species $CO_{2(aq)}$ and $H_2CO_{3(aq)}$ (Moran, 2010). The constant K_0 depends on temperature, pressure, salinity and the ionic composition of the water. The increase of the temperature favours the gas phase and shifts reaction (2) to the left, as will an increase in salinity of the electrolyte solution, whereas increasing pressure results in higher solubility of CO_2.

$$CO_{2(g)} = CO_{2(aq)} \quad (2)$$

Moreover, the presence of different salts in the solution reduces the solubility of CO_2 in the solution and this decrease is known as the "salting out effect" (Yasunishi & Yoshida, 1979; Al-Anezi et al., 2008). Adding large quantities of $MgCl_2$ creates a network of hydrogen bonds with H_2O molecules that shifts the equilibria (2) and (3) toward the left.

$$CO_{2(aq)} + H_2O_{(l)} = H_2CO_{3(aq)} \quad (3)$$

This reduction in the solubility of CO_2 would seem to significantly reduce the efficiency of the process. In fact, with a more concentrated solution of magnesium chloride, the residence time of CO_2 is enhanced in the aqueous medium because of a reduced tendency to produce $CO_{2(g)}$. At equivalent CO_2 contents, the more saline the solution, the less effective is the loss of the gas phase. The low degree of CO_2 degassing results in a longer-term availability of carbonic ions to react with Mg ions to form stable carbonate minerals (Mignardi et al., 2011).

3.1.2 Stability of nesquehonite

Nesquehonite occurs as a low-temperature carbonate in alkaline soils, in cave deposits and as a weathering product of ultramafic rocks.

Nesquehonite is a monoclinic hydrated carbonate of Mg and its structure consists of infinite flat ribbons of corner-sharing MgO_6 octahedra along the **b** axis of the crystal, which is the fiber axis, linked by hydrogen bonds. Within the chains, CO_3 groups link three MgO_6 octahedra by one edge and two common corners. The Mg atoms are in a distorted coordination, and each atom is coordinated by two H_2O ligands; one free H_2O molecule is located between the chains (Giester et al., 2000).

In order to evaluate the appropriateness of a reaction of CO_2 with Mg chloride solutions as a process for storing carbon dioxide and wastewater, the thermal behaviour of nesquehonite was investigated *in situ* using real-time parallel-beam X-ray powder diffraction. Nesquehonite appears to be stable up to 373 K (Ballirano et al., 2010) suggesting that its storage as "sequestering medium of CO_2 and reject brines" remains stable under the temperature conditions that prevail at the Earth's surface. At temperature above 373 K the process of thermal decomposition of nesquehonite (*via* intermediate hydrated magnesium carbonate phases) ultimately produces magnesite in the range 423-483 K. This sequence involves the formation of carbonate minerals thermodynamically more stable than nesquehonite, assuring the stable storage of these hazardous materials for millions of years.

According to information in the literature, no appreciable CO_2 release will occur if Mg hydrate carbonates are leached in solution having $pH > 2$. Because the pH of acid rain is unlikely to be below 2.5, the possible release of CO_2 from sites of nesquehonite storage due to acid rain should be unimportant.

3.1.3 Nesquehonite could be an ideal starting point for a manmade and environmentally safe supply of carbonate

The sequestration of CO_2 and the disposal of reject brines *via* carbonation produces a solid material that can be utilized directly in many products such as acoustic panels, non structural panels, insulation, or agglomerated in concrete as a manmade aggregate.

Because nesquehonite growths as fibrous and acicular crystals (Fig. 3), it improves the microstructures of manufacturing materials.

Fig. 3. SEM images of synthetic nesquehonite

3.2 Challenge 2: The role of reject brines to contribute to the abatement of the massive discharge of animal waste

The treatment of waste from animal farming represents a considerable problem from environmental and economic points of view. Indeed, the uncontrolled discharge of these wastes is generally regarded as the primary source of pollutants responsible for eutrophication of rivers and lakes.

Numerous approaches to P and N removal, including chemical precipitation, biological removal, crystallization, tertiary filtration, and ion exchange (Morse et al., 1998) are currently being applied.

The crystallization of P and N as struvite (magnesium ammonium phosphate, $MgNH_4PO_4{\cdot}6H_2O$) is one of the option most promising to remove both P and N. It has been applied in various countries for the treatment, for example, of swine waste (Burns et al., 2001; Nelson et al., 2003) and agro-industrial effluents (Altinbas et al., 2002).

The precipitation of struvite and other complex phosphates in wastewater treatment plants was already identified in the last century, but the development of an *ex situ* methodology based on the direct synthesis of these mineral phases for the treatment of both saline and animal waste has not been explored so far.

3.2.1 Reject brines and animal waste disposal within phosphate minerals

Our research group recently approached the problem of phosphorus and nitrogen removal from animal waste by developing a process of mineral synthesis using solutions involving both slurries and reject brines. In particular, the process was applied to a mixture composed by anaerobic lagoon-based swine slurry and simulated reject brine. The concentration of P and NH_4-N in the slurries ranged from about 160 to 200 mg L^{-1} and 600 to 2150 mg L^{-1}, respectively; whereas the reject brine was a magnesium chloride solution ($MgCl_2{\cdot}6H_2O$) having 32 g L^{-1} of Mg. The Mg^{2+}:PO_4^{3-} molar ratio was adjusted to 1:1. The suitable range of pH for the phosphates precipitation (~9) in our experimental conditions was obtained by adding ammonia solution (25% NH_3). The solution was thoroughly mixed using a magnetic stir plate at room temperature. The reaction rate was rapid, with solid product formation almost complete in few minutes.

The XRD pattern of the precipitates was found to be in good agreement with that of struvite (Fig. 4). Well-formed crystals (Fig. 5) are greyish; their formation occurred according to reaction (4).

$$Mg^{2+} + NH^{4+} + PO_4^{3-} + nH_2O \rightarrow MgNH_4PO_4 \cdot 6H_2O \quad (4)$$

However, this equation is a simplification of the reaction involved in struvite precipitation.

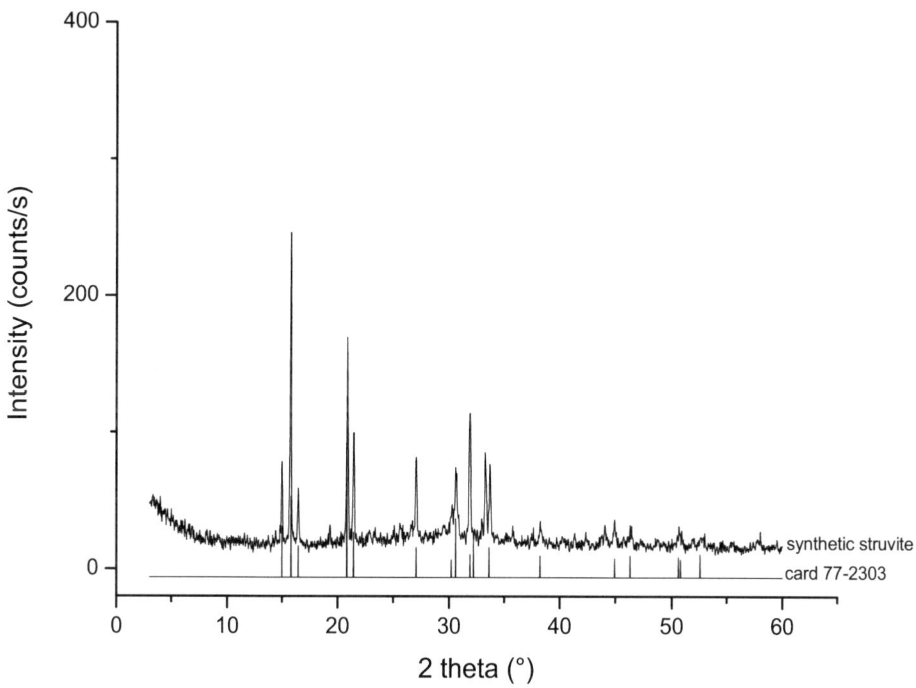

Fig. 4. XRD pattern of synthetic struvite compared with the strongest lines recorded in JCPDS card 77-2303

Indeed, the crystallization of struvite is controlled by pH, temperature, supersaturation of the reactants, and also the presence of calcium in the solution. However, the low concentration of calcium in the treated slurry did not affect negatively the formation of struvite in our experiments.

The synthesis of struvite efficiently removed P (never less than ~75% in fifteen experiments) and as the molar ratio NH^{4+}/PO_4^{3-} is >> 1, the precipitation of struvite left an ammonium chloride–H_2O mixture as a residual solution. This solution can be used economically to prepare crystalline ammonium chloride.

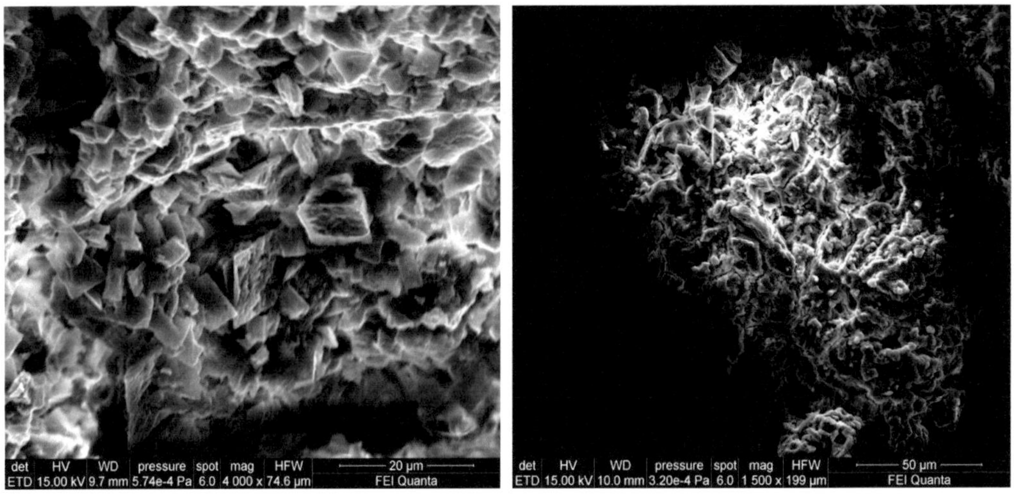

Fig. 5. SEM images of synthetic struvite

3.2.2 Struvite formation in the management of reject brines and animal waste

The synthesis of struvite can be considered as a suitable alternative for the treatment of animal waste and reject brines as it is a relatively inexpensive and effective technology. The results of our experimental work along with those reported in the literature, suggest that the abatement of P and NH_4-N *via* mineralization is technically and operationally feasible. However, as the composition of animal waste in term of P and NH_4-N concentrations is extremely variable, the treatment *via* struvite formation must be carefully adapted to the actual composition of the waste to be managed. Indeed, on the basis of the NH_4-N/P molar ratio, in an ideal solution having a ratio 1:1, the process would achieve the almost complete abatement of these contaminants. Moreover, as slurries with low concentration of P also exist, other P-bearing wastes for the formation of struvite should be involved in the process. In this way, alternative P sources include streams from corrosion treatment of iron metals, domestic sewage, and detergent and agricultural effluents (Verma et al., 2007).

Where high concentrations of Ca ions are present in the solution, calcium phosphate will preferentially precipitate, and cause a reduction in the yield of struvite. However, calcium phosphates also remove P and can be used for agricultural applications, just as struvite.

The synthetic phosphates produced using the method described above represent environmentally friendly resources, as their formation is based on the re-use of by-products whose regulatory standards for their discharge are becoming very stringent. They are also economically important as fertilizers.

3.2.3 Synthetic struvite as a resource

The abatement of P and N *via* mineralization produces a solid material, i.e., struvite, which offers numerous advantages. It is an excellent fertilizer compared with those normally used (Ghosh et al., 1996). Its agronomic qualities include: low solubility in water, resulting in the prolonged release of nutrients during the growing season, without danger of burning roots of crops treated, highly effective source of nutrients (P, N, and Mg) for plants, and the addition of possible supplements as micronutrients.

Global reserves of high quality phosphate ore are forecast to be exhausted by the end of the century, resulting in increasing prices and environmental impacts from the processing of ore reserves of lesser quality or more difficult access. Therefore, the possible recovery of P from wastewater could become an increasingly appealing opportunity (Gaterell et al., 2000).
Another incentive for the treatment of slurries through the formation of struvite is the important reduction in the volume of sludge to be discharged.

4. Future research

The results of our experimental work have highlighted the potential of reject brines as a source of several elements to be involved in processes contributing to the development of environmental technologies.
Challenge 1 - Future research should focus on the efficiency of the carbonation process starting from multi-elemental solutions, simulating the composition of reject brines now being produced industrially. In particular, we shall investigate the carbonation process using brines with elevated concentrations of sodium and aluminium for the synthesis of dawsonite, $NaAl(CO_3)(OH)_2$. This hydrated carbonate occurs in nature as aggregates of acicular to fibrous crystals in several sedimentary sequences and in various depositional environments (e.g., associated with eruptive rocks, derived but not directly associated with magmatic rocks, and related to hydrothermal mineralization; Baker et al., 1995; Ferrini et al., 2003; Sirbescu & Nabelek, 2003).
Dawsonite has been successfully synthesized as a way to deal with the etching waste streams of the aluminium anodizing industry, in order to reduce the emissions to the environment and also to recover a useful and marketable mineral resource (Alvarez-Ayuso & Nugteren, 2005).
Most investigators dealing with the reactions involved in the injection of CO_2 into deep sedimentary formations hypothesize the formation of various carbonate minerals, including dawsonite. Therefore, any information about the mechanisms and kinetics of the formation of this carbonate in aqueous solution should be utilized to determine the potential of a process aimed at sequestering CO_2 involving Na-enriched reject brines and Al-bearing wastewater.
Challenge 2 – To assess the possible use of synthesized struvite as an agricultural fertilizer, an evaluation of its heavy metal content (e.g., Pb, Zn, Cu, Cd, Cr, Fe, and Ni) will be of critical importance. The presence of such pollutants in fertilizers is strictly regulated and excessive amounts can result in the fertilizer being banned from use in agriculture.
A rather limited number of previous studies on the quality evaluation of struvite in terms of heavy metal content exist in the literature. Therefore, our research will be focused on the synthesis of struvite using for the reactions slurries having various concentrations of toxic metals with the objective to test the maximum concentrations of these elements that can be hosted in the structure of struvite.

5. Acknowledgments

Financial support by Sapienza Università di Roma is acknowledged. Additional funding was provided by the Ministry of Education, University and Research (MIUR) through a National Research Program (PRIN 2006045331 001). The authors thank T. Coppola and S. Stellino for the assistance in the ICP-AES and X-ray diffractometer laboratories, respectively.

6. References

Abdul-Wahab, S.A., Al-Weshahi, M.A. (2009). Brine management: substituting chlorine with on-site produced sodium hypochlorite for environmentally improved desalination processes. *Water Resources Management*, 23, pp. 2437-2454

Ahmed, M., Shayya, W.H., Hoey, D., Al-Handaly, J. (2001). Brine disposal from reverse osmosis desalination plants in Oman and the United Arab Emirates. *Desalination*, 133, pp. 135-147

Al-Anezi, K., Somerfield, C., Mee, D., Hilal, N. (2008). Parameters affecting the solubility of carbon dioxide in seawater at the conditions encountered in MSF desalination plants. *Desalination*, 222, pp. 548-571

Altinbas, M., Ozturk. I., Aydin, A.F. (2002). Ammonia recovery from high strength agro-industry effluents. *Water Science Technology*, 45, pp. 189-196

Alvarez-Ayuso, E., Nugteren, H.W. (2005). Synthesis of dawsonite: a method to treat the etching waste streams of the aluminium anodising industry. *Water Research*, 39, pp. 2096-2104.

Aneja, V.P., Chauhan, J.P., Walker, J.T. (2000). Characterization of atmospheric ammonia emissions from swine waste storage and treatment lagoons. *Journal of Geophysical Research*, 105, pp. 11535-11545

Baker, J.C., Bai, G.P., Hamilton, P.J., Golding, S.D., Keene, J.B. (1995). Continental-scale magmatic carbon dioxide seepage recorded by dawsonite in the Bowen-Gunnedah-Sydney Basin system, eastern Australia. *Journal of Sedimentary Research*, 65, pp. 522-530

Ballirano, P., De Vito, C., Ferrini, V., Mignardi, S. (2010). The thermal behaviour and structural stability of nesquehonite, $MgCO_3 \cdot 3H_2O$, evaluated by *in situ* laboratory parallel-beam X-ray powder diffraction: New constraints on CO_2 sequestration within minerals. *Journal of Hazardous Materials*, 178, pp. 522-528

Burns, R.T., Moody, L.B., Walker, F.R., Raman, D.R. (2001). Laboratory and *in situ* reductions of soluble phosphorus in swine waste slurries. *Environmental Technology*, 22, pp. 1273-1278

Conrad, R., Klose, M. (1999). Anaerobic conversion of carbon dioxide to methane, acetate and propionate on washed rice roots. *FEMS Microbiology Ecology*, 30, pp. 147-155

Doyle, J.D., Parsons, S.A. (2002). Struvite formation, control and recovery. *Water Research*, 36, pp. 3925-3940

ESCWA (1993). Water desalination: the experience of GCC countries. Regional Symposium on Water Use ad Conservation, Report E/ESCWA/NR/1993/WG.1/WP.10. Amman, Jordan

ESCWA (2009). Role of Desalination in Addressing Water Scarcity 2009, 2009 November Available from http://www.escwa.un.org/information/publications/edit/ upload/sdpd-09-4.pdf

Ferrini, V., De Vito, C., Mignardi, S. (2009). Synthesis of nesquehonite by reaction of gaseous CO_2 with Mg chloride solution: Its potential role in the sequestration of carbon dioxide. *Journal of Hazardous Materials*, 168, pp. 832-837

Ferrini, V., Martarelli, L., De Vito, C., Çina, A., Deda T. (2003). The Koman dawsonite and realgar-orpiment deposit, northern Albania: inferences on processes of formation. *Canadian Mineralogist*, 41, pp. 413-427

Freund, P., Ormerod, W.G. (1997). Progress toward storage of carbon dioxide. *Energy Conversion and Management*, 38, pp. 199-204

Friedmann, S.J. (2007). Geological carbon dioxide sequestration. *Elements*, 3, pp. 179-184

Gaterell, M.R., Gay, R., Wilson, R., Lester, J.N. (2000). An economic and environmental evaluation of the opportunities for substituting phosphorus recovered from wastewater treatment works in existing UK fertilizer markets. *Environmental Technology*, 21, pp. 1067-1084

GHD (2003). Desalination in Queensland. Department of Natural Resources and Mines, Final Report, July 2003

Ghosh, G.K., Mohan, K.S., Sarkar, A.K. (1996). Characterisation of soil fertilizer P reaction products and their evaluation as sources of P for gram (*Cicer arietinum* L.). *Nutrient Cycling in Agroecosystems*, 46, pp. 71-79

Giester, G., Lengauer, C.L., Rieck, B. (2000). The crystal structure of nesquehonite, $Mg(CO_3) \cdot 3H_2O$, from Lavrion, Greece. *Mineralogy and Petrology*, 70, pp. 153-163

Gunter, W.D., Wiwchar, B., Perkins, E.H. (1997). Aquifer disposal of CO_2-rich greenhouse gases: extension of the time scale of experiment for CO_2-sequestering reactions by geochemical modelling. *Mineralogy and Petrology*, 59, pp. 121-140

Haga, K. (1998). Animal waste problems and their solution from the technological point of view in Japan. *Japan Agricultural Research Quarterly*, 32, pp. 203-210

Hamed, O.A. (2005). Overview of hybrid desalination systems - Current status and future prospects. *Desalination*, 186, pp. 207-214

Hansen, L.D., Dipple, G.M., Gordon, T.M., Kellett, D.A. (2005). Carbonated serpentinite (listwanite) at Atlin, British Columbia: a geological analogue to carbon dioxide sequestration. *Canadian Mineralogist*, 43, pp. 225-239

Holloway, S. (1997). An overview of the underground disposal of carbon dioxide. *Energy Conversion and Management*, 38, pp. 193-198

Holloway, S., Pearce, J.M., Hards, V.L., Ohsumi, T., Gale, J. (2007). Natural emissions of CO_2 from the geosphere and their bearing on the geological storage of carbon dioxide. *Energy*, 32, pp. 1194-1201

Hsu, J.H., Lo, S.L. (2001). Effect of composting on characterization and leaching of copper, manganese, and zinc from swine manure. *Environmental Pollution*, 114, pp. 119-127

Huijgen, W.J.J., Witkamp, G.J., Comans, R.N.J. (2006). Mechanisms of aqueous wollastonite carbonation as a possible CO_2 sequestration process. *Chemical Engineering Science*, 61, pp. 4242-4251

Hwang, I.S., Min, K.S., Choi, E., Yun, Z. (2005). Nitrogen removal from piggery waste using the combined SHARON and ANAMMOX process. *Water Science and Technology*, 52, pp. 487-494

Jondreville, C., Revy, P.S., Dourmad, J.Y. (2003). Dietary means to better control the environmental impact of copper and zinc by pigs from weaning to slaughter. *Livestock Production Science*, 84, pp. 147-156

JWR & DTF (2005). Water Innovation Symposium, 2005 October 17-21, Available from http://www.coloradowaterquality.com/ro/docs/water_innovation_symposium.pdf

Kalyuzhnyi, S., Fedorovich, V., Nozhevnikova, A. (1998). Anaerobic treatment of liquid fraction of hen manure in UASB reactors. *Bioresource Technology*, 65, pp. 221-225

Kessler, J., Zogg, M., Bachler, E. (1994). Phosphor, kupfer und zink im schweinetrog (Phosphorus, copper and zinc in pig slurry). *Agrarforschung*, 1, pp. 480-483

Köttner, M. (2001). Biogas in agriculture and industry potentials, present use and perspectives, In: *International Biogas and Bioenergy Centre of Competence*, Available from http://crest.org/discussiongroups/resources/biomass/biogas/ BIOGASMK.pdf

Lackner, K.S. (2002). Carbonate chemistry for sequestering fossil carbon. *Annual Review of Energy and the Environment*, 27, pp. 193-232

Lackner, K.S. (2003). A guide to CO_2 sequestration. *Science*, 300, pp. 1677-1678

Lackner, K.S., Butt, D.P., Wendt, C.H. (1997). Progress on binding CO_2 in mineral substrates. *Energy Conversion and Management*, 38, pp. 259-264

Lackner, K.S., Butt, D.P., Wendt, C.H., Ziock, H.J. (1998). Mineral carbonates as carbon dioxide sinks, LANL Internal Report, LA-UR-98-4530

Lackner, K.S., Wendt, C.H., Butt, D.P., Joyce, E.L., Sharp, D.H. (1995). Carbon dioxide disposal in carbonate minerals. *Energy*, 20, pp. 1153-1170

Lattemann, S., Höpner, T. (2008). Environmental impact and impact assessment of seawater desalination. *Desalination*, 220, pp. 1-15

Le Corre, K.S., Valsami-Jones, E., Hobbs, P., Parsons, S.A. (2009). Phosphorus recovery from wastewater by struvite crystallization: a review. *Critical Reviews in Environmental Science and Technology*, 39, pp. 433-477

Loughrin, J.H., Szogi, A.A., Vanotti, M.B. (2006). Reduction of malodorous compounds from a treated swine anaerobic lagoon. *Journal of Environmental Quality*, 35, pp. 194-199

Mallin, M.A., (2000). Impacts of industrial animal production on rivers and estuaries. *American Scientist*, 88, pp. 26-37

Marsh, G., Bates, J., Haydock, H., Hill, N., Clark, C., Freund, P. (2002). The 6th International Conference on Greenhouse Gas Technologies in Kyoto, Japan, 1. (4 October 2002).

Maugin, G., Corsin, P. (2005). Concentrate and other waste disposals from SWRO plants: characterization and reduction of their environmental impact. *Desalination*, 182, pp. 355-364

Metz, B., Davidson, O., de Coninck, H., Loos, M., Meyer, L. (Eds.) (2005). *Carbon dioxide capture and storage*, IPCC Special Report, Cambridge University Press, Cambridge, UK, pp. 1-433

Mezher, T., Fath, H., Abbas, Z., Khaled, A. (2010). Techno-economic assessment and environmental impacts of desalination technologies. *Desalination*, 266, pp. 263-273

Mickley, M. (2010). State of the Science White Paper, Brackish groundwater concentrate management. Prepared for New Mexico State University (NMSU) and Consortium for High Technology Investigations on Water and Wastewater (CHIWAWA)

Mickley, M., Hamilton, R., Gallegos, L., Truesdall, J. (1993). Membrane concentration disposal. *American Water Works Association Research Foundation*, Denver, Colorado

Mignardi, S., De Vito, C., Ferrini, V., Martin, R.F. (2011). The efficiency of CO_2 sequestration via carbonate mineralization with simulated wastewaters of high salinity. *Journal of Hazardous Materials*, 191, pp. 49-55

Miri, R., Chouikhi, A. (2005). Ecotoxicological marine impacts from seawater desalination plants. *Desalination*, 182, pp. 403-410

Mohamed, A.M.O., Maraqa, M., Al Handhaly, J. (2005). Impact of land disposal of reject brine from desalination plants on soil and groundwater. *Desalination*, 182, pp. 411-433

Moral, R., Perez-Murcia, M.D., Perez-Espinosa, A., Moreno-Caselles, J., Paredes, C., Rufete, B. (2008). Salinity, organic content, micronutrients and heavy metals in pig slurries from South-eastern Spain. *Waste Management*, 28, pp. 367-371

Moran, D. (2010). Carbon dioxide degassing in fresh and saline water. I: Degassing performance of a cascade column. *Aquacultural Engineering*, 43, pp. 29-36

Morita, T., Robinson, J., Adegbulugbe, A., Alcamo, J., Herbert, D., Rovere, E.L., Nakicenovic, N., Pitcher, H., Raskin, P., Riahi, K., Sankovski, A., Sololov, V., Vries, H.J.M., Dadi, Z. (2001). Greenhouse gas emission mitigation scenarios and implications, In: *Climate Change 2001: Mitigation, Contribution of Working Group III to the Third Assessment Report of the Intergovernmental Panel on Climate Change,* Metz, B., Davidson, O., Swart, R., Pan, J. (Eds.), pp. 115–166, Cambridge University Press, Cambridge (UK)

Morse, G., Brett, S., Guy, J., Lester, J. (1998). Review: Phosphorus removal and recovery technologies. *Science of the Total Environment*, 212, pp. 69-81

Nelson, N.O., Mikkelsen, R.E., Hesterberg, D.L. (2003). Struvite precipitation in anaerobic swine lagoon liquid: effect of pH and Mg:P ratio and determination of rate constant. *Bioresource Technology*, 89, pp. 229-236

O'Connor, W.K., Dahlin, D.C., Rush, G.E., Dahlin, C.L., Collins, W.K. (2002). Carbon dioxide sequestration by direct mineral carbonation: process mineralogy of feed and products. *Minerals and Metallurgical Processes*, 19, pp. 95-101

Parson, E.A., Keith, D.W. (1998). Fossil fuels without CO_2 emissions. *Science*, 282, pp. 1053-1054

Robertson, A. (2004). Development of concepts concerning the genesis and emplacement of Tethyan ophiolites in the Eastern Mediterranean and Oman regions. *Earth-Science Reviews*, 66, pp. 331-387

Seifritz, W. (1990). CO_2 disposal by means of silicates. *Nature*, 345, pp. 486

Sirbescu, M.L.C., Nabelek, P.I. (2003). Dawsonite: An inclusion mineral in quartz from the Tin Mountain pegmatite, Black Hills, South Dakota. *American Mineralogist*, 88, pp. 1055-1060

Soong, Y., Fauth, D.L., Howard, B.H., Jones, J.R., Harrison, D.K., Goodman, A.L., Gray, M.L. (2006). CO_2 sequestration with brine solution and fly ashes. *Energy Conversion and Management*, 47, pp. 1676-1685

Szogi, A.A., Vanotti, M.B., Stansbery, A.E. (2006). Reduction of ammonia emissions from treated anaerobic swine lagoons. *Transactions of the American Society of Agricultural and Biological Engineers*, 49, pp. 217-225

Vanotti, M.B., Millner, P.D., Hunt, P.G., Ellison, A.Q. (2005). Removal of pathogen and indicator microorganisms from liquid swine manure in multi-step biological and chemical treatment. *Bioresource Technology*, 96, pp. 209–214

Vanotti, M.B., Szogi, A.A., Hunt, P.G., Millner, P.D., Humenik, F.J. (2007). Development of environmentally superior treatment system to replace anaerobic swine lagoons in the USA. *Bioresource Technology*, 98, pp. 3184-3194

Verma, M., Brar, S.K., Tyagi, R.D., Surampalli, R.Y., Valero, J.R. (2007). Industrial wastewaters and dewatered sludge: rich nutrient source for production and formulation of biocontrol agent, *Trichoderma viride*. *World Journal of Microbiology & Biotechnology*, 23, pp. 1695-1703

Wadman, W.P., Sluijsmans C.M.J., de la Lande Cremer, L.C.N. (1987). Value of animal manures: changes in perception, In: Animal manure on Grassland and Fodder Crops. Fertilizer or Waste? Van der Meer, H.G. et al. (Eds.), pp. 1-16, Martinus Nijhoff Pub., Dordrecht, The Netherlands

Winters, H., Isquith, I., Bakish, R. (1979). Influence of desalination effluents on marine ecosystems. *Desalination*, 30, pp. 403-410

Xu, T., Apps, J.A., Pruess, K. (2004). Numerical simulation of CO_2 disposal by mineral trapping in deep aquifers. *Applied Geochemistry*, 19, pp. 917-936

Yasunishi, A., Yoshida, F. (1979). Solubility of carbon dioxide in aqueous electrolyte solutions. *Journal of Chemical & Engineering Data,* 24, pp. 11-14

Yetilmezsoy, K., Sapci-Zengin, Z. (2009). Recovery of ammonium nitrogen from the effluent of UASB treating poultry manure wastewater by MAP precipitation as a slow release fertilizer. *Journal of Hazardous Materials*, 166, pp. 260-269

5

Rotary Pressure Exchanger for SWRO

Zhou Yihui, Bi Mingshu and Liu Yu
School of Chemical Machinery, Dalian University of Technology, China

1. Introduction

A rotary pressure exchanger (RPE) is a kind of fluid energy recovery equipment which is based on the positive displacement principle. The key components of RPE include a rotor with several circular ducts, two end covers and one sleeve. At any time during operation, half the ducts are exposed to the high pressure fluid and the other half are exposed to the low pressure fluid. There are two indispensable requirements for RPE when used in SWRO: (1) The concentration of fresh seawater at high pressure outward pipe should not be so increased that it will have bad effect on the reverse osmosis. (2) The high pressure system and low pressure system in RPE should be completely separated, otherwise the leakage will happen and lead to the energy recovery efficiency decreases dramatically.

The former one depends on the stable liquid piston which is formed and moves reciprocally in the ducts of rotor. Zhou[1,2] illustrates the dynamics mixing process in the ducts. And the transition process and conditions from initial mixing zone to liquid piston were explained through 2D numerical simulation. The fluid's radial velocity in the duct was neglected during rotation.

The latter requirement depends on the end face seal between rotor and end cover. The face seal of RPE is very difficult for the following reasons: (1) the fluid is seawater which is incompressible and nearly inviscid and (2) the clearance between rotor and end cover has to be existed because it is an indispensable requirement for rotor starting up and rotating. Until now there are no detailed research reports on face seal mechanism of RPE being published.

In this chapter, the primary objective is presented as following :

- The study on the mixing process and the effects of RPE parameters on the liquid piston on the basis 3D model, which will describe the dynamics mixing more accurately and closer to the real operation conditions.
- The study on the face seal with the aim of getting the real seal formation in the clearance of RPE through theoretical analysis and numerical simulation. The results will guide the seal design of RPE experimental equipment. And the effects of clearance on the pressure, leakage and energy recovery are also studied.
- The effects of clearance on energy recovery and mixing rate under different flow and pressure conditions will be studied through RPE energy recovery experiments.

2. 3D Numerical simulation on dynamics mixing in ducts

The 3D model of RPE has a better view on the mixing process between different concentration fluids. The model description is firstly put forward. Then the computational results are to be illustrated and discussed. Finally, the effects of velocity and rotor speed on the mixing will be discussed.

2.1 Geometry model

In order to verify the validity of model the geometry dimension is in accordance to the real PX equipment[3]. Fig.1 shows the solid model of rotor and endcovers. The length of rotor is 200mm and diameter is 120mm with 12 ducts.

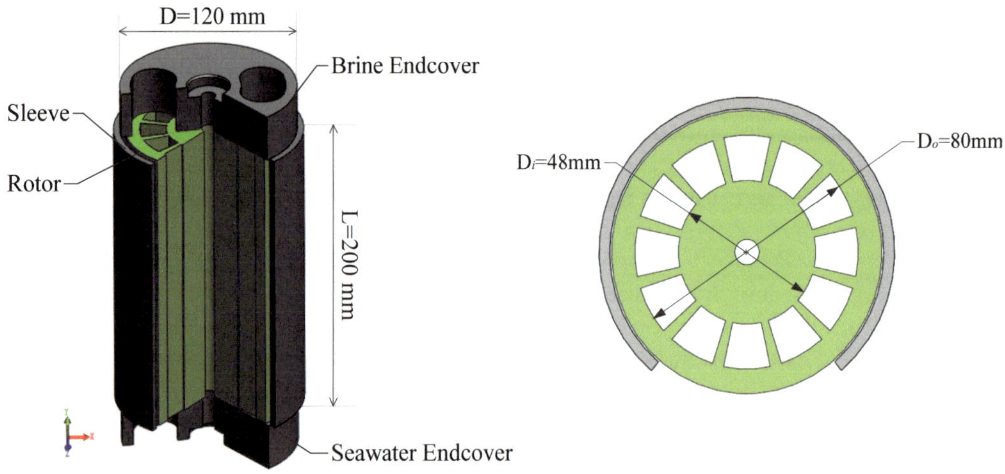

Fig. 1. Geometry model of 3D mixing simulation

2.2 Governing equations and boundary conditions

The numerical simulation is carried out under cylindrical coordination. And the governing equations are[4],

Continuity equation

$$\frac{\partial \rho}{\partial t} + \nabla \cdot (\rho \vec{v}) = 0 \tag{1}$$

Momentum equation[5]

$$\frac{\partial}{\partial t}(\rho \vec{v}) + \nabla \cdot (\rho \vec{v}\vec{v}) = -\nabla P + \nabla \cdot (\bar{\bar{\tau}}) + \rho \vec{g} \tag{2}$$

Where P is the static pressure, $\bar{\bar{\tau}}$ is the stress tensor (described below), and $\rho \vec{g}$ are the gravitational body force.

The stress tensor $\bar{\bar{\tau}}$ is given by

$$\bar{\bar{\tau}} = \mu \left[(\nabla \vec{v} + \nabla \vec{v}^T) - \frac{2}{3} \nabla \cdot \vec{v} I \right] \tag{3}$$

Where μ is the molecular viscosity, *I* is the unit tensor

Specifies equation

$$\frac{\partial}{\partial t}(pY_i) + \nabla \cdot (\rho\vec{v}Y_i) = -\nabla \cdot \vec{J_i} + R_i \tag{4}$$

Where R_i is the net rate of production by chemical reaction and according to the assumption that there is no chemical reaction during the mass transfer.

$$R_i\text{=}0$$

In turbulent flows,

$$\vec{J_i}\text{=} -\left(\rho D_{i,m} + \frac{\mu_t}{Sc_t}\right)\nabla Y_i - D_{T,i}\frac{\nabla T}{T} \tag{5}$$

Where Sc_t is the turbulent Schmidt number,

$$Sc_t\text{=}\frac{\mu_t}{\rho D_t}$$

Where μ_t is the turbulent viscosity and D_t is the turbulent diffusivity. The default Sc_t is 0.7.
Turbulence model:
Turbulent kinetic energy:

$$\frac{\partial}{\partial t}(pk) + \frac{\partial}{\partial x_i}(pku_i) = \frac{\partial}{\partial x_j}\left[\left(\mu + \frac{\mu_t}{\sigma_k}\right)\frac{\partial k}{\partial x_j}\right] + G_k + G_b - \rho\epsilon - Y_M \tag{6}$$

Where G_b is the generation of turbulence kinetic energy due to buoyancy and Y_m represents the contribution of the fluctuating dilatation in compressible turbulence to the overall dissipation rate,
Turbulent kinetic energy simplified to:

$$\frac{\partial}{\partial t}(pk) + \frac{\partial}{\partial x_i}(pku_i) = \frac{\partial}{\partial x_j}\left[\left(\mu + \frac{\mu_t}{\sigma_k}\right)\frac{\partial k}{\partial x_j}\right] + G_k + \rho\epsilon \tag{7}$$

Turbulent energy dissipation:

$$\frac{\partial}{\partial t}(\rho\epsilon) + \frac{\partial}{\partial x_i}(\rho\epsilon u_i) = \frac{\partial}{\partial x_j}\left[\left(\mu + \frac{\mu_t}{\sigma_\epsilon}\right)\frac{\partial \epsilon}{\partial x_j}\right] + C_{1\epsilon}\frac{\epsilon}{k}(G_k + C_{3\epsilon}G_b) - C_{2\epsilon}\rho\frac{\epsilon^2}{k} \tag{8}$$

Where

$$\mu_t = \rho C_\mu \frac{k^2}{\varepsilon}$$

The empirical constants for the RNG k–e model are assigned as following [6]:

$$C_{1\varepsilon} = 1.44, C_{2\varepsilon} = 1.92, C_\mu = 0.09, \sigma_K\text{=}1.0, \sigma_\varepsilon = 1.3$$

Boundary conditions
Non-slip boundary condition is used on the inner wall of the duct. The standard wall function method is applied to simulate the flow in the near-wall region. The mass-flow-inlet and pressure-outlet boundary condition are used on the high-pressure brine inlet, low-pressure feed water inlet and high-pressure feed water outlet, low-pressure brine outlet.

And the rotor speed is given by cell zone conditions. Fig.2 shows the computation (a) and grid model (b) ready for the simulation.

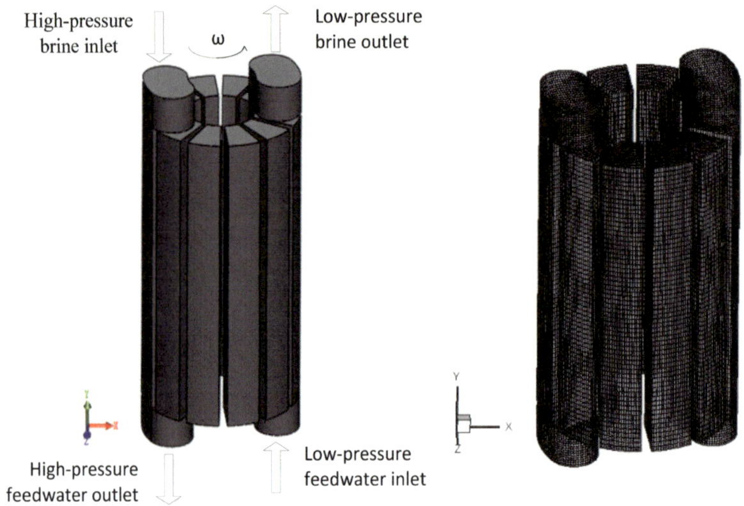

(a)Computation model (b) Grid model

Fig. 2. Geometry model in simulation

2.3 Results and discussion

2.3.1 Verification of 3D model

Fig.3 illustrates the average concentration on the high pressure out surface compared with the ones under real operation conditions[5]. The divergence between those is less than 5%which verifies the numerical model is accurate for the study.

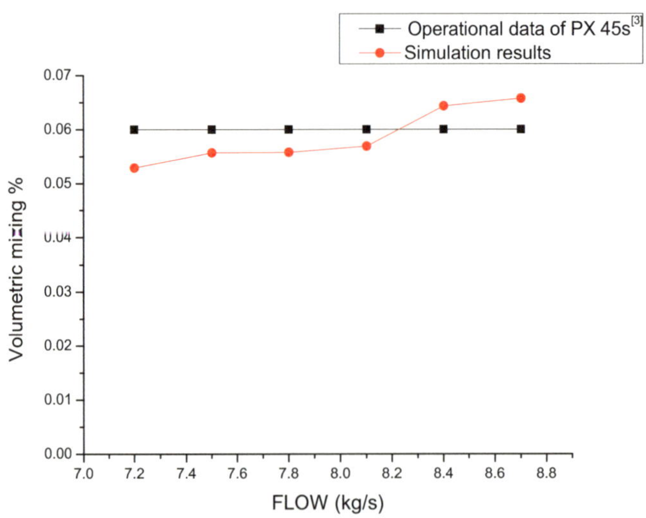

Fig. 3. Comparison between numerical result and real data on high pressure out surface concentration

2.3.2 Liquid piston in the ducts

Fig.4 shows the concentration distribution in all the ducts of rotor. It could be clearly seen that the liquid piston formation in the ducts during rotation. Fig.5 shows the movement of liquid piston separates the fresh water and brine to prevent from over-mixing with each other.

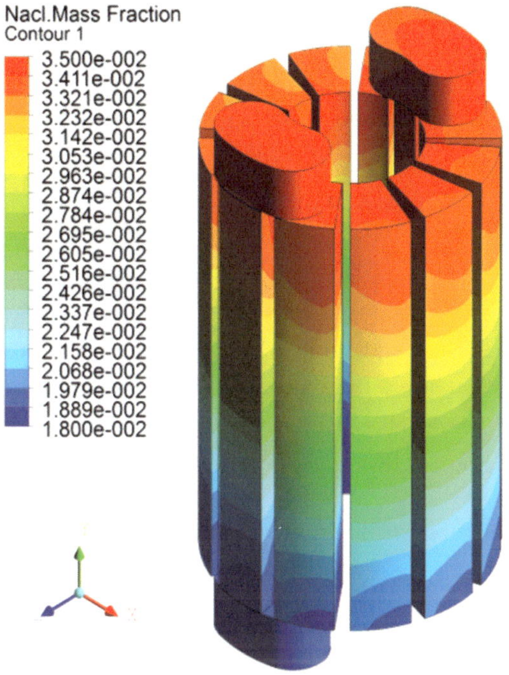

Fig. 4. Concentration distribution in the ducts of rotor

Fig. 5. Liquid piston in the ducts of rotor

2.3.3 Parameters affecting mixing rate

Many parameters affect the mixing rate, in which flow-in velocity and rotation speed are the most important ones. When the rotor was driven by the flow the rotor speed increased with increasing of flow rate linearly. Consequently, the distance of liquid piston moving in the duct almost keep constant because of the effects of velocity and rotor speed on the mixing cancelled with each other[8,9]. While in this chapter, the velocity and rotor speed are studied as independent parameters.

The mixing rate of fresh seawater and concentration stream is calculated according to Eq.(9) and salinity of samples are obtained through electrical conductivity[1].

$$\text{Volumetric Mixing Rate}=\frac{HP_{out}-LP_{in}}{LP_{out}-LP_{in}} \tag{9}$$

Fig.6 illustrates the relationship between flow-in velocity and volumetric mixing rate when the rotor speed $\omega = 1200\text{rpm}$. It could be seen from Fig.6 that the mixing rate increases with the increasing of flow-in velocity. And if the mixing rate should be controlled less than 6% the maximum flow-in velocity should not more than 2m/s according to the geometry model and rotor speed.

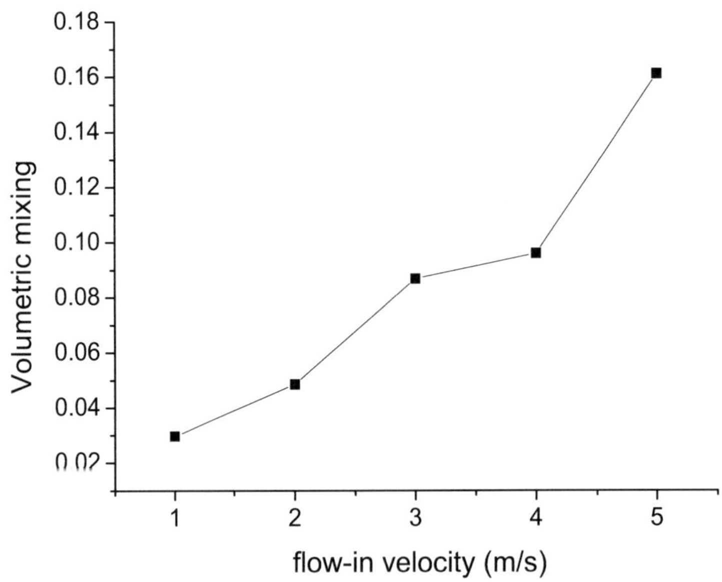

Fig. 6. Relationship between flow-in velocity and mixing rate

Fig.7 is the relationship between rotor speed and mixing rate when the flow-in velocity $v = 2.5\text{m/s}$. It could be seen that with the increasing of rotor speed the mixing rate decreases firstly, and then almost be stable when rotating from 600rpm to 1400rpm. The results show that the mixing rate will be out of control if the rotor speed is low. However, the rotor speed has little effect on the mixing rate once it is more than 800rpm.

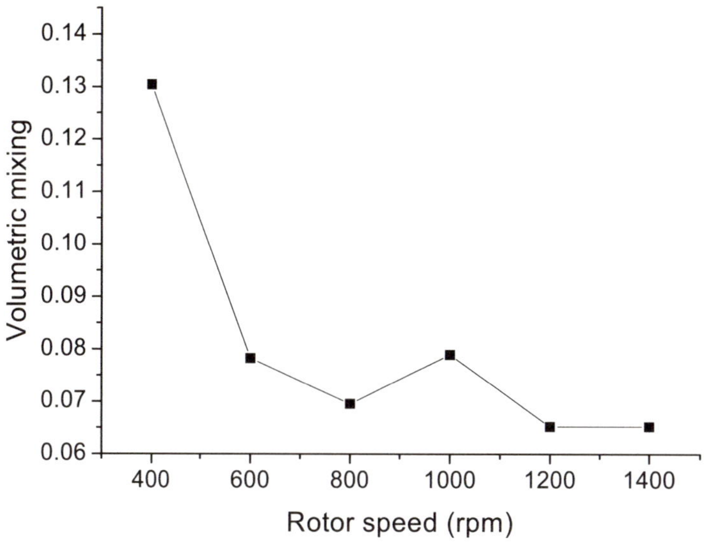

Fig. 7. Relationship between rotor speed and mixing rate

3. Face seal of RPE

Fig.8 shows the seal structure of assembled RPE which is composed of rotor, end covers and sleeve. A set of rings made of 316L is used to adjust clearance in experiment. The center shaft is used to assembly the above components and it will not affect the face clearance because there is no radial contact between center shaft and seal components. The materials of rotor and end covers are cemented carbide and the planarity and parallelism are ensured through precious surface finishing. The clearance can be adjusted by changing different rings from 0.02mm to 0.13mm.

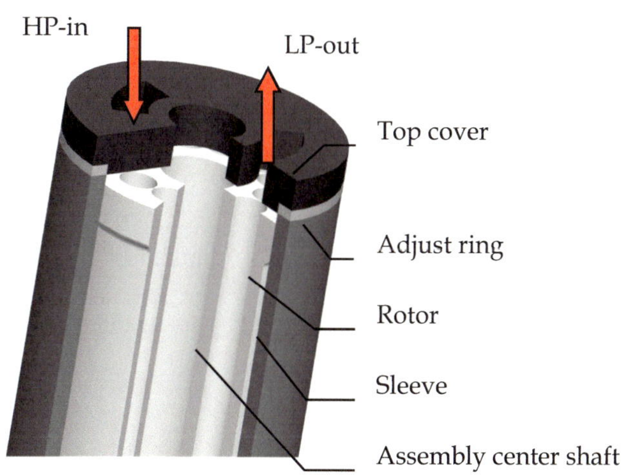

Fig. 8. Face seal structure of RPE

Fig.8 is the front view of axial RPE. The fluid distributors on the end-covers are designed to release the flow shock caused by the suddenly contact between high pressure fluid and low pressure ones. The fluid enters the clearance through seal surface, which is the most significant difference from typical mechanical seal.

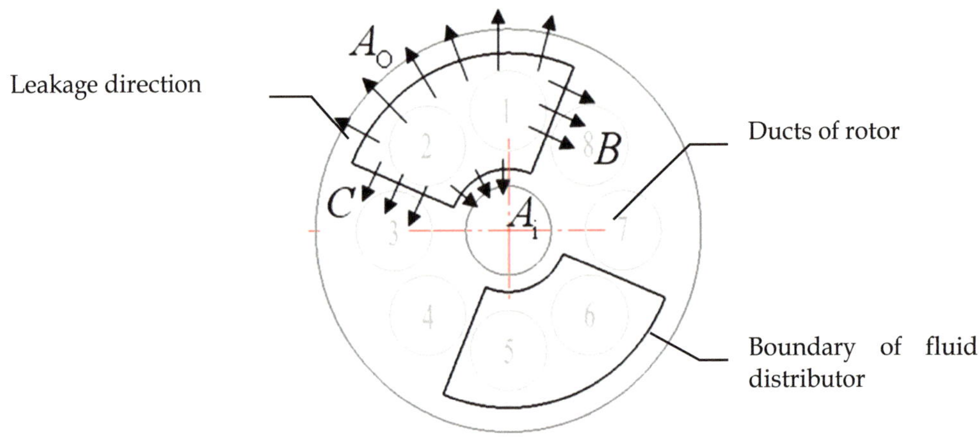

Fig. 9. Flow direction in clearance

Fig. 9. is the top view of RPE and shows the boundary of fluid distributor. Number1~8 are the ducts of rotor when rotating. The arrows illustrate the leakage after fluid entering the clearance whose direction is normal to the boundary

Generally speaking, the face seal structure of RPE has its own special characteristics:

1. There is constant flow crossing seal surface into the seal clearance.
2. There are leakages both at radial and at tangential direction from high pressure side to low pressure side.
3. The shear flow caused by relative rotating could be parallel or perpendicular to pressure-driven flow caused by pressure gradient. Moreover, when these two kinds of flows are parallel they could be at same direction or at opposite direction.

Therefore, face seal of RPE is much more complicated than traditional mechanical seal technology. However, the study on it in this paper could be done from both theoretical analysis and experiments ways.

3.1 Numerical simulation on face seal of RPE

3.1.1 Hypothesis

There are several hypotheses for the face seal numerical simulation.

1. The velocity at the thickness direction $\upsilon_Z = 0$
2. The high pressure fluid entering the clearance will flow at the direction normal to the boundary of fluid distributor to low pressure side.
3. There is no thermal effect during face sealing.

3.1.2 Geometry model and boundary conditions

Fig.10 is the geometry model of clearance between rotor and end cover with arrows illustrate the flow direction in hypothesis. Fig.11 is the grid display.

Fig. 10. Geometry model of clearance (not to scale)

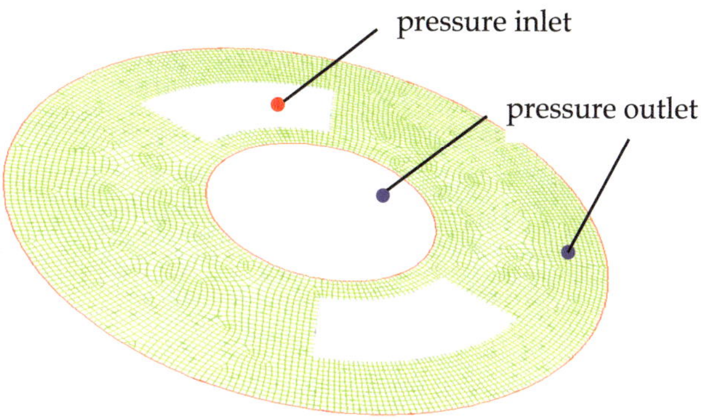

Fig. 11. Gridded geometry model

The red surfaces in Fig.10 are defined as pressure-inlet and $p = 6.0\text{MPa}$. The blue surfaces are defined as pressure-outlet and $p = 0.2\text{MPa}$.

3.1.3 Governing equations

The numerical simulation is carried out under cylindrical coordination. And the governing equations are[4,5],

Continuity Equation $$\frac{1}{r}\frac{\partial(r\upsilon_{\mathrm{r}})}{\partial r}+\frac{1}{r}\frac{\partial(\upsilon_{\theta})}{\partial\theta}=0 \qquad (10)$$

Momentum Equation

Radial direction:

$$\rho\left(\frac{\partial\upsilon_r}{\partial t}+\upsilon_r\frac{\partial\upsilon_r}{\partial r}+\frac{\upsilon_\theta}{r}\frac{\partial\upsilon_r}{\partial\theta}-\frac{\upsilon_\theta^2}{r}\right)=\rho f_r-\frac{\partial p}{\partial r}+\mu\left[\frac{\partial}{\partial r}\left(\frac{1}{r}\frac{\partial}{\partial r}(r\upsilon_r)\right)+\frac{1}{r^2}\frac{\partial^2\upsilon_r}{\partial\theta^2}-\frac{2}{r^2}\frac{\partial\upsilon_\theta}{\partial\theta}\right] \qquad (11)$$

Tangential direction:

$$\rho\left(\frac{\partial\upsilon_\theta}{\partial t}+\upsilon_r\frac{\partial\upsilon_\theta}{\partial r}+\frac{\upsilon_\theta}{r}\frac{\partial\upsilon_\theta}{\partial\theta}+\frac{\upsilon_\theta\upsilon_r}{r}\right)=\rho f_\theta-\frac{1}{r}\frac{\partial p}{\partial\theta}+\mu\left[\frac{\partial}{\partial r}\left(\frac{1}{r}\frac{\partial}{\partial r}(r\upsilon_\theta)\right)+\frac{1}{r^2}\frac{\partial^2\upsilon_\theta}{\partial\theta^2}-\frac{2}{r^2}\frac{\partial\upsilon_r}{\partial\theta}\right] \qquad (12)$$

The energy equation is not necessary because of hypothesis (3).

3.2 Simulation results and discussion

3.2.1 Verification of model validity

Firstly, the validity of simulation model was verified through comparison with experimental results. Fig.12 is the comparison between simulation results and experimental ones.
The maximum difference between simulation results and experimental ones is $0.02m^3/h$ and the minimum one is $0.004m^3/h$, which ensured the model valid to guide the experiments and engineering design.

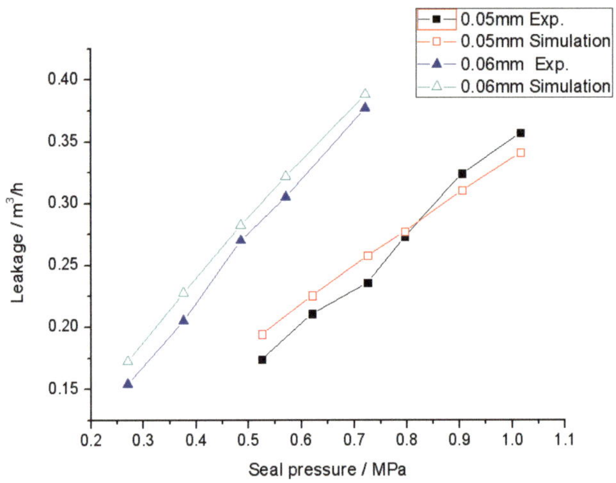

Fig. 12. Comparison between simulation results and experimental ones

3.2.2 pressure distribution in clearance

Fig.13 shows the liquid film pressure distribution in clearance.

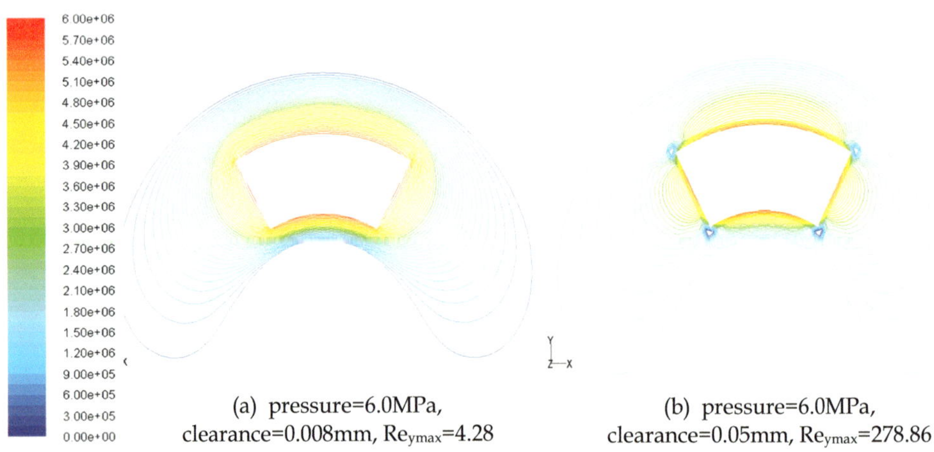

(a) pressure=6.0MPa, clearance=0.008mm, Re_{ymax}=4.28

(b) pressure=6.0MPa, clearance=0.05mm, Re_{ymax}=278.86

Fig. 13. Pressure distribution in clearance

Fig.13 illustrates the totally different form of face seal at different clearance. It could be seen that pressure is well sealed when laminar flow in the clearance shown in (a). However, the

Reynolds number increased with the increasing of clearance and the turbulence occurred at the boundary which makes the seal pressure decreased dramatically shown in (b) and means the seal has been destroyed.

3.2.3 Effects of rotor speed on leakage

Fig.14 shows the variation of leakage with rotor speed at 0.02mm and seal pressure is 6.0MPa. It could be seen that the leakage has little changed with the rotor speed increasing from 600rpm to 1600rpm, which means the rotor has no effects on leakage. The results were accordance with the theoretical computations and could be explained that the rotor speed has no effects on flux crossing boundary A and its effects on boundary B and C cancelled each other. Therefore the rotor speed has no effects on total leakage.

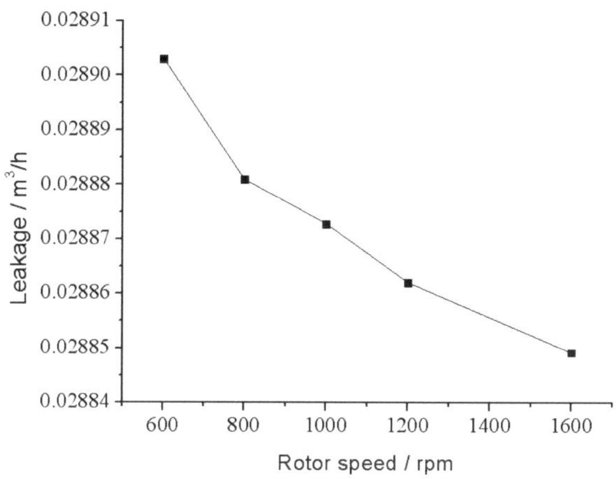

Fig. 14. Relationship between leakage and rotor speed

3.2.4 Effects of clearance on seal pressure and pressure

Fig.15 shows the relationship between clearance and seal pressure at different flux.

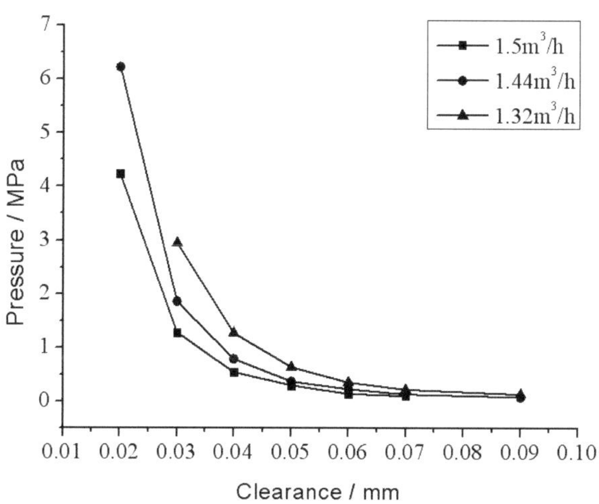

Fig. 15. Relationship between clearance and pressure

It could be seen from Fig.15 that the seal pressure decreased dramatically with the increasing of clearance especially when the clearance is less than 0.03mm. Under different flux when the clearance is larger than 0.04mm the seal pressure will be less than 1MPa which can not meet the requirements of SWRO operation. Therefore, the clearance should be controlled to as minimum as possible while keeping rotor rotating, otherwise the more flux is needed to makeup leakage loss.
Fig.16 shows the relationship between clearance and leakage at different seal pressure.

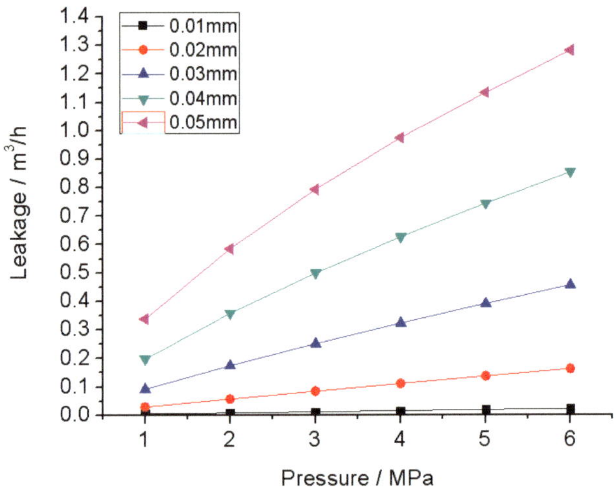

Fig. 16. Relationship between clearance, pressure and leakage

It could be seen that the leakage increases more rapidly at larger clearance than that does at smaller one. If the clearance of RPE could not be controlled under 0.01mm the pressure requirements for SWRO operation have to be met at the cost of increasing flux, which consequently leads to the flow unbalance into and out of RPE.

4. Energy recovery experiment with RPE

4.1 Design of noncontact face seal structure

The face seal structure of RPE in the experimental RPE is shown in Fig.17. The face seal structure is composed of rotor, end covers and sleeve. A set of rings with different thickness is used to adjust clearance between rotor and end covers in RPE.
In Fig.17, h_1 and h_2 are the height of rotor and sleeve respectively. h_3 is the thickness of adjust ring. h_4 and h_5 are the assembly height of top and bottom end cover respectively.
The clearance is controlled in the order of 10^{-2} mm and to 0.02mm as its minimum through adjustable ring with different thickness during energy recovery experiments. When the clearance is less than 0.02mm the rotor of RPE will not start driven by jet flow under experimental conditions. During SWRO energy recovery experiments, the clearance is adjusted to minimum at first step as long as the rotor could start. And then mixing rate, face leakage and energy recovery efficiency are evaluated through salinity, pressure and flux data.

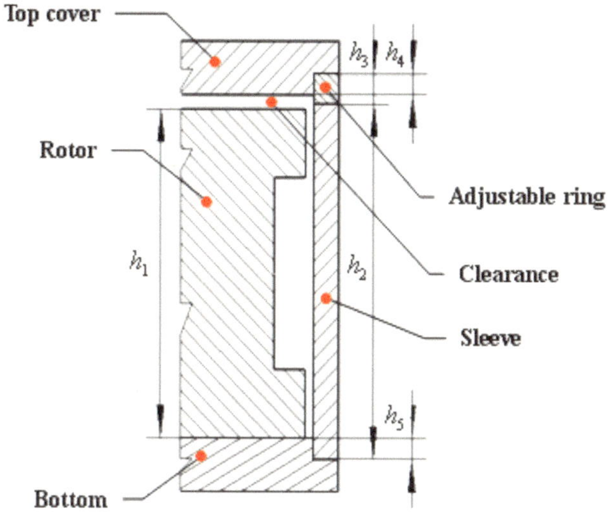

Fig. 17. Noncontact seal structure of RPE unit

4.2 RPE unit

Fig.18 is the RPE unit in the experimental study. The materials of rotor, end covers and sleeve are carefully chosen to counteract corrosion of NaCl solution. The face materials of rotor and end covers are cemented carbide and the planarity and parallelism are ensured through precious surface finishing.

The rotor is driven by external jet flow to start up. The rotor speed could be adjusted through changing the orifice diameter of jet.

Fig. 18. RPE unit driven by fluid

4.3 Design of SWRO experimental system

Fig.19 is the scheme of RPE energy recovery experiment system with seawater reverse osmosis (SWRO). The capacity of SWRO is $1.2m^3/h$ and the salt rejection is over 99% and recovery rate is 40%. The purity of production water is monitored by conductivity. The pressure in front of the reverse osmosis is adjusted by throttle valve (V5) and monitored by pressure gauge (P5).

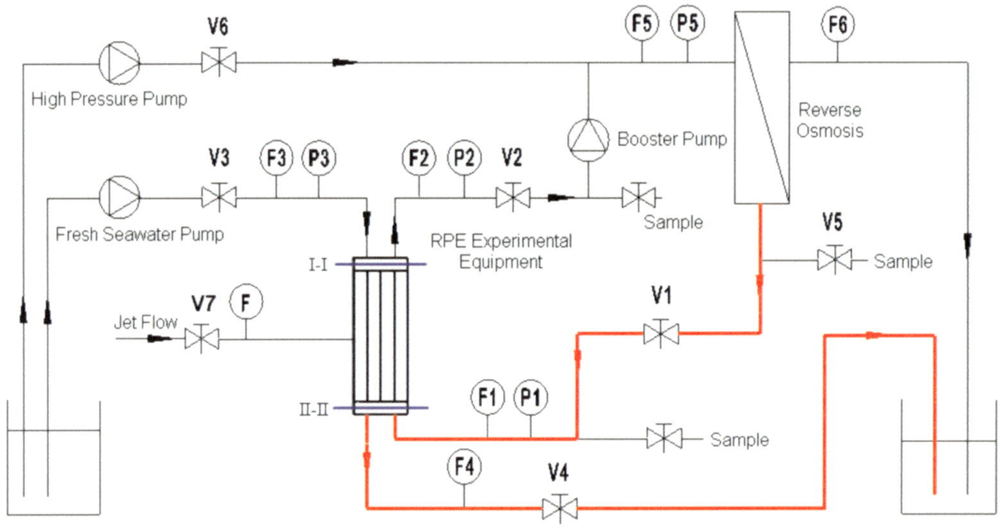

P— pressure gauges or sensors F— flux meters or sensors V—Valve

Fig. 19. SWRO experimental system scheme

The pressure and flow rate are measured and recorded through pressure sensors and flow meters at the pipes from and to the RPE experimental unit. Moreover, there are three sampling points which are used to salinity measurement during experiments. The rotor speed is adjusted through orifice diameter. The configuration of valves is designed to make face seal performance, mixing rate, energy recovery and their relationship being studied.

4.4 Experimental procedure

According to Fig.19 the V1~V4 are initially closed. Firstly the SWRO system starts and operation pressure can be adjusted through V7. Secondly, rotor of RPE driven by jet flow starts and then the booster pump and V1~V4 open to assemble the RPE into the SWRO system.

4.5 Experimental results and discussion

4.5.1 Face seal performance

In this experiment, tap water is used as experimental fluids. Fig.20 shows the characteristics of face seal at different clearance in experiments. It could be seen that the face seal has better performance at smaller clearance than larger one.

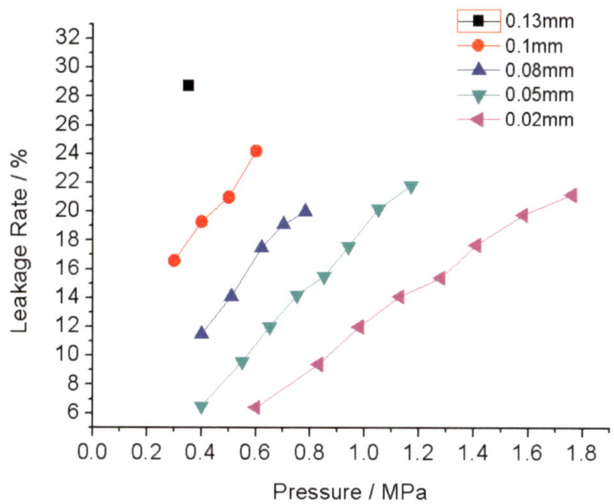

Fig. 20. Relationship between system pressure and leakage rate at different clearances

Fig.21 shows the relationship between pressure and clearance. It is clear that with the increasing of clearance the system pressure deceases and leakage increases dramatically. And when the clearance is larger than 0.1mm the leakage will so unbearable that the SWRO system has lost operation.

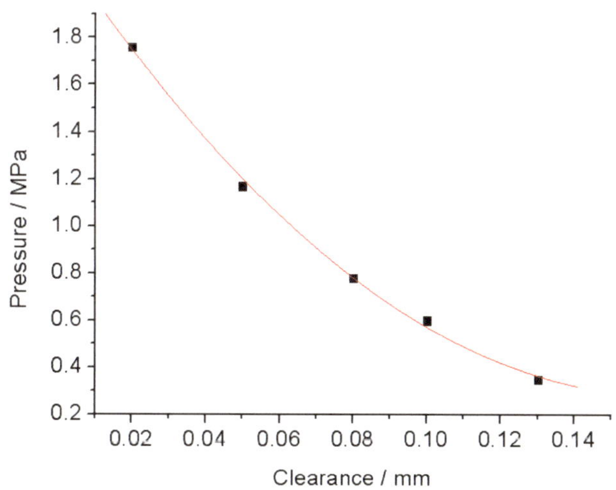

Fig. 21. Relationship of system pressure and clearance

Fig.22 shows the relationship between energy recovery efficiency and clearance. It could be seen that the energy recovery efficiency decreases with the increasing of clearance.
The face seal experiments demonstrate that its performance plays a very important role in the operation of RPE. The larger clearance makes the leakage increase and the SWRO lost normal operation point.

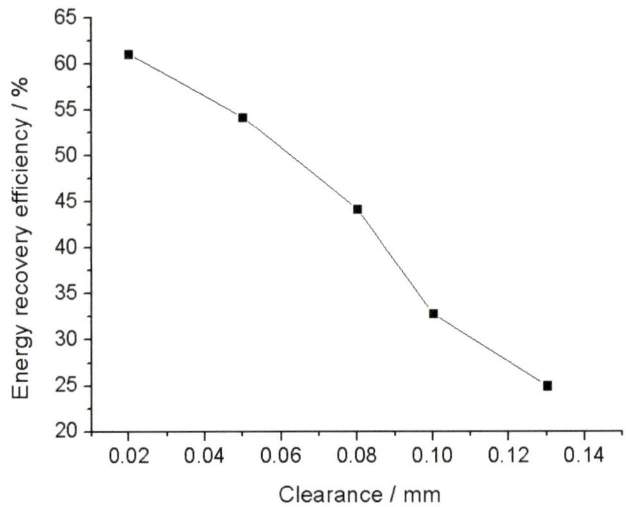

Fig. 22. Relationship of clearance and energy recovery efficiency

4.5.2 Mixing rate and energy recovery experiments

In this experiment, the NaCl solution is used to replace real seawater as experimental fluid. The main purpose of this part is to study characteristics of mixing during energy recovery process.

The experimental results are listed in Tab.1.

Experiments	Item	Low pressure inlet	High pressure outlet	High pressure inlet	Low pressure outlet	Mixing Rate	Energy recovery efficiency
1	Pressure (MPa)	0.1	0.53	0.83	0		
	Flow Rate (L/min)	20.5	20	21.4	22		
	Electric Conductivity (mS/cm)	47.3	42.1	25.9	26.3		
	Salinity (%)	4.79	4.71	2.21	2.26	3.16%	53.5%
2	Pressure (MPa)	0.1	0.87	1.15	0		
	Flow Rate (L/min)	22.5	20	21.4	24.1		
	Electric Conductivity (mS/cm)	47.3	42.1	25.9	26.3		
	Salinity (%)	4.79	4.71	2.21	2.26	3.16%	64.8%
3	Pressure (MPa)	0.1	1.07	1.41	0		
	Flow Rate (L/min)	24.5	20.4	22	26.3		
	Electric Conductivity (mS/cm)	46.9	42.2	26.1	26.4		
	Salinity (%)	4.76	4.71	2.21	2.26	2%	65.3%
4	Pressure (MPa)	0.1	1.42	1.76	0		
	Flow Rate (L/min)	25.5	20.6	23.5	27.4		
	Electric Conductivity (mS/cm)	47.2	42.2	23.4	26.4		
	Salinity (%)	4.79	4.71	2.24	2.24	3.14%	66.7%

Table 1. Experiments of energy recovery with different fluid

It could be seen from the Tab.1 that the mixing rate is well controlled less than 3.5%. However, because the clearance is only 0.02mm the serious leakage deteriorates the face seal performance. The seal pressure decreased dramatically and the energy recovery efficiency was not satisfactory.

5. Conclusion

The effective control of mixing rate in the ducts of rotor is very important to the stable operation of SWRO. The 3D mixing model is setup in the paper and the conclusions are as following:

1. The liquid piston is formed and moves reciprocally in the ducts. And the fresh seawater is separated by the liquid piston to be prevented from over-mixing with brine.
2. The flow-in velocity and rotor speed are the most important parameters affecting mixing rate. The mixing rate will increase with increasing of flow-in velocity.
3. The mixing rate will almost keep stable even the rotor speed increases once the rotor speed is more than 800rpm.

At the same time, excellent face seal plays very important role in separating the high pressure system from low pressure of RPE in order to ensure the normal operation of SWRO. The characteristics of face seal and relationship between clearance, pressure and leakage are studied through theoretical analysis, numerical simulation and experimentally studied in the paper. The conclusions are as following:

1. Laminar flow in clearance makes the face seal reliable. When the turbulent flow occurs the pressure gradient and face seal are destroyed.
2. Rotor speed has no obvious effects on the leakage in face seal process.
3. The seal pressure decreases and leakage increases dramatically with the increasing of clearance. The clearance should be controlled less than 0.01mm for the SWRO operation.
4. The pressure decreases and leakage increases dramatically with the increasing of clearance. The clearance in RPE is controlled to 0.02mm as minimum. Under such clearance the mixing rate of different concentration NaCl solutions is less than 3.5%. However the larger clearance leads to deterioration of face seal, building up pressure and energy recovery efficiency. Although the highest energy recovery efficiency is just 66.7% the experimental study clearly illustrates the face seal between rotor and end cover is the MOST important factor in RPE energy transfer process.

6. References

[1] Zhou Yihui etc. Numerical simulation on dynamic mixing process in duct of RPE, Desalination and Water Treatment,2009.1

[2] Zhou Yihui. Theoretical and experimental study on energy recovery process of RPE in SWRO(D),Dalian University of Technology,2010

[3] User Manual PX45s, Energy Recovery Inc. USA 2006

[4] ANSYS FLUENT 12.0 user's guide FLUENT Inc., 2006

[5] G. K. Batchelor. An Introduction to Fluid Dynamics. Cambridge Univ. Press, Cambridge, England, 1967

[6] ANSYS FLUENT 12.0 user's guide(section 12.4.2, section 7.5, section7.10, section 7.13.1). FLUENT Inc., 2006,

[7] B. E. Launder and D. B. Spalding. Lectures in Mathematical Models of Turbulence. Academic Press, London, England, 1972.
[8] R.L. Stover, Development of a fourth generation energy recovery device. Desalination, 165 (2004) 313–321
[9] Victor L. Streeter, E. Benjamin Wylie, Keith W. Bedford. Fluid Mechanics (9th edition) [M].McGraw Hill Publish
[10] John D. Anderson, Jr. Computational Fluid Mechanics-The Basics with Applications [M].McGraw Hill PublishC. Fritzmann, J. Löwenberg, T. Wintgens, T. Melin. State-of-the-art of reverse osmosis desalination. Desalination 216 (2007) 1–76

Part 2

Distillation

6

Flexibility Study for a MSF by Monte Carlo Simulation

Enrique Tarifa, Samuel Franco Domínguez,
Carlos Vera and Sergio Mussati
Universidad Nacional de Jujuy,
Universidad Nacional del Litoral, CONICET,
Argentina

1. Introduction

Traditionally, processes and controllers are designed sequentially. Firstly, the process configurations (structures) and parameters are designed to satisfy the economic objectives, such as maximum profit or minimum operational costs. The designs are based on steady state models, and they are subjected to the operational constraints. Then, the controllers are designed, with a focus on rejecting the possible effects of external disturbances and process uncertainties, and achieving the desired dynamic performance. This approach carries a risk in that it may end up choosing the cheapest process design that can prove difficult to control. It may also miss out a slightly less economic but easier to control design, the one that might be more profitable in the long run (Weitz & Lewin, 1996).

Operability properties of a process determine how process dynamics affect the quality of a process control design. These include flexibility, controllability, optimality, stability, selection of measurements and manipulated variables. The flexibility is defined as 'the ability to maintain the process variables within the feasible operational region, despite the presence of uncertainties' (Grossman et al., 1983). Flexibility is often considered simultaneously with the economic objectives and hence the optimality issue is raised. As a consequence, flexibility studies are dominated by numerous optimization strategies. Those studies aim at the determination of flexible operational spaces and flexibility measurements. The analysis generally involves two complementary tasks, the calculation of the flexibility index and the flexibility test.

Operational flexibility is an important issue when designing and operating a chemical plant. Very often, flexibility is concerned with the problem of ensuring a feasible steady-state operation over a variety of operating uncertainties. To quantify how flexible a process is many metrics have been developed. Grossmann et al. (1983) first introduced the flexibility index *FIG* which quantifies the smallest percentage of the uncertain parameters' expected deviation that the process can handle. Another metric named resilience index *RI* was adopted by Saboo et al. (1985). These two measurements -*FIG* and *RI*– require identification of the nominal point, which must be located within the feasible region. These measurements however only take the critical uncertainty into account. This may cause serious flexibility under-estimation or neglect the ability of the process to handle other process uncertainties. To solve this problem, Pistikopoulos and Mazzuchi (1990) proposed an index called stochastic flexibility, *SF*, that is determined from the probability distribution of all the uncertain parameters. Although *SF* accounts for the chance that the process can operate feasibly, the probability distribution of all the uncertain parameters may not be available at the design stage. Even though the probability distributions can be obtained, the calculation of *SF* is usually tedious. To avoid this difficulty, Lai and Hui (2007) proposed the index *FIV*. This was calculated as the size ratio of the feasible space to the overall space bounded by the expected limits of the uncertain parameters. The feasible space is the subspace of the overall space in which the uncertain parameters can be feasibly handled. The index *SF* and *FI* belong to the interval [0, 1]; a higher value means a higher flexibility.

In this work several flexibility indexes for a multi-stage flash (MSF) desalination plant were estimated. To mimic the plant operation a stationary simulator was developed, and the determination of the feasible space was carried out with Monte Carlo simulation (Metropolis & Ulam, 1949; Rubinstein & Kroese, 2007). This approach does not involve an optimization model, but only a simulation one; hence the implementation is more simple and robust than other approaches. Finally, the proposed method yields additional information besides the flexibility indexes, and the relevance of this additional information shows the potential of this approach.

2. Mixing tank modelling

The strategy of the proposed flexibility study will be introduced by using a simpler system, a mixing tank, as shown in Fig. 1. In this system, stream F_1, water at 25 °C, and stream F_2, water at 80 °C, are mixed to yield a new stream at 52 °C. The pressure at the valve V_1 input is 1.115×10^5 Pa. The volumetric flow rate of F_2 is 0.02 m^3/s. The tank is open and discharges to the atmosphere; its diameter is 1.5 m and the maximum allowed liquid level is 2 m. The controller CT (a P+I controller) controls the tank temperature *T*, and the controller CL (a P controller) controls the tank level *L*; the respective set points are 52 °C and 1 m.

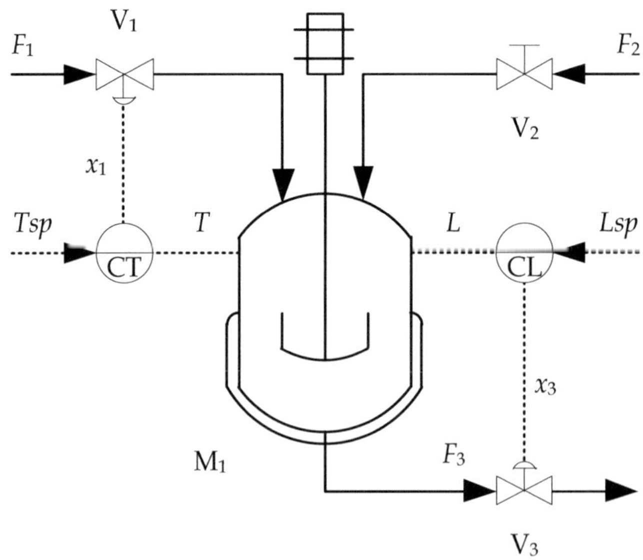

Fig. 1. Mixing tank.

The dynamic model for the mixing tank comprises the following equations:

$$\frac{dL}{dt}=\frac{F_1+F_2-F_3}{A} \tag{1}$$

$$\frac{dT}{dt}=\frac{F_1(T_1-T)+F_2(T_2-T)}{AL} \tag{2}$$

$$\frac{dAi_{\mathrm{T}}}{dt}=e_{\mathrm{T}} \tag{3}$$

$$e_{\mathrm{T}}=T-Tsp \tag{4}$$

$$x_1=Ab_{\mathrm{T}}+K_{\mathrm{T}}\left(e_{\mathrm{T}}+\frac{1}{\tau i_{\mathrm{T}}}Ai_{\mathrm{T}}\right) \tag{5}$$

$$e_{\mathrm{L}}=L-Lsp \tag{6}$$

$$x_3=Ab_{\mathrm{L}}+K_{\mathrm{L}}\,e_{\mathrm{L}} \tag{7}$$

$$F_1=C_{v1}\,x_1\sqrt{P_1-P_0} \tag{8}$$

$$F_3=C_{v3}\,x_3\sqrt{\rho g L} \tag{9}$$

Table 1. presents the list of variables used in the previous model. The parameters of controller CT are Tsp = 52 °C, Ab_{T} = 0.5, K_{T} = 0.05 °C^{-1} and τi_{T} = 30 s; the parameters of controller CL are Lsp = 1 m, Ab_{L} = 0.5 and K_{L} = 20 m^{-1}. The valve parameters are $C_{v1} = 4.039\times10^{-4}$ m$^{3.5}$/kg$^{0.5}$ and $C_{v3} = 8.078\times10^{-4}$ m$^{3.5}$/kg$^{0.5}$.

τi: integral time constant (s).
A: cross sectional area (m^2).
Ab: controller bias.
Ai: integral effect of CT (s °C).
AL: output of CL.
C_v: valve flow coefficient (m$^{3.5}$/kg$^{0.5}$).
e_{L}: controller error of CL (m).
e_{T}: controller error of CT (°C).
F: volumetric flow rate (m^3/s).
K_{L}: gain of CT (°C^{-1}).
K_{T}: gain of CT (m^{-1}).
L: level (m).
Lsp: set point of CL (m).
P: pressure (atm).
T: temperature (°C).
t: time (s).
Tsp: set point of the CT (°C).
x: valve opening.

Table 1. Variables of the mixing tank model.

The steady model is obtained from the dynamic one by setting to zero all the derivative terms. The resulting model contains 9 equations and 9 unknowns (F_1, F_3, T, L, e_T, x_1, e_L, x_3 and Ai_T). If T, e_T, e_L and Ai_T are removed, the model can be reduced to the following equations:

$$F_1 = F_2 \frac{(T_2 - Tsp)}{(Tsp - T_1)} \tag{10}$$

$$x_1 = \frac{F_1}{C_{v1}\sqrt{P_1 - P_0}} \tag{11}$$

$$F_3 = F_1 + F_2 \tag{12}$$

L and x_3 are calculated from the two following equations:

$$x_3 C_{v3} \sqrt{\rho g L} - F_3 = 0 \tag{13}$$

$$Ab_L + K_L (L - Lsp) - x_3 = 0 \tag{14}$$

3. Standardization of variables

In the proposed method every variable has to be standardized as follows:

$$\delta X = \frac{X - Xn}{\Delta Xn} \tag{15}$$

where X is the actual value of the variable, Xn is the nominal value that was considered for the variable during the system design, and ΔXn is the half-band of acceptable variability for the variable. As it can be deduced, δX is a dimensionless value that belongs to the open interval (-1, 1) under normal conditions, and it takes the null value at the nominal condition. This study considers two set of variables. The first set, called D, is formed by the disturbances; the second one, called Y, is formed by the other process variables (i.e. all the variables of the process, disturbances not being included). In the case of the mixing tank, the selected disturbances are F_2 and T_2; whereas the selected process variables are x_1 and x_3.

4. Overall and feasible spaces

According to the above definitions, the overall space bounded by the expected limits of the uncertain parameters (Lai & Hui, 2007) can be defined as:

$$\left|\delta X_j\right| \le 1 \quad \forall j \in \mathrm{D} \tag{16}$$

On the other hand, the feasible space in which the uncertain parameters can be feasibly handled (and where the process is operable) can be defined as:

$$\left|\delta X_j\right| < 1 \quad \forall j \in \mathrm{Y} \tag{17}$$

In order to determine whether the process is operable or not in a given point of the overall space, it is convenient to define the operability index in the following way:

$$Io = \max_{j \in Y}\left(\left|\delta X_j\right|\right) \tag{18}$$

While *Io*, the maximum observed deviation of the process variables, belongs to the interval [0, 1) the process is operable because the deviations of all process variables are lower than their respective acceptable variabilities. This index is used to determine the feasible space, which is formed by all the process states with *Io* belonging to the interval [0, 1).
Fig. 2 shows the overall space for two disturbances, D_1 and D_2. A square circumscribed about a circle of radius *r* and centre (0, 0) is also shown in that figure. That square is called the maximum square if it is the largest square that can be defined in the feasible space.

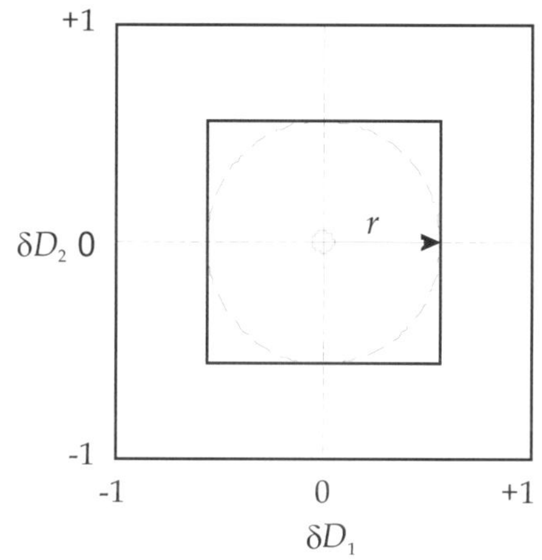

Fig. 2. Overall space and maximum square.

The probability that the disturbances yield a point inside the maximum square depends on the probability distributions associated to the disturbances. If every disturbance follows the uniform distribution (Fig. 3) that probability has the distribution plotted in Fig. 4. Conversely, if every disturbance follows the triangular distribution (Fig. 5) that probability has the distribution shown in Fig. 6. Both distributions (in Fig. 4 and Fig. 6) were obtained by Monte Carlo simulation with a sample of 10000 points (Metropolis & Ulam, 1949; Rubinstein & Kroese, 2007).

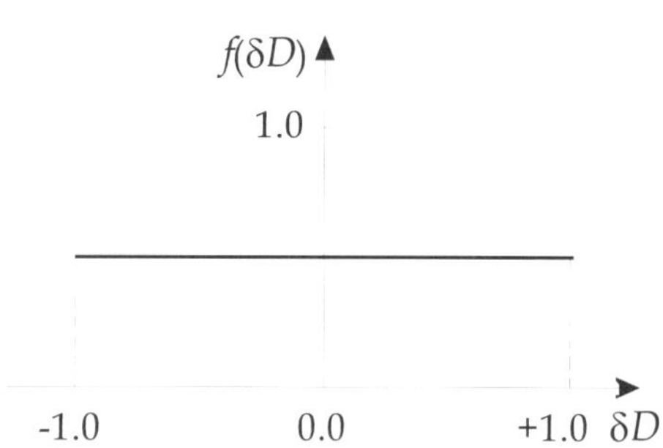

Fig. 3. Uniform distribution for δD.

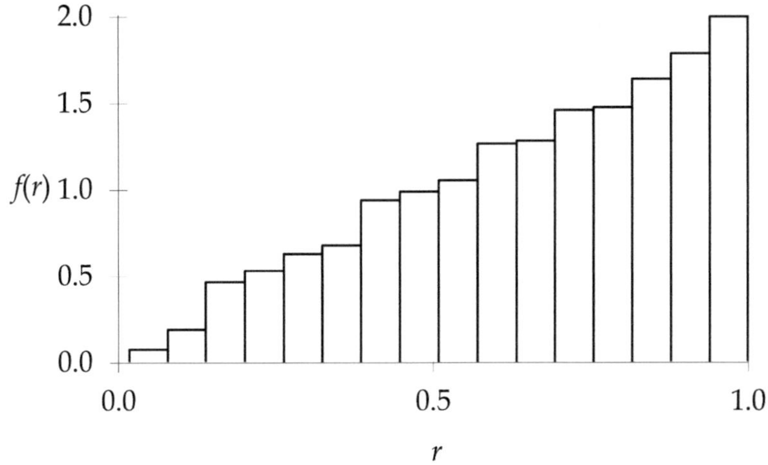

Fig. 4. Probability density distribution of r for uniform distribution of δD. The median is 0.71.

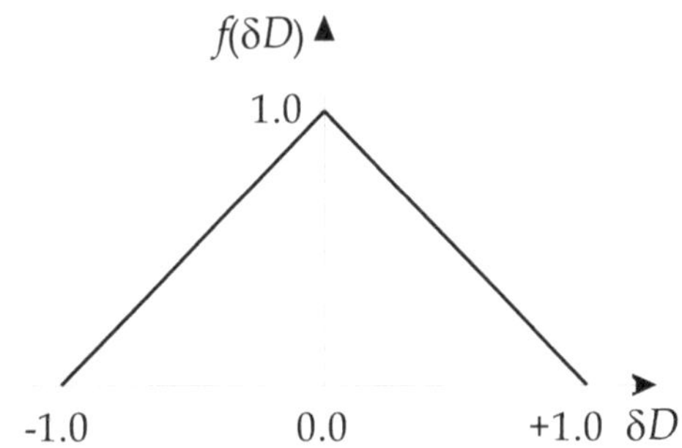

Fig. 5. Triangular distribution for δD.

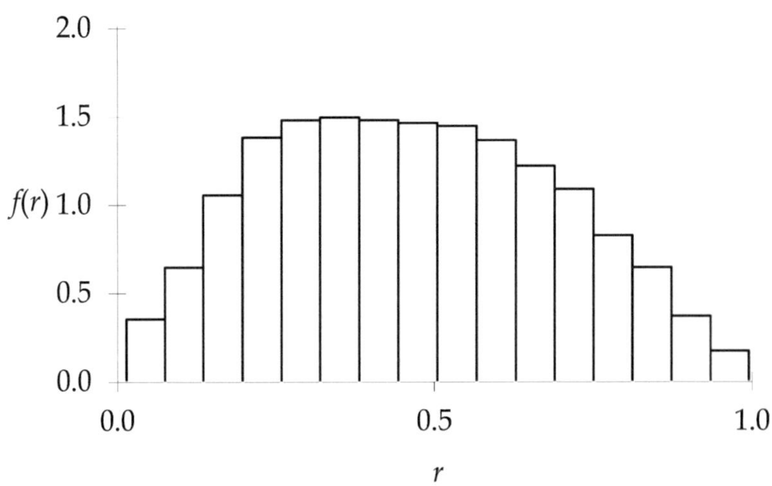

Fig. 6. Probability density distribution of r for triangular distribution of δD. The median is 0.46.

For both the uniform and triangular distributions, it is possible to determine analytically the respective probability density function -pdf- $f(r)$ and the cumulative distribution function -cdf- $F(r)=\int_0^r f(t)dt$ (Rose & Smith, 2002) associated to the maximum hypercube of dimension n (maximum square if n = 2). If every disturbance follows the uniform distribution, the corresponding pdf and cdf for the maximum hypercube of dimension n are:

$$f(r)=nr^{n-1} \tag{19}$$

$$F(r)=r^n \tag{20}$$

If every disturbance follows the triangular distribution, the corresponding pdf and cdf for the maximum hypercube of dimension n are:

$$f(r)=2n\left(r^2+2r(1-r)\right)^{n-1}(1-r) \tag{21}$$

$$F(r)=\left(r^2+2r(1-r)\right)^n \tag{22}$$

Then, the probability that the combination of disturbances were constrained to the maximum square defined by r is equal to $F(r)$.
If every disturbance follows a normal distribution (Fig. 7) the probability of the disturbances yielding a point inside the maximum square has the distribution plotted in Fig. 8. The distribution was obtained by Monte Carlo simulation with a sample of 10000 points. In contrast to the case of the uniform and triangular distributions, it is not possible to obtain analytical expressions of $f(r)$ and $F(r)$ when the disturbances are normally distributed. For this reason, Monte Carlo simulation was again used to determine $F(r)$ (Fig. 9). A good approximation for $F(r)$ is achieved by data regression, this approximation has a maximum error of 0.04, and is equal to $F(r)=\exp\left(-3.6109(1-r)^{3.7664}\right)$.

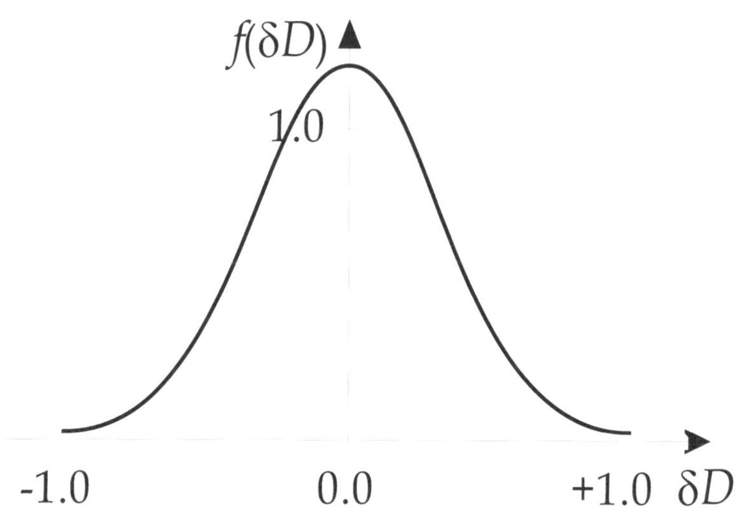

Fig. 7. Normal distribution for δD with μ = 0, $\sigma = 1/3$.

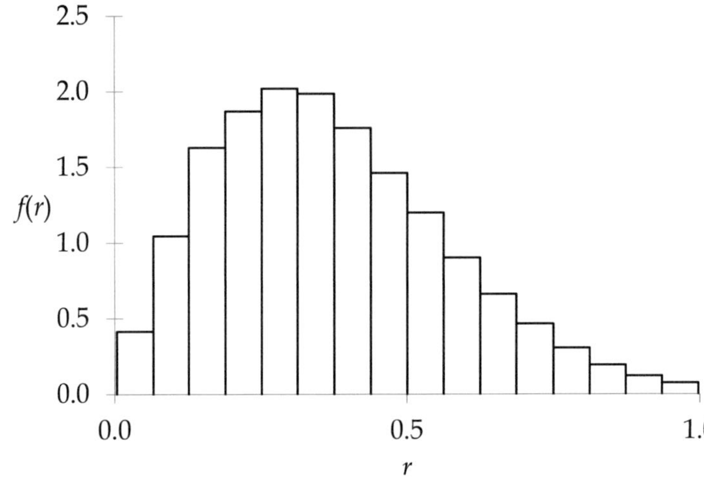

Fig. 8. Probability density distribution of r for normal distribution of δD. The median is 0.35.

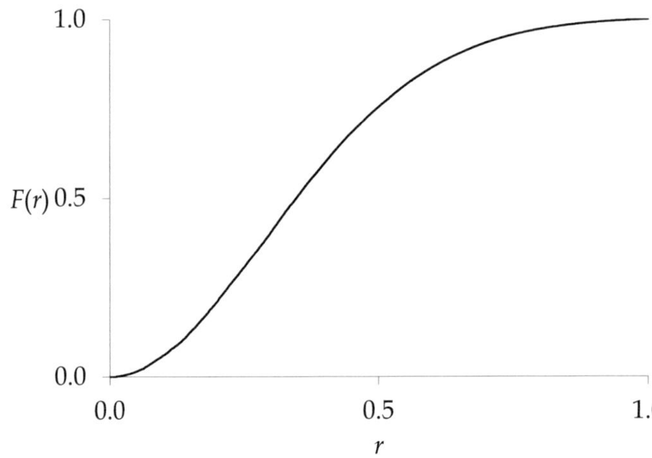

Fig. 9. Cumulative distribution function of r for normal distribution of δD.

The maximum hypercube is useful because it makes it easy to verify whether a given operating point is inside it. This is a sufficient condition to guarantee the process operability. If the point is outside the maximum square, a deeper analysis is needed such as the outlined below.

5. Flexibility study of the mixing tank

Table 2 presents the nominal values and variability half-bands adopted in the case of the mixing tank. Fig. 10 shows the corresponding feasible space (without shadow). Several simulations were run to obtain that figure. In each simulation, a particular combination of disturbances was generated, equations (10)-(14) were solved, and the operability index Io was calculated to determine the operability of that point. When Io was lower than 1, that point was marked as belonging to the feasible space.

According to Fig. 10, the process is not operable for T_2 below 52 °C (the set point value of controller CT), a constant limit. However, the upper limit for T_2 is a function of F_2. The radius r of the inscribing circle into the maximum square is equal to 0.42. By using the corresponding cfd, it can be estimated that the maximum square covers 18% of the possible cases if every disturbance has uniform distribution. The covered cases are 44% when every disturbance follows the triangular distribution.

	F_2 (m³/s)	T_2 (°C)	x_1	x_3
Xn	0.02	80	0.5	0.5
Δ*Xn*	0.01	40	0.5	0.5

Table 2. Nominal values and half-bands of acceptable variability for the case of the mixing tank.

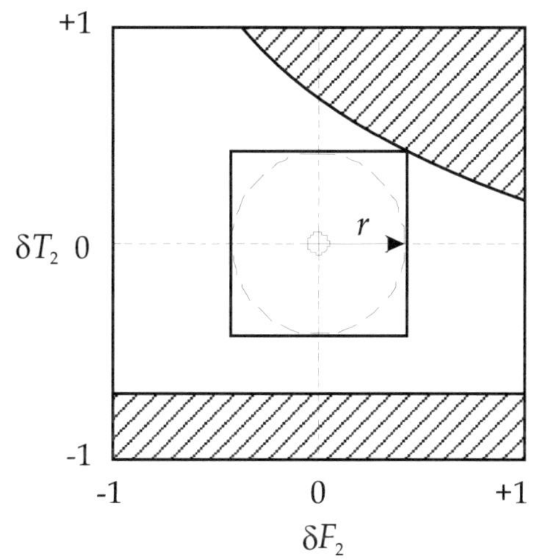

Fig. 10. Mixing tank example. Feasible space of operation (without shadow), r = 0.42.

Fig. 11 shows the values adopted by the process variables Y inside the feasible space. Several simulations were run to obtain that figure (the same ones used to determine the feasible space). In each simulation, a particular combination of disturbances was generated, and the corresponding values of the process variables were calculated. If the process state thus generated belonged to the feasible space –*Io* belonging to the interval [0, 1)–, δx_1 and δx_3 were added to the plot as 'pair *n*'.

According to Fig. 11, the most critical variable is x_1 because it reaches the limits of acceptable variability. More precisely, δx_1 reaches the value 1, which means x_1 reaches its highest allowed value. Therefore, the feasible space can be expanded by acting on the sector supervised by the CT controller; e.g. increasing the size of the valve V1 or decreasing the temperature T_1. In fact, it is evident from eq. (11) that it is possible to reduce x_1 (the critical variable) by only increasing C_{v1} (the size of valve V1) and without changing the other variables.

From the results of Fig. 11, it can also be deduced that the reduction of x_1 will cause that some process states become now feasible states, augmenting in this way the feasible space. This is one of the several conclusions that can be obtained from that figure and demonstrates one advantage of the proposed method. To prove that the previous conclusion was correct, a new test was made and the new feasible space was determined after multiplying C_{v1} by two. The new feasible space thus obtained was bigger than the original one. The same conclusion was again obtained though it required many more simulations. The usefulness of the information presented by Fig. 11, a contribution of the proposed method, cannot be overemphasized.

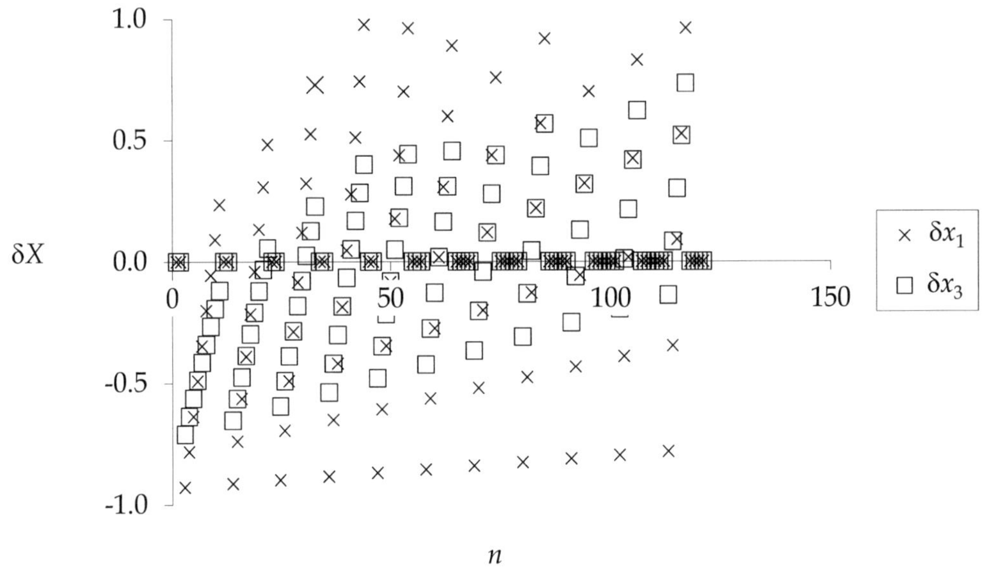

Fig. 11. Mixing tank example. Process variables in the feasible space.

Fig. 12 shows the process states corresponding to the representative points of the overall space. These points were explored by the simulations used before for defining the feasible space. The figure is a plot with parallel coordinates; this is a common way of visualizing high-dimensional geometry and analysing multivariate data. To show a set of points in an n-dimensional space, a backdrop is drawn consisting of n parallel lines, typically vertical and equally spaced. A point in that n-dimensional space is represented as a polyline with vertices on the parallel axes; the position of the vertex on the i-th axis corresponds to the i-th coordinate of the point. In this work, the set of represented points correspond to the studied process states, i.e. every plotted polyline represents a particular steady state. The vertical axis represents the studied variable, e.g. Fig. 12 has axis for F_2, T_2, x_1 and x_3.

In Fig. 12 every point of the overall space is depicted by a line linking the values corresponding to all the considered δX. For the sake of clarity, the vertical axis corresponding to the process variables are not drawn. Among those points only those with absolute values of δx_1 and δx_3 lower than 1 are operable and hence belong to the feasible space. That figure also shows that x_1 is the most critical variable. Moreover the strong effect of T_2 over x_1 is evident: decreasing T_2 is more risky than increasing it. x_1 is indeed the most critical variable because it has more values out of the (-1, 1) interval. The effect of T_2 on x_1 provokes that the polylines associated to an increase of T_2 are also associated to an increase of x_1. An equivalent but weaker relation exists between the decrease of T_2 and the decrease of x_3. This conclusion is just one of the many that be inferred from Fig. 12. This demonstrates another advantage of the proposed method.

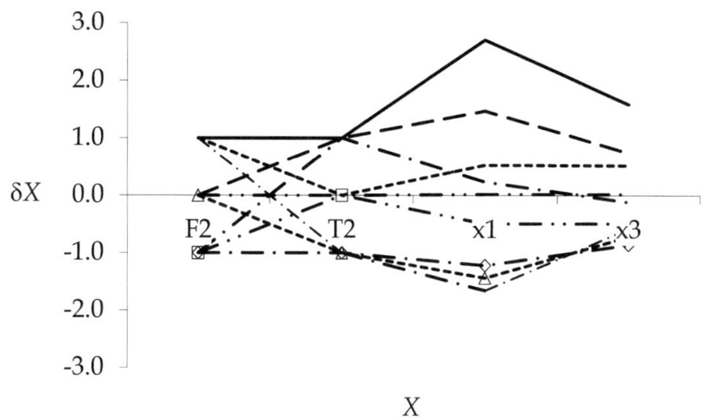

Fig. 12. Mixing tank example. Representative points of the overall space.

6. Flexibility indexes

Different kinds of indexes have been defined for representing the process flexibility. In general, the better the index the more complex is its calculation. That's why at the first stages of the process design simpler indexes should be used. Conversely, at the final stages of the design the most exact indexes should be calculated. In line with this reasoning several indexes were evaluated for the mixing tank. Those indexes are presented below, ordered by increasing quality and complexity.

The first index is *Iv*. It is equivalent to *FIG* (Grossmann et al., 1983) while $FIG \leq 1$. *Iv* is defined as:

$$Iv = r \tag{23}$$

where *r* is the radius of the circle tangential to the maximum square.

The second index is *Ic*, which is defined as the ratio of the size of the maximum hypercube of dimension *n* (maximum square if $n = 2$) to the size of the overall space:

$$Ic = r^n \tag{24}$$

The third index is *Ir*, which is defined as the size ratio of the feasible space to the overall space –it is equivalent to *FIV* (Lai & Hui, 2007)–. All the above indexes belong to the interval [0, 1], and the value 1 represents the maximum flexibility.

The indexes *Iv* and *Ic* are conservative because they take into account a subspace (i.e., the maximum square or hypercube) of the whole feasible space. They also rely on geometric ratios between the feasible space and the overall space. This is adequate when every disturbance follows the uniform distribution. However this is also a limitation of *Ir*. For other distributions, it becomes necessary to define additional indexes. The fourth defined index would be *Pc*, the probability that the disturbances yield a point inside the maximum square (or hypercube). This depends on the probability distributions of the disturbances. Equations (20) and (22) are used to calculate *Pc* in the cases of uniform and triangular distributions, respectively. For other distributions, Monte Carlo simulation can be instead used, as was previously outlined for the normal distribution (Fig. 7-Fig. 9). The fifth and last index is *Pr*, the probability of the disturbances yielding a point inside the feasible space. It is equivalent to *SF* (Pistikopoulos & Mazzucchi, 1990), which also depends on the disturbances probability distributions. *Pr* is the index most difficult to calculate. In this work *Pr* was calculated by Monte Carlo simulation with 10000 samples.

Table 3 shows all the defined indexes calculated for the mixing tank example. For the calculation, it was assumed that the disturbances probability distributions were triangular. When the disturbances produced a point into the feasible space, the process was by definition considered operable at those conditions. Therefore, the higher the probability of the disturbances yielding a feasible point the higher the probability of the process of being operable -i.e., the process is more flexible–. From all the indexes presented in Table 3 the more realistic and useful is *Pr,* which represents the probability that the process be operable. While the value 0.88 may prove enough for some experts, this can be improved by expanding the feasible space as it was done when studying Fig. 11 and Fig. 12.

Iv	*Ic*	*Ir*	*Pc*	*Pr*
0.42	0.18	0.68	0.44	0.88

Table 3. Flexibility indexes of the mixing tank.

As stated before almost all the indexes, except *Pr,* are conservative. The values reported in Table 3 show how conservative these indexes can be. For example, *Ic* assigns a value of 0.18 to the mixing tank, suggesting a rather poor flexibility; *Pr,* in contrast, assigns a much better value. The difference between these values is a strong evidence of the convenience of utilizing *Pr* for flexibility studies.

7. MSF modelling

The strategy proposed for performing flexibility studies was applied to the analysis of a MSF desalination plant (Fig. 13). This kind of plant has a series of flash units (stages) where sea water is evaporated to obtain distilled water. The plant has *N* stages; the first *M* ones belong to the recovery section and the remaining ones belong to the rejection section. There are also six P+I controllers that set the operating conditions for the heater, the feed, the recycle and the level of the last stage.

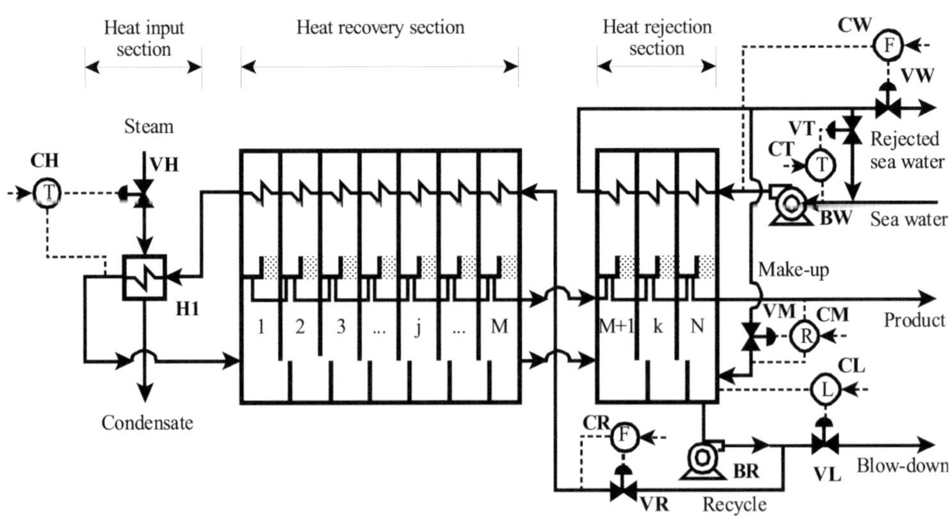

Fig. 13. Simplified scheme of the studied MSF plant.

For the dynamic model of the MSF plant an early model developed by Tarifa and Scenna (2001) was used. This comprises a set of ordinary differential equations and a set of algebraic equations. The steady model was obtained by setting to 0 every derivative term. The model parameters were adjusted to represent the system studied by Thomas et al. (1998). This system has 15 stages in the recovery section and 3 stages in the rejection section. Table 4 shows the adopted operating conditions.

Sea water:
Tsw = 28 °C
Xsw = 51500 ppm
Vapour:
Pvh = 0.937 atm
Controller set points:
T0s = 90 °C
Ls = 0.6 m
Wcws = 14800 tn/h
Tcws = 33 °C
Rmus = 4.6
Wbs = 14380 tn/h

Table 4. MSF plant. Operating conditions.

8. Flexibility study of the MSF plant

The disturbances considered were the seawater temperature *Tsw* and the seawater salinity *Xsw*. These variables show a wide range of variability and they have large effects on the operation of MSF plants (Tanvir & Mujtaba, 2006). The studied process variables were those presented in Table 5.

AL: output of CL
AWmu: output of CM
AT0: output of CH
ATcw: output of CT
AWb: output of CR
AWcw: output of CW

Table 5. MSF plant. Process variables selected for the study.

Table 6 shows the nominal values and half-bands of variability adopted for the MSF plant. For each process variable, its half-band of variability ΔXn was set equal to 80% of the corresponding nominal value *Xn*.

	Tsw (°C)	*Xsw*	*AL*	*AWmu*	*AT0*	*ATcw*	*AWb*	*AWcw*
Xn	30	0.0515	0.35	0.42	0.54	0.32	0.53	0.22
ΔXn	5	0.0165	0.28	0.34	0.43	0.26	0.42	0.18

Table 6. MSF plant. Nominal values and half-bands of acceptable variability.

Fig. 14 shows the feasible space (without shadow) of operation of the MSF plant. That feasible space was obtained by simulation, a process that proved time consuming due to the model complexity. The simulations were performed taking samples in steps of 0.20 for both δXsw and δTsw. The radius r of the circle inside the maximum square was equal to 0.40. According to the simulations the process was not operable for values of Tsw higher than 33 °C (constant limit, also the set point value of the CT controller). On the other side, the lower limit for Tsw was a function of Xsw. The upper limit of Tsw corresponded to a value of r of 0.60. Therefore the actual radius belonged to the interval [0.40, 0.60). Taking the worst case, r becomes equal to 0.40. By using the corresponding cfd it was estimated that the maximum square covered 16% of the possible cases when every disturbance had uniform distribution. The covered cases were 41% when every disturbance followed the triangular distribution.

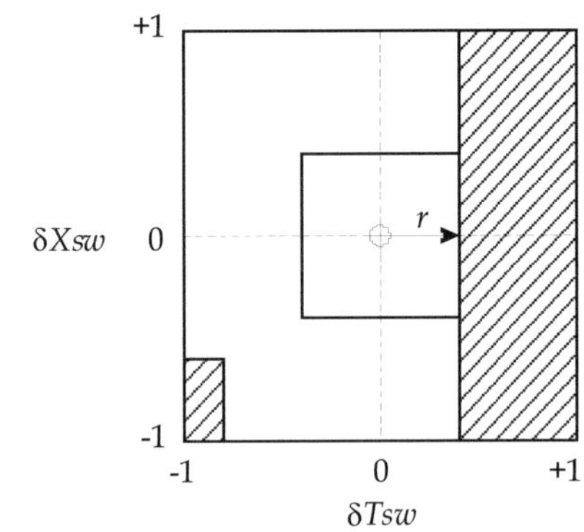

Fig. 14. MSF plant. Feasible space (without shadow) for $r = 0.40$.

Fig. 15 presents the values adopted by some of the analysed process variables inside the feasible space. The remaining ones were not plotted because their changes were not meaningful. The most critical variable are $ATcw$ and $AWcw$ because they reach the limits of acceptable variability; therefore, the feasible space can be expanded by acting on the sectors supervised by the controllers CT and CW (e.g. increasing the size of corresponding valves).

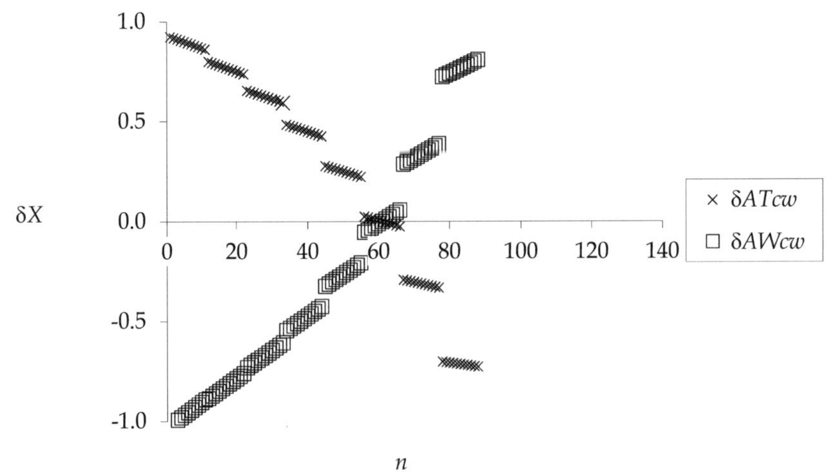

Fig. 15. MSF plant. Process variables in the feasible space.

Fig. 16 shows representative points of the overall space of operation of the MSF plant. Every point is depicted by a line linking the values corresponding to all the considered δX. Of those points, only those with absolute values of $\delta ATcw$ and $\delta AWcw$ lower than 1 are operable. The figure also shows that both process variables, *ATcw* and *AWcw*, are critical. Moreover, the strong effect of *Tsw* on *ATcw* and *Awcw* is evident. Increasing *Tsw* values are more risky than decreasing ones.

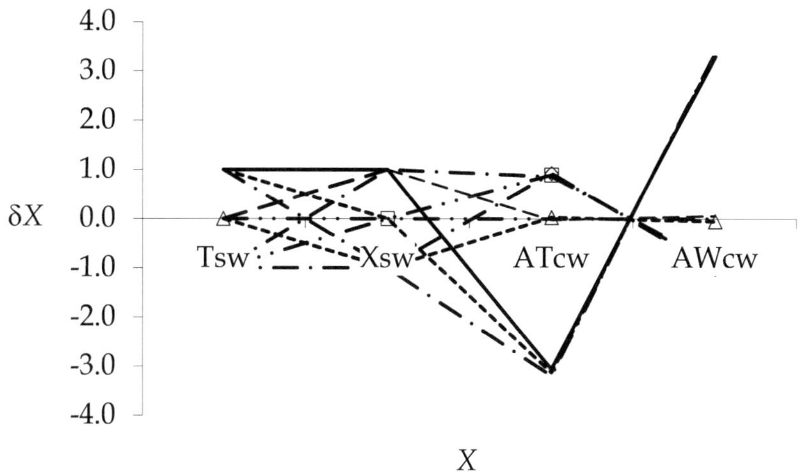

Fig. 16. Representative points of the overall space for the MSF plant.

Finally, Table 7 shows the previously defined indexes as calculated for the case study of the MSF plant. In the calculation triangular distributions for the disturbances were again assumed.

Iv	*Ic*	*Ir*	*Pc*	*Pr*
0.40	0.16	0.68	0.41	0.82

Table 7. MSF plant flexibility indexes.

9. Conclusion

A strategy for performing a flexibility study was presented and it was applied to an introductory simple example and to a complex case study of a multistage flash plant. The strategy begins with the development of a steady-state model of the analysed process. Next, the main disturbances and process variables are identified. Those variables are then properly standardized. The feasible space is determined by simulation. At this point, a set of indexes can be evaluated and the process flexibility estimated.

The proposed methodology allows to calculate the probability associated to the feasible space. It also enables the identification of the critical variables of the process; which can then be modified at the implementation level (changing equipment design parameters) in order to increase the flexibility. The inverse problem can also be considered, i.e. determining the effects on the process flexibility produced by a modification of some components or parameters of the plant. Then, the new feasible space, its new associated probability and the involved costs will together establish the convenience of such modification.

10. Acknowledgement

The authors wish to acknowledge the financial support of the Consejo Nacional de Investigaciones Científicas y Técnicas CONICET (Argentina) and Universidad Nacional de Jujuy UNJu (Argentina).

11. References

Grossmann, I.E., Halemane, K.P. & Swaney, R.E. (1983). Optimization Strategies for Flexible Chemical Processes. *Computers & Chemical Engineering*, Vol. 7, No. 4, pp. 439-462, ISSN 0098-1354

Lai, S.-M. & Hui, C.-W. (2007). Measurement of plant flexibility. *Computer Aided Chemical Engineering*, Vol. 24, pp. 189-194, ISSN 1570-7946

Metropolis, N. & Ulam, S. (1949). The Monte Carlo Method. *Journal of the American Statistical Association*, Vol. 44, No. 427, (September 1949), pp. 335-341, ISSN 0162-1459

Pistikopoulos, E.N. & Mazzucchi, T.A. (1990). A Novel Flexibility Analysis Approach for Processes with Stochastic Parameters. *Computers & Chemical Engineering*, Vol. 14, No. 9, (September 1990), pp. 991-1000, ISSN 0098-1354

Rose, C. & Smith, M.D. (2002) *Mathematical Statistics with Mathematica*, Springer-Verlag, ISBN 9780387952345, New York, USA

Rubinstein, R.Y. & Kroese, D.P. (2007) *Simulation and the Monte Carlo Method* (second edition), John Wiley & Sons, ISBN 9780470177945, Hoboken, USA

Saboo, A.K., Morari, M. & Woodcock, D.C. (1985). Design of Resilient Processing Plants-VIII: A Resilience Index for Heat Exchanger Networks, *Chemical Engineering Science*, Vol. 40, No. 8, pp. 1553-1565, ISSN 0009-2509

Tanvir, M.S. & Mujtaba, I.M. (2006). Neural network based correlations for estimating temperature elevation for seawater in MSF desalination process. *Desalination*, Vol. 195, No 1-3, (August 2006), pp. 251-272, ISSN 0011-9164

Tarifa, E.E. & Scenna, N.J. (2001). A Dynamic Simulator for MSF Plants. *Desalination*, Vol. 138, No. 1-3, (September 2001), pp. 349-364

Thomas, P.J., Bhattacharyya, S., Patra, A. & Rao, G.P. (1998). Steady state and dynamic simulation of multi-stage flash desalination plants: A case study. *Computers & Chemical Engineering*, Vol. 22, No. 10, (September 1998), pp. 1515-1529, ISSN 0098-1354

Weitz, O. & Lewin, D.R. (1996). Dynamic Controllability and Resiliency Diagnosis Using Steady State Process Flowsheet Data. *Computers & Chemical Engineering*, Vol. 20, No. 4, (April 1996), pp. 325-335, ISSN 0098-1354

7

A Computational Model Study of Brine Discharges from Seawater Desalination Plants at Barka, Oman

H.H. Al-Barwani and Anton Purnama
Department of Mathematics and Statistics, Sultan Qaboos University, Muscat, Sultanate of Oman

1. Introduction

Oman is situated at the south-east of the Arabian Peninsula at the entrance to the Arabian Gulf, and its coastline stretches 1700 km along the Gulf of Oman in the north to the Arabian Sea in the south. Most of the population lives in the north-eastern coastal areas and in the capital area of Muscat. The climate of Oman is typically described as a tropical hyper-arid, with two distinct seasons: winter and summer. The winter period extends from late November to April, during which rains at irregular intervals occur. However, based on 27 years of rainfall data from 1977 (Kwarteng et al., 2009), the annual mean rainfall for the whole country is 117 mm. Hot weather with high humidity is experienced in the coastal areas during the summer months. The mean air temperature in northern Oman varies between 32 ^{o}C to 48 ^{o}C from May to September, and between 26 ^{o}C to 36 ^{o}C from October to April. The mean wind speeds range between 2 and 3.5 m/s, with high winds encountered during the summer months.

Desalinated water has been used in Oman since 1976 when the Al-Ghubrah (co-location) power and seawater desalination plant using a thermal technology of multi-stage flash (MSF) was first commissioned in Muscat. To meet continuously growing water demand due to population growth and economic and social development and to reduce the reliance on groundwater resources, by 1999 the Al-Ghubrah plant had seven MSF desalination units installed. The first seawater desalination unit installed had a capacity of 22,750 m^3/d, and the other six MSF units each have a capacity of 27,000 m^3/d. Desalinated water usage in Oman is expected to increase further in the future, due to new industrial and tourism-related developments.

Desalination plants extract large volumes of seawater and discharge hot, hypersaline brine back into the marine environment. Therefore, the main concern of continuous brine discharges has been the potential impact upon the salinity of seawater (and possible thermal stress for discharges from MSF plants), and the resultant effects to marine communities around discharge outlets. Other occasional discharges from the plants include corrosion products, toxic antifoulants and antiscalants used in maintaining plant infrastructure (Roberts et al., 2010). As brine discharges are often denser and heavier than receiving marine waters, the brine streams tend to sink and spread further along the seabed than at the surface, and thus it has a greater exposure to benthic organisms than that of pelagic and planktonic organisms.

In order to demonstrate the compliance with the regulations for discharging brines from modern and recently operated Barka desalination plants in the Omani marine environment (MRMEWR, 2005), CORMIX simulations are carried out for two scenarios: scenario I uses temperature as a measure of the positively buoyant plume concentration to simulate the previous heated brine discharges from Barka I plant (up to November 2009), and scenario II uses salinity as a measure of the negatively buoyant plume concentration to simulate the new dense brine discharges from the combined Barka I and II plants (after 2009). In contrast to scenario I, where the simulated plume rises towards the surface, the simulated brine plume for scenario II should sink and stay at the seabed. Due to inherent uncertainty in the input data, sensitivity analysis were also carried out for scenario II using both submerged single port and multiport discharges by varying the ambient current velocity (to evaluate the effect of uncertainty in sea conditions), the brine discharge density (to evaluate the effect of uncertainty on the brine characteristic) and the brine flow rate (to evaluate the effect of uncertainty on the desalination plant's operation).

Fig. 1. Seawater desalination plants in the Gulf of Oman

2. Outlook for seawater desalination in Oman

To manage the future water demands in Oman, the Oman Power and Water Procurement (OPWP) company was established in 2005, and one of its responsibilities is to procure the production of desalinated water in conjunction with electricity and to ensure the adequacy of generation of resources for new desalination capacity. Based on the annual report issued by OPWP in 2009, the forecasted total demand for the desalinated water is expected to increase from 259.5 million m^3 in 2010 to 405.2 million by 2016, an average annual increase of 13% per year. OPWP has forecasted also that by 2016 (Table 1), the additional 32,000 m^3/d of water desalination capacity is needed.

The desalinated water in the capital Muscat area is supplied mainly by Al-Ghubrah and Barka plants (Fig. 1). Due to proximity to demand, availability of land and infrastructure, both Al-Ghubrah and Barka plants sites are the preferred locations for additional power and seawater desalination capacity. The site for the Barka plants has been designated as the location for the construction of up to four (co-location) power generation and seawater desalination plants. Barka I was the first plant to be built, operated in 2003 and owned by the private sector, AES Barka. The power plant has a capacity of 427 MW and the (MSF) seawater desalination plant has a capacity of 91000 m^3/d. Adjacent to Barka I, Barka II plant was recently commissioned in November 2009, and owned by the private sector, SMN Barka Power company; the power plant with a capacity of 685 MW and the seawater desalination plant using a membrane technology of reverse osmosis (RO) with a capacity of 120000 m^3/d.

	2010	2011	2012	2013	2014	2015	2016
Forecasted water demand (in thousand m^3/d)	**711**	**782**	**850**	**925**	**1020**	**1094**	**1110**
Al-Ghubrah East plant	205	205	205	161	138	138	138
Al-Ghubrah West plant				135	135	135	135
Barka I + II plants	211	211	211	211	211	211	211
Barka III plant					180	180	180
Sohar plant	150	150	150	150	150	150	150
Sohar II plant					135	135	135
Sur I + II plants	80	80	80	80	80	80	80
Sur III plant						23	23
Other smaller plants	14	16	19	22	22	22	22
Salalah (MSF) plant		68	68	68	68	68	68
Total desalination capacity (in thousand m^3/d)	**660**	**730**	**733**	**827**	**1119**	**1142**	**1142**

Table 1. Projected water demand and total seawater desalination capacity in Oman

From 2012 onward, the reduction in desalination capacity is mainly due to the retirement of old MSF desalination units at the Al-Ghubrah plant. In order to meet the future water demands, OPWP has launched a new scheme to expand/redevelop the Al-Ghubrah plant by splitting the existing plant site into two parts. The existing plant will be called Al-Ghubrah East, and the new independent water and power project Al-Ghubrah West with a total desalination capacity of 135,000 m^3/d will be built and operated by a private sector by 2013. A new additional of independent water and power project at Barka with a total desalination capacity of 180,000 m^3/d will also be built and operated by a private sector by 2014.

Sohar (co-location) power and seawater desalination plant was the second plant to be built, operated in 2007 using MSF technology and owned by the private sector, Sohar Power company. To meet the additional water desalination requirements, a new independent power and water project at Sohar with a total desalination capacity of 135,000 m^3/d will be built and operated by a private sector by 2014.

Neither of the RO seawater desalination plants at Sur are co-generation plants and they are owned by Public Authority for Electricity and Water. The first plant with a capacity of 12,000 m^3/d was built in 1993, and the second plant with a capacity of 68,000 m^3/d was built in 2009. A new desalination plant at Sur with a total capacity of 23,000 m^3/d will need to be built by 2015 in order to meet the future water demand.

Finally, the first Salalah (co-location) power and seawater desalination plant with a total desalination capacity of 68,000 m^3/d will be built, and operated by a private sector by 2011. However, to meet the future water demands, a new additional seawater desalination plant with a capacity of 80,000 m^3/d will need to be built by 2016.

3. Environmental standards for effluent discharges into Omani coastal waters

Brine discharges from coastal seawater desalination plants in Oman are regulated according to the Omani ministerial decision No. 159/2005 (MRMEWR, 2005) on "promulgating the bylaws to discharge liquid waste in the marine environment", where the term liquid waste is defined as "any liquid containing environmental pollutants discharged into the marine environment from land or sea sources". To regulate such discharges, Article 5 states that "no liquid waste shall be directly or indirectly discharged in the marine environment without obtaining prior license". The license is issued by the Department of Inspection and Environment Control and depends on the following conditions as further measures to reduce the potential environmental impacts. Firstly, as stated in Article 7, the desalination plant operators must undertake to reuse or recycle the liquid waste, or destroy hazardous contents of such waste, or mitigate it by using proper environmental treatment.

Secondly, the plant operators have to provide a detailed description and the description of the characteristics of the liquid waste (Article 8) and the quality of liquid waste has to conform to the discharge limits specified in Annex 1 (Article 9). Besides the discharge limits, a circular mixing zone of 300 m in diameter around the outfall is specified as the initial mitigating area. Within the mixing zone, no marine life at the seabed may be destroyed (Article 14). Beyond the mixing zone, the ambient water temperature must not be increased by more than 1 ^oC (weekly average), and the average ambient salinity must not be changed by more than 2 ppt (daily average). A mixing zone can conceptually be defined as a limited area or volume of water where initial dilution of a discharge takes place.

Thirdly, the plant operators have to also provide information about the discharge location, such as physical, chemical and biological characteristics of seawater and recreational areas or other usages of the concerned shoreline (Article 10). The outfall pipes must not be installed less than one meter from the lowest tide line (Article 11). The temperature of liquid waste at the discharge point should not exceed 10 ^oC over the temperature of the water surrounding the seawater intake, if any (Article 2). The discharge pipes must be located in a place where it is impossible for the waste plume to hit corals and seaweed at the bottom (Article 3).

For the selection of the discharge site and the construction of the outfall, information about wind speed and direction for one month, low and high tide currents in an area of 1 km around the outfall and the average sea depth in the same area should be included (Article 13). Multiport diffuser pipes are recommended to be installed in order to improve the brine dilution. Finally, for those violating any of these regulations, the penalties of the Environment Protection and Pollution Control Law shall be applied (Article 17).

4. Cornell mixing zone model expert (CORMIX) system

CORMIX (www.cormix.info) is a United States Environmental Protection Agency (US EPA) approved software system for the analysis, prediction, and design of marine outfall mixing zones resulting from continuous point source discharge of aqueous pollutants into diverse water bodies (Doneker & Jirka, 1990; Akar & Jirka, 1991; Jirka & Akar, 1991; Jirka & Doneker, 1991; Del Bene et al., 1994; Jones et al., 1996). It employs an easy-to-use rule-based expert system to screen input data and check for consistency, analyze regulatory requirements, and select the appropriate hydrodynamic computational model to simulate the physical mixing processes likely to be present for flow within a given discharge-environment interaction, ranging from internally trapped plumes, buoyant plumes in uniform density layers, and sinking of negatively-buoyant plumes. Boundary interaction, upstream intrusion, buoyant spreading, and passive diffusion in the far field are also considered.

The hydrodynamic classification schemes in CORMIX system use the length scale concepts, as a measure of the influence of each potential mixing process due to momentum flux and buoyancy of the discharge in relation to boundary interactions, to predict steady-state mixing zone characteristics and plume geometry. Boundary interaction analysis on mixing processes from laboratory and field experiments provide a rigorous and robust expert knowledge base that distinguishes among the many hydrodynamic flow patterns that may occur. For every flow class, CORMIX assembles and executes a sequence of appropriate hydrodynamic computational modules to predict the trajectory and dilution characteristics of a complex flow. Efficient computational algorithms provide simulation results in seconds for mixing zone problems with space scales of meters to kilometers and time scales of seconds to hours. The governing equations and formulations used in CORMIX system have been reported elsewhere (Jirka & Akar, 1991; Jirka & Doneker, 1991; Mendez-Diaz & Jirka, 1996; Doneker et al., 2004; Jirka, 2004; Jirka, 2006).

CORMIX is designed to analyze water quality criteria within regulatory mixing zones and has been successfully applied by engineers and environmental scientists to the design and monitoring of wastewater disposal systems in ocean, rivers, lakes, and estuaries, and it is also recognized by regulatory authorities for environmental impact assessment (US EPA, 1999). Extensive comparison with available field and laboratory data (including the negatively buoyant discharges) has shown that the CORMIX system predictions on plume dilutions and concentrations (with associated plume geometries) are reliable for the majority of cases (Roberts & Toms, 1987; Pun & Davidson, 1999; Kang et al., 2000; Roberts & Tian, 2004; Etemad-Shahidi & Azimi, 2005; Azimi et al., 2005; Kikkert et al., 2007). CORMIX simulation outputs include contemporary three dimensional plume and diffuser visualizations, design recommendations, flow class descriptions and reporting oriented on mixing zone analysis and regulatory compliance (Doneker & Jirka, 2001; Doneker & Jirka, 2007). CORMIX model has also been used to simulate heated brine discharges (Del Bene et al., 1994; Alameddine & El-Fadel, 2007), and the results of CORMIX simulations of dense brine discharges from the submerged single port and multiport diffusers are reported here.

Parameter	Scenario I	Scenario II	
Outfall type	Single	Single	Multiport
Ambient (unbounded coastal environmental)			
Velocity of the currents (m/s)	0.2	0.3	
Depth at discharge (m)	9	9	
Bottom slope (degree)	-	0.8944	
Wind speed (m/s)	2.5	2.5	
Temperature (°C)	24	24	
Salinity (ppt)	36	36	
(Uniform) Density (kg/m³)	1024.4	1024.4	
Brine discharge			
Distance to shoreline (m)	576.5	576.5	
Diffuser length (m)	-	-	91
Number of ports	1	1	36
Port height (m)	1	1	
Port diameter (m)	0.7	0.7	
Vertical angle (degree)	10	10	
Horizontal angle (degree)	90	90	
Flow rate (m³/s)	0.474	0.942	
Temperature (°C above ambient)	9	6	
Salinity (ppt above ambient)	2	13	
(Uniform) Density (kg/m³)	1022.91	1032.27	
Mixing zone			
WQS = Water quality standard (°C)	1	-	
WQS = Water quality standard (ppt)	-	2	
RMZ = Regulatory mixing zone (m)	150	150	
ROI = Region of interest (m)	1000	1000	

Table 2. Input data for the CORMIX simulations

5. Model simulations of brine discharges from Barka plants

Barka is situated on the coastal plain near to the Gulf of Oman, an agricultural area that has been suffering from seawater intrusion due to excessively pumped out groundwater usage for irrigation (Purnama et al., 2003). The surface land features can be described as flat with a sandy strip parallel to the coastline approximately 300 m inland. Ground surface elevations vary typically from 1.5 to 5 m above mean sea level. From the regional bathymetry of the Barka coastline (National Hydrographic Office, 2008), the gradient is about 1:140 up to the 10 m depth contour and about 1:220 between the 10 m and 30 m depth contours. Near the plant site, the water depth reaches 5 m at about 500 m offshore and 10 m at about 2000 m offshore. The general flow in the ocean currents in the Gulf of Oman is in the anticlockwise direction, that is, along the northern coast of Oman, the predominant coastal current moves southeastward from the Gulf of Oman (Purnama & Al-Barwani, 2006), with maxima at spring tides of up to 0.42 m/s. Tides in the Gulf of Oman have a strong diurnal component with a spring-tide range of 2.6 m or more (National Hydrographic Office, 2008).

There are two power generation and seawater desalination plants currently operated at Barka: Barka I was commissioned in 2003, and Barka II in 2009. There are four intake and marine outfall systems constructed as part of Barka I plant facilities. So far, only two seawater intake and marine outfall systems are used for both Barka I and Barka II plants. The intake system consists of four parallel pipes of 1.2 km in length and a diameter of 2.2 m. The pipes are spaced 2 m apart, buried under the seabed (not visible on the surface) and the intake structure opens at 1.5 m above the seabed. Barka I plant withdraws seawater from a depth of 10 m up to a maximum rate of 67500 m^3/h, and Barka II plant at a rate of 59000 m^3/h for cooling purposes. To avoid the circulation of concentrated brine discharges to the intake system, the outfall discharge point is constructed at a distance of 800 m from the intake point (Abdul-Wahab, 2007).

The marine outfall systems are designed to discharge the combined brine reject (and other effluents) from the desalination plants and the once through condenser cooling water system from the power plant. It also comprises of four parallel pipes angled at 62 degrees to the beach, each with a diameter of 2.5 m, buried at 5 m below the seabed and spaced equally at 4.8 m apart. The 62.4 m long staged multiport diffuser, consisting of nine ports, are installed at the end of each outfall pipe. The diffusers are arranged in two nested V shapes, and each pair diverges at an angle of 30 degrees on either side of the outfall pipelines. The two internal pipes of length 653 m have its end at a depth of 9 m, while the other two shorter external pipes of length 582 m end at a depth of 8.4 m. Each port with a diameter 0.7 m opens up at 1 m above the seabed, and the jet brine stream is discharged up at an angle of 10 degrees above the horizontal. Barka I plant discharges up to a maximum flow rate of 61500 m^3/h, and Barka II plant discharges up to a maximum flow rate of 60600 m^3/h.

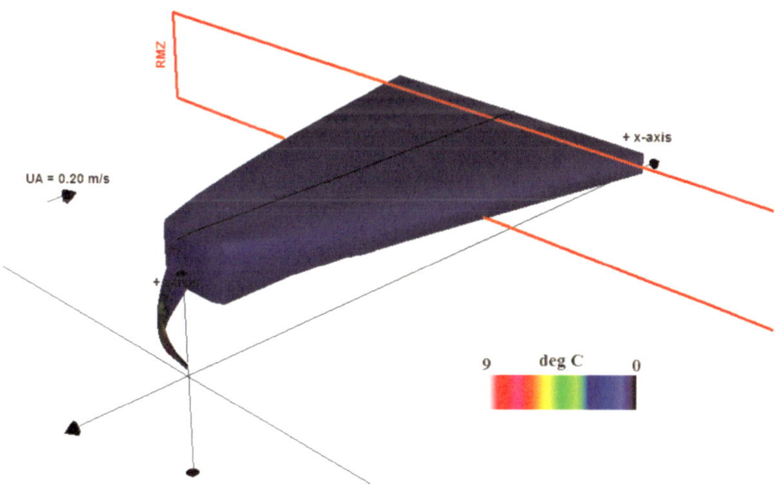

Fig. 2. Scenario I brine plume from the single port discharge

The mixing processes of brine discharges into marine coastal waters are significantly influenced by the effluent density, which is controlled by salinity and temperature of the effluent (Bleninger & Jirka, 2008). The brine of MSF desalination plants is extremely hot, and must be diluted with cooling water from the power generation plants. As a result, the brine discharges are lighter than the receiving seawater. In contrast, the brine of RO desalination plants is higher in salinity, and such discharges are heavier than the receiving waters. During 2009, the physical properties of brine discharge from Barka plants are changing from the previously positively buoyant plume (from Barka I only) to a negatively buoyant plume (from combined Barka I and II). In order to assess the compliance with the regulations for discharging effluents in the Omani marine environment, CORMIX v6.0 simulations are carried out for two scenarios of brine discharges from the Barka plants. The input data are summarized in Table 2. The outfall discharge point for the single port and the midpoint of diffuser line for the multiport are located at 576.5 m offshore, and 9 m below the sea surface. Since the height of the ports is 1 m above the seabed, according to CORMIX, this is a deeply submerged discharge.

Outfall type	Single port
Temperature (in °C)	0.2
Plume dilution	43.9
RMZ centerline (in m)	$x = 150$, $y = 10.85$, $z = 9$
RMZ plume dimensions (in m)	Half-width = 21.60 Thickness = 2.41

Table 3. The RMZ characteristics of scenario I

5.1 Scenario I : Heated brine discharges from Barka I plant (up to November 2009)

This scenario is used to simulate the brine discharges of 9 °C above ambient on an unbounded flat bed coastal environment with a surface heat loss rate of 30 W/m²/°C. The origin is located at the seabed, directly below the outfall, and 576.5 m offshore. The x-axis points downstream, the y-axis points to left (in the flow direction), and the z-axis points upward. As the effluent density 1022.91 kg/m³ is less than the surrounding ambient density 1024.4 kg/m³, the effluent plume is positively buoyant and will rise to the surface. The maximum permissible temperature limits set by the Omani government is WQS = 1 °C above ambient within the RMZ of 150 m radius from the outfall (MRMEWR, 2005).

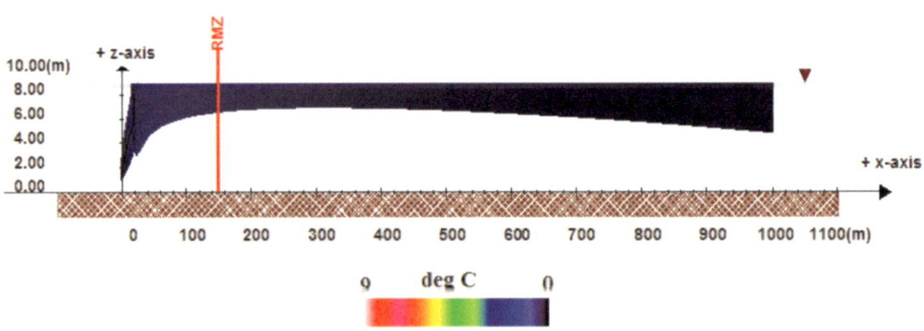

Fig. 3. Side view of scenario I brine plume from the single port discharge

CORMIX system classified the motion of the brine plume from the single port discharge as the flow class H2 near bottom (horizontal), positively buoyant flows in a uniform density layer (Jirka & Doneker, 1991; Doneker & Jirka, 2007). The jet-like plume is weakly deflected by the ambient current into the flow direction, and the bent-over brine plume rises towards the surface (Fig. 2). After reaching the surface within 17 m downstream, the plume spreads laterally due to buoyant ambient spreading, and its thickness (measured vertically) is observed to be decreasing initially (Fig. 3), the lower part of the plume reaches its maximum height within 300 m downstream. Further downstream, as the ambient flow is the dominating mixing mechanism, the passive brine plume grows back in thickness, and it eventually becomes vertically fully mixed beyond the ROI, 1000 m downstream as the bottom plume attaches to the seabed; after that the plume collapses and continues to spread due to passive ambient mixing in uniform ambient.

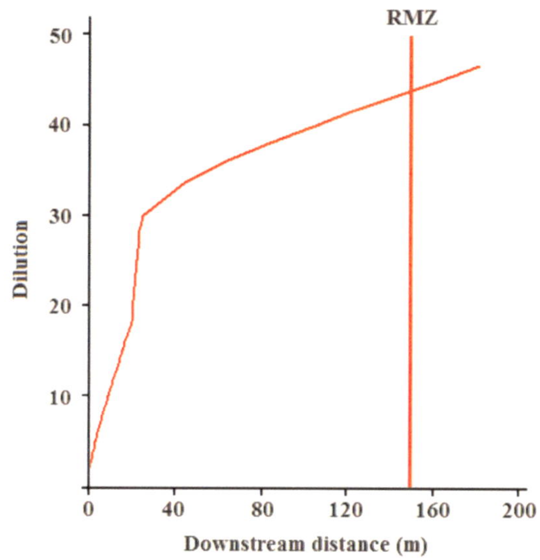

Fig. 4. Brine plume dilution of scenario I

The specified WQS has been met within 8 m downstream, and the plume characteristics at the edge of the RMZ are presented in Table 3. The mixing performance (concentration reduction) is measured in the CORMIX system as brine plume dilution, which is defined as the ratio of initial brine concentration (above ambient) at the outfall discharge point to that at a given location. The increasing values of hydrodynamic brine plume dilution within the RMZ are shown in Fig. 4, and a dilution of 271 is obtained at the end of the ROI. The CORMIX simulation results show that the overall temperature rise at the edge of the RMZ is slightly more than 0.2 °C (above ambient), which is well below the maximum permissible limits by the Omani government. This appears to be in agreement with the field survey conducted in the vicinity of Barka I plant (Abdul-Wahab, 2007), and also with the heavy metals analysis of the bottom sediments sample collected (Abdul-Wahab & Jupp, 2009), which is consistent with the assumption that the brine plume rises to the surface.

5.2 Scenario II : Dense brine discharges from Barka plants (after 2009)

This scenario is used to simulate the brine discharges of salinity 13 ppt above ambient (set to a concentration of 100% above ambient) on the uniformly sloping beach coastal environment with slope 0.9 degree. In contrast to scenario I, the effluent density 1032.27 kg/m^3 is denser and heavier than the surrounding ambient density 1024.4 kg/m^3, the effluent plume is negatively buoyant and it will tend to sink at the seabed. The origin is located at the sea surface directly above the submerged outfall, and 576.5 m offshore. The maximum permissible salinity limits set by the Omani government is WQS = 2 ppt above ambient within the RMZ of 150 m radius from the outfall (MRMEWR, 2005).

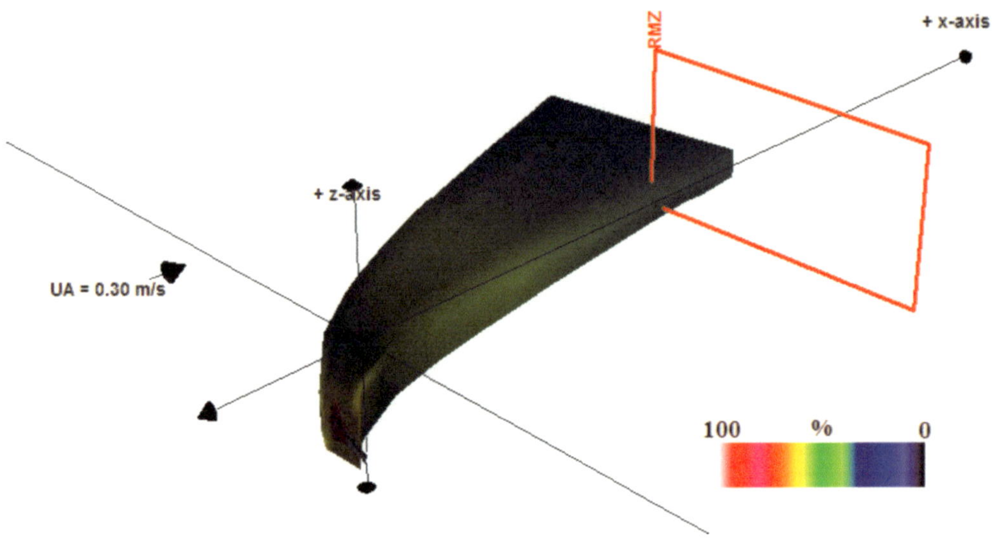

Fig. 5. Scenario II brine plume from the single port discharge

CORMIX system classified the brine plume motion from the single port discharge as the flow class NH4 near bottom (horizontal), negatively buoyant flows in a uniform density layer (Jirka, 2004; Doneker & Jirka, 2007). As the brine stream is released from the outfall, the ambient current strongly deflects the jet-like plume trajectory into the current direction, inducing higher dilution. The bent-over brine plume slowly descends toward the sloping bed, it impinges on the seabed within 1.3 m downstream, and the plume remains at the seabed due to its negative buoyancy. Impingement is a complex three-dimensional process, with forward, lateral, and partially reverses spreading, until a bottom density current is formed. The height of the top part of the plume slowly rises and reaches its maximum within 26 m downstream, and the concentration distribution becomes relatively uniform across the plume width and thickness (Fig. 5). Thereafter, it continues to spread laterally due to bottom density current while it is being advected by the ambient current (Bleninger & Jirka, 2008). In the absence of ambient stratification, the brine plume will propagate down the slope until it reaches the ROI (Fig. 6), and thus, the potential benthic impact due to brine plume should be of concern.

Outfall type	Single port	Multiport
Plume concentration	0.6 ppt	0.056 ppt
Plume dilution	21.6	234
RMZ centerline (in m)	$x = 150$, $y = 45.50$, $z = -9.69$	$x = 150$, $y = 1.62$, $z = -9.03$
RMZ plume dimensions (in m)	Half-width = 14.24 Thickness = 2.99	Half-width = 47.03 Thickness = 9.01

Table 4. The RMZ characteristics of scenario II

Again, the specified WQS has been met within 7 m downstream, and the plume characteristics are presented in Table 4. The increasing values of brine plume dilution with the RMZ are plotted in Fig. 7, and a dilution of 30.9 is obtained at the edge of the ROI for the single port discharge. The CORMIX simulation results show that the overall salinity increase is less than 0.6 ppt (above ambient) at the edge of the RMZ, which is well below the maximum permissible limits by the Oman government.

For the submerged multiport, the brine discharge volume is distributed over 36 ports, which are placed in the alternating arrangement perpendicular to the diffuser line, and the average spacing between the individual ports is 2.6 m. Due to the lateral merging and interactions of adjacent brine plumes, forming a dynamically equivalent two-dimensional plume as if the discharges are made from a two-dimensional slot diffuser and that the downstream plume behaviour after merging is independent of the port arrangement. CORMIX system classifies the motion of the brine plume from the multiport discharge as the flow class MNU13 near bottom, negatively buoyant flows in a uniform density layer (Jirka, 2006; Doneker & Jirka, 2007). As CORMIX assumes the complex merging process of the individual jets from each port as a plume releases from a long slot discharge, the merging jet-like plume is rapidly deflected by the strong ambient current, and due to instability, it immediately becomes vertically fully mixed as it leaves the multiport, descending toward the sloping bottom. After it is attached to the seabed, it continues to spread laterally due to the bottom density current and in the absence of ambient stratification will proceed down the slope until it has reached the ROI. As shown in Fig. 7, in comparison with the single port, a ten-fold brine plume dilution is achieved for the multiport discharges at the edge of the RMZ.

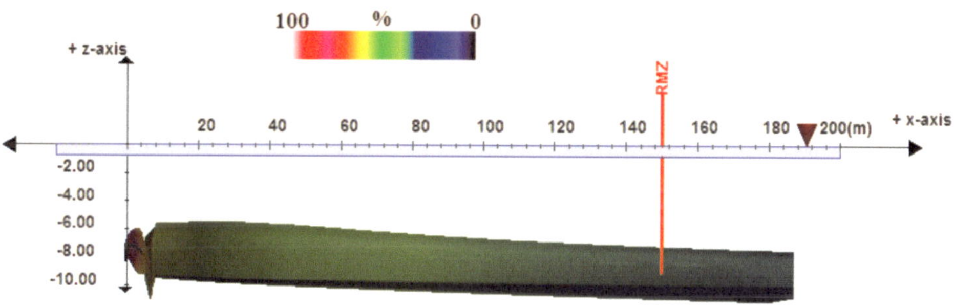

Fig. 6. Side view of scenario II brine plume from the single port discharge

6. CORMIX iterative simulations

CorSens is the CORMIX sensitivity analysis tool that generates a sensitivity study case to address model performance due to inherent uncertainty in the input data (Doneker & Jirka, 2007; Alameddine & El-Fadel, 2007). Therefore, instead of the tedious repetition of manual data entry, it automatically increments input data to analyze mixing zone conditions. Sensitivity studies are also motivated by the fact that there are no user-adjustable parameters for model calibration within CORMIX system. The basis for this restriction is that normal variations in ambient conditions are likely to have greater influence over mixing zone behavior than a model parameter to obtain a desired result. Only the scenario II simulations are carried out to represent dense brine discharges from Barka plants after 2009. Firstly, to evaluate the effect of uncertainty in sea conditions, simulations were conducted by varying the ambient current velocity (a coastal environment parameter) while keeping the other input parameters the same as the base (Section 5.2) simulation specified in Table 2. The results summary of reducing ambient velocity from 0.45 m/s to 0.05 m/s are presented in Table 5, and for the single port discharges, it is observed that flow class changes from

NH4 to a new class NH5: momentum dominated bottom-attached jet motion (Jirka, 2004; Doneker & Jirka, 2007). For calm sea conditions, i.e. the ambient velocity smaller than 0.25 m/s, the discharge induced momentum flux dominates the flow, and due to instability, the brine plume becomes attached to the sloping seabed and vertically fully mixed as it leaves the port. The ambient stratification also occurs when the ambient velocity is less than 0.1 m/s, and for the smallest value of 0.05 m/s, the stratification occurs within 29 m downstream. CORMIX also predicts the occurrence of localized recirculation into the jet-plume like regions, thus blocking the ambient current and reducing the plume dilution.

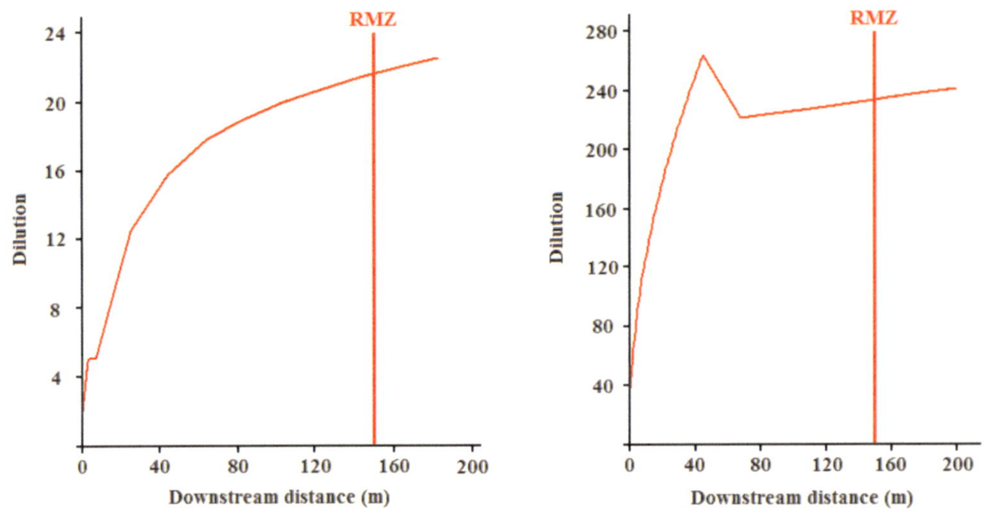

Fig. 7. Brine plume dilution of scenario II: single port (left) and multiport (right) discharges

At the edge of the RMZ, plume dilution values are much greater than the base dilution for the single port discharges, and for the ambient velocity 0.01 m/s, dilution is more than three times the base value. Therefore, the uncertainty in sea conditions may result in the overall salinity increase to be less than the base value of 0.6 ppt (above ambient).

For the multiport discharges, it is observed that by reducing the ambient velocity, the plume dilution decreases. The bottom density current develops along the diffuser line due to the continuous inflow of mixed buoyant water becoming stronger, reducing the mixing rate and thus the plume dilution. The flow class changes from MNU13 to MNU2 (Jirka, 2006; Doneker & Jirka, 2007) occur when the velocity value equals to 0.05 m/s, where the momentum flux is now weak relative to the buoyancy flux. The ambient stratification also occurs when the ambient velocity is less than 0.1 m/s, and for the smallest value of 0.05 m/s, the bottom stratification occurs within 23.44 m downstream. It is found that the uncertainty in sea conditions may result in the overall salinity increase of 0.4 ppt (above ambient) at the edge of the RMZ.

Next, the simulations were conducted by varying the discharge density (an effluent parameter), which reflect the uncertainty on the brine characteristic. Salinity and temperature directly influence the density of the effluent. The results summary of reducing discharge density from 1034.27 kg/m^3 to 1026.27 kg/m^3 are given in Table 6, showing that there are no flow class changes for the single and multiport discharges. At the edge of the

RMZ, plume dilution increases to more than double for the single port discharges since the stronger momentum with relatively weak buoyancy controls the flow class NH4, but the same effect is minor for the multiport discharges. Thus, the uncertainty on the brine characteristic may result in the overall salinity increase being less than 0.65 ppt (above ambient).

Outfall type	Single port			Multiport		
Ambient velocity (m/s)	Flow Class	Dilution	Salinity (ppt)	Flow Class	Dilution	Salinity (ppt)
0.45	NH4	29.7	0.438	MNU13	349	0.038
0.4	NH4	26.8	0.486	MNU13	311	0.042
0.35	NH4	23.8	0.545	MNU13	272	0.048
0.3	NH4	21.6	0.602	MNU13	234	0.056
0.25	NH5	80.9	0.161	MNU13	196	0.066
0.2	NH5	84.1	0.155	MNU13	158	0.082
0.15	NH5	82.9	0.157	MNU13	101	0.129
0.1	NH5	69.2	0.189	MNU13	72.2	0.181
0.05	NH5	62.7	0.207	MNU2	36.0	0.360

Table 5. CorSens results summary on the ambient velocity at the edge of the RMZ

Outfall type	Single port			Multiport		
Discharge density (kg/m^3)	Flow Class	Dilution	Salinity (ppt)	Flow Class	Dilution	Salinity (ppt)
1034.27	NH4	20.1	0.645	MNU13	234	0.0555
1032.27	NH4	21.6	0.602	MNU13	234	0.0555
1030.27	NH4	25.4	0.512	MNU13	234	0.0556
1028.27	NH4	32.6	0.399	MNU13	233	0.0558
1026.27	NH4	46.3	0.281	MNU13	232	0.0559

Table 6. CorSens results summary on the discharge density at the edge of the RMZ

Lastly, the simulations were carried out by varying the brine flow rate (a discharge parameter), which reflect the uncertainty on the plant's operation. During the winter months, Barka I plant operates with 60% load by shutting down one of the three desalination units for maintenance. The results summary of increasing discharge rate from 0.542 m^3/s to 1.242 m^3/s are presented in Table 7. For the single port discharges, the brine flow rates lower than the base rate lead to smaller plume dilution due to less momentum flux controlling the flow class NH4. The flow class changes to NH5 occur at higher flow rates than the base rate, and thus, larger momentum flux produces higher plume dilution. However in contrast, for the multiport discharges there are no flow class changes, and plume dilution decreases. It is found that the uncertainty in the brine discharge rate may result in the overall salinity increase of 0.9 ppt (above ambient) at the edge of the RMZ.

Outfall type	Single port			Multiport		
Discharge flow rate (m^3/s)	Flow Class	Dilution	Salinity (ppt)	Flow Class	Dilution	Salinity (ppt)
0.542	NH4	16.2	0.802	MNU13	405	0.0321
0.642	NH4	17.6	0.737	MNU13	343	0.0380
0.742	NH4	19.0	0.685	MNU13	297	0.0438
0.842	NH4	20.4	0.638	MNU13	262	0.0497
0.942	NH4	21.6	0.602	MNU13	234	0.0555
1.042	NH5	23.0	0.566	MNU13	212	0.0614
1.142	NH5	49.8	0.261	MNU13	193	0.0672
1.242	NH5	51.5	0.252	MNU13	178	0.0731

Table 7. CorSens results summary on the discharge flow rate at the edge of the RMZ

7. Conclusion

CORMIX simulations for submerged single port and multiport discharges were carried out for two scenarios to assess the compliance of brine discharge from Barka plants within the regulations for discharging effluents in the Omani marine environment (MRMEWR, 2005). The results show that, up to November 2009, the overall temperature rise due to heated brine discharges from Barka I plant is found to be around 0.2 °C (above ambient) within the regulatory mixing zone of 150 m radius from the outfall. Similarly, since November 2009, the overall salinity increase due to dense brine discharges from the combined Barka I and II plants is found to be less than 1 ppt (above ambient). These values are well below the maximum permissible limits set by the Omani government, which are respectively 1 °C and 2 ppt above ambient.

It is also observed that there is a change in the Barka brine discharge characteristic, from a positively buoyant plume to a negatively buoyant plume. In contrast to the positively buoyant plume that will rise towards the surface, the new negatively buoyant plume will sink and attach to the sloping seabed and then spreads due to the bottom density current downslope, and therefore the potential exposure of benthic organisms should also be monitored and investigated further.

Like other numerical models, the CORMIX system has several inherent limitations. One major limitation results from the representation of the coastal environment as the unbounded rectangular channel, and as a uniformly sloping cross-section channel for concentrated brine discharges, where the current velocity is assumed to be uniform. Another limitation is the flow classification system (without even starting a numerical computation) based on hydrodynamic criteria using significant length scale analysis to predict mixing processes and its subsequent dilution in the receiving water environment. Thus, as shown in Table 5, a small change in an input parameter may result in a different CORMIX flow class leading to marked discontinuities in the plume dilutions.

8. Acknowledgements

The authors would like to thank the Middle East Desalination Research Center for its financial support of the project on the environmental planning, prediction and management of brine discharges from desalination plants (http://www.brinedis.net.ms). The authors also very much indebted to Drs. T. Bleninger and R.L. Doneker for their guidance and technical assistance in CORMIX.

9. References

Abdul-Wahab, S.A. (2007). Characterization of water discharges from two thermal power/desalination plants in Oman. *Environmental Engineering and Science*, Vol.24, pp. 321–337.

Abdul-Wahab, S.A. & Jupp, B.P. (2009). Levels of heavy metals in subtidal sediments in the vicinity of thermal power/desalination plants: A case study. *Desalination*, Vol.244, pp. 261–282.

Akar, P.J. & Jirka, G.H. (1991). *CORMIX2: An expert system for hydrodynamic mixing zone analysis of conventional and toxic submerged multiport diffuser discharges*. Technical Report EPA/600/3-91/073, US EPA, Environmental Research Laboratory, Athens, Georgia, USA.

Alameddine, I. & El-Fadel, M. (2007). Brine discharge from desalination plants: a modeling approach to an optimized outfall design. *Desalination*, Vol.214, pp. 241–260.

Azimi, A.H.; Azizi, F. & Etemad-Shahidi, A. (2005). A new formulation to estimate initial dilution of dense jet flow. *Proceedings of the XXXI IAHR Congress*, pp. 4241-4248, Seoul, Korea.

Bleninger, T. & Jirka, G.H. (2008). Modelling and environmentally sound management of brine discharges from desalination plants. *Desalination*, Vol.221, pp. 585-597.

Del Bene, J.V.; Jirka, G.H. & Largier, J. (1994). Ocean brine disposal. *Desalination*, Vol.97, pp. 365–372.

Doneker, R.L. & Jirka, G.H. (1990). *CORMIX1: An expert system for hydrodynamic mixing zone analysis of conventional and toxic submerged single port discharges*. Technical Report EPA/600/3-90/012, US EPA, Environmental Research Laboratory, Athens, Georgia, USA.

Doneker, R.L. & Jirka, G.H. (2001). CORMIX-GI systems for mixing zone analysis of brine wastewater disposal. *Desalination*, Vol.139, pp. 263–274.

Doneker, R.L.; Nash, J.D. & Jirka, G.H. (2004). Pollutant transport and mixing zone simulation of sediment density currents. *Journal of Hydraulic Engineering*, Vol.130, pp. 349-359.

Doneker, R.L. & Jirka, G.H. (2007). *CORMIX user manual: A hydrodynamic mixing zone model and decision support system for pollutant discharges into surface waters*. MixZon Inc., Portland, Oregon, USA. (available at http://www.mixzon.com and http://www.cormix.info)

Etemad-Shahidi, A. & Azimi, A.H. (2005). Performance of CORMIX model applied to dense jet flow. *Proceedings of the XXXI IAHR Congress*, pp. 4211-4218, Seoul, Korea.

Jirka, G.H. & Akar, P.J. (1991). Hydrodynamic classification of submerged multiport diffuser discharges. *Journal of Hydraulic Engineering*, Vol.117, pp. 1113-1128.

Jirka, G.H. & Doneker, R.L. (1991). Hydrodynamic classification of submerged single port discharges. *Journal of Hydraulic Engineering*, Vol.117, pp. 1095-1112.

Jirka, G.H. (2004). Integral model for turbulent buoyant jets in unbounded stratified flows. Part 1: Single round jet. *Environmental Fluid Mechanics*, Vol.4, pp. 1-56.

Jirka, G.H. (2006). Integral model for turbulent buoyant jets in unbounded stratified flows. Part 2: Plan jet dynamics resulting from multiport diffuser jets. *Environmental Fluid Mechanics*, Vol.6, pp. 43-100.

Jones, G.R.; Nash, J.D. & Jirka, G.H. (1996). *CORMIX3: An expert system for the analysis and prediction of buoyant surface discharges*. Technical Report CR818527, De Frees

Hydraulics Laboratory, School of Civil and Environmental Engineering, Ithaca, New York, USA.

Kang, S.W.; You, S.H. & Na, J.Y. (2000). Near-field mixing characteristics of submerged effluent discharges into Masan Bay. *Ocean Research*, Vol.22, pp. 45-56.

Kikkert, G.A.; Davidson, M.J. & Noakes, R.I. (2007). Inclined negatively buoyant discharges. *Journal of Hydraulic Engineering*, Vol.133, pp. 545-554.

Kwarteng, A.Y.; Dorvlo, A.S. & Vijaya Kumar, G.T. (2009). Analysis of a 27-year rainfall data (1977–2003) in the Sultanate of Oman. *International Journal of Climatology*, Vol.29, pp. 605–617.

Mendez-Diaz, M.M. & Jirka, G.H. (1996). Buoyant plumes from multiport diffuser discharges in deep coflowing water. *Journal of Hydraulic Engineering*, Vol.122, pp.428-435.

Ministry of Regional Municipalities, Environment and Water Resources (MRMEWR). (2005). *Ministerial Decision No. 159/2005: Promulgating the bylaws to discharge liquid waste in the marine environment*. Minister's Office, Muscat, Sultanate of Oman.

National Hydrographic Office. (2008). *Oman Maritime Book 2009*. Royal Navy of Oman, Muscat, Sultanate of Oman.

Oman Power and Water Procurement. (2009). *OPWP's 7-year Statement (2010-2016)*. Muscat, Sultanate of Oman. (available at http://www.omanpwp.co.om/7yearstatement-1016.pdf)

Pun, K.L. & Davidson, M.L. (1999). On the behavior of advected plumes and thermals. *Journal of Hydraulic Research*, Vol.37, pp. 519-540.

Purnama, A.; Al-Barwani, H.H. & Al-Lawatia, M. (2003). Modeling dispersion of brine waste from a coastal desalination plant. *Desalination*, Vol.155, pp. 41–47.

Purnama, A. & Al-Barwani, H.H. (2006). Spreading of brine waste discharges into the Gulf of Oman. *Desalination*, Vol.195, pp. 26–31.

Roberts, D.A.; Johnston, E.L. & Knott, N.A. (2010). Impacts of desalination plant discharges on the marine environment: A critical review of published studies. *Water Research*, Vol.44, pp. 5117-5128.

Roberts, P.J.W. & Toms, G. (1987). Inclined dense jets in a flowing current. *Journal of Hydraulic Engineering*, Vol.114, pp. 323-341.

Roberts, P.J.W. & Tian, X. (2004). New experimental techniques for validation of marine discharge models. *Environmental Modelling and Software*, Vol.19, pp. 691–699.

US Environmental Protection Agency (1999). *Review of the D-CORMIX model*. Science Advisory Board Report EPA-SAB-EC-99-011, Washington DC, USA. (available at http://www.epa.gov/science1/ec99011.pdf)

8

Brine Outfalls: State of the Art

Daniela Malcangio and Antonio F. Petrillo
Department of Water Engineering and Chemistry,
Technical University of Bari
Italy

1. Introduction

With the growth of urbanisation, industrialisation and intensive agricultural practices, all superficial and underground, inland and marine bodies of water have become the repository for large quantities of every type of substance extraneous to the natural aquatic environment. In particular, the use of the sea as a dumping ground for a wide variety of waste has become a well-established practice worldwide over recent years. Increasingly greater stress is placed on the potential of these waters to dilute and disperse a greater quantity of waste pollutants created by continually growing populations. However, reliance on the self-purification capacity of the sea must be constantly monitored and not overestimated in order to protect the marine environment.

This is the point of departure for the author's desire to synthesize the complex phenomenon of the diffusion of discharge in the marine environment: a process which is governed by the hydrodynamic conditions of miscible fluids with different proprieties and which primarily depends on typical aspects of the environment which are often overlooked.

Particular attention will be paid to a certain type of discharge, called brine or concentrate, which is the salty water obtained from desalination plants that produce fresh water by separating dissolved salts from saline water. This waste water is generally disposed of through a pressurized conduit which carries the brine from the desalination plant to the shoreline. At this point, a specific submerged pipe, either buried or laid along the natural seabed, completes the discharge process. In order to guarantee the correct functioning of the discharge system, it is a good idea to provide the discharge conduit with a piezometric water supply tower of the appropriate dimensions. It must have a planimetric size sufficient for the water in transit and a sufficient height to guarantee that the system function under pressure all the way to its opening into the sea, and the entry of the wastewater into the sea must be carried out at a speed that favours the dilution of the brine. A loading tank will further ensure regular flow in the case of sudden oscillations in pressure in the underground conduit. The brine discharge outfall pipe should start at the piezometric tower, situated adjacent to the tank for the accumulation of the desalted water at the desalination plant. The submerged conduit should start from the shoreline, as previously mentioned, and be suitably coated so as to guarantee an effective passive protection against corrosion, and possibly thickened, if not buried, to create optimal mechanical protection and, at the same time, render the system perfectly stable on the seabed. At the head of this outfull pipe there is usually a diffuser, preferably with various shafts in decreasing diameter to guarantee a variable range of internal flow speeds so as to avoid possible deposits (low speed), and to safeguard the conduit against the effects of erosion (high speed). This design technique minimises the otherwise common phenomenon of sedimentation inside the diffuser itself, especially in low capacity systems. Discharge openings, called nozzles, are set into each branch of the diffuser in the appropriate size both to allow for a speed of effluence adequate for diffusion and arranged with a discharge speed and direction that will make it possible to avoid the overlapping of the plume rise at a short distance from the nozzles.

The theoretical excursus is performed in the discharged salt fluid direction considering the global system outfall pipe-diffuser-sea and focusing on the following different phases of the diffusion process:

1. the brine waste conveyance in the outfall pipe;
2. its release in the sea through the diffuser;
3. primary dilution, which occurs in the so-called near field and depends primarily on discharge characteristics, such as the density difference between the concentrated brine and the seawater (the buoyancy flux), as well as on the momentum flux, the flow rate, the outfall geometry and the seabed depth;
4. natural dilution in the far field which is further away, considered as a result of the diffusion and mixing processes produced by the sea currents and waves, i.e. the environmental conditions and turbulence.

As integral part of the diffusion process, it muts be mentioned in the lead the seawater desalination treatment, which will not be analyzed in this work. However, it is important to rememeber that nowadays two desalination processes are prevailing: the so-called multistage flash distillation (MSF) and the reverse osmosis (RO) making up about 89% of the total capacity (Glade, 2005). A sharp distinction in brine volume exists between the two desalination processes. RO plants have a conversion ratio from 20 to 50%, while MSF plants have lower recovery rates (10-20%) because of being additionally mixed with cooling water (Goebel, 2005). Thus, the effluent flow rate is 4-5 times higher for thermal desalination than for RO processes referring to the same amount of produced fresh water (Niepelt et al., 2008).

The rest of the aforementioned processes will be analysed later, following the direction of the movement of the brine (and then of the mix) within the system. All of the processes, which, in actuality, can be seen for any type of discharge, are analysed under the various aspects which characterize submerged outfalls: sanitary, construction, technological, environmental, managerial and, primarily, due to the professional background of the author, hydraulic.

A few practical mathematical modelling examples will be presented to support the planning of the brine outfall pipes carried out during the course of the research by the Department of Water Engineering and Chemistry at the Laboratory of Research and Experimentation for the Defence of the Coasts (LIC), both of which are affiliated with the Technical University of Bari, Italy.

The authors intention is to report on the state of the art in waste water discharge in general and brine waste in particular, without intending to offer a single solution to the many problems linked to such a complex issue. We would also like to underline that many of the size values cited in the following paper refer to Italian scientific and technical studies and, therefore, are closely correlated with the characteristics of the Mediterranean Sea as a repository body of water and the national regulations regarding discharges.

2. The sea ecosystem

To understand the impact of the possible alterations of the sea ecosystem as a result of the pollutants discharged into it, one must be familiar with its salient ecological characteristics. First of all, its considerable vastness, considering that about 70% of the earth's surface is covered with water; its depths, which are home to different animals and plant life at different depths; the continuity of its sections, in the face of variations of temperature, salinity and depth; and finally, its ionic neutrality, which makes it a chemical system with high buffering capacities, able to resist variations in pH and remain an alkaline environment.

Another determining factor in the life of the sea ecosystem is its currents which, with their different horizontal and vertical movements, help to mix the waters. There are various types of currents, depending on the marine weather conditions that cause them. There are currents caused by the tangential action of the wind on the surface of the sea; currents in the shallows caused by the wave movement; gradient currents caused by differences in horizontal pressure, which in turn are caused by changes in atmospheric pressure; tidal currents generated by the vertical movements of the marine surface, as a result of astronomical forces; and finally, density currents, caused by variations in the temperature and/or salinity of the water.

The sea is populated by an enormous quantity of biotic factors, for example microscopic animal and plant organisms which form plankton, invertebrates, vertebrates, algae and marine plants. The complexity of pelagic life, in number of species and variety in the food chain, depends on environmental factors, the most important of which are light and the salinity of the waters. In particular, the former determines a subdivision of seas and oceans in regions, with different environmental conditions and, therefore, different species of flora and fauna. The two most important regions are the photic region, which reaches 200 m below the surface and in which photosynthesis takes place, allowing flora to exist, and the pelagic region, below the photic region, where it is only possible to find fauna.

Considering the sea ecosystem as the final container for discharges has, for millennia, essentially been due to the size of the diluter-diluted relationship, which creates a means for dispersal of practically infinite capacity. In truth, its ability to self-purify is determined by the metabolic activity of bacteria, microscopic organisms which participate in the formation of plankton. If, however, the substances extraneous to the marine environment, but discharged into it, are of such a quality and quantity to counter the homeostatic power of the marine ecosystem and damage the cleansing power of its organisms, leading to a permanent alteration of the ecological parameters, the sea system can be defined as polluted. With an extreme range of types of water pollutants from something as ubiquitous as waste heat to deathly chemicals, environmental strategies must obviously related to the substance to be disposed of. For example, the strategy of wide dispersal is suitable only for heat and natural organic materials, which are to be re-assimilated in the global ecosystem. Trace metals in small amounts and non-toxic compounds can be dispersed in large bodies of water if the resulting increases in the background concentrations are minimal, but the strategies of containment or prevention of the discharge of the pollutant is by far preferable for persistent organic chemicals as well as for trace metals.

The polluting agents not included in the processes of transformation of materials and energy which take place within the marine ecosystem, come from the land, the sea and the air. The polluting sources coming from the land can, in broad terms, be detected in the currents which very often transport a considerable quantity of pollutants spilled along river channels: direct cloacal sewage (not passed through a purification plant), civil, industrial and agricultural waste, brine discharged as waste in the desalination process, as well as the cooling waters of electrical power plants. There are also indirect discharges such as soil washout waters or the discharge of solid waste accumulated in the territory and purposely sunk in the sea (macro-pollution). Other sources of pollution that end up in the sea include combustible oils discharged from the petrol tanker washing and the accidental loss of hydrocarbons. Although these last are important and have a great impact on public opinion, accidental spills of hydrocarbons are only a small part of all discharges in the sea. In fact, most discharges are caused by routine operations, such as washing cisterns, discharge of ballast waters, sewage and bilge.

Therefore, pollutants can be released into the environment through point and non-point sources. A point source is the discharge from a structure that is specifically designed for the outflow of waste water, for example from some industrial process or municipal sewerage system. This kind of source has been the target of most of the laws and regulations for water pollution control. The accidental spill of oil from a ship and the release of radioactive wastes from a power plant can also be considered as point sources. On the other hand, non-point sources are defined as widely distributed points where pollutants are introduced into the hydrologic cycle. In such cases, water treatment is usually not feasible. Examples are the runoff of salts used for de-icing highways in winter, soil erosion, acid rainfall and street drainage.

Alterations to the ecosystem's state of health, due to an inferable type of wide-ranging pollutants, are chemical, physical and biological in nature. The chemical-physical characteristics of the sea that influence the self-purification processes are, for example, hydrodynamics, water volume, temperature, thermal stratification, salinity, type of substratum and clearness. The biological properties which play a role and influence these same self-purification processes even more strongly, as previously mentioned, are organisms such as bacteria and fungi, including heterotrophics, and planktonic algae (phytoplancton) and benthic algae (phytobenthos) among the autotrophics. The aforementioned properties act on the pollutants introduced into the marine environment with specific mechanisms, based on processes which take place during the self-purification of the sea system. These chemical-physical processes include reactions of chelation, neutralisation, salification, as well as phenomena of absorption, entrapment, dilution and dispersion. The aforementioned organisms, with their metabolism and mineralisation of the organic substance, define the biological processes essential for purification, together with the photosynthetic processes. In truth, these last mechanisms can paradoxically worsen the pollution of the marine ecosystem when the quantity of the organic substance is too great to be handled, causing phenomena of anoxia and, therefore, the death of the organisms.

The environmental damage that derives from the highly complex possibility of pollution of the recipient water body are at times irreparable, in part as a result of the duration of the repercusions. Thus, there can be:

- damages to natural marine resources, in particular the death of organisms due to eutrophication of the waters as a result of the excess of nutrients carried to the sea by fluvial and/or urban discharges;
- danger for public health, as a result of bathing in polluted waters, ingestion of contaminated raw shellfish, and, although to a lesser degree, the inhalation of sea spray in the presence of high microbial concentrations;

- degradation of recreational activities connected with sea bathing, as a result of aesthetic, but more importantly, hygienic damage;
- impairment of maritime activities, in primis fishing and mussel farming.

In particular, the desalination of sea water can have an impact on five basic environmental domains (Einav et al., 2002):

1. Use of the coastal strip, which instead of being destined exclusively for tourism and/or recreation is also used for industrial activity.
2. Impact on the aquifer. The increased salinity of coastal waters intensifies the problem of saline intrusion into the underground aquifers near the shoreline. A remedy to this potential effect of the discharge of salt waters on the coastal waterbed is the creation of longer pipes into the sea which carry the discharge farther out from the coast. Therefore, this is an intervention strictly linked to the planning stage of the discharge system.
3. Noise pollution, especially in the case of reverse osmosis, which requires high pressure pumps which often generate noise. To remedy this inconvenience it is a good rule to position the plants at the correct distance from inhabited areas.
4. Intense use of energy, which in truth has an indirect effect on the environment, causing, for example, an increase of atmospheric or thermal pollution.
5. The impact on the marine environment, essentially due to the discharge of brine in the sea, and its intensity depend on environmental factors and hydro-geological characteristics of the sea, such as currents, tides, temperature, bathymetry, salinity and density, as well as changes in them. The living organisms that populate the sea environment, both flora and fauna, suffer the greatest consequences. The main effects on the marine biota can be varied and are essentially related to the increase in the concentration of salt. Some animals and plants resist up to certain salinity values and even improve their productivity, while other species succumb in particular conditions. Generally, the larvae and young individuals, especially of invertebrates, are more sensitive than the adults to changes in salinity. Moreover, because of the presence of the brine discharge, fish migrate offshore, so endangering their chances of survival because of the longer distances covered and the wider and wilder marine environment. Hopner &Windelberg (1996) divide marine habitats into 15 different classes on the basis of their sensitivity to the effects of desalination plants. A possible model for the scale of sensitivity is the one formulated by d'Ozouville et al. (1981) on the occasion of the AMOCO CADIZ (Brittany, France) accident, in 1978. According to the hierarchy defined by d'Ozouville, the most appropriate sites for the construction of desalination plants are on the shores of the ocean, characterised by high energy capable of favouring mixing the waters, while he advises against discharging brine near rocks, where the presence of marine organisms and biota is very high. Particularly, the effect of brine on the marine environment is mainly evident in the vicinity of the discharge pipe, where its release creates a "salty desert". In fact, the high specific weight of the salt concentration and the presence of chemicals accumulated during the pre-treatment of the desalination processes, especially for reverse osmosis plants, necessitate brine mixing and increase the risk of potential damage to local flora and fauna. In addition, for desalination plants that use evaporation processes, the effects of temperature increase must also be considered.

With particular reference to dumping waters with a high concentration of salt from desalination plants into the sea, Talavera & Riuz (2001) have reported a specific evaluation of their environmental impact.

The treatment of all the parameters and mechanisms that determine the quality of water, presented in this paragraph, is indispensable when discussing discharges, since the correct location of the submerged outfalls and, above all the point of exit, a definition of the limits for tolerance of the discharge and an evaluation of its impact on the marine environment and the possible damage that it may receive, depend on the chemical-physical and biological characteristics of the recipient body of water. For example, it is not surprising that, in the planning stages, as a parameter of reference for the sizing of a discharge conduit for sewage, the colimetric index is used. This is a microbiological parameter that measures the quantity of bacteria and, specifically, total coliforms and fecal coliforms, estimating the MPN (Most Probale Number). Instead, for the planning of brine discharge conduits, the system is generally sized on the basis of technical data about the flow of the discharge of the desalination process of sea water. In addition to that flow, the brine discharge system can be sized taking into consideration the effluent, or, alternatively, the water that, for management reasons or possible anomalies, by-passes the desalination process. Therefore, besides the verification of the ordinary functioning of the system, the flow to be calculated can be the one which corresponds to the worst possible conditions, i.e. the by-pass of the entire flow of seawater at the entry to the desalination plant.

It is also necessary to consider the legal situation together with the biological criteria, which are aimed at regulating what enters a body of water, determining the acceptable limits needed to protect the water in general, and with the multidisciplinary research, carried out in concert with technicians and ecologists both during the phase of design and that of verification.

Veltri & Maiolo (1992) define two types of discharge acceptability criteria. One is an absolute standard, objectively simpler and more economical in terms of time and cost because it does not require particular and specific investigation of water quality through the acquisition of basic general data, and, therefore, assumes more restrictive standards. The other is a relative criterion, which sets variable limits for discharges on the basis of the quality of the recipient body of water and the destination for its use.

The law on the protection of waters and management of water resources in Europe is the EU Water Framework Directive 2000/60/CE. The goal of Directive 2000/60/CE is to institute a complete and shared European framework for the protection of superficial internal waters, transition waters, coastal and underground waters, underlining the concept of protection through the future objective of the reduction and elimination of discharges of dangerous substances. In particular, regarding the inspection of discharges, the combined approach of point and diffuse sources is indicated in Article 10 and is based on the contextual application of emissions and the best techniques available to check the pollution provoked by point sources (the discharges) and the application of controls and the best environmental practice to contrast the pollution from diffuse sources.

There is no specific law regarding the discharge of salty waters from desalinators located near the coastline. This is, perhaps, because the concentration of salt water, the brine, is considered to be a natural compound and not a pollutant. Unfortunately, these assumptions can not be confirmed in toto. In fact, brine is a liquid product with characteristics similar to seawater, but with approximately double the salt concentration. However, what renders the discharged waters more polluting is the presence of possible chemical products used in the phases of desalinating the seawater, especially in systems using reverse osmosis (Morton et al., 1996). In the case of plants that use the principle of evaporation, account must also be made of their higher temperature (for example, see the study by Anschutz et al., 1999).

3. Outfull pipes

The most important examples of submerged outfall pipes were built from the 1930s onward, like, for example, the cast iron conduit with flexible joints laid in 1930 to cross the New Orleans ship canal. The types, sizes and number of these projects increased rapidly over time, passing from siphons for crossing natural waterways to aqueducts to supply water to islands, to large gas and oil pipelines, conduits for capturing and releasing cooling waters for thermal power plants and industries, and on to the discharge pipes for brine from desalination plants, which we are discussing here.

The sizing of an outfall pipe is done keeping in mind all the factors which condition its functioning. In general, for the choice of the course, the length of the conduit, the depth of the discharge and the characteristics of its ending section, generally made of a diffuser, it is important to consider the morphology of the seabed, the biotics present and the sea weather characteristics of the area concerned. The evaluation of the location is additionally linked to an analysis of possible damage to the marine ecosystem (see considerations made in paragraph 1). Near the seabed the sea is least able to dilute pollutants, as will be further discussed later and defined as "initial dilution", this creates an even greater risk to marine life considering that this zone is ecologically richer. This area is at risk for the sedimentation of the substances transported by the effluent in the conduit. For this reason, it is necessary to carry out a correct analysis, particularly a biological analysis, of the seabed before choosing the point of outlet. During the design phase, the study of the plant component (phytobenthos) can be carried out with the "synecological" method (study of the community), which is performed directly in loco by researcher-divers, and in special cases of the suspected introduction of toxic substances, with the "autoecological" method (study of single organisms), slower than the first since it requires a laboratory for the analysis of samples. The detailed study of the course also requires the study of bathymetric profiles, the lateral configuration of the seabed, the consistency of the seabed's superficial strata, obtainable with surface echosounders used by a swimmer, or by direct observation submarines. All of the above serves to further identify the characteristics of the possible location for the conduit, like, for example, possibly uneven surfaces or the impossibility to build an underground trench.

For a coastal area characterised by an assigned bathymetry, an assigned sea weather climate and an assigned chemical-physical water regime, the correct length for the outfall pipe depends on the other fundamental fact for planning, the effluent flow, maximum and minimum, to be transported far from the seashore, or rather, on the number of inhabitants served, for sewage discharges.

The choice of the diameter of the submerged discharge conduit is made using economic criteria, which take into consideration the cost of supply and laying the conduit in the sea, the capitalisation of the annual energy expenditure, and the expenditure necessary for the acquisition and installation of the electrical pumps.

The design of an outfall pipe requires a series of calculations aimed at determining the project's minimum construction requirements. Among these, the sizing of the conduit is of fundamental importance, as is the resulting definition of the thickness of its walls. Considering that the diameter of a pressurised pipe linked to its functional and hydraulic requirements and, therefore, known a priori, the construction requirements for the conduit are the following:

- the state of the circumferential stress caused by the internal design pressure must be less than the limit established by the applicable law;

- the condition of sectional collapse as a result of the external pressure must coincide, as much as possible, with the maximum acceptable combined state of stress, established by the applicable law.

Note that while the first requirement is essentially related to safety, the second tends to optimise the use of materials, avoiding redundancy in the less critical condition of collapse. However, the choice of the diameter depends primarily on economic criteria, which consider for example the cost of the pipe supply and laying in the sea, and the cost for the purchase and the installation of the electro-pumps.

Of great importance is the evaluation of the type of material to be adopted for the outfall pipe because not only do the conduits determine the velocity of effluence and, above all, the laying procedure and, therefore, the impact on the biotics, but they also determine the duration of the project itself. After shrewd application of interior and exterior surface protection against the chemical aggressions of the sea and the effluent, steel discharge conduits are generally preferred given the fact that they are heavier than the other materials commonly used. Keep in mind that protection carried out with cement mortar casings, if of the passive type, or induction currents and zinc or graphite anodes, if of the active type, increase the costs of the conduits, but, at the same time, render them technologically competitive on the market. There are still some problems for small to average sized pipes because the restoration of the continuity of internal protection near soldered joints is unfeasible. In cases like this, it is a good idea to provide for steel tubes with rapid socket joints and elastic gaskets. In the past, cast iron tubes were very commonly used. Their limited mechanical resistance and extreme fragility was improved with the introduction of spheroidal cast iron, while in the specific case of discharge, conduits were protected internally with cement plaster and weighed down with an external casing in reinforced concrete. The use of plastic materials, like, for example, high density polyethylene (PEAD), has shown itself to be particularly competitive over time for medium-small diameter conduits. PEAD is known for its resistance to the aggressiveness of the marine environment and the transport of fluids, for its lack of roughness, for its flexibility and lightness which facilitate moving it on land and at sea while laying it. However, because of this material's lightness, ballast is necessary, for example with flexible materials, with cast iron collars, with concrete rings, and, to further protect the pipes from the action of the waves, these conduits are generally buried in special trenches. If that is impossible due to the nature of the seabed, the conduit is encompassed in a concrete box. Reinforced polyester with fibreglass (PRFV) is one of the most recent materials to be developed. It is used for tubes of large diameter, and is of better quality than PEAD. It is particularly heavy, so it does not need ballast. It is, however, always a good idea to bury these tubes in special trenches to protect them from the action of the waves and currents.

The cost of the material is, actually, less than that of the installation of the pipelines and, therefore, the choice falls to the material that is easiest to install. The criteria for the choice of methods of launching and laying the tubes depend on local situations, on the depth to be reached, on the type of seabed, on the material the conduit is made of and the type of joints, as well as the type of protection planned for the tubes (Figs. 1-2). Generally, flexible tubes with limited diameters are laid down with techniques derived from those used for deepsea cables. For rigid pipes, instead, the procedure for launching can be divided into three categories: the pull on the seabed, the pull of floating pipe and, finally, the launching of the pontoon with various methods.

Fig. 1. Launch of a submerged pipe with junction in the sea.

Fig. 2. Launch of a submerged methan pipe. The buoys along the launched lenght are in full view.

Following the various evaluations that must be made during the design phase of an outfall pipe, like the identification of the best plan and the choice of the safest and most economic solution from the engineering and technical point of view, it is necessary to carry out a series of verifications regarding stability, structural resistance and preservation over time of the initial conditions. The most common stresses that conduits are subject to are generally due to the following causes:

- chemical aggressions, external (sea environment) and internal (transported fluid);
- biological activity (living organisms);
- mechanical actions (wave and current movement, local phenomena of erosion);
- fishing activities (especially in the area around the diffusers).

A great number of types of protection exist to defend outfall pipelines on the seabed from these risk factors, if necessary following those already provided for the tubes before their launching. The most effective type of protection remains that of burial in a trench, as previously mentioned, with adequate coverage.

Continual monitoring, from the various phases of construction up to the phase of use of the conduits, is indispensable for guaranteeing a long life, and now easier than ever thanks to modern inspection techniques.

4. Diffusers

The ending point of an outfall pipe is generally made of a diffuser, also called a distributor, which plays a fundamental role in the functioning of the entire discharge system. In fact, the job of this device is to separate the effluent fluid to be disposed of in the marine environment into several jets, so as to favour the process of dilution which then takes place in the sea, and to reduce local phenomena of turbidity and deoxygenation.

The diffuser usually consists of sections of tubing of declining size, which at times are subdivided into two parallel sections to form a Y, but which still have a cross-section larger than the sum of the areas of all the openings present on the lateral surface. These openings, called nozzles, are positioned in alignment along a single generatrix (generally the upper one), or in two symmetrical generatrices, so that the jets of at least one row shoot up in the direction of the undersea current, which at times is reversed in comparison with the dominant one. These are then oriented in such a way as to make the jet shoot up vertically, slightly inclined off the vertical, or horizontally. This last example is often used for the disposal of effluent from urban treatment plants, to create curved density jets, which tend to rise to the surface in longer trajectories and, so, favour dilution. An excellent treatment of the various types of diffusers (Fig. 3) was made by Jirka & Akar (1991).

Regarding the arrangement of the diffusers in the design, to make sure that the dominating marine currents exercise a correct dispersion of the effluent in the recipient water body, the diffuser is generally placed orthogonally to them, as well as to the nearby coast. Instead, in a situation where there is no prevalent direction of the dominating current, the aforementioned Y-shaped diffusers are used.

Generally, a single diameter d is assigned to the n nozzles, so as, so as to evenly divide the overall flow and guarantee the correct functioning of the full section, in order to avoid the intrusion of seawater into the tubes. It is advisable to contain the diameter of the nozzles to favour the initial dilution, but within certain limits to avoid occlusion due to marine deposits.

In truth, at times it is possible to carry out a premixing process in which the seawater is pulled into the end section of the diffuser on purpose. In this way, there are two dissimilar effluents, the tout court fluid discharge and the premixeddischarge, which is denser than the first and in part already undergoes a type of dilution in the conduit. Experiments have shown that the beneficial effects on the initial dilution provided by this methodology are minimal and, in any case, not comparable with the negative effects, like the less effective functioning of some nozzles, the entry of marine organisms and the deposit of solid substances which increase the roughness of the tubes.

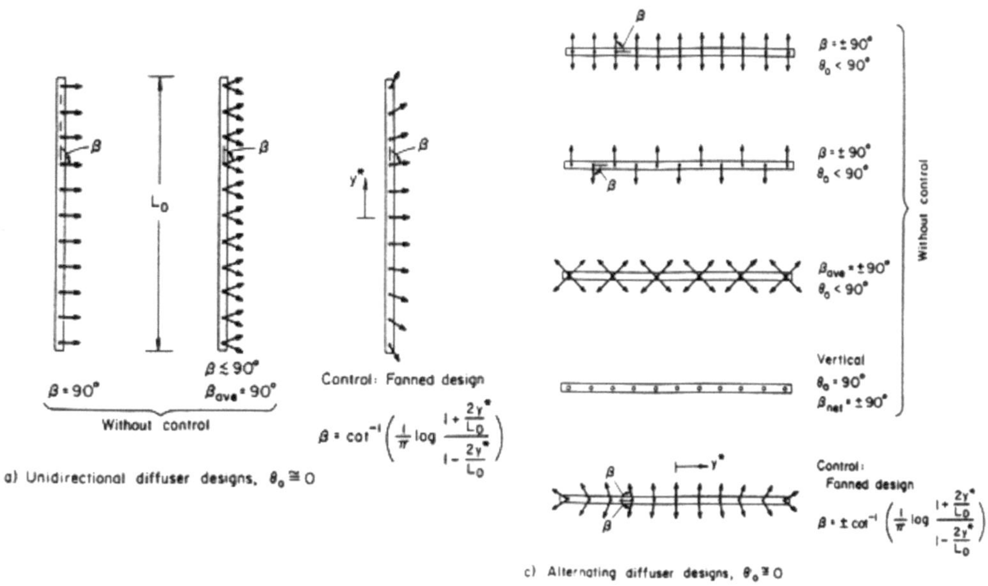

Fig. 3. Different types of unidirectional (left) and alternating (right) diffusers.

The hydraulic load at the exit point of an outfall pipe is not equal to the head *H*, but considering the contracted section of the effluent of density ρ_0 greater or lesser than the density of the sea ρ_a, depending on the type of discharge, the relative piezometric level is superior to the hydrostatic level *H* of a quantity $H' = H(\rho_a - \rho_0)/\rho_0$, having placed the balance of the pressures inside and outside the edge of the same section. In general, it can be assumed that the load does not vary before and after each single opening, so, ignoring the kinetic height, it is possible to obtain the simplified expression of the piezometric difference along the entire diffuser, by means of an iterative calculation. The increase of the load (increase of the piezometric level) at one point and before the same point can be seen as a way of indicating conditions of anomalous circulation in the whole system, due to the entry of seawater in the final nozzles of the diffuser. Another way to proceed is to attribute to the kinetic load in the orifices 10-20% of the available difference in gravity systems, or of the overall dissipated load (difference plus prevalence of the pump) in the case of pumping systems.

An important aspect of the sizing of a diffuser lies in considering the dilution of the first phase *Si* to be a fundamental parameter, which will be covered in detail in the following section, to which is assigned a prefixed value defined as "acceptable" on the basis of the characteristics required both by the seawater near the diffuser and by the inshore water. Besides this parameter, other sizes come into play in the sizing of the diffuser, such as the total discharge load, a component of the cross current and the difference of density between seawater and effluent. Once these measures have been designated as project data, together with the velocity of effluent from the diffuser, it is possible to determine the length of the diffuser as well as the diameter and arrangement of the openings which guarantee the prescribed value of dilution *Si*. In principle, if greater dilution is desired, the length of the diffuser should be increased instead of increasing the velocity of effluent, so that the distance between the nozzles is great enough to guarantee reciprocal contact at the lower border of the mixing field

In any case, during the design phase it is important to anticipate an effluent velocity V_0 of at least 0.6- 0.8 m/s, to guarantee the self-cleaning of the diffuser and annul the possibility of deposits, but, above all, to ensure a value of the densimetric Froude number:

$$F = \frac{V_0}{\sqrt{\frac{\rho_a - \rho_0}{\rho_0} gd}} \tag{1}$$

much greater than the unit, since the initial dilution along the jet depends on this. The latter is defined as:

$$Si = \frac{\rho_0 - \rho_a}{\rho - \rho_a} \tag{2}$$

in which ρ_0, ρ and ρ_a represent respectively the densities of the effluent in the effluent section, in the generic section of the jet, and of the sea, which generally must be superior to 60-80. To avoid the erosion of the submerged outfall pipe, it is also a good idea to guarantee a maximum value of the velocity of effluence for the single nozzles, no greater than 2.5 m/s. Equally important is that the jets not interfere with each other during their upward flow, which takes place at a short distance from the discharge. To do this, the interaxis distance between the nozzles is generally equal to one third of the trajectory, which is usually considered as equal to three quarters of the head *H*.

To end the specific discussion of the hydraulic design of diffusers, it is worth reasserting the importance of the verification of their functioning with timely monitoring techniques, and at times even to compensate for possible design errors. In truth, an issue presented by the installation of these devices is that, differently from the existing conduits, these can not be protected with trenches; since they lack any coverage for the tubes they need adequate protection from fishing activities. For example, in the case of a spill of effluents rich in substances nutritious for fish species, the zone around the diffuser becomes an attractive site for fishing boats, which can cause serious damage to the diffusers with their anchors and trawl nets. Expedients aimed at protecting the outlet can be of the passive type, like signal buoys and steel nets around the diffuser, and of the active type, like concrete blocks equipped with hooks to break fishing nets.

5. Diffusion and dilution processes

The subject of diffusion, interpreted as "turbulent transport associated with dilution," is a highly complex topic and has a consolidated history in terms of research and both theoretical and experimental analysis. The term dilution, first expressed in (2), can be considered as the reduction of the concentration, in comparison with the initial concentration, of a specific effluent introduced into a recipient body of water. In truth, in the literature there are a series of formulas with which to identify the total dilution of a generic discharge, which depends essentially on the various phases in which the process itself can be divided, namely:

1. first phase or initial dilution, near the outlet into the marine environment;
2. second phase or following dilution, in proximity to the free surfaces.

The two phases take place in two distinct zones. In particular, the first happens in the so-called near field, it has a prevalently vertical development in which transport and mixing depend on the intensive magnitude in correspondence to the exit outlet (density, velocity of the fluid and turbulence index). Instead, the second phase takes place in the far field, in a practically horizontal direction, along which transport and mixing depend on the surrounding movement field (e.g. sea, wind and density currents) and on the characteristics parameters of the discharge (flow rate, average dilution). In the particular case of sewage discharge, a third phase must be taken into consideration, that of bacterial decline, which takes place at the end in the far field, and which will not be treated in this paper.

The various density differences between the effluent and the receiving water represented by the buoyancy flux causes different flow characteristics of the discharge. When the waste water density is higher than the sea water density, the dense effluent flow has the tendency to fall as negatively buoyant plume. In the contrary case, the effluent is distinguished by a neutral to positive buoyant flux causing the plume to rise. Figure 4 illustrates the typical behaviour of positively or negatively buoyant jets discharging into the receiving water through a submerged single port, for the particular case of brine discharge. The same Figure 4 underline also the different behaviour of the salt buoyant jet depending on the desalination processes from which it comes.

Hereafter, a detailed analysis will be made of the phenomenon of dilution, starting with the first phase, during which the aforementioned process of diffusion takes place. The current that carries up the effluent dumped into the sea, which has a different density than the sea itself, is indicated with the term buoyant jet, because this fluid dynamic structure has all of the characteristics of the current itself. In the exit section a zone, called zone of flow establishment (ZFE), is created. It contains a conical nucleus with constant density and velocity and is the starting point for the jet's turbulence structure. Then there is the so-called zone of established flow (ZEF), in which the field of turbulent flow reaches the axis of the jet and the velocity and concentration (the inverse of the dilution) diagrams assume the typically Gaussian flow, with decreasing values both from the axis of the jet towards the periphery and along the axis itself, as the distance from the source increases. A reference parameter for the identification of these two jet development zones is the densimetric

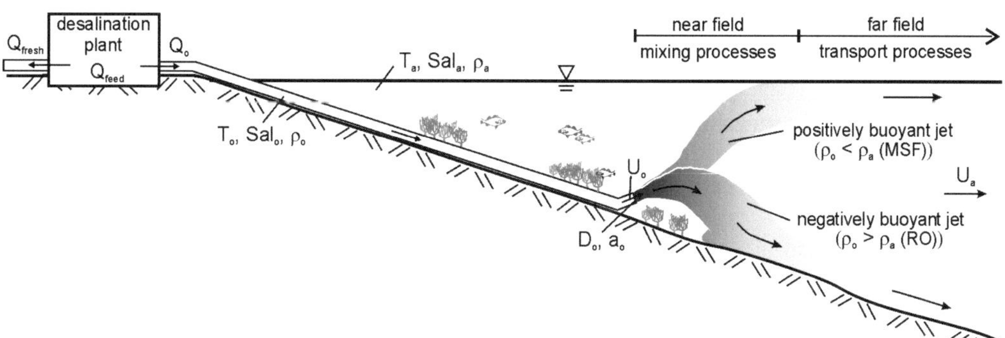

Fig. 4. Brine discharge characteristics of desalination plants (by Niepelt et al., 2008).

Froude number F defined in (1), which represents the relationship between the forces of inertia and the forces of buoyancy.
Using this parameter, it is also possible to formulate the initial dilution:

$$Si = Si\ (F', z/d') \tag{3}$$

where F' is the Froude number referring to the contracted section of diameter $d' = 0.61d$, where d is the diameter of the single nozzles, and z is the depth. In the preceding section, the importance of the determination of the initial dilution, in its acceptable value, was highlighted as being indispensable, for example, for the sizing of the diffuser. It was also pointed out that the initial dilution increases with the increase in the number of orifices that the diffuser has, as long as the distance between them is great enough to guarantee that the jets do not overlap, which would, on the contrary, reduce its dilution.
The active forces which push the jet to rise through the overlying seawater are the momentum flux at the outlet, and the buoyancy flux due to the difference of density between effluent and sea. In its permanently delayed movement, the sea water is pulled into the jet, thus expanding it. This process is known as entrainment, which Pedersen (1986) defines as "the diffusion of a fluid characterised by a field of turbulent flow within an environmental fluid in nonturbulent flow." Therefore, in conditions of dynamic equilibrium, the forces of inertia and buoyancy are opposed by the resisting forces which act on the surface of separation between the jet and the recipient body of water and are the cause of elevated velocity gradients in the directions transversal to the main flow. This causes, in part because of the effect of the density gradient, coherent hydrodynamic structures, or toroidal vortexes which transport heat, momentum flux and mass, and, therefore, favour the diffusion of the effluent in the surrounding marine environment and the intrusion of seawater into the jet. Their presence was detected by the analysis of velocity surveys carried out in both recent and past experimental laboratory models, with both pioneering and advanced measuring techniques.
All researchers agree that the conditions of flow in the marine environment and the turbulent agitation in the initial section of the jet favour diffusion considerably, although for this last aspect it is difficult to quantify the turbulence intensity which, clearly, depends fundamentally on the form of the nozzles. There is a series of recent studies on the effect of turbulence on the diffusion process of the jet. One of particular interest, for its use of various approaches, is by Malcangio (2004).
Then, the presence of seasonal effects must be pointed out: for example, in the summer, since the temperature of deep seawater is lower than the surface temperature, the gradient of negative density is enhanced along the vertical, which is generally always present. In such circumstances, the jet, which during its upward rise varies in density due to the effect of progressive dilution, can remain entrapped under the thermocline, and be unable to reach the surface, since the structure of the jet deteriorates in a formless cloud of increasingly large volume. Often this phenomenon is considered desirable, as it makes it possible to obviate the impairment of qualitative standards linked to aesthetic factors, as in the case of sewage effluent.
For more details on jets, consult for example the pioneering work of Rajaratnam (1976), Fisher et al. (1979), Papanicolau & List (1988), as well as the vast literature by Abraham (e.g. 1960, 1963, 1965).
As the jet moves away from the nozzle, the cloud of effluent, which we will now call "mix," expands horizontally and tends to dissipate as a result of the dilution of the second phase,

Ss. This dilution takes place thanks to sea weather agents which are often unrecognised, and so are presented as a complex mixed process. They are of a small entity in comparison with the initial phase, but nonetheless can not be neglected. In terms of size, this process can be expressed in the relationship:

$$Ss = Ss\ (\beta, x/b) \tag{4}$$

in which β is the opening angle of the sewage plume, expressed as:

$$\beta = \frac{12 \cdot \varepsilon}{U \cdot b} \tag{5}$$

where ε is the coefficient of turbulent diffusivity computable through experimental interpretations, U the current velocity, b the diffuser length, while x represents the horizontal x-axis evaluated in the direction of the flow hypothesised as the normal axis of the diffuser.

Keeping in mind that in the sea the action of the currents generated a continual exchange of seawater mixed with the discharge, the mixing field above the diffuser assumes a thickness of h so that

$$Q\ Ss = U \cos\alpha\ h\ b \tag{6}$$

where Q is the overall flow rate of the effluent exiting from the diffuser, Ss the dilution at the entry of the accumulation cloud created over the discharge, $U\cos\alpha$ is the velocity of the current in the direction parallel to the axis of the diffuser and b is the length of the diffuser itself. This formula makes it possible to verify the risk of nutrient accumulation and, as a consequence, the rise of eutrophic processes as a result of an insufficient exchange of the recipient water mass. Therefore, it is wise to make reference to the maximum discharge of effluent introduced by the diffuser into the marine environment and the minimum current velocity. Note, however, that the conditions which are created between the effluent cloud rate and that removed by the currents are not stationary, since the cloud tends to gradually increase in thickness and concentration with a reduction of the intensity of currents or an increase in the discharge, and viceversa.

Therefore, transport to the far field is essentially due to sea weather agents, such as currents and wave motion, as well as the different density between effluent fluid and the recipient water body. Also in this phase of dilution there is turbulent agitation of the fluid mix and, therefore, a highly complex process is generated which has a strong influence on the quality of inshore waters.

In the literature there are various mathematical expressions of the various phases of dilution, and, in general, the overall dilution is indirectly evaluated as their product. However, it is always important to consider individually the various phases which the process of dilution is divided into, their singularity and specificity should never be forgotten if the goal is to obtain a complete, well-structured result in the analysis of the global conduit-distributor-sea system.

A different approach that can be adopted for the analysis and anticipation of discharge transport and dilution phenomena in the sea uses mathematical models which make it possible to resolve the equations that govern these phenomena numerically. As a result of the complexity of a process as composite as dilution, both in the near and far fields, there is a tendency to use zone models, which examine the single processes on a more local scale. For matters in the near field, reference is made to the traditional mathematical and experimental type of formulation, described in the same section, which supply the initial dilution and, therefore, the distribution of the concentration of the effluent, which in this phase assumes the structure of a buoyant jet. Regarding the phenomena of the far field, often a coincidence can be found between the reference equations and those that govern any hydraulic phenomenon, which are the classic Navier Stokes equations. In truth, these equations lack a specific identification with the particularity of the phenomenon of circulation of pollutants in coastal waters. The results obtained are, actually, quite far from being a real answer to the prediction of phenomena of transport and diffusion, if they are not accompanied by a physical comprehension of the mechanisms of coastal circulation, characterised by time-space scales which are quite various and complex. These mechanisms, which supply any numerical model with the boundary conditions, together with an adequate experimental basis, are able to render a numerical simulation able to understand the fundamental characteristics of the physical issues at play. For this reason, along with the traditional equations of Navier Stokes and of continuity, which express the principle of the conservation of the momentim and mass which govern flow, it is also important to take into consideration the equations which express transport, descriptive of the advective and dispersive processes, and the biochemical equations which consider chemical and biological factors which reduce the mass of the polluting mix.

The resolution of such a mass of equations, both when they are adopted in their complete forms and when simplifications are allowed, is possible with diagrams of numerical integration, to finite differences or to finished elements, together with the definition of the boundary conditions. It is fundamental to underline the importance of thorough field survey to correctly validate the simulation models of the diffusion of pollutants in the sea, as well as the estimation of the physical parameters of the simplified models, such as the density of the effluent and the recipient body, the drift velocity of the current, and the dispersion coefficient.

Applicable examples of mathematical models of the dispersion into the sea of brine discharge from desalination plants are shown in section 7.

6. Wave motion and current effects on dilution

The realistic situation facing any fluid introduced by an outfall pipeline into the marine environment includes the presence of currents in the recipient water body. Setting aside the effect of the hydrodynamic actions wielded by the flow fields on the discharge conduit, in this paper we will simply underline the effect that the currents and waves have on the process of diluting the solute. Regarding this, as mentioned in the previous section, the diffusion phenomena are increased by the presence of these motions, therefore in the hydraulic design of the pipe and the diffuser a condition of calm sea or weak superficial currents should be considered a cautionary situation. Instead, in the definition of the position and depth of the conduit and the diffuser it is very important to take into consideration the flow and the currents; the flow so as to position the diffuser far enough offshore from the area of the reefs, and the currents to ensure that the prevalent direction of diffusion be toward the open sea.

Speaking of currents, the following is a classification of types, on the basis of what causes them:

- convection currents, due to differences in density and only found in particular conditions in large basins like the Mediterranean, which in turn generate coastal currents;
- drift currents, determined directly by the action of the wind, which moves the surface strata of liquids in the same direction, also creating an action of uplift which is accompanied by ebb currents;
- compensation currents, generated by the encounter of a marine current with the shoreline and everything that makes it not uniform;
- tide currents, due, as the name itself indicates, to the effect of the tide, generally reduced in closed basins, like for example the Mediterranean.

Currents and wind work, in particular on the medium-term effects, due for example to the introduction of hydrocarbons, oils and fats, heavy metals, differently than on the short-term localised effects around the point of the introduction of the waste. To correctly be able to evaluate the effect of currents on hydrodynamic dispersion in the far field it is important to identify the resultant of the velocity vector in direction and intensity. There are different measurement techniques, which can be distinguished essentially by the depth to which they refer. For surface measurements, buoyant bodies are left to drift on the surface and, today, also equipped with sensors that indicate their position, like modern equipped buoys, while for deep measurements current meters are generally used.

Additionally, it is necessary to make a distinction between currents depending on the depth at which they take place. Therefore, there are submarine currents, to be enumerated in the group of convection currents since they are due to variations in density, which distribute themselves parallel to the coast and can be considered useful for the phenomenon of dilution because of their diluting effect on spilled wastewater, and there are surface currents, due, above all, to the wind, which instead can play a harmful role if they carry back to shore a mix that has not yet been completely diluted.

The beneficial effect of currents on the diffusion of discharges in the sea is well known in the literature. A valid example is seen in the experiences of Platten & Keffer (1968), who arrived at the important conclusion that intrusion, as defined in the preceding section, from the (vertical) jets introduced into a fluid field in flow in a direction transversal (horizontal) to their axis, is one order larger in size than jets in still water. When considering a uniform current directed in the same direction as the jet (horizontal), Lee & Nerville-Jones (1987) propose two different expressions for initial dilution, depending of the prevalence of the effect of buoyancy on the effect of the current, and viceversa. In the first case:

$$Si = k_1\,(A/d)^{5/3}\,F^{-1/3} \tag{7}$$

where k_1 = 0.36, and where A represents the greatest height of plume rise, while in the second situation the result is:

$$Si = k_2\,(A/d)^{5/3}\,F''/F \tag{8}$$

where F'' is the Froude number relative to the current velocity. It has been seen experimentally that with these expressions, and in special conditions, values of initial dilution that are 3 to 5 times greater than those corresponding to the condition of a calm fluid environment can be obtained.

Let us now focus on the effect of the wave flow field of the fluid environment on the phenomena of diffusion of the waste dumped into it. In these conditions, the process of the mixing and diffusion of the solutes benefits from a considerable advantage, as Chin (1987)

also affirms, "waves are almost always present in oceans and seas, a proper consideration of their beneficial effect in the dilution of underwater discharges can serve for design economy."
To identify a subsector of reference of origin for the characteristics of the wave flow which transfers the inshore energy contents, it is necessary to carry out an analysis of the frequencies of occurrence and the classification of the sea weather events which have historically happened during the period of observation and in the area being studied. This analysis is fundamental, for example, in the choice of the planimetric device for the outfall pipes, in those cases where it is impossible, because of the geological nature of the seabed, to completely bury them. Therefore, it is advisable to place the axis of the pipe parallel to the maximum nodal incidence of the wave motion, so as to offer it the section of least resistance. As a result of the increasing deformation and rotation of the front closer to the shallows, it is indeed preferable to provide the conduit with a planimetric layout that varies with the depth, according to the evolution of the orthogonals to the wavefront.
In truth, the effect of the wave motion does not always manifest itself for the entire depth of the recipient body of water. This depends, according to Airy's theory, on parameters such as the measured depth in comparison with the average level of the calm sea h and the wavelength L, therefore leading to the following cases:

- $h/L>0.5$ refers to deep waters, in which the effect of the wave flow extends to the depth of $L/2 < h$;
- $0.05<h/L<0.5$ refers to intermediate waters, in which the effect of the wave extends for the entire depth but with variable intensity;
- $h/L<0.05$ refers to shallow waters, in which the effect of the wave motion extends for the entire depth but uniformly.

Although Chyan &Hwung (1993) showed scant knowledge of the action of the wave motion on the diffusion of the jets, in contrast with a rich literature on discharges in the sea with regards to the processes of intrusion and interaction with the currents, Ger (1979) and Chin (1987) had already experimentally demonstrated that wave allows for the improvement of dilution of the solute transported by a jet. Therefore, consider a horizontal jet introduced into a recipient water body in the presence of a monochromatic wave motion. The initial dilution that it undergoes can be expressed by means of the following functional relationship:

$$Si = f(d, V_0, g', h', \theta, a, T, L) \tag{9}$$

where V_0 is the exit velocity of the jet, g' is the acceleration of reduced gravity defined as

$$g' = \frac{\rho_a - \rho_0}{\rho_0} g; \tag{10}$$

h' is the depth of the seabed in correspondence with the plume, θ is the angle between the exit direction of the jet and the propagation direction of waves, finally, a, T and L respectively represent the amplitude, period and length of the wave. Following the criteria of adimensionalisation proposed by Chin (1987), by means of following simplifications which allow for the grouping of the quantities indicated in (9) they are rendered adimensional, the author goes so far as to represent the physical process of average dilution Si on the surface of a horizontal jet of a solution of set initial density and velocity, introduced into a field of wave motion, by means of the following relationship:

$$\frac{S_i}{S_0} = 1 + 6.15\left(\frac{L_Q}{Z_M}\right). \qquad (11)$$

In (11), *Si* represents the average surface dilution in the case of a calm recipient water body, L_Q the distance starting from the nozzle beyond which the port has an effect on the flow and Z_M the distance from the nozzle beyond which the wave flow prevails over the initial momentum in jet behaviour. Following his experiments, Chin (1987) observed that, as a global result of the interaction between the wave flow field and the jet, there is a spraying effect at the level of the nozzle and an oscillating movement during the rising phase (Mossa & Petrillo, 2001; Mossa 2004). Other experiments were carried out analogously (e.g. Fischer et al., 1979), and all reached the conclusion that the presence of a wave flow field determines initially greater dilution values, even doubled, in comparison with those of calm in the surrounding fluid.

7. Mathematical modelling case studies

The results obtained by means of the theoretical analysis of the dilution in the sea of a brine discharge coming from a desalination plant are effective for a comparison with those obtained from mathematical models. The result of this comparison has often been positive, thereby it would seem profitable to join theoretical studies with both mathematical and experimental analyses. In this paragraph two practical applications of relative mathematical models will be briefly illustrated, already seen elsewhere (Malcangio & Petrillo, 2009; Malcangio & Petrillo, 2010), both carried out at the Department of Water Engineering and Chemistry at the Technical University of Bari, Italy. Both the area under investigation are environmentally vulnerable because of their dryness and lack of natural watercourses. Therefore, underground water resources have been intensely exploited to meet the growing demand for water, in particular through extraction from wells. Moreover, as the replenishment of the aquifer is very low due to the shortage of rainfall, coastal areas of South Italy face the critical problems of seawater intrusion.

Considering the first case study (Malcangio & Petrillo, 2009), in order to avoid the aforementioned problems of water lack, a coastal desalination plant has been planned in a coastal zone in Brindisi, southern Italy, and not constructed yet. One special feature of the seabed concerned is that of the presence of a protected vegetative species, the oceanic Posidonia, which is an endemic plant that forms prairies of Posidonia, also called seaweed fields (although they are not formed by seaweed). This prairies are the most important ecosystem of the Mediterranean, equivalent to forests in land ecosystems, as they are also a great source of biodiversity being the habitat of numerous vegetable and animal species, some of which are in danger of extinction. Figure 5 shows a map of the existing biocoenosis and a preliminary overview of the intake and outfall pipe location in the area under assessment.

The initial part of the study focused on the hydrodynamics of this area, and then on the current generated by several environmental factors (wind, tide, etc.) and simulated by mean a 3D mathematical model. Then, the virtual brine discharge was implemented by the same commercial code as a punctual salt source for several simulations, and the resulting spread of the brine plume at different layers were analyzed. The aim of this work was to find both the most suitable position and length of the brine outfall pipe in order to minimize the

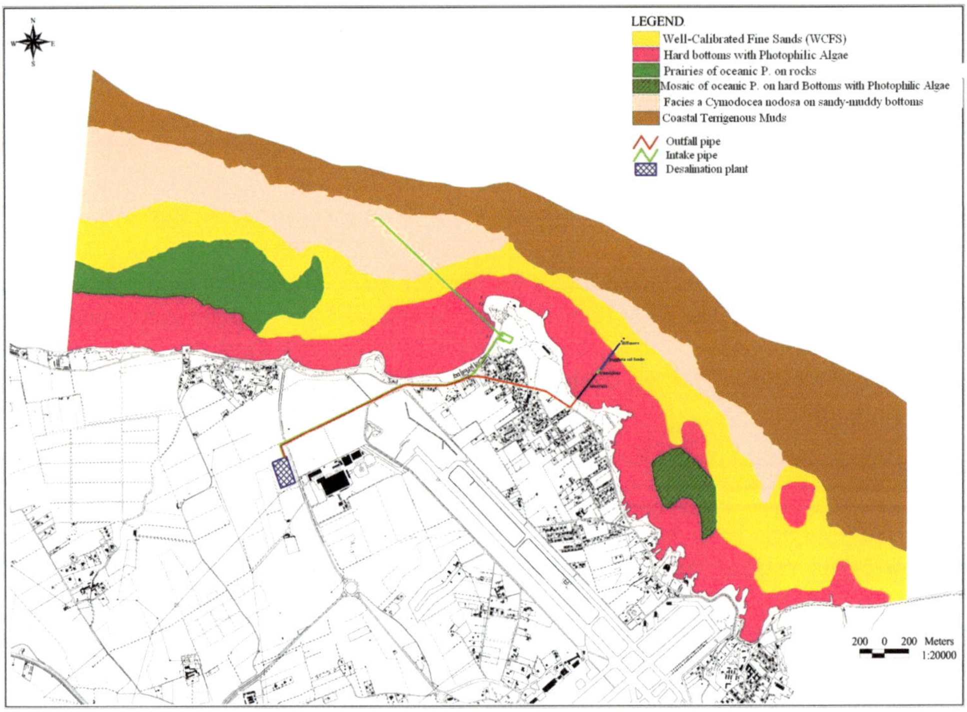

Fig. 5. Biocoenosis of the area under study and preliminary identification of the desalination intake and outfall pipes site.

impact of salt concentration release on the protected flora species in the area. The results of the simulations allowed the authors to conclude that between the two simulated project hypotheses for the outfall pipe length (605m and 770m), the best choice would be that of the shorter pipe taking into account both environmental and economic considerations.

As second test case, the authors analyzed the optimal site for an outfall pipe originating from a seawater desalination plant in Bari, South Italy, at the planning stage (Malcangio & Petrillo, 2010). As in the previous case, the study focused on the dilution of the system brine discharge in the sea determined by a punctual salt source, considering the salinity as a tracer in recognizing the brine discharge path. The same 3D mathematical model, which was utilized in the former work here summarized and previously calibrated in the same zone of interest by a measurement survey, was used to simulate brine dilution in the sea. Several simulations in barotropic conditions were performed, in order to test the interaction of the real variation in wind direction and intensity on the dilution of the brine discharge, and more importantly the best location for the brine outfall with the least environmental impact for wildlife, particularly in an area which has already been damaged by the presence of treated waste water disposal.

In Figure 6a the salinity map obtained for the test which simulated the single treated waste water and the real situation typical in a mean annual range is plotted, together with the salinity maps achieved at the end of runs where the same meteo-climatic conditions were simulated but with the added detail of brine outfall, 60m (fig. 6b) and 30m distant respectively (fig. 6c). Qualitatively, it can be deduced from their comparisons that the planning choice to locate a brine outfall pipe close to the real outfall pipe for treated waste water is environmentally more suitable than the more economically favorable option which would place the brine outfall in front of the desalination plant and about 1330m from the other outfall pipe. But which distance between 60m and 30m is not so evident.

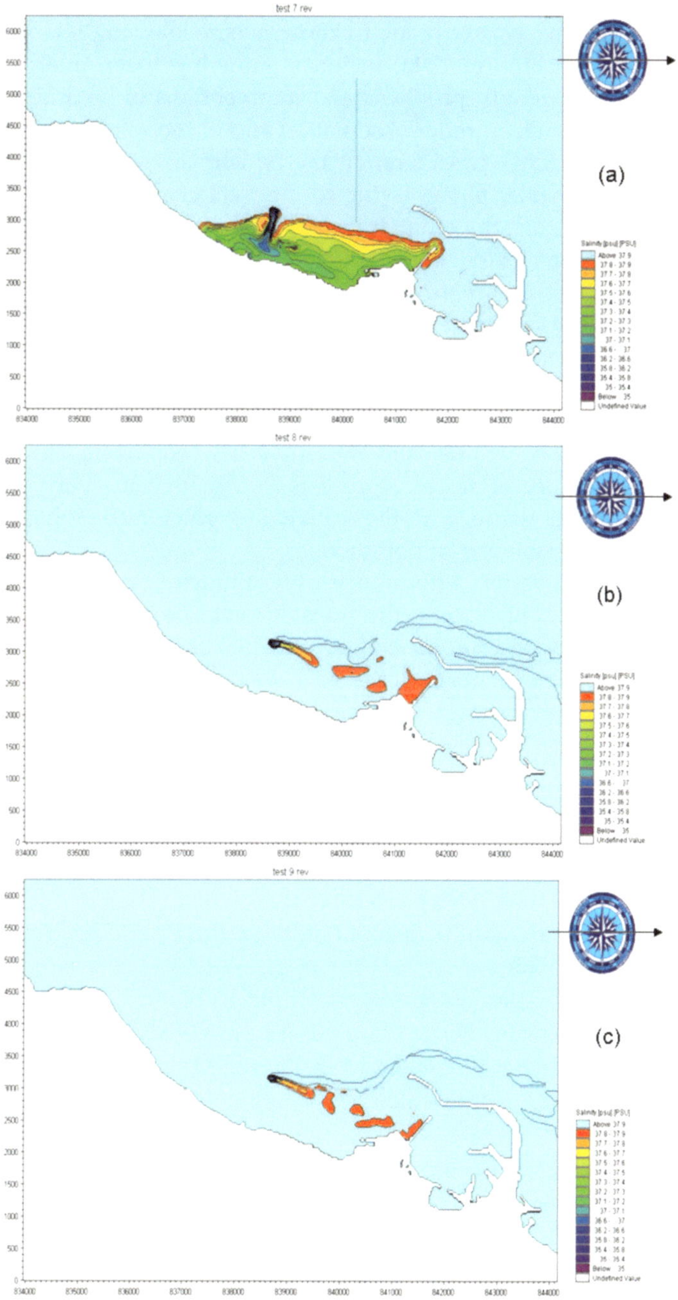

Fig. 6. Superficial salinity map (a) when the single treated waste water release is simulated, (b) with the further contribution of the brine outfall, far 60m and (c) 30m.

8. Conclusion

Protection of the marine ecosystem is one of the main environmental issues of our times. In fact, it is a common practice to subject bodies of water to the constant introduction of discharges, many of which are polluting in nature. In the field of water resource management, to protect means, above all, to know, where knowing refers to a process that, beyond the simple acquisition of data, includes gathering them, interpreting them, but, above all, using data to identify problems and interventions to be undertaken on a short timescale both in the ecosystem under observation and in those that interface with it. All of this is done while taking into consideration the aquatic ecosystem (water, sediment and biotics) and not just the water alone, trying to intersect chemical, physical and biological parameters so as to determine the ecological state of bodies of water, dictating precise rules for the development of monitoring.

In truth, to truly protect the marine environment prevention must prevent first, and therefore, taking into consideration all of the components of the outfall pipe-diffuser-sea system that have been defined and analysed in this paper, it is important to make sure, starting at the design phase of any discharge system, that each of these components be thoroughly evaluated in terms of environmental impact. It is essential to underline this point because of the too frequent use and abuse that has been made, and is still commonly carried out, on deep bodies of water, since it is known that the very dimensions of the masses of waters receiving waste, with the sea weather phenomena inherent to them, carry out a positive role in the resolution of pollution.

It is hoped that this discussion, although it only alludes to or touches on the issue of discharge of waters with high concentrations of salts deriving from the desalination process, within a much more complex and heterogeneous issue like marine pollution due to discharges in general, can in any case be cause for reflection on a concept that is essential for every person: respect for a resource like the sea is indisputably the best lesson to be learned.

9. References

Abraham, G. (1960). Jet diffusion in liquid of grater density, *Journal of Hydraulic Division*, Vol. 86, No. HY6, June 1960, pp. 1-13.

Abraham, G. (1963). Jet diffusion in stagnant ambient fluid, In: *Delft Hydraulics Laboratory*, Pub. N. 29, July, 1963.

Abraham G. (1965). Horizontal jets in stagnant fluid of other density, *Journal of Hydraulic Division*, ASCE, Vol. 91, No. HY4, July 1965, pp. 139-150.

Anschutz, P.; Blanc, G.; Chatin, F.; Geiller, M. & Pierret, M.C. (1999). Hydrographic changes during 20 years in the brine-filled basins of the Red Sea, *Deep-Sea Research I*, Vol. 46, pp. 1779-1792.

Chin, D.A. (1987). Influence of surface waves on outfall dilution, *Journal of Hydraulic Engineering*, Vol. 113, N. 8, Paper No. 21710, ASCE, New York, pp. 1005-1017.

Chyan., J.M. & Hwung, H.H. (1993). On the interaction of a turbulent jet with waves, *Journal of Hydraulic Research*, Vol.31, N. 6, pp. 791-810.

d'Ozouville, L.; Bernè, S.; Gundlach, E.R. & Hayes, M.O. (1981). Evolution de la pollution du littoral Breton par les hydrocarbures de l'AMOCO CADIZ entre Mars 1978 et Novembre 1979. In: *AMOCO CADIZ*, Conséquences d'une

pollution accidentelle par les hydrocarbures, Centre National pour l'Exploitation des Océans, Paris, 1981.
Einav, R. ; Harussi, K. & Perry, D. (2002). The footprint of the desalination processes on the environment, *Desalination*, Vol. 152, pp. 141-154, ISSN 0011-9164.
Fischer, H.B.; List, E.J.; Koh, R.C.Y.; Imberger, J. & Brooks, N.H. (1979). *Mixing in inland and coastal waters*, Academic Press, New York.
Ger, A.M. (1979). Wave effects on submerged buoyant jets, *Proc. 18th Congress Int. Ass. Hydr. Res.*
Glade, H. (2005). Design of seawater distillation plants, *DME Seminar Introduction to Seawater Desalination*, Berlin, Germany, June 20th, 2005.
Goebel, O. (2005). Markets and desalination technologies in brief, *DME Seminar Introduction to Seawater Desalination*, Berlin, Germany, June 20th, 2005.
Höpner, T. & Windelberg, J. (1996). Elements of environmental impact studies on coastal desalination plants, *Desalination*, Vol. 108, pp. 11-18, ISSN 0011-9164.
Jirka, G.H. & Akar, P.J. (1991). Hydrodynamic classification of submerged multiport diffusers discharges, *Journal of Hydraulic Engineerimg*, Vol. 117, N. 9, pp. 1113-1128.
Lee, J.H.W. & Nerville-Jones, P. (1987). Sea outfall design - Prediction of initial dilution, *Proc. Inst. Civ. Eng.*, Part 1.
Malcangio, D. (2004). Modelling of wastewater discharges in a turbulent environment, *PhD thesis*, Politecnico di Bari, Italy.
Malcangio, D. & Petrillo, A.F. (2009). Desalination brine discharge modelling as a support in planning decisions, *Proc. 33rd IAHR Congress - Water Engineering for a Sustainable Environment*, Vancouver, Canada, August 9-14, 2009, ISBN 978-94-90365-01-1.
Malcangio, D. & Petrillo, A.F. (2010). Modeling of brine outfall at the planning stage of desalination plants, *Desalination*, Vol. 254, pp. 114-125, ISSN 0011-9164.
Morton, A.J.; Callister, I.K. & Wade, N.M. (1996). Environmental impacts of seawater distillation and reverse osmosis processes, *Desalination*, Vol. 108, pp. 1-10, ISSN 0011-9164.
Mossa, M. & Petrillo, A.F. (2001). The effects of waves on the jets of a sewage outfall diffuser, *Meeting on Coastal Zone Management in the Mediterranean Region*, Izmir, Turkey, April 26 –May 1, 2001.
Mossa, M. (2004). Experimental study on the interaction of non-buoyant jets and waves, *Journal of Hydraulic Research*, IAHR, Vol. 42, N. 1, pp. 13-28.
Niepelt, A.; Bleninger, T. & Jirka, G.H. (2008). Desalination brine discharge modeling - Coupling of Hydrodynamic Models for Brine Discharge Analysis, *Proc. MWWD & IEMES 2008*, Croazia, Cavtat (Dubrovnik), Ooctober 27-31, 2008, ISBN 978-9944-5566-3-7, CD-ROM.
Papanicolau, P.N. & List, E.J. (1988). Investigation of round vertical turbulent buoyant jets, *Journal of Fluid Mechanics*, Vol. 195, pp. 314-391.
Pedersen, F.B. (1986). *Environmental Hydraulics: stratified flows – lecture notes on coastal and estuarine studies*. Springer Verlag, Berlin, Germany.
Platten, J.L. & Keffer, J.F. (1968). Entrainment in deflected axisymmetric jets at various angles to the stream, *Tech. Rep. 6808*, Dpt. of Mech. Eng., Univ. of Toronto.

Rajaratnam, N. (1976), *Turbulent jets*. Elsevier Scientific Publishing Company.

Talavera, J.L.P. & Ruiz, J.J.Q. (2001). Identification of the mixing processes in brine discharges carried out in Barranco del Toro beach - South of Gran Canaria (Canary Islands), *Desalination*, Vol. 139, pp. 277-287, ISSN 0011-9164.

Veltri, P. & Maiolo, M. (1992). Environmental Aspects in the Use of Sea Outfalls: a Sensitive Analysis. Marina Technology, *Proc. Second International Conference*, Edited by W.R.Blain, Computational Mechanics Publication, Southampton, March 1992.

Part 3

Nano, Ultra and Microfiltrations

9

The Influence of Electrochemical Properties of Membranes and Dispersions on Microfiltration

Petr Mikulášek and Pavlína Velikovská
University of Pardubice
Czech Republic

1. Introduction

Membrane filtration processes are widely used in various water and wastewater treatment applications. Crossflow microfiltration is a pressure driven membrane process, which has been shown effective in a great number of processes, including the removal of colloidal inorganic and organic solids.

The main factor limiting the application of crossflow microfiltration and other pressure-driven membrane processes is flux decline due to membrane fouling and concentration polarization. The steady value of permeate flux depends on various process variables such as cross-flow velocity, kind of membrane (pore size and chemical composition), the Reynolds number, shear stress and shear rate at the membrane surface in the membrane system and on physico-chemical particle-particle and particle-membrane interactions (Broussous et al., 2000; Huisman et al., 1999; Narong & James, 2006; Velikovská & Mikulášek, 2007). One way of characterizing these particle-particle and particle-membrane interactions is the knowledge of the ζ-potential of both particle and membrane. This electrostatic characterization of membranes is a useful way to predict and interpret the performance of microfiltration process. The magnitude of the ζ-potential gives the information of the stability of the system. Near the isoelectric point - IEP (the value of pH, where the charge and therefore ζ-potential of particles are equal to zero (Takagi & Nakagaki, 2001)) the system is unstable and the particles tend to flocculate. Therefore, the stability of the particles and the particle-membrane system could affect the separation process. Many studies showed that permeate flux, *J*, can be easily changed by pH, kind of added salt and salt concentration of the microfiltration dispersion (Baik & Seung, 2010; Gustafsson et al., 2000; Martín et al., 2003; Moritz et al., 2001; Mullet et al., 1997; Nazzal & Wiesner, 1994; Oo & Song, 2009).

Baik & Seung (2010) investigated surface charge properties of bentonite colloids to study their colloidal stability in a solution as a function of the pH and ionic strength. The results of ζ-potential measurements for the bentonite colloid showed that the bentonite colloids were stable at lower ionic strengths of 0.01 and 0.001M $NaClO_4$ but instable at a higher ionic strength of 0.1M $NaClO_4$ within the whole studied pH range.

Faibish et al. (1998) studied the effect of electrostatic double layer interaction on permeate flux decline and deposit cake formation in crossflow membrane filtration of colloidal suspensions. They found that the rate of flux decline is strongly dependent on solution ionic strength and, to a much lesser degree, on solution pH (for the investigated pH range of 6.1-10.0). Variations in flux decline rate with solution ionic strength are especially significant as the particle size decreases. For given physical and chemical conditions, the cake layer porosity increased with decreasing particle size, while cake permeability decreased with decreasing particle size.

Franks (2002) investigated the influence of Li^+, Na^+, K^+ and Cs^+ on the surface properties of silica particles. Hofmeister series (Cs^+, K^+, Na^+, Li^+) order ions from the least hydrated ions to the most hydrated ions. He found that the adsorption sequence on the silica surface follows the Hofmeister series ($Cs^+>K^+>Na^+>Li^+$) with Cs^+ adsorbing in greater quantities than Li^+. The measurements of ζ-potential indicate that adding of these ions has influence on the magnitude of the ζ-potential. Obtained results show that Cs^+ produces lower magnitude than Li^+. The results also showed that there exists a shift in the IEP to higher pH values as the salt concentration is increased. This indicates that the alkali cations are adsorbing to the silica surface in quantities greater than required for charge neutralisation. The greatest shift in the IEP is found to follow the greatest shift to least IEP shift $Cs^+>K^+>Na^+>Li^+$.

TiO_2 belongs to the mineral oxide. It is well known that if the surface of a mineral oxide is exposed to water, it becomes hydrated. The presence of water causes forming of the hydroxide layer containing MOH groups on the particle surface (M stands for a cation such as Al^{3+}, Zr^{4+}, Si^{4+} or Ti^{4+}). These amphoteric MOH surface groups are capable to dissociate when the surface gets in contact with polar liquids. The dissociation of amphoteric surface groups strongly depends on the pH of the solution and causes positive or negative charge of the surface.

Many authors (Elzo et al., 1998; Greenwood & Kendall, 2000; Gustafsson et al., 2000; Kosmulski et al., 1999; Kosmulski, 2002) showed that the charge of surface of mineral oxides particles consists of two parts. One part is the particles charge due to the dissociation of the ionisable groups of particles, and the other part is the particles charge affected by the electrolytes in the bulk solution.

Elzo et al. (1998) studied the charge effects on inorganic membrane performance on a cross-flow microfiltration of silica particles. The microfiltration experiments showed that the permeate flux is affected significantly by the pH, the salt concentration and the salt valency. Steady-state flux was higher at high pH and low salt concentration. In contrast, the flux was lowered at high salinity, low pH and in the presence of a bivalent salt such as $CaCl_2$. The higher fluxes were obtained when particles were highly charged, i.e. when strong repulsive interactions between the particles occurred.

Greenwood & Kendall (2000) studied surface of alumina particles in the presence of KCl. They found that there exists a shift of the isoelectric point to higher values of pH when the concentration of KCl increases. This shift may be due to some adsorption of potassium ions at the alumina interface. Authors also found out that the ζ-potentials were smaller in magnitude in the suspensions where the salt was present. This is due to screening of the surface charge by the electrolyte. The same results were shown by Gustafson et al. (2000). They studied the influence of pH and NaCl on the ζ-potential of anatase dispersions and they also found that the isoelectric point is shifted to higher pH values with increasing salt concentrations. The results also showed that addition of electrolyte affects thickness of double layer and the magnitude of ζ-potential. The magnitude of ζ-potential will decrease with increasing concentration of added electrolyte.

Kosmulski (2002) studied the electrochemical properties of alumina in the presence of high concentration of certain 1-1 electrolytes. His results showed that the shift in the IEP induced by different salts comply with the following rules: Salts with large cations (Cs^+) have a rather insignificant and anion independent effect. Salts with small cations (Li^+, Na^+) have a significant and anion-dependent effect. Salts with small anions (Cl^-) have a rather insignificant and cation-dependent effect.

Nyström et al. (2001) investigated the influence of Na^+, Ca^{2+}, Ba^{2+} and La^{3+} on the ζ-potential of calcite dispersions. The obtained results showed that adding salt influences the electrostatic interactions between the particles and leads to a change in the shear-induced aggregation. For Na^+ a pure electrostatic screening was observed. The two divalent cations, Ca^{2+} and Ba^{2+}, showed a similar adsorption behavior. Up to a certain concentration they caused an increased ζ-potential, as a result of specific adsorption, after which it decreased. From the results it is obvious that the monovalent sodium shows the weakest and the trivalent lanthanum the strongest effect.

Zhao et al. (2005) studied the influence of pH and different concentrations of NaCl on microfiltration of TiO_2 suspensions. They found that the steady flux increased with increasing ionic strength and decreasing pH. This was mainly due to the influences of ionic strength and pH on the dispersity of the TiO_2 suspensions-that was confirmed by the results of particle size of dispersion TiO_2.

For the microfiltration of particulate dispersions, surface properties of both dispersions and membranes vary with the solution environments. Although considerable work on the effects of pH and ionic strength have been reported, contradicting results have been obtained due to the complicate interactions between related substances in such separation system. Moreover, the influence of multivalent salts on membrane performance has received little or no attention. In fact, different inorganic electrolytes are always encountered in the treatment of industrial dispersions by membrane technology (Mulder, 2000).

The aim of present study was to show that pH, kind of salt and ionic strength are very important parameters that significantly affect the particle size of the dispersion and electrochemical properties of dispersion and membrane and by this way influence the microfiltration process at all. Attention was given to the ζ-potential measurements of the membrane surface and anatase dispersion under different solution environments. These measurements were combined with a series of cross flow microfiltration experiments at various pH values, salt concentrations and with NaCl, $MgCl_2$, $AlCl_3$, Na_2SO_4 and Na_3PO_4 electrolytes. Due to measurements of the permeate flux and the ζ-potential of the system particle-particle the filtration behavior was analyzed.

2. Experimental

2.1 Materials

The membranes used in the filtration experiments were tubular asymmetric multi-layered membranes (Terronic, a.s., Czech Republic). They consist of a thin layer deposited on the internal surface of the alumina support. The characteristics of the membrane are: length, $L = 0.25$ m; internal diameter, $d_i = 6 \times 10^{-3}$ m; nominal pore size, $d_p = 0.091$ μm; mean membrane resistance, $R_m = 1.17 \times 10^{12}$ m^{-1} (obtained from water flux measurements) and membrane surface area, $A_m = 43.35 \times 10^{-4}$ m^2. A new membrane was used in each experiment, and before the run the pure water flux was measured with deionised water.

The microfiltration experiments were performed with an aqueous dispersion of anatase obtained from Precheza a.s., Přerov, Czech Republic. The mean diameter of particles of pure dispersion (without acid or electrolyte) was 310 nm; however, the distribution of particles was very wide.

Selection of electrolytes used in this study was application-oriented. All chemical used were analytical grade and the solutions were prepared with ultra-pure water of conductivity less than 1 μScm^{-1}.

For investigating the effect of ionic strength on the performance of the ceramic membrane, NaCl solutions at different ionic strengths of 0.001 and 0.01M were prepared, respectively. To investigate the influences of different valent cations, NaCl, $MgCl_2$ and $AlCl_3$ solutions were used at an ionic strength of 0.001 and 0.01M. Na_2SO_4 and Na_3PO_4 solutions at an ionic strength of 0.001 and 0.01M were prepared for investigating the influence of different valent anions on the performances of the ceramic membrane. The pH of the solution was adjusted with NaOH and HCl solutions.

2.2 Experimental methods

2.2.1 Separation experiments

The microfiltration experimental apparatus used is shown schematically in Figure 1.

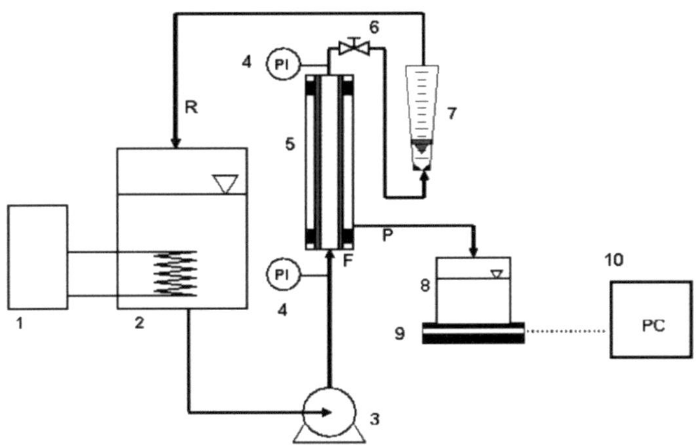

Fig. 1. Schematic diagram of crossflow microfiltration system
1 - tempering system with thermostat, 2 - stock tank, 3 – pump, 4 - sensor of pressure, 5 - membrane module, 6 - needle valve, 7 – rotameter, 8 – beaker for permeate, 9 – scale, 10 - output on PC. F – feed, P – permeate, R – retentate.

The circulating loop was constructed of stainless steel and contained a three litre feed suspension reservoir, pump and the membrane module. This loop was also equipped with a pressure and flow monitoring system. The crossflow velocity was controlled using the speed of the pump and the operating pressure (and hence transmembrane pressure) in the loop was controlled by a needle valve at the module outlet; both parameters were varied independently. The loop was also equipped with a temperature regulating system.

In the experiments anatase suspensions were used with a concentration of 5% by weight at various pH and with or without electrolyte NaCl. During all the tests, the filtration was run at a constant crossflow velocity of 1 m s^{-1} and a constant pressure of either 100 or 200 kPa.

Before each filtration experiment the resistance of membrane was measured. The membrane was placed into the membrane module, and deionised water was circulated in the test loop at a pressure 100 kPa and a velocity of 1 m s^{-1} for about 30 minutes. During this time a stable value of the permeability of the membrane to water was observed. From the measured permeate flow rate, the membrane resistance R_m was calculated from:

$$R_m = \frac{\Delta P}{\mu_p J_w} \tag{1}$$

where μ_p is dynamic viscosity of the permeate, J_w is the permeate flux for deionised water and ΔP is the transmembrane pressure.
After this measurement, the dispersion was placed into the reservoir and the filtration was started, the operating pressure as well as feed velocity was adjusted by the regulation system. The stock dispersion was kept at a constant temperature of 25°C. The flux through membrane was measured by weighing permeate and timing the collection period (by use of a balance interfaced with a computer). Both the retentate and permeate were recirculated back into the reservoir. Therefore, the concentration in the recirculation loop remained virtually constant.
The experiment was stopped after obtaining a constant permeate flux. The adjustment of pH value was performed 18 hours before the filtration experiments, to allow stable equilibration conditions to be reached in the dispersion (that is, an equilibrium state of particle-particle interactions). After each set of experiments the unit and the membranes were rinsed with deionised water and the pure water flux was measured again under condition of initial test until the steady state was attained. From this value was the membrane resistance was calculated again. The fouling tendency of the membrane can be calculated from the difference between the two resistances (that is the "before" and "after" filtration resistances). The total filtration resistance R_t was calculated from:

$$R_t = \frac{\Delta P}{\mu_p J_\infty} \tag{2}$$

where J_∞ is the steady state permeate flux. The difference between the total filtration resistance and the membrane resistance is filter cake resistance R_c.

2.2.2 Particle size and ζ–potential measurements

The particle size and ζ-potential measurements were carried out on the instrument Zeta PALS (Brookhaven Instruments Corp., USA).
The stock dispersion was diluted to the concentration of 1 kg m^{-3} using permeate, in order to retain the same solution conditions (pH, salt concentration) that existed during filtration. The ζ- potential of the particles was calculated from a measurement of the electrophoretic mobility using:

$$\zeta = \frac{\mu \upsilon}{\varepsilon_o \varepsilon_r} \tag{3}$$

where υ is the mobility of the particles, μ is the dynamic viscosity and ε_o and ε_r are permittivity of vacuum and relative permittivity of the liquid.
One membrane was ground in the oscillatory mill. This powder was dispersed in permeate by the help of ultrasonic bath and than the ζ-potential of this sample was measured. The sample concentration was 1 kg m^{-3}, too.

3. Results and discussion

3.1 The pH dependence of the ζ–potential of TiO_2 particles in the presence of various electrolytes

The Fig. 2 shows the pH dependence of the ζ-potential of TiO_2 particles in the presence of NaCl, $MgCl_2$, $AlCl_3$, Na_2SO_4 and Na_3PO_4 (at two ionic strengths of 0.001 and 0.01M), and without any salt. Fig. 3 shows the same pH dependence of the ζ-potential of TiO_2 in the presence of the rest of studied electrolytes - Na_2SO_4 and Na_3PO_4 (at two ionic strengths of 0.001 and 0.01M).

As shown in the Fig. 2 and Fig. 3, the surface charge of the particles is strongly dependent on the pH, kind of salt and ionic strength of the electrolyte solution. With the increase of the pH from 2 to 8, the ζ-potential of the TiO_2 particles decreased and became negative at the isoelectric point (IEP); exception is $AlCl_3$ for which there is no IEP and the ζ-potential is always positive.

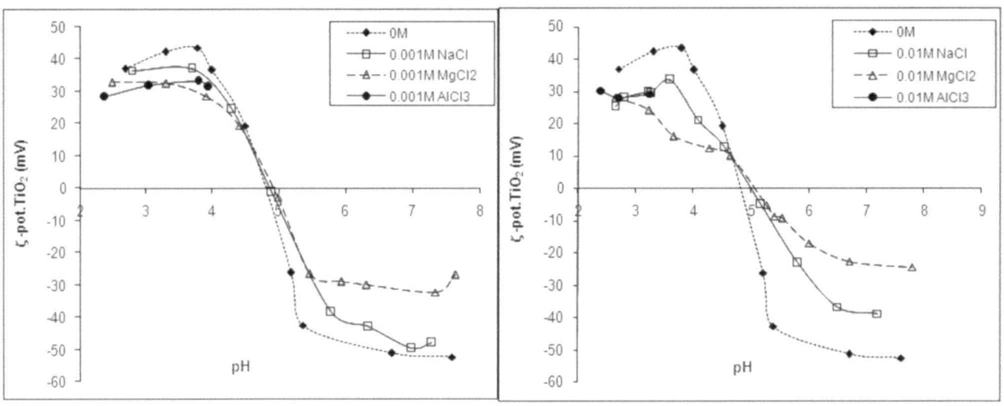

Fig. 2. The pH dependence of the ζ-potential of TiO_2 particles in the NaCl, $MgCl_2$ and $AlCl_3$ solution under different ionic strengths

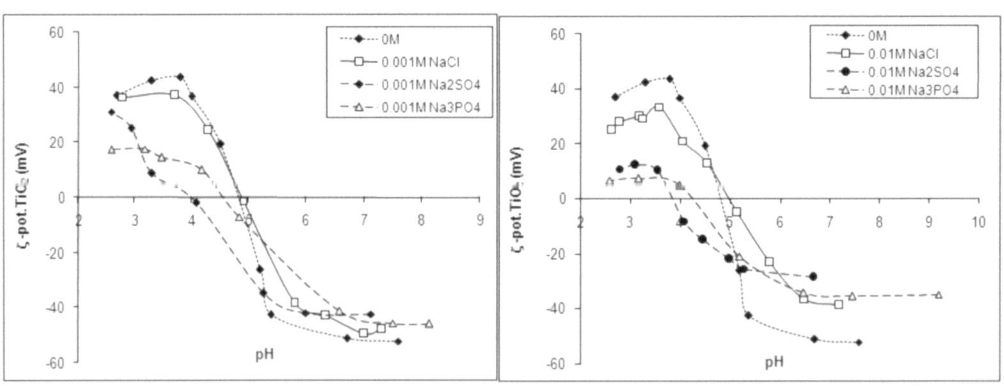

Fig. 3. The pH dependence of the ζ-potential of TiO2 particles in the NaCl, Na2SO4 and Na3PO4 solution under different ionic strengths

From the curves shown in Fig. 2 and Fig. 3 it is obvious that the ζ-potential decreased with increasing ionic strength. These results may be explained by a decrease in the effective thickness of the diffuse layer as the ionic strength increases. From the results it can be also found that IEP of TiO_2 dispersion without any salt and with addition of NaCl is nearly the same for both the concentrations 0.001 and 0.01M. This fact indicated that there is absence of a specific adsorption of Cl^- and Na^+ ions on the particle surface.

From the curves shown in Fig. 2 and Fig. 3 it is apparent, that in comparison with the results obtained for TiO_2 dispersion without any salt the adding of NaCl and $MgCl_2$ induces the shift of IEP to higher values of pH, while the adding of Na_2SO_4 and Na_3PO_4 causes the shift to lower pH values. The results indicate the specific adsorption of cations in the case of NaCl and $MgCl_2$, respectively the specific adsorption of anions in the case of Na_2SO_4 and Na_3PO_4.

For electrolytes NaCl and $MgCl_2$ and both of the studied concentrations of added electrolyte the magnitude of ζ-potential of TiO_2 was found in the order $MgCl_2 <$ NaCl. The difference in the magnitude of ζ-potential is in the area pH>pH(IEP) caused by the adsorption of cations. The diameter of these cations as different, and therefore exists the difference in hydration of these cations. The bigger and therefore poorly hydrated ion (Mg^{2+}) adsorbs in greater quantity to the TiO_2 surface than the smaller and well-hydrated ion (Na^+) and by this way Mg^{2+} causes bigger compression of the diffuse layer than Na^+ and produces lower magnitude of ζ-potential of TiO_2 particles. At the same time the valence of Mg^{2+} is two time bigger than in the case of Na^+. From this is obvious, that the same amount of Mg^{2+} cation affects the magnitude of ζ-potential more than Na^+ cation.

In the area where is the pH<pH(IEP), the difference in the magnitude of ζ-potential is caused by the anions. The concentration of Cl^- anions is in the case of $MgCl_2$ two times higher than in the case of NaCl. The adding of $AlCl_3$ has the same effect as the addition of $MgCl_2$.

In the area of pH where is the pH<pH(IEP), the magnitude of ζ-potential of TiO_2 particles is influenced by adsorption of the anions. The bigger and poorly hydrated anion PO_4^{3-} adsorbs in greater quantity to the surface of TiO_2 than small and well-hydrated anion Cl^- and produces lower magnitude positive ζ-potential. This result was found for both studied concentrations of the electrolytes (see Fig. 3).

In the area where the pH>pH(IEP), the magnitude of ζ-potential is affected by adsorption of cations. For 0.001M concentration of added electrolyte (see Fig. 3) the magnitude of ζ-potential was nearly the same for all kind of electrolytes. For 0.01M concentration of NaCl, Na_2SO_4 and Na_3PO_4 was found that in comparison with the results obtained for TiO_2 dispersion without any salt the adding of NaCl, Na_2SO_4 and Na_3PO_4 produces lower magnitude of ζ-potential of TiO_2 particles.

3.2 The pH dependence of the particle size of TiO_2 particles in the presence of various electrolytes

The dependences of diameter of TiO_2 particles on the value of pH are shown in Fig. 4 and Fig. 5.

From the curves shown in Fig. 4 and Fig. 5 it is obvious that the diameter of TiO_2 particles increased with decreasing value of pH. By the comparison of results obtained for 0.001M and 0.01M concentration of electrolytes and was found that increasing ionic strength of the added electrolyte causes increase in the particle diameter.

For all studied systems there was found a step change in the particle size. In all cases these step change of particle size happened near the value of pH corresponding to the IEP.

The size of the particles tend to increase in the order NaCl < $AlCl_3$ < <$MgCl_2$ < Na_2SO_4 < Na_3PO_4. These results may be explained by adsorption of the counter ions into the Stern layer of the TiO_2 particle.

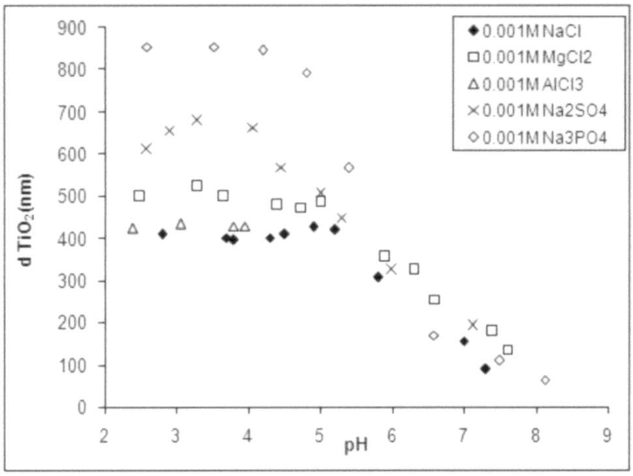

Fig. 4. Particle diameter of TiO_2 as a function of pH and kind of electrolyte; electrolyte concentration 0.001M

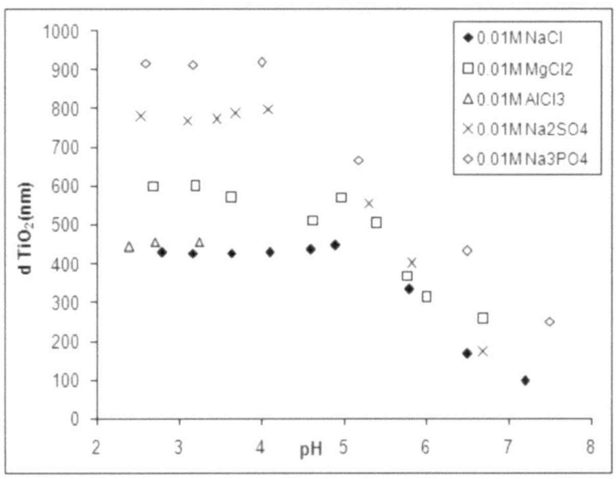

Fig. 5. Particle diameter of TiO_2 as a function of pH and kind of electrolyte; electrolyte concentration 0.01M

3.3 Steady value of the permeate flux as a function of pH, kind of electrolyte and ionic strength

To investigate the influences of pH, kind of electrolyte and its concentration a set of microfiltration experiments was carried out. All experiments were performed at the same conditions. The cross-flow velocity was $1 m.s^{-1}$, the temperature was set to 25°C. The transmembrane pressure was 100 or 200kPa. Fig. 6 shows the results from the cross-flow microfiltration experiments as a function of pH for 0.001M concentration of electrolytes NaCl, $MgCl_2$ and $AlCl_3$. The studied transmembrane pressure was 100 and 200kPa. Fig. 7 shows the same dependencies this time for the 0.01M concentration of mentioned electrolytes.

As can be seen from Fig. 6 and Fig. 7, all the dependencies had a similar trend. The values of steady permeate flux decreased as the pH value increased in a pH range of 2 to 8. All dependencies have maximum permeate flux near the isoelectric point of dispersion. From Fig. 6 and Fig. 7 it is obvious, that the steady flux increased with increasing ionic strength of NaCl, $MgCl_2$. Exception is $AlCl_3$; increasing concentration of $AlCl_3$ caused decrease in the steady of permeate flux.

From the curves shown in Fig. 6 and also from Fig. 7 it is apparent, that increase in the value of transmembrane pressure causes growth of the steady value of permeate flux.

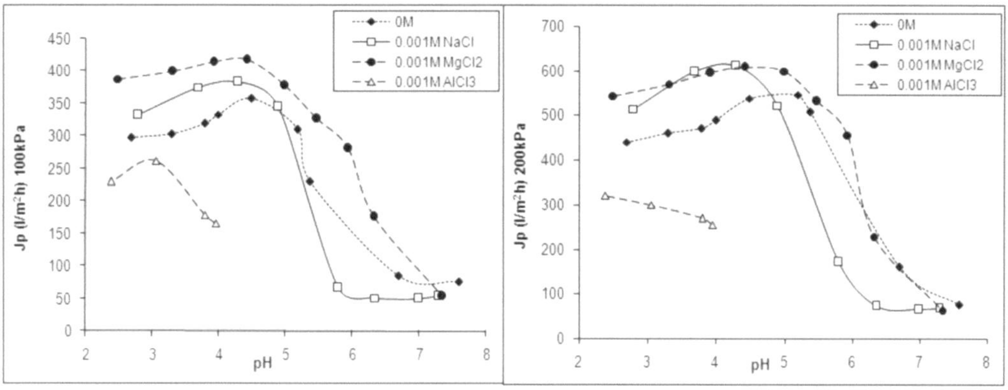

Fig. 6. Effect of pH, transmembrane pressure and 0.001M concentration of NaCl, $MgCl_2$ and $AlCl_3$ on steady permeate flux

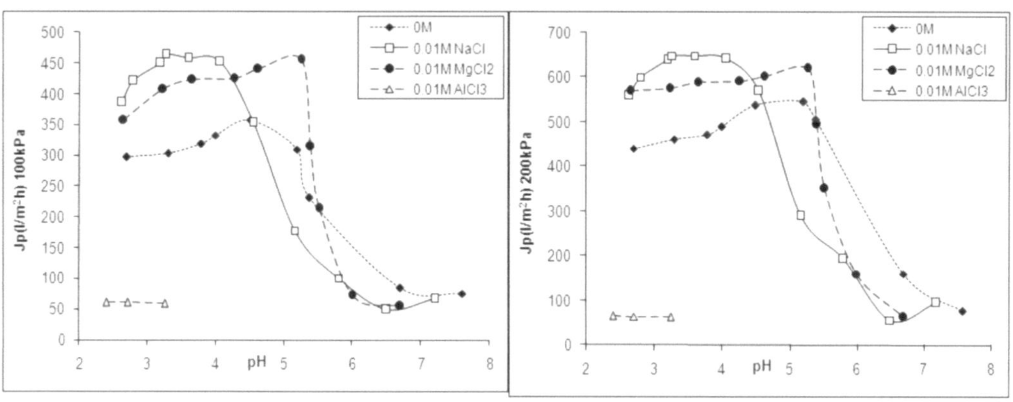

Fig. 7. Effect of pH, transmembrane pressure and 0.01M concentration of NaCl, $MgCl_2$ and $AlCl_3$ on steady permeate flux

The last phenomenon resulted from Fig. 6 and Fig. 7 is the effect of kind of the electrolyte on the steady value of permeate flux. Adding of the NaCl and $MgCl_2$ in comparison with the results obtained for TiO_2 dispersion without any salt produces higher values of steady permeate flux. For the 0.001M concentration and 100kPa transmembrane pressure value the maximum of steady value of permeate flux observed with the $MgCl_2$ is higher than that observed with NaCl electrolyte. In all other cases (0.001M NaCl and $MgCl_2$ 200kPa; 0.01M NaCl and $MgCl_2$ 100 and 200kPa) the maximum steady value of permeate flux was higher in the presence of NaCl.

The growth of the steady permeate flux in the presence of electrolyte is caused by the specific adsorption of counter ions on the particle surface what results in less repulsive forces between the particles and in increasing size of the particles. Both of these factors improve the conditions for effective microfiltration.

The Fig. 8 shows the results from the cross-flow microfiltration experiments as a function of pH for 0.001M concentration of electrolytes NaCl, Na_2SO_4 and Na_3PO_4. The studied transmembrane pressure values were 100 and 200kPa. The Fig. 9 shows the effect of pH and transmembrane pressure for 0.01M concentration of electrolytes NaCl, Na_2SO_4 and Na_3PO_4. The same way as the curves shown in Fig. 6 and Fig. 7, all the dependencies from Fig. 8 and Fig. 9 had a similar trend. The values of steady permeate flux decreased as the pH value increased in a pH range of 2 to 8. All dependencies have maximum permeate flux near the isoelectric point of dispersion.

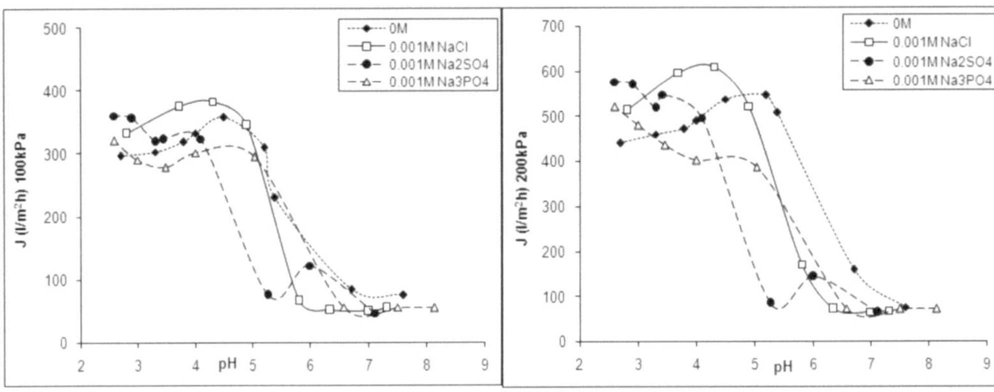

Fig. 8. Effect of pH, transmembrane pressure and 0.001M concentration of NaCl, Na_2SO_4 and Na_3PO_4 on steady permeate flux

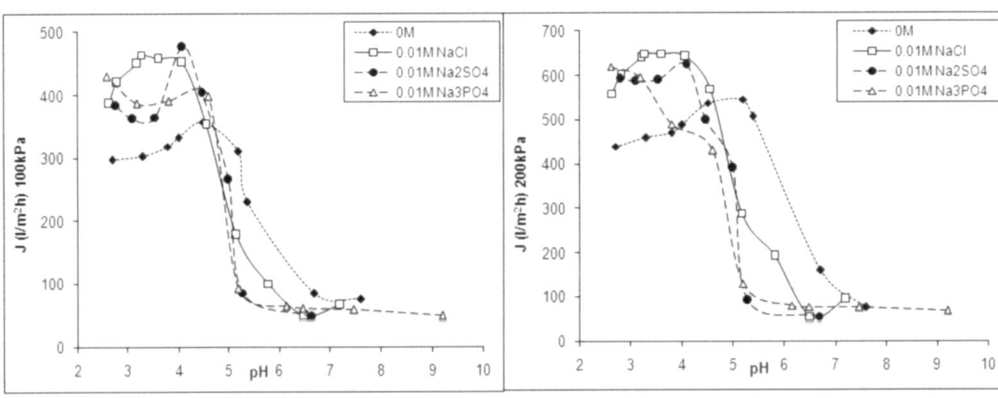

Fig. 9. Effect of pH, transmembrane pressure and 0.01M concentration of NaCl, Na_2SO_4 and Na_3PO_4 on steady permeate flux

From Fig. 8 and Fig. 9 it is obvious, that in comparison with the results obtained for TiO_2 dispersion without any salt addition of NaCl produces higher values of steady permeate flux no matter what the concentration of NaCl or the value of the transmembrane pressure is. Addition of electrolyte Na_2SO_4 and Na_3PO_4 caused the growth of permeate flux only in narrow range of pH values. This range of pH values for 0.001M electrolytes Na_2SO_4 and Na_3PO_4 is from pH 2 to pH 3 and for 0.01M concentration of Na_2SO_4 and Na_3PO_4 from pH 2

to pH 5. From this result it is obvious that the 0.01M concentration of electrolyte affected the microfiltration process much more than the concentration of 0.001M.
This result is in contrast to that observed with microfiltration of silica particles (Elzo et al., 1998). The authors found that high permeate fluxes were obtained at low salt concentrations and they attributed this event to the strong repulsion between the silica particles at low salt concentrations. Such a contradiction may be attributed to the different nature of the particles. There was no flocculation of silica particles in their work. However, TiO_2 particles flocculated easily as the solution environment changed in our study.
As mentioned before the growth of the steady permeate flux in the presence of electrolyte is caused by the specific adsorption of counter ions on the particle surface what results in less repulsive forces between the particles and in increasing size of the particles.

3.4 Modeling of the value of permeate flux as a function of ζ–potential of TiO_2 particles

The dependence of the values of permeate flux as a function of ζ-potential of TiO_2 particles was modeled by the following equation reported in the literature (Doran, 1995):

$$y = y_0 + y_\infty \cdot e^{a \cdot e^{b(x - x_0)}} \tag{4}$$

Which was because of the modeling of the values of permeate flux as a function of ζ-potential of TiO_2 particles arranged into the following form:

$$J = J_D + J_H \cdot e^{a \cdot e^{b(\zeta - zeta_o)}} \tag{5}$$

where J is steady value of permeate flux [l m^2h^{-1}],
J_D - parameter [l m^2h^{-1}],
J_H - parameter [l m^2h^{-1}],
a - parameter [-],
b - parameter [mV^{-1}],
ζ - ζ-potential of TiO_2 particles [mV],
$zeta_o$ - parameter [mV^{-1}].

For single electrolyte and its concentrations, by the transmembrane pressure difference value of 100 and 200kPa, were by the minimizing the function $((J_{exp.} - J_{cal.})/J_{exp.})^2$ observed values of the parameters J_D, J_H, a, and b ($J_{exp.}$ are the experimental values of steady permeate flux, $J_{cal.}$ are values of permeate flux calculated by model described by Eq. (2)).
The resulting values of the parameters J_D, *a*, *b*, *and* J_H are presented in Table 1.

	100 kPa	200 kPa
J_D[$lm^{-2}h^{-1}$]	55.10	55.10
a [-]	-10.55	-10.55
b [mV^{-1}]	-0.41	-0.41
J_H[$lm^{-2}h^{-1}$]	340.1	507.9

Table 1. The resulting values of parameters J_D, *a*, *b* a J_H (Eq. (5)) as a function of transmembrane pressure

While the parameters J_D, *a*, *b* and J_H are independent of the kind of electrolyte and its concentration and theirs values changed only with the change of the value of transmembrane pressure, the parameter *zeta*$_o$ is very dependent on the kind and

concentration of the electrolyte. The kind and the concentration of the electrolyte can be express by the ionic strength.
The values of parameter $zeta_o$ as a function if ionic strength of dispersions are presented in Table 2. The parameter $zeta_o$ is independent on the value of transmembrane pressure and therefore for the transmembrane pressure value of 100 and 200kPa are the values of parameter $zeta_o$ the same.
The dependence of the value of parameter $zeta_o$ on the ionic strength of TiO_2 dispersion is possible to describe by following Eq. (6):

$$zeta_o = -8496\, I^2 + 850.8\, I - 43.24 \quad (6)$$

where $zeta_o$ is parameter [mV^{-1}],
I - ionic strength of dispersion of TiO_2 [mol/l].

	c [mol/l]	I [mol/l]	$zeta_0$ [mV]
without electrolyte	0	0	-47.36
NaCl	0.001	0.001	-41.18
$MgCl_2$	0.001	0.003	-39.95
Na_2SO_4	0.001	0.003	-39.95
Na_3PO_4	0.001	0.006	-36.24
NaCl	0.01	0.01	-35,64
$MgCl_2$	0.01	0.03	-25.94
Na_2SO_4	0.01	0.03	-25.94
Na_3PO_4	0.01	0.06	-22.52

Table 2. Values of the parameter $zeta_o$ for different ionic strength

The deviations between the experimental results and values of steady permeate flux calculated by using Eq. (5) (for the values of the parameters presented in Tables 1 and 2) for chosen electrolyte and its concentration are presented in Tables 3 and 4.

	100 kPa		200 kPa	
$\zeta(TiO_2)$	$J_{exp.}$ [$lm^{-2}h^{-1}$]	$J_{cal.}$ [$lm^{-2}h^{-1}$]	$J_{exp.}$ [$lm^{-2}h^{-1}$]	$J_{cal.}$ [$lm^{-2}h^{-1}$]
25.20	386.3	395.2	559.4	562.9
28.00	420.6	395.2	600.5	562.9
29.90	450.0	395.2	640.3	562.9
29.40	464.4	395.2	649.0	562.9
33.56	458.3	395.2	649.0	562.9
21.18	452.0	395.2	643.0	562.9
12.70	353.8	395.2	569.8	562.9
-4.90	178.2	395.2	288.9	562.9
-23.14	100.6	374.9	192.6	532.7
-36.60	49.7	55.1	55.3	55.1
-38.70	66.3	55.1	94.0	55.1

Table 3. Comparison of experimental and calculated values of steady permeate flux as a function of ζ-potential of TiO_2 dispersion in the presence of 0.01M NaCl

The deviations between the experimental results and values of the steady permeate flux calculated by using Eq. (5) (for the values of the parameters presented in Tables 1 and 2) are caused by the fact, that the model reflects parameters as the ζ-potential of dispersion, kind and concentration of electrolyte (by using the ionic strength). The model does not reflect the parameters like the particle size of the dispersion, ζ-potential of the membrane or physico-chemical particle-particle and particle-membrane interactions.
Based on the results presented in Tables 3 and 4 (and the others comparison of experimental and calculated values of steady permeate flux, which are not presented in this work) is possible to say that the model represented by Eq. (5) presents good correlation between experimental and prediction values of steady permeate flux.

	100 kPa		200 kPa	
$\zeta(TiO_2)$	$J_{exp.}$ $[lm^{-2}h^{-1}]$	$J_{cal.}$ $[lm^{-2}h^{-1}]$	$J_{exp.}$ $[lm^{-2}h^{-1}]$	$J_{cal.}$ $[lm^{-2}h^{-1}]$
31.16	360.8	395.2	576.0	562.9
25.34	356.1	395.2	570.2	562.9
8.64	319.3	395.2	517.8	562.9
-1.87	323.8	395.2	496.5	562.9
-34.91	74.6	145.3	83.5	189.7
-41.99	121.8	55.1	144.5	55.1
-42.42	45.7	55.1	63.7	55.1

Table 4. Comparison of experimental and calculated values of steady permeate flux as a function of ζ-potential of TiO_2 in the presence of 0.001M Na_2SO_4

4. Conclusion

The present study is focused on the study of the influence of electrochemical properties of dispersions and membranes on the microfiltration of model dispersion of TiO_2 on asymmetric tubular α-Al_2O_3 microfiltration membranes.
The permeate flux of the ceramic membrane in the microfiltration of TiO_2 suspension was affected significantly by the particles environment including pH and ionic strengths. The steady-state flux increased with increasing ionic strength in the presence of electrolytes, NaCl, Na_2SO_4 and Na_3PO_4, and decreasing pH. This was mainly due to the influences of ionic strength and pH on the dispersity of TiO_2 suspensions. The results showed that the influences of inorganic salts on the microfiltration of TiO_2 dispersion were related to changes in the surface properties of both the ceramic membrane and TiO_2 particles.
The results of the experiments show that the ζ-potential of particles as well of membrane surface have important effects on the permeate flux. This is especially pronounced during microfiltration of the dispersion when it is close to its isoelectric point, when the value of permeate flux increased to about fivefold the value of non-treated dispersion. This is due to the particle interactions; close to the isoelectric point the dispersion had a tendency towards instability, the particle aggregation was evident. It resulted in either deposition of a lower thickness of cake or of one with a higher porosity, thereby leading to higher values of permeate fluxes and lower filter cake resistances.
From the presented results the following general conclusions may be drawn:

- ζ-potential of the TiO_2 particles and also ζ-potential of membrane surface decreased with increasing ionic strength;
- Addition of electrolytes caused the shift of the value of the IEP (to lower pH values in case of Na_2SO_4 and Na_3PO_4 and to higher values of pH in the case of NaCl and $MgCl_2$; addition of $AlCl_3$ did not cause any shift, because there is no IEP and the value of ζ-potential of membrane and particles is always);
- The shift of IEP to lower pH values indicate the specific adsorption of anions, while the shift of IEP to higher values of pH indicate specific adsorption of cations;
- The diameter of TiO_2 particles increased with decreasing value of pH;
- Increasing ionic strength of the added electrolyte caused increase in particle diameter (because of the adsorption of counter ions on the particle surface);
- For all studied systems the steady permeate flux decreased as the pH value increased and the maximum permeate flux was near the isoelectric point of dispersion;
- The addition of electrolytes NaCl and $MgCl_2$ in concentration of 0,001 and 0,01M produces higher values of steady permeate flux in comparison with the results obtained for TiO_2 dispersion without any salt;
- Addition of electrolytes Na_2SO_4 and Na_3PO_4 in the concentration of 0.001M caused the growth of permeate flux only in narrow range of pH values (pH 2 - 3);
- Addition of electrolytes Na_2SO_4 and Na_3PO_4 in concentration of 0.01M caused the growth of steady values of permeate flux in the range of pH 2 - 5.

The results showed that the influences of inorganic salts on the microfiltration of TiO_2 dispersion were related to changes in the surface properties of both the ceramic membrane and TiO_2 particle. The influence of solution environment on the dispersity of the TiO_2 dispersion played a major role in the permeate flux. When the particles were well dispersed, changes in the surface charge of the membrane would be important in determining membrane performance. Electroviscous effects are more complicated both on membrane and cake layer. Further investigations are deserved.

5. Acknowledgement

This project was financially supported by Ministry of Education, Youth and Sports of the Czech Republic, Project MSM0021627502.

6. References

Baik, M.H. & Seung, Y.L. (2010). Colloidal stability of bentonite clay considering surface charge properties as a function of pH and ionic strength. *Journal of Industrial and Engineering Chemistry*, Vol.16, No.5, (September 2010), pp. 837-841, ISSN 1226-086X

Broussous, L; Schmitz, P.; Boisson, H.; Prouzet, E. & Larbot, A. (2000). Hydrodynamic aspects of filtration antifouling by helically corrugated membranes. *Chemical Engineering Science*, Vol.55, No.21, (November 2000), pp. 5049-5057, ISSN 0009-2509

Doran, P.M. (1995). *Bioprocess Engineering Principles.* Academic Press, ISBN 0-12-220856-0, London, Great Britain

Elzo, D.; Huisman, I.; Middelink, E. & Gekas, V. (1998). Charge effects on inorganic membrane performance in a cross-flow microfiltration process. *Colloids and Surface A-Physicochemical and Engineering Aspects*, Vol.138, No.2-3, (July 1998), pp. 145-159, ISSN 0927-7757

Faibish, R.; Elimelech, M. & Cohen, Y. (1998). Effect of interparticle electrostatic double layer interactions on permeate flux decline in crossflow membrane filtration of colloidal suspensions. *Journal of Colloid and Interface Science*, Vol.204, No.1, (August 1998), pp. 77-86, ISSN 0021-9797

Franks, G.V. (2002). Zeta potentials and yield stresses of silica suspensions in concentrated monovalent electrolytes: isoelectric point shift and additional attraction. *Journal of Colloid and Interface Science*, Vol.249, No.1, (May 2002), pp. 44-51, ISSN 0021-9797

Greenwood, R. & Kendall, K. (2000). Effect of ionic strength on the adsorption of cationic polyelectrolytes onto alumina studied using electroacoustic measurements. *Powder Technology*, Vol.113, No.1-2, (November 2000), pp. 148-157, ISSN 0032-5910

Gustafsson, J.; Mikkola, P.; Jokinen, M. & Rosenholm, J.B. (2000). The influence of pH and NaCl on the zeta potential and rheology of anatase dispersions. *Colloids and Surface A-Physicochemical and Engineering Aspects*, Vol.175, No.3, (December 2000), pp. 349-359, ISSN 0927-7757

Huisman, I.H.; Tragardh, G. & Tragardh, Ch. (1999). Particle transport in crossflow microfiltration - II. Effects of particle-particle interactions. *Chemical Engineering Science*, Vol.54, No.2, (January 1999), pp. 281-289, ISSN 0009-2509

Kosmulski, M.; Gustafsson, J. & Rosenholm, J.B. (1999). Ion specificity and viscosity of rutile dispersions. *Colloid and Polymer Science*, Vol.277, No.6, (June 1999), pp. 550-556, ISSN 0303-402X

Kosmulski, M. (2002). Confirmation of the differentiating effect of small cations in the shift of the isoelectric point of oxides at high ionic strengths. *Langmuir*, Vol.18, No. 3, (February 2002), pp. 785-787, ISSN 0743-7463

Martín, A.; Martínez, F.; Malfeito, J.; Palacio, L.; Prádanos, P. & Hernández, A. (2003). Zeta potential of membranes as a function of pH. Optimization of isoelectric point evaluation. *Journal of Membrane Science*, Vol.213, No.1-2, (March 2003), pp. 225-230, ISSN 0376-7388

Moritz, T.; Benfer, S.; Árki, P. & Tomandl, G. (2001). Influence of the surface charge on the permeate flux in the dead-end filtration with ceramic membranes. *Separation and Purification Technology*, Vol.25, No.1-3, (October 2001), pp. 501-508, ISSN 1383-5866

Mulder, M. (2000). *Basic Principles of Membrane Technology*, 2nd Ed., Kluwer Academic Publishers, ISBN 0-7923-4248-8, Dordrecht, The Netherlands

Mullet, M.; Fievet, P.; Reggiani, J.C. & Pagetti, J. (1997). Surface electrochemical properties of mixed oxide ceramic membranes: Zeta-potential and surface charge density. *Journal of Membrane Science*, Vol.123, No.2, (January 1997), pp. 255-265, ISSN 0376-7388

Narong, P. & James, A.E. (2006). Effect of the ζ-potential on the micro/ultrafiltration of yeast suspensions using ceramic membranes. *Separation and Purification Technology*, Vol.49, No.2, (April 2006), pp. 149-156, ISSN 1383-5866

Nazzal, F.F. & Wiesner, M.R. (1994). pH and ionic strength effects on the performance of ceramic membranes in water filtration. *Journal of Membrane Science*, Vol.93, No.1, (August 1994), pp. 91-103, ISSN 0376-7388

Nyström, R.; Lindén, M. & Rosenholm, J.B. (2001). The influence of Na^+, Ca^{2+}, Ba^{2+} and La^{3+} on the ζ-potential and the yield stress of calcite dispersions. *Journal of Colloid and Interface Science*, Vol.242, No.1, (October 2001), pp. 259-263, ISSN 0021-9797

Oo, M.H. & Song, L. (2009). Effect of pH and ionic strength on boron removal by RO membranes. *Desalination*, Vol.246, No.1-3, (September 2009), pp. 605-612, ISSN 0011-9164

Takagi, R. & Nakagaki, M. (2001). Characterization of the membrane charge of Al_2O_3 membranes. *Separation and Purification Technology*, Vol.25, No.1-3, (October 2001), pp. 369-377, ISSN 1383-5866

Velikovská, P. & Mikulášek, P. (2007). The influence of Cl^-, SO_4^{2-} and PO_4^{3-} ions on the ζ-potential and microfiltration of titanium dioxide dispersions. *Separation and Purification Technology*, Vol.58, No.2, (December 2007), pp. 295-298, ISSN 1383-5866

Zhao, Y.; Zhang, Y.; Xing, W. & Xu, N. (2005). Influences of pH and ionic strength on ceramic microfiltration of TiO_2 suspensions. *Desalination*, Vol.177, No.1-3, (June 2005), pp. 59-68, ISSN 0011-9164

10

Nanofiltration and Low Energy Reverse Osmosis for Advanced Wastewaters Treatment

Gamal Khedr
National Research Centre, Cairo, Egypt

1. Introduction

Nanofiltration (NF) and low energy reverse osmosis (LERO) are pressure driven, selective membrane separation techniques that progressively acquire new fields of application in treatment of surface, ground and wastewater (WW) (Deraniyagala, 2006) in view of their continuous development. Trends of their development are lower energy requirements, higher efficiency and specificity solute rejection, longer working lifetime at steady performance, and consequently higher process cost effectiveness, in addition to promotion of environmental safety. In fact, according to the results presented in the present chapter, NF is expected to replace, in the near future, several conventional techniques of WW treatment (Khedr, 2004, 2008, 2009).

NF membranes are generally characterized by relatively high surface charge (Childress et al, 2000), usually negative, and pores in the range of nanometer diameter (Jiang et al, 2004). In contradiction with the usually stated oversimplified description of NF membranes as intermediate type between loose RO membranes and tight ultrafiltration ones, the permeation performance of these membranes is mainly determined by their surface charge density, pores diameter and distribution, surface roughness and fouling susceptibility depending on the simultaneously working mechanisms of solute species size exclusion and electrostatic interaction (Thanuttamavong et al, 2002).

In general, NF membranes have high permeate flux and rather high rejection of polyvalent ions at low or moderate rejection of TDS. Therefore, they have lower opposite osmotic pressure and, consequently, lower operation pressure and lower energy consumption than RO under the same operation conditions (Khedr, 2008). The moderate salt rejection in NF also maintains the stability of the water and prevents it from being aggressive to network piping.

Efficient rejection of organics by NF and LERO was reported in a wide range of applications in wastewater treatment, food, chemical and pharmaceutical industries, for purification of industrial effluents and minimizing waste discharge. NF was also applied for rejection of natural organic matter which imparts color (Jensen & Thorsen, 1995). Efficiency of NF in removal of organics, suspended solids, colloids, microorganisms, and hardness components enabled its application for pre-treatment of sea water prior to SWRO or MSF in hybrid desalination processes. Such pre-treatment promoted the percent recovery of the subsequent RO step to 65% instead of 40% and lowered the fouling potential of RO membranes so as to lower the maintenance rate (Hassan et al, 1998, 2000; khedr, 2004).

Rejection of organic compounds, (micropollutants) is mainly controlled by their physicochemical properties, membrane material, and characteristics like porosity, pore diameter, and surface charge. Compounds with higher molecular weight show higher rejection. However, other factors as static and polar effects, degree of dissociation and adsorption play an important role in the rejection mechanism (Hofman et al, 1995).
Selection of adequate membrane process for decontamination of WW or groundwater depends on the level of contamination and the possible parallel requirement of desalination. In view of the operation pressure, energy consumption in membrane processes increases in the order NF < LERO < Conventional RO.

2. Evaluation of performance of selective membrane methods, NF and LERO in treatment of WWs and contaminated groundwater in comparison with the conventional methods

Several recent WWs treatment problems revealed the low efficiency or failure of conventional methods and suggested the search of more adequate methods of higher efficiency and wider validity ranges. As an example the radioactive contamination of groundwater common to certain areas in USA and the Middle East at levels that are dangerous for drinking is due, in fact, to ultratrace concentrations of radioisotopes as Ra^{2+} in mixture with other dissolved salts of more than 10 orders of magnitude greater concentration and may be of similar chemical nature as Ca^{2+} which would farther complicate the separation process (Clifford, 1991).
As for removal of heavy metal cations (HMCs) contamination of some surface, groundwaters and industrial WWs, granular activated carbon (GAC) was reported to be inefficient adsorbent in view of the higher temperature and pH dependence of the cation uptake (Gabaldon & Gonzalez, 2000). NF is expected to enable higher separation efficiency at lower cost.
On the other hand, the increasingly stringent norms of environmental protection render the disposal of the industrial or municipal WWs or sludge a major factor in the selection of industrial or municipal waste treatment processes.
The following results are part of a long term investigation (Khedr, G, 2004, 2007) that aims to evaluate the removal of contaminants by the modern cross flow membrane methods NF and LERO, in single or hybrid processes, in comparison with some of the common conventional methods e. g. chemical precipitation, softening, ion exchange resin (IER), and Coagulation/Settling/Filtration, and to determine the optimum process conditions for promotion of separation efficiency, cost effectiveness and environmental safety.
The main undesired classes of contaminants in industrial and municipal WWs required to be removed prior to reuse of water or to safe environmental disposal are:

- Organic compounds as dyes, fertilizers, food products ...... etc.
- Inorganic pollutants as HMC's, hardness components and/or total salinity.
- Trace radioactive isotopes.
- Colloidal matter either inorganic as colloidal silica or iron, or organic like humic substances.
- Dissolved gases like CO_2, H_2S, SO_2 or the radioactive Rn 222.
- Microorganisms and their life products.

During the comparative evaluation of performance of selective membrane methods and the subsequent consideration of their application problems, our discussion will be concentrated on the first four classes of the above mentioned pollutants. Dissolved gasses are not removed by neither the selective desalination membranes nor the mentioned conventional techniques of WWs treatment. They are mostly removed by aeration, sometimes followed by chemical scrubbing or by chemical treatment like oxidation. Rejection of colloidal substances will be considered under organic or inorganic pollutants. Microbiological contamination cannot be removed by the conventional methods of separation. In membrane processes, feed water is usually sanitized in order to avoid biofouling. In fact, NF and LERO membranes reject a high percent of microorganisms, however, they cannot be considered as absolute barrier for water sterilization. Our results of high temperature RO of WWs, to be discussed later on, have shown remarkable control of biological growth under certain conditions without interference with the rejection of the other water contaminants (Khedr, G, 1997).

2.1 Testing details

A mobile pilot unit was designed so as to enable conduction of NF and LERO runs over a wide range of operation conditions, Fig (1), feed pressures, flow rates, and pretreatment steps. Percent recovery ranged from 85-90 at feed pressures of 6 and 8 bar for NF and RO, successively. Both permeate and reject streams were re-circulated back to the feed tank which was thermostated at 25°C. Feed pH was adjusted to the range 7.5 to 8 except upon comparing membrane results with those of hot lime softening (HLS) where the high pH of the softening method was adopted. Ionic concentrations were determined by ICP-AES (Perkin- Elmer, Boston, USA). The radioisotopes Ra $(NO_3)_2$ and UO_2Cl_2 were supplied and analyzed by radio counting curtsey from the Regional Middle East Centre of Radioactive Isotopes (Cairo, Egypt).

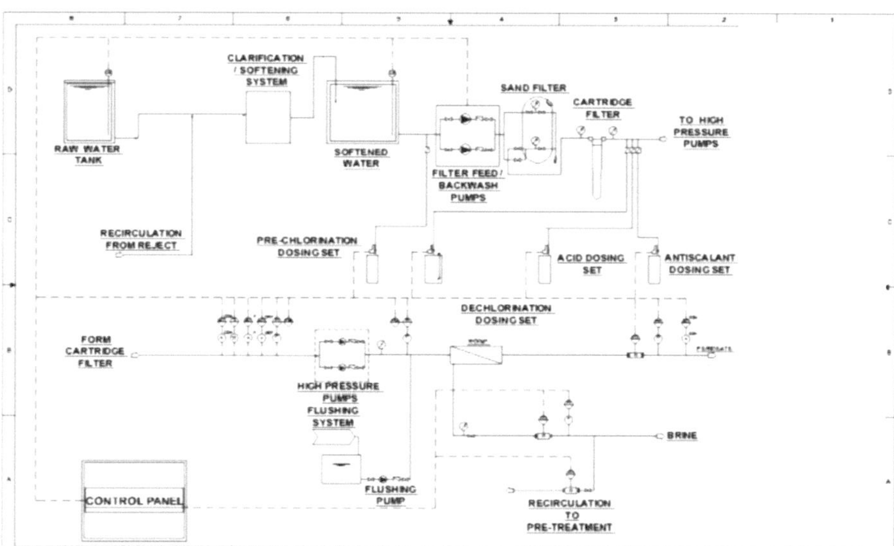

Fig. 1. Schematic representation of the mobile NF + RO pilot unit

A laboratory experimental system having six test cells with circular turbulent agitation at the level of surface of membrane coupons was used in a test circuit which consisted of a low pressure pump, pressure gauge, cartridge filter, flowmeter and thermostated feed tank. Membrane samples were stored dry and thoroughly rinsed with deionized water before use. They were compacted in distilled water at 120 psi, prior to testing, until steady flux is obtained, then conditioned by soaking in the testing solution for one hour. The testing feed pressures ranged from 80 to 100 psi for NF experiments and from 100 to 120 psi for LERO ones. Tangential cross flow velocity ranged from 0.005 to 0.1 m/s and feed flux from 120 to 720 L/m^2.d. The synthetic contaminated feed water used was prepared by dissolution of Cu SO_4 to give 5.5 ppm Cu^{2+}, $CaCl_2$ to give 57.3 ppm Ca^{2+}, $FeCl_3$ to give 15.8 ppm Fe^{3+}, Na_2 SO_4 to give 88.6 ppm Na^+ and KSO_4 to complete SO_4 $^{2-}$ concentration to 163.4 ppm. The same procedure was used to prepare the other feed water compositions. Thin-film composite NF (HL4040F) and LERO (AK4040F) membranes of polyamide chemistry (GE/Osmonics) were used throughout the present study in coupons for cell testing and in 4" elements for pilot testing.

2.2 Performance of NF and LERO in removal of HMCs and radionuclides

Fig. 2 shows the results of treatment by RO, LERO, and NF of synthetic contaminated water containing beside usual drinking water components, Cu^{2+} as HMC as well as trace concentrations of the radionuclides Ra^{2+} and UO_2^{2+}. Rejection of radionuclides is calculated according to:

$$\text{Radionuclide \% rejection} = \frac{(\text{pCi}/\text{L})\ \text{feed} - (\text{pCi}/\text{L})\ \text{permeate}}{(\text{pCi}/\text{L})\ \text{feed}} \times 100$$

Only modest rejection is observed for monovalent species as is expected with RO and particularly with NF, however, rather strong rejection is recorded for polyvalent ions. LERO and NF permeate showed decontamination till less than the maximum contaminant level (MCL), for drinking water according to the norms of the US, Environmental Protection Agency i.e. 5 pC_i/l for combined radium (Ra_{226} + Ra_{228}) and 20 ppb for uranium 238. The observed order of percent rejection of the solution components agrees with the expected mixed control of both Donnan and diffusion effects for a porous charged membrane (Hilal et al, 2004).

Polyvalent radionuclides, whether cationic or anionic, were strongly rejected despite their presence as trace concentrations in mixture with tremendously higher ionic concentrations of similar chemical nature, which indicates the absence of interference with rejection of the contaminants. Still higher rejection is measured for LERO, specially for monovalent species and TDS.

2.3 Effect of ion background concentration on NF and LERO contaminant rejection

Fig. 3 shows the performance of NF rejection of the TDS, Na^+, Cu^{2+}, and Ra^{2+} from a mixture solution of $CuSO_4$ and NaCl to which a trace of $Ra(NO_3)^2$ was added. The test runs were conducted over a wide range of background feed TDS, at constant contaminant cation concentration of 25 mg/l Cu^{2+} and 124 pCi/L Ra^{2+}.

Cation rejection decreases with concentration due to increase in concentration polarization at membrane solution interface (Khedr, 2000).This effect is much more pronounced for Na+ where, a rather linear decrease of rejection is observed. Since the monovalent ions represent the main component in the feed water, the NF rejection of TDS decreases practically parallel to that of Na^+.

As for divalent cations, on the other hand, stronger rejection is observed which concords with Donnan effect considerations (Albino & Donald, 1988). It is interesting to notice that the rejection of radioactivity, which belongs to a very low chemical concentration, is practically complete and independent of the background concentration over the investigated range.

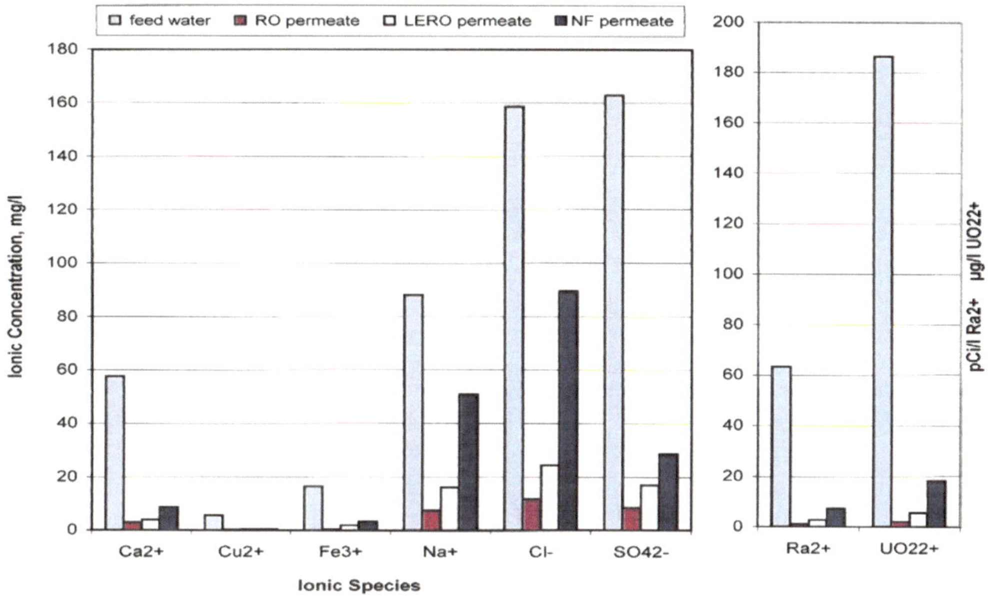

Fig. 2. Comparison of performance of membrane methods RO, LERO, and NF in treatment of contaminated feed water

These results confirm that the rejection of a trace HMC or radioactive isotope by NF is not subject to any screening or interference by the co-present much higher concentration ionic background in water. Similar behaviour was obtained with LERO membranes. However, in this case the recorded general higher rejection and lower decline of TDS and Na^+ rejection with concentration are due to the higher selectivity of the RO membrane. Radium rejection of the order of 98% is recorded.

According to the results of Figs (2, 3), NF would be the optimum method for rejection of the considered contaminants from drinking water sources. When parallel water desalination is required in addition to the removal of HMCs or radionuclides, on the other hand, LERO is the candidate technique. Contaminant rejection takes place parallel to that of other salts in the same system without using additional equipment or process modification.

NF may also be applied as the first step in a hybrid membrane process (NF/RO). Results have shown in such case that the last polishing RO step, while it improved a lot the quality of the final product water so as to attain drinking water norms, it added only slightly to the process cost since it was performed at lower pressure and high recovery. In fact, analysis of **NF permeate confirmed that pretreatment by NF upgraded waste water quality through efficient removal of colloids, suspended solids, hardness components, microorganisms and organics (Khedr, 2008).**

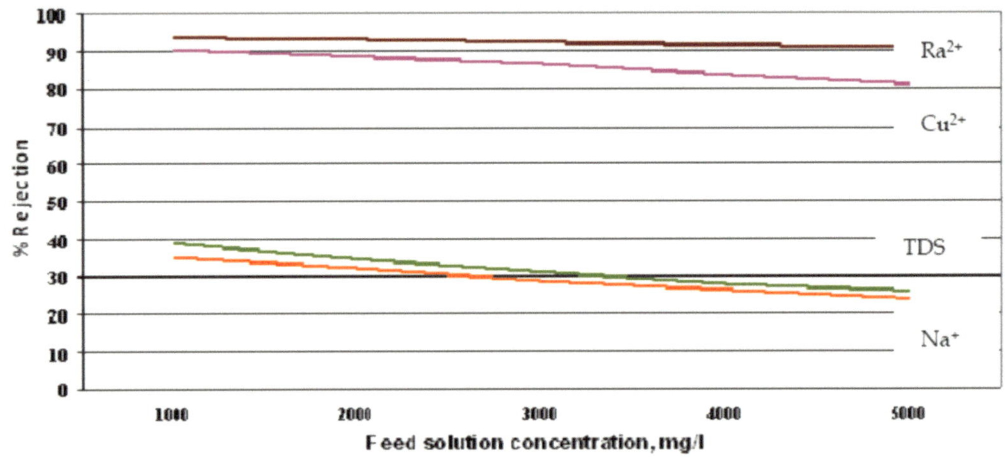

Fig. 3. Variation of rejection of Cu^{2+} , trace of Ra^{2+}, Na^{+} and TDS with solution concentration in LERO

2.4 Removal of the contaminant ions by IERs

Synthetic water samples contaminated by HMCs and radionuclides were treated by chelating strong cation and anion exchange resins. Samples of product water were periodically collected from the resin columns for chemical analysis and radioactive count. Removal of contaminant ions of the order of 85 to 95% was recorded. However, results revealed the following drawbacks of this method as compared to membrane methods:

Strong interference of the polyvalent ions with ion exchange of the contaminant ions was invariably observed. This is particularly pronounced for ions of similar physico-chemical behaviour as e.g. interference of scale forming cations as Ca^{2+} with rejection of Ra^{2+}, both belonging to alkaline earth metals. Comparison of separation of Ra^{2+} in absence and in presence of Ca^{2+} confirmed that affinity of CER is much higher for the heavier cation, however, this is partially counteracted by the much higher concentration of Ca^{2+}. Since parallel Ca^{2+} removal can't be avoided, a bigger system is required to realize the required efficient contaminant removal.

The high affinity of the resin to the HMCs or the radioisotopes makes the subsequent resin regeneration quite difficult. 10% NaCl regenerant did not remove more than 23% of the retained uranium. Slightly better regeneration was observed by successive elution by 4% NaCl followed by 1N HCl as was previously advised (Sorg, 1988). However, neither process was totally effective in removing all of the uranium or radium which continued to accumulate after each run on the IER. Use of higher concentration NaCl solution and/or increasing the regeneration time improved only slightly the percent regeneration. This can be attributed to the relatively large size and poor hydration of Ra^{2+} ion which would locate it in relatively inaccessible hydrophobic regions of the resin phase. Results of the present work showed that chelating resins are more efficient for removal of polyvalent contaminant ions than conventional resins; however, they are also more difficult to regenerate. Large quantities of regenerant solution would be required if 100% removal of uranium should be approached.

According to Fig (4), at the same acid pH of feed water containing 70 μg/l U 238, the H^{+} form of the CER enabled 93% removal compared to 68% for the Ca^{2+} form under the same

conditions. At intermediate pH of 6.0 lower removals was observed. At pH 7.6, AER showed much higher removal. **On the other hand, over the whole pH range NF showed higher and rather steady rejection behavior, slightly lower at pH 6.0**.
The observed high pH – dependence of U removal by IERs is due to the high reactivity of U and its affinity to form complexes to give various ionic species (Sorg, 1991). In acid Cl^- solution, UO_2Cl_2 is the prevailing form of U where the uranyl cation UO_2^{2+} is highly stable and exchanges at such high rate with H+ ions. In slightly acidic pH in the range from 5 to 6.5 neutral carbonate molecule is forward, $UO_2CO_3^{\circ}$. In near neutral pH values, on the other hand, the uranyl ion forms stable complexes with - CO_3^{2-} which is common to drinking water supplies, the predominant anion species is $UO_2\ (CO_3)_2^{2-}$ and $UO_2\ (CO_3)_3^{4-}$. This explains the observed rather complete removal of uranium by AER and their difficult regeneration at such pH. These results confirm and add to the previous consideration that IER would probably not be practical for decontamination of drinking water supplies (Lee et al, 1982).
Separation of contaminant ions by IERs is not steady. It decreases with time with start of contaminant break through near the resin exhaustion.
Important amounts of concentrated brine, or other regeneration chemicals, are required for periodical resin regeneration (which is seldom complete). The contaminated regenerant solution has then to be disposed of.

2.5 Removal of the contaminant ions by softening and chemical precipitation

2.5.1 Removal of HMCs and radioactive isotopes by NF and LERO in comparison with Hot Lime Softening (HLS)

Rejection of the poisonous HMC Cd^{2+} and a trace of Ra^{2+}, in presence of Na^+, Ca^{2+}, Cl^-, SO_4^{2-}, HCO_3^- and CO_3^{2-} was investigated in comparison with their removal by the conventional (HLS). In order to compare the performance of these methods under similar experimental conditions, the dosing of lime and soda ash was adjusted so as to realize almost the same level of Ca^{2+} rejection by the NF. In Fig (5), for each species, the first column gives the initial concentration, the second gives the result after NF treatment and the third gives the result after HLS, independently realized.

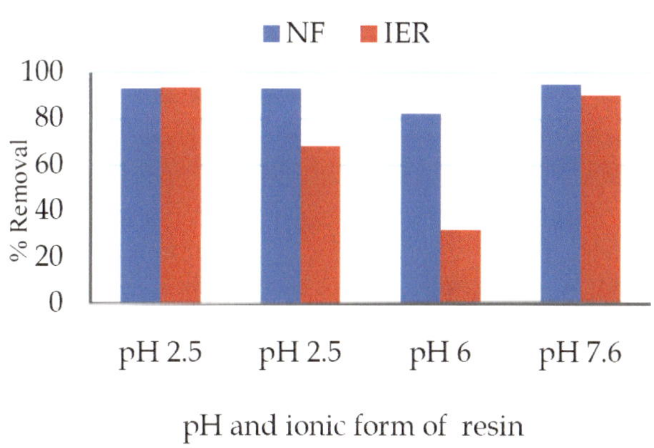

Fig. 4. Comparison between Removal of U238 by NF and IERs

While NF led to lowering the concentration of all the species to different extents and consequently to lowering of TDS, HLS lowered only the concentration of hardness components Ca^{2+}, Ra^{2+}, and HCO_3^- and of HMC. On other hand, HLS raised Na^+ and CO_3^{2-}

concentrations. Concentration of the anions which are not included in softening reactions like Cl^- and SO_4^{2-} were not modified. These modifications are directly explained by the softening reactions of HLS.

NF has, therefore, the advantages over HLS of (1) parallel general water desalination, (2) absence of continuous chemical dosing stoichiometric to the hardness rejection which attains several tens of tons per day in big HLS softening facilities, (3) absence of contaminated solid sludge formation which requires periodical disposal, (4) mechanism do not include slow steps like settling, (5) do not require subsequent filtration, NF filtrate being crystal clear, (6) the relative ease of operation of membrane processes, (7) NF do not require such sophisticated installations with dosing silos and agitated reactors.

Further (7), as for the process cost, results of Fig (5) show that removal of other hardness components by HLS cannot be avoided. In order to remove the radionuclide contamination to the desired low level, most of hardness components parallel to Ra^{2+} should simultaneously be removed. Such interference results in a higher rate of chemicals consumption in higher capacity softening installations, and leads to a higher amount of contaminated sludge. NF application, on the other hand, is shown to lead to co-rejection of all these components without interference. It will have, therefore, lower cost.

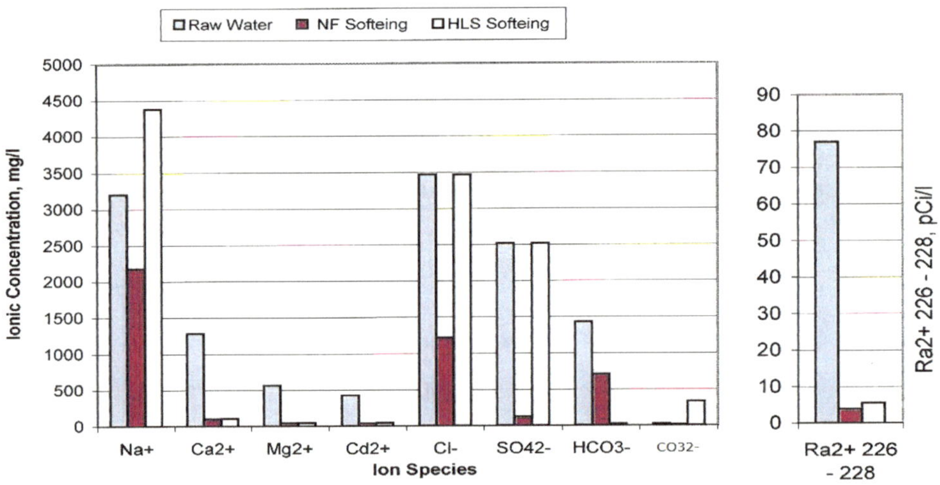

Fig. 5. Comparison between performance of NF and HLS in removal of HMC and radioisotopes.

2.5.2 Separation of HMCs and radionuclides by chemical precipitation

The most common method used is precipitation of contaminant polyvalent cations as hydroxides by means of e.g. lime dosing. According to Fig (6) pH adjustment is critical for optimal precipitation of most of cations and specifically for amphoteric ones like Cr^{3+} and Zn^{2+}, where a slight increase in pH results in a remarkable decrease in precipitation efficiency by dissolution as CrO_4^{2-} or ZnO_2^{2-}. For the purpose of comparison, the separation by NF on Fig (7) was realized independently for each cation at the same concentration of 75 ppm in mixture solution with Na Cl at the same TDS from (Na_2CrO_4 + NaCl), ($CrCl_3$ + NaCl), and ($ZnCl_2$ + NaCl) mixture solutions. Steady rejection of HMCs is obtained over the whole pH range for both cationic and anionic forms. Preferential rejection is even more pronounced for

anions than for cations. CrO_4^{2-} showed higher rejection than Cr^{3+} despite the higher ionic charge, also rejection of Zn^{2+} increased in the high pH range of the line Zn^{2+} / ZnO_2^{2-}. This behavior was previously observed and attributed to the coulombic repulsion with the residual negative charge on the polyamide membrane surface (Khedr, 2000).
According to Fig (6), in case of water contaminated by more than one HMC the possible pH of co-precipitation will be at much higher solubility than the theoretical minimum value for each cation. In such case, NF and LERO enabled efficient co-rejection of contaminants, Fig (2), without measurable interference. Membrane rejection is also much more feasible than the recently proposed two-step precipitation of HMCs, first as hydroxide then as sulphide (Cort, 2005), each step being followed by independent coagulation, flocculation settling and filtration.
As for chemical precipitation of radioisotopes, in the inhand work, only limited testing was preformed to deposit Ra^{2+} through dosing of $BaCl_2$ according to:

$$Ba^{2+}_{(excess)} + Ra^{2+}_{(trace)} + SO_4^{2-} = Ba\,(Ra)\,SO_4\,(s)$$

This reaction is reported to be highly specific to Ra^{2+} separation and leads to its quantitative deposition (Clifford, 1991). However, our results showed its deposition in a finely divided form which clogs in part to the walls and piping of the reactor and settling equipment causing strong radioactive contamination of the system. Also since such deposit includes highly concentrated radionuclide, it represents a waste disposal problem. In fact, it should be buried in a standard radioactive waste disposal facility with all the related potential environmental risks.
The NF radioactive waste, on the other hand, is dissolved in the reject stream. Upon drying in e.g. the evaporation ponds it is diluted in the solid state by the co-rejected, non radioactive salts and compounds so that the radioactivity of the deposit is usually lower than the level that imposes burying in a "low-level radioactive waste disposal site" i.e. 1-3 nCi/cm^3 (Hahn, 1991). Instead, contaminated NF or RO reject streams have several selections for safe environmental disposal. As per the level of contamination they may be discharged to storm sewers, surface water bodies or sanitary sewers, or blended with sanitary effluent streams so that the final stream radioactivity becomes lowered to the trace state.

2.5.3 Removal of the contaminant ions by coagulation, flocculation, and settling

In order to compare the results of rejection by membrane methods to those of conventional coagulation, laboratory testing was conducted using the same synthetic contaminated feed water under the previously described testing conditions. Jar tests were conducted using both ferric and aluminum coagulants over the dosing rate of 1 to 20 mg//l for various feed pH values covering both acid and alkaline ranges. Results of Fig (7) showed wide variation of behavior between different contaminant ions. As an example while Cu^{2+} and Pb^{2+} were removed at 95% and 97%, successively, Hg^{2+} was only 52% removed, and Cr^{3+} and Ra^{2+} were not removed.
Efficient uranium removal by coagulation was observed, however, the extent of removal is function of water pH and selected coagulant. While ferric chloride led to 82% removal over the pH range of 6 to 10, higher removal of 95% was obtained in presence of alum coagulant at pH 10. The sensitivity to pH in case of coagulation is due to its influence on the nature of the ionic form of U as was previously mentioned.
According to these results separation of such contaminant ions by conventional coagulation does not show a general trend of behavior. It is too much dependent on water composition,

pH, valency state of ionic contaminants, type and dose of coagulant, as well as other process variables. **In order to assess the validity of this method for a specific application, evaluation testing has to be conducted.**

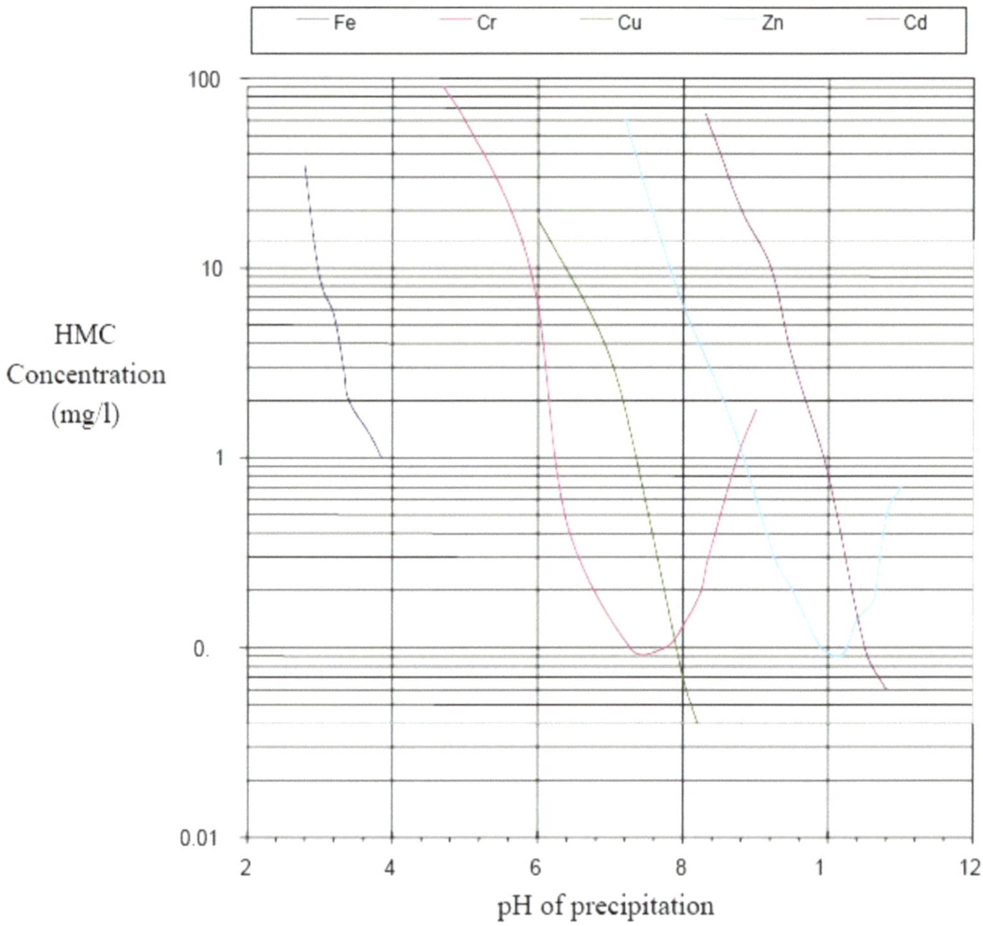

Fig. 6.

Rejection by NF and LERO, on the other hand Fig (7), is quite predictable and is mainly determined by the characteristics of the contaminant ions. Only slight effect of pH on NF separation is recorded. Membrane methods, again, do not include slow steps like settling, and do not require specific chemical dosing or subsequent filtration.

3. Problems encountered in NF and LERO treatment of WWs

The main application problems of these techniques are the phenomena of membrane fouling (khedr, 2000) where undesired rejected components accumulate on and/or in membrane and block its surface and would interfere with its function. Organics, microorganisms and their life products lead to development of colloidal fouling (Rizwan et al, 2007), organic fouling (Elimelech & Li, 2004), and biofouling (Koek et al, 2003). Saturation of inorganic sparingly soluble salts leads to scale deposition (Bertels et al, 2008). These phenomena are far of being cases of simple mechanical surface blocking. In fact, they interfere with membrane rejection mechanism.

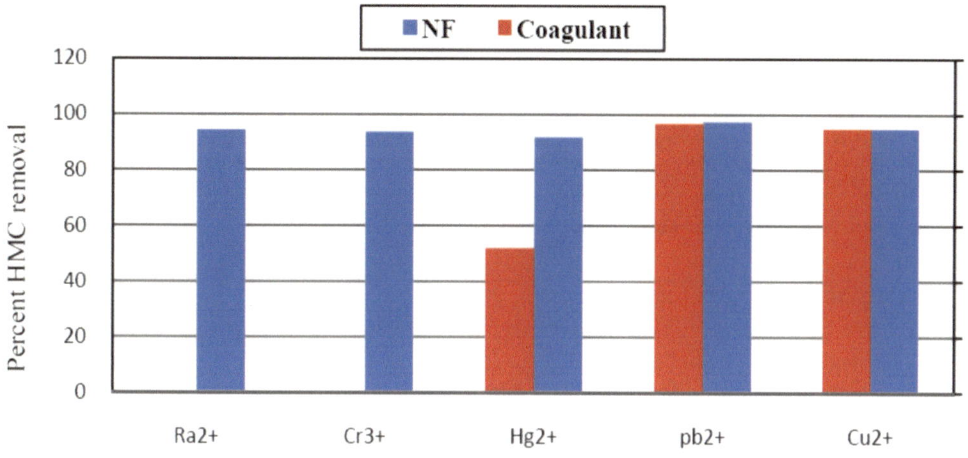

Fig. 7. Comparison between removal of HMCs and Radioisotopes by NF and by coagulation.

Fouling lowers both process efficiency and cost effectiveness through decline of the permeate rate, repeated shutdown for cleaning which would shorten the membrane lifetime, increasing the energy consumption, and modifying the rejection selectivity which decreases the product quality. Only limited works were reported about the potential interference between the various forms of fouling and its effects on NF performance (Khedr, 2000; Elimelech & Li, 2004).

Another difficulty encountered in NF system design which promotes fouling is due to its quite low operation pressure. A balanced flux throughout the system stages is sometimes difficult to achieve and maintain. Too high permeate rate would be obtained in the first stage which leads to too low brine flow rate in the last ones. This increased the potential of organic fouling in first stages and scale deposition in the last ones (Khedr 2008, 2009).

In order to characterize the development of complex fouling on NF membrane and the possible effect of interaction between the different forms of fouling, results of periodical inspection of membrane surface during advance of fouling by SEM and analysis of fouling deposit by energy dispersive X-ray and Fourier Transform Infra Red Spectroscopy, were correlated to the measured parallel decline of NF permeation and solute rejection.

Fouling tests were conducted using water from one of the collection streams of wastewater of the industrial city of Riyadh, KSA; typical analysis is given in Table 1. Composition of this stream remained rather steady over the duration of the experimental work.

Membranes used in investigation of high temperature NF are of quality characterized by high temperature resistance (Duratherm Membrane Series, Osmonics/GE). In this case, permeation rates and rejection values are measured in absence of temperature gradient i.e. they are free from effect of thermoosmosis. The reported changes in membrane performance are reversible and not related to polymer degradation due to high temperature.

Component	Concentration ppm	Parameter	Value
Ca^{2+}	122.8	TDS	1543.7 ppm
Mg^{2+}	12.5	pH	7.6
Na^{+}	424.8	TOC	64 ppm
K^{+}	17.0	Bacterial count	Too numerous to count
NH_4^{+}	12.8		
NO_3^{-}	46.0		
Cl^{-}	630.0		
SO_4^{2-}	160.4		
SiO_2	22.5		

Table 1. Typical Raw Wastewater Analysis.

3.1 Fouling behavior of NF membranes

Fig (8) shows the decline of NF product rate with time for the three stages of the NF system in absence of antiscalant. Different rates of decline are observed for different stages. While the first and second stages started by a gradual rate of decline which increased remarkably after the first 20 hours of operation, the flow rate of the third stage suffered a rather rapid decline to practical blocking after a much shorter time. The first stage then showed some increase of permeation to compensate for the decrease of the last one. This behavior is typical to scaling of the third stage membranes. Decline of first and second stage permeation rate indicates another form of fouling that accumulates on the membrane surface as e.g. organic and/or biofouling. The chemical analysis of the spent cleaning solutions confirmed the mentioned nature of the fouling film.

In presence of antiscalant (Genesysis International, DS, UK), on the other hand, usual decline in permeation was recorded for the three stages of the system, Fig (9).

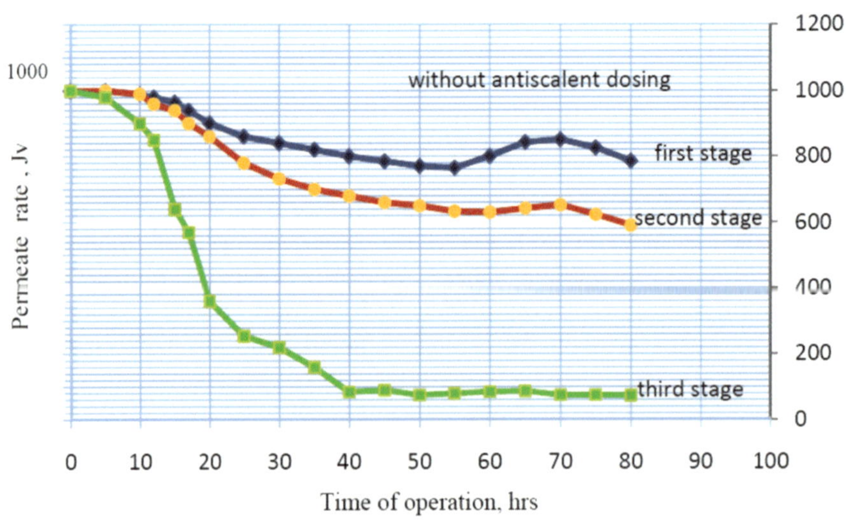

Fig. 8. Variation of permeate rate with time for the three stages of the NF system

In order to verify the relative importance of organic fouling and biofouling in formation of the fouling film similar test series were conducted in absence then in presence of UV

radiation at the level of inlet to cartridge filtration (Trojan UV Swift ECT, FL, and USA). In absence of UV radiation a much faster rate of decline of permeation is observed particularly after an incubation period of 14 days. Our results have also shown that: the use of phosphate-based antiscalent as hexametaphosphate contributed to increase of biofouling rate, specially in the high temperature condition of summer time (ambient temperature of 30 to 35 °C) higher rates of bacteria development and fouling film thickening were observed than in the case of non-phosphate based, polymeric antiscalents.
Upon testing of granular activated carbon filters for lowering the raw water content of organics, those filters became themselves contaminated with bacteria and acted as substrate for bacterial growth and source of additional contamination of feed water.
Correlation between decline in permeation and progress of fouling is confirmed by SE micrographs realized for membrane coupons collected at different time intervals over the test duration, as well as through analysis of the deposit accumulated on membrane surface.

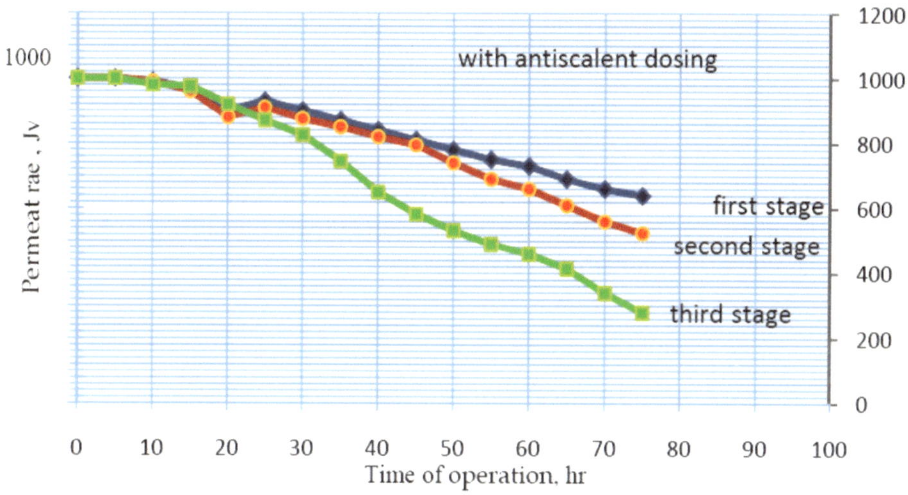

Fig. 9. Variation of permeate rate with time for the three stages of the NF system

3.2 Chemical composition of the fouling film

Table (2) shows the chemical analysis of the fouling deposit. Elemental composition of the film shows sulphur on the fresh membrane surface which would belong to the polysulphone supporting film of the membrane. The main elements in the deposit are phosphorous (P) and Calcium (Ca). P would be due to adsorption of the phosphate-based antiscalant in the fouling film. HMP may also become hydrolyzed to orthophosphate, which would deposit calcium phosphate scale. Phosphates may also be fixed by bacteria. On the other hand, calcium detected in the fouling film may be part of metal-organic complex created by interaction of the divalent cation with organic molecules. This was reported to result in highly compacted fouling layer and substantial flux decline (Seidel & Elimelech, 2002; Cho et al, 2000).
Since raw water did not contain iron, its presence in the deposit is attributed to corrosion of steel or stainless steel. In fact, a trace of some heavy metal cations e.g. in sewage water, can induce a rather severe galvanic corrosion of steel in view of cathodic depolarization through an autocatalytic displacement mechanism (Khedr, 1989).

As for the organic components detected in the film without being present in feed water, proteinacious and hydrocarbon derivatives would be metabolic products of bacteria. (proteinacious amide and long chain hydrocarbon ester). Polysulphone and polystyrene would be contribution removed from membrane surface and membrane separators along with the accumulated deposit. Analysis of samples of the fouling film and bacterial count confirmed the progressive accumulation of organics and bacteria cells in case of organic/biofouling parallel to the measured decline of permeation rate. In case of UV exposure a lower film material is accumulated which is practically free from growth and propagation of bacteria cells. While organic fouling accumulates on membrane surface since the beginning of each run, an incubation period is required for biological fouling to contribute effectively to the decline of permeation together with a higher rate of organics accumulation in the film.

3.3 Forms of membrane fouling

Membrane fouling is usually classified into different types which are attributed to presence of certain components in the feed water (Ridgway, 1988).

- Inorganic or crystalline fouling due to deposition of sparingly soluble salts of Ca, Mg, Sr and Ba or of SiO_2.
- Organic fouling, due to adsorption on membrane surface of various dissolved organic compounds.
- Particulate fouling, due to accumulation of suspended particles as silt or clay.
- Colloidal fouling, due to colloidal organic or inorganic components.
- Biofouling, due to microorganisms and their metabolic products.

Sample	Major Elements	Minor Elements	< Minor < 1%
Fresh membrane surface	–	S	
case 1 (deposit on membrane surface)	Ca, P	Si, Fe	Ti, Al, Mg
case 2 (deposit removed from membrane surface)	Ca, P	Si, Fe	Ti, Al, Mn

Concentrations are estimated from spectral lines intensities.

Table 2. Analysis of fouling deposit (Energy dispersive X-ray results)

However, such consideration of individual forms of membrane fouling oversimplifies the situation observed on NF or LERO membranes in applications which include rejection of inorganic, organic and biological contaminants from waste waters. In fact, most of, the cases of membrane fouling include combinations of these forms or even all of them in one and the same film (Khedr, 2000). The present work considered the potential interaction between these forms of fouling and its possible influence on the fouling rate, NF performance decline, and recovery upon cleaning.

3.4 Mechanism of development of fouling film

Through surface inspection by Scanning Electron Microscopy (SEM) of periodically collected membrane samples during the progress of membrane fouling, analysis of fouling deposit by EDX-ray and FTIR-Spectroscopy, and correlation to the parallel decline of permeate rate and solute rejection, the following sequence of steps was observed to determine the mechanism of development of fouling on membrane surface:

3.4.1 Colloidal fouling

The fouling film starts to form invariably under all operational conditions by the primary accumulation on membrane surface of organics, silt, and colloidal particles of organic and inorganic nature that escape filtration. These particles are entrained and adhere to membrane surface by convective forces. They may also be fixed on it by forces of electrostatic or chemical nature, e.g.:

- Through interaction between the surface charge of the membrane and the stabilizing layer of charges of the colloidal particles.

In fact, zeta potential measurements of NF membranes have shown that membrane fouling susceptibility is higher the higher the measured negative surface charges (Elimelech & Li, 2004).

- And/or the interaction between the polyvalent cations like iron or aluminum accumulated by membrane rejection in the diffusion layer with the stabilizing charge of colloidal particles, which leads to the neutralization of the latter and their consequent deposition on membrane surface.

3.4.2 Primary scale deposition

It takes place if the dosing of acid or antiscalant is unadequate to cover the supersaturation of the scale forming salts or compounds or if the related system design parameters are exceeded i.e. if NF is conducted at too slow brine flow or too high recovery.
Results confirmed what was previously observed (Khedr, 2008, 2009) that if the balance of flows in the NF testing is not precisely controlled according to system design values, fouling and scaling susceptibilities of membranes are enhanced. On the other hand, organics may interfere with scale deposition. Even at slightly positive LSI values, scale formation was not observed. In fact, surface adsorbed of organics may act as scale inhibitors (Bartels, 2008).
In presence of bacterial contamination of the feed water, membrane fouling may be further complicated by the parallel development of biofouling:

3.4.3 Formation of biofouling film

Results confirmed that particularly in hot summer time the contribution of biofouling to the decline of permeation of NF is quite more important than the other forms of fouling in absence of scale deposition. Fig (10) shows the SE micrographs of surface of membrane samples periodically collected at 3-days intervals during the progressive development of biofouling. Biofilm starts by adsorption of bacteria cells on membrane followed by film growth.

3.4.3.1 Adsorption of bacteria

A small fraction, less than 1% of bacteria cells (biocolloids) in the feed water adsorbs on the membrane to cover about 10 to 15 % of its surface. Such low adsorption is attributed to electrostatic repulsion between the residual negative charge on the thin film polyamide NF membrane and that on the bacteria cells at near neutral pH values. Adsorption is assumed to take place through intermediate connection through the extracellular polymer substance (EPS), life product of bacteria. Bacteria cells may reside on or near the membrane surface under force of permeation drag until the extruded EPS ensures strong attachment.
Electrostatic repulsion is, therefore, overcome by a combination of chemical bonding and brownian displacements of the bacteria cells. Factors like cross flow rate, permeate rate, water solutes, suspended matter, and membrane surface characteristics would also contribute to the rate of adsorption of microorganisms (Hoek et al, 2003). Possible effect of membrane surface roughness on the extent of bacterial adsorption was not evaluated in the present work.

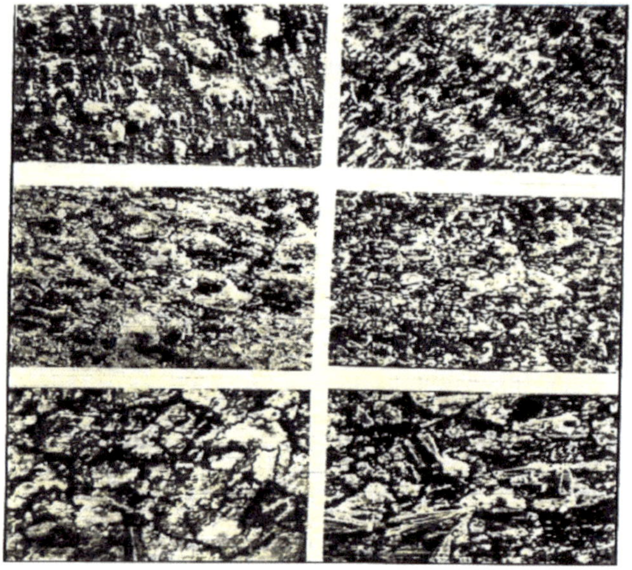

Fig. 10. SEM's of membrane samples showing the progressive accumulation of biofilm over 15 days of NF

3.4.3.2 Propagation of biofilm

The adsorbed bacteria were observed to progressively cover the membrane surface by the life products. Full surface coverage by a layer of several microns thick was obtained in about 15 days at an ambient temperature of 32° C. Rate of propagation of biofilm was observed to depend on the continuous exposure to contaminated feed water. The faster film growth and performance decline in this case is due to the role played by the slime gel matrix of the EPS of bacteria. It favors further organic and biofouling through entrapping of the organic molecules, colloidal particles, suspended particles and bacteria cells carried by the feed WW stream which we considered as (secondary adhesion). **This is an example of interaction between the different forms of membrane fouling. In fact, in contact with uncontaminated stream, the initial biofilm showed a much slower rate of growth.**

Addition of organic biocides could control the biological growth, (DOW Antimicrobial 7287 and Permatreat PC-11, Nalco). However, sometimes organic slimes were detected on the membrane surface which would be due to interaction between organic biocides with the present polyvalent cations (Yuan & Zydney, 1999; Cho et al, 2000).

3.4.3.3 Fouling film denaturation

Up to this stage in the development of NF and LERO membrane fouling results of chemical cleaning showed successful film removal and recovery of initial membrane performance. The optimum combination of mechanical and chemical washing conditions resulted in dissolution of the binding organic matrix of the film, killing of bacteria and washing these components off the membrane surface.

However, if cleaning is not timely conducted or is not successful, the accumulating fouling film undergoes, with time, a progressive change in nature. It becomes increasingly more dense and sticky, more adherent to membrane surface, and of lower permeability. Such modifications may be attributed to several factors which are either physico-chemical like film compaction due to thickening under pressure, or increase of hydrophobic character of the EPS of the microorganism upon interaction with membrane surface, and/or biological e.g. through the development of anaerobic bacterial reaction. The characterstic black coloration of the fouled membrane elements is thought to be due to the tendency of bacteria, at certain depth in the fouling film, to become anaerobic due to the lack of oxygen and the prevailing reducing conditions. SO_4^{2-} reducing bacteria lead to the production of the black ferric sulphide in the presence of ferric ions. Only limited reference is reported in the literature (Paul & Abanmy, 1990; Kaakinen & Moody, 1984) about such modification of the nature of fouling film.

3.4.4 Secondary scale deposition

The fouling film is assumed to act as a secondary membrane parallel to the original one. The fouled membrane, therefore, will have **a higher transmembrane pressure i.e. feed-permeate ΔP**, and in view of the partial blocking of the feed flow channels through the NF membrane element the fouled membranes will also have **higher transelement pressure, i.e. feed-brine ΔP**. The hydrodynamic conditions of the feed flow at the vicinity of the membrane surface would become further impaired if the fouling deposit causes blocking of the feed spacer in contact with membrane surface. This, together with the mentioned increase in ΔP values, would induce severe localized concentration polarization at the membrane surface (Khedr & Varoqui, 1981) with the consequent deposition of inorganic scale which we considered as (secondary scale deposition). In fact, in the advanced stages of membrane fouling, inorganic scale crystal aggregates were detected in the fouling film. **Again, this represents an example of interaction between the different forms of membrane fouling which further complicates the structure and composition of the fouling film and causes higher decline in NF performance.**

3.5 Control of NF membrane fouling

Results of Fig (11) show the decline of permeation rate under fouling conditions and the success of prompt cleaning to recover the initial performance (alkaline clean followed by acid clean) which was not the case with retarded cleanings Trails were conducted to control the membrane fouling in NF through:

3.5.1 Periodical backwash of membrane coupled with intermittent chlorination

In this case the choc dosing of chlorine for 5 minutes every 6 hours of nanofiltration together with periodical backwash of the membranes at a higher flow rate and a lower pressure could control the development of organic/biofouling as shown by Fig (11). Chlorination of feed stream is accompanied by dechlorination through dosing of a reducing agent (MBS) at a dose higher than stoichiometric for chlorine reduction in order to protect the polyamide thin film of membrane from destruction through oxidation/dissolution. Success of intermittent chlorination is usually attributed to (Baker & Dudley, 1998) either overcoming the increase in secretion of life products by microorganisms that takes place in presence of continuous dosing of chlorine, or to the decrease in availability of nutrients for bacteria through feeding on non-viable bacteria or on the products of decomposition by oxidation of complicated organics which otherwise cannot be digested by microorganisms.

However, chlorine-tolerant polymers have been prepared and adapted for the production of performant RO membranes (Nita et al, 1993; Glater et al, 1994). This will enable the introduction of residual chlorine in the feed stream up to membrane surface and killing of bacteria without risk of membrane damage. The role of chlorination will also include oxidation and depolymerization of the EPS secreted by bacteria leading to the destruction of the fouling film matrix and consequently to its detachment by the cross flow stream.

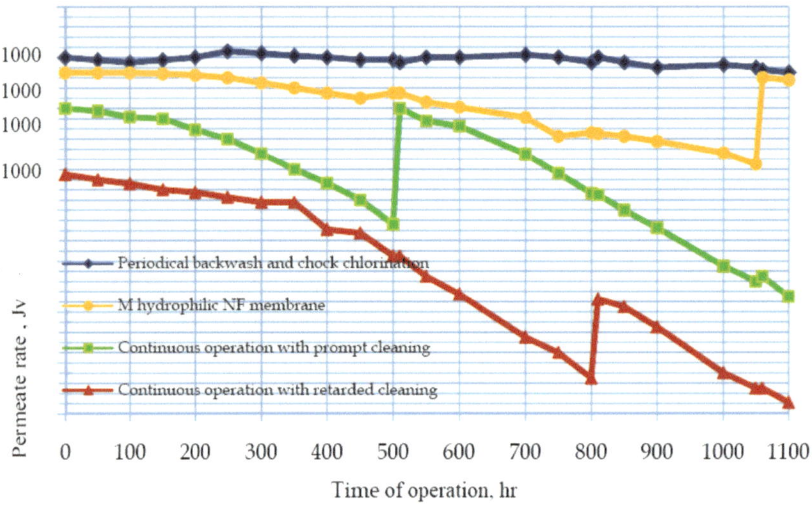

Fig. 11. Decline of product rate for various NF membranes upon fouling and remedy

3.5.2 Use the hydrophilic membranes

The use of the M membrane, based on modified PAN, previously supplied by Osmonics Corp. (currently under GE) which is characterized by high hydrophilicity, with the same fouling wastewater and under the same operation conditions, enabled to realize steady NF rate without decline of permeation and with largely spaced cleaning requirements as shown by Fig (11). Cleaning of hydrophilic membranes is shown to recover easily original performance.

When compared to conventional PAN membrane, the M membrane has a very low contact angle with water drop than that of other polymer membrane surface (Nicolaisen, 2002) i.e. much stronger hydrophilic character. This explains the much lower fouling rates observed with this membrane in view of the suppressed affinity towards adsorption of organics and colloids, including bacteria cells on the hydrophilic membrane surface. No sanitization was required for control of biofouling. In fact, much lower bacteria adhesion and bioflouling were detected by SE microscopy and chemical analysis of film on M membrane surface.

3.5.3 Effect of high temperature operation on NF & LERO membrane fouling

In case of hot WW streams direct treatment by NF or RO without cooling is of both technical and economic interests. Beside acceleration of permeate rates at lower pressures i.e. saving of energy, HTNF/RO save installation of cooling facilities as well as their required chemical dosing and maintenance costs.

Thermal acceleration of Jv is attributed to the increase of water diffusivity and increase of efficiency of sweep of fouling film by the cross flow feed stream in view of decreased viscosity. Under these conditions our pervious results (Khedr & Quadri, 1997) have shown that the hydraulic permeability varies with temperature according to an Arrhenius type relation:

$$C_{HPT} = C_{HPO} \exp [- Eo / RT] \quad (1)$$

Where the C_{HPO} and C_{HPT} are the membrane hydraulic permeabilities at initial temperature, and at temperature T, respectively.
Eo is the activation energy of the hydraulic permeability.
The product flux will vary because of these viscosity changes according to

$$J_{VT}=J_{V25}\,1.03^{(t-25)} \tag{2}$$

The temperature correction factor of the product rate at a given temperature "t" referred to the value at 25 ° C is

$$TFC=[J_{V25} / J_{VT}]=1.03^{(25-t)} \tag{3}$$

It is worthy to clarify here that the activation energy Eo, equation (1), has no kinetic implication. The Arrhenius equation models the change in hydraulic permeability with the change in temperature, however, it does not indicate that an initial energy barrier should be exceeded in order to activate a change in the hydraulic permeability.

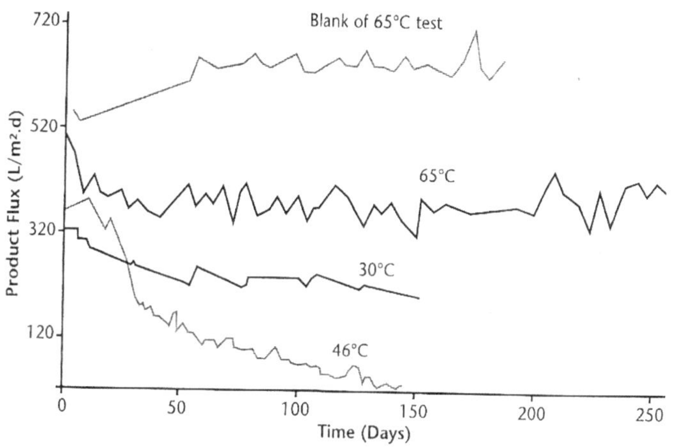

Fig. 12. Variation of permeate flux with time at various feed temperatures

It was observed that with mixed bio/organic fouling the acceleration of permeation rate upon rise of temperature (25 – 40° C) was leveled by the parallel acceleration of biofouling. At still higher temperature (45 -55 ° C) the influence of enhanced biofouling became predominant and progressive decline of permeation was obtained accompanied by formation of a thicker fouling film. However, at still higher temperatures, fig (12), 55 - 65 °C, the permeate flux remained steady as function of time at a value lower than that obtained in absence of fouling (blank line). In fact, at such temperatures bacteria cells are not killed and when temperature was lowered biofouling growth was reactivated. According to SE micrographs the minor film with minor bacteria adsorption and growth detected at such high temperatures suggests that bacterial growth and multiplication were just inhibited.
Increase of operation temperature, Fig (13), on the other hand, caused a rapid and rather linear decline of rejection of mono-monovalent salt while the di-divalent one showed a much higher rather steady rejection over a wide temperature range, followed by a slow decline. Rise of temperature enhances back diffusion of salt through membrane to the dilute side. This effect may be assisted, in part, by the viscous drag of salt ions due to interference with the accelerated product water flux, in the same direction, according to the friction

model of ion and water transport through selective membranes (Jansson & Boesen, 1975) It is evident that both factors i.e. back diffusion of ions and influence of viscous drag by the permeate flux exert higher effect on the lower rejection monovalent ions.
As for the rejection of organics investigated under exposure to UV radiation in order to avoid the contribution from biofouling, lower organic fouling was observed with rise of temperature in view of decreased adsorption of organics on membrane surface as well as lowered viscosity of the fouling film which facilitated its detachment by cross flow feed stream. TOC rejection showed a primary decline with temperature rise followed by increase of rejection at higher temperatures. Analysis of permeate and reject streams showed that the initial decline is mainly the case for low molecular weight organics, while the subsequent rise is only apparent due to the additional dilution of the permeate stream due to the enhanced water permeation.

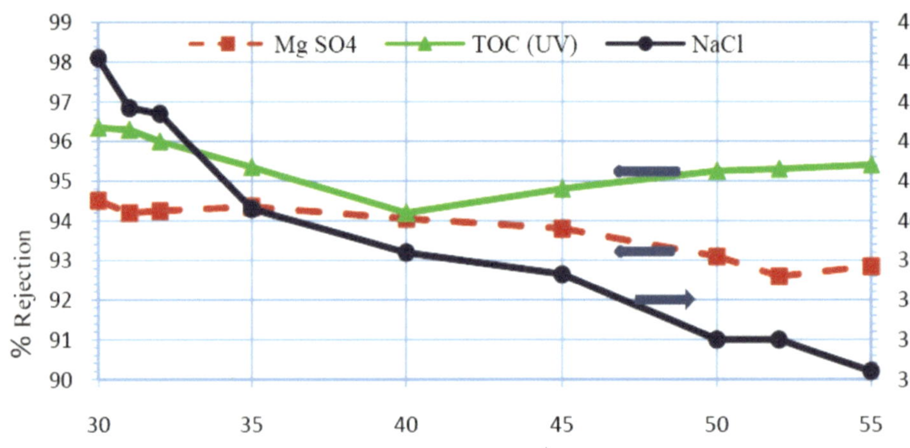

Fig. 13. Variation of NF rejection of various components with temperature

These results reveal the improvement of Nanofiltration and Low Energy Reverse Osmosis process efficiency for high temperature WW streams or groundwater through rise of permeation rate at lower pressure, better control of organic and biofouling, and increase of rejection of organics without significant loss of rejection of polyvalent contaminant ions (heavy metal cations, hardness components, and radioactive isotopes).

4. Conclusion

1. In view of the general advantages of NF & LERO of low energy consumption, cost effectiveness over a wide range of water salinities and plant capacities, and particularly their efficient rejection of wide variety of contaminants, it was decided to evaluate their performance in treatment of WW problems in comparison with the conventional methods applied in this respect. The general results of the long-term **investigation suggest that the low energy membrane processes are candidate to replace several conventional technologies of WW treatment.**
2. NF & LERO showed the following advantages for the separation of HMCs, hardness components, and radioactive isotopes from WWs and drinking water

sources over IER, chemical precipitation, softening and coagulation/ settling/ filtration methods:

2.1. Higher rejection of contaminants which include anions, cations and uncharged components with co-rejection of various trace contaminants without masking or interference.

2.2. No critical feed pH adjustment is required. Rejection is practically not pH dependent.

2.3. They are continuous processes and membranes do not require subsequent regeneration. Their mechanisms have no slow steps like settling.

2.4. They lead to general feed desalination and do not require subsequent filtration.

2.5. They do not require continuous chemical dosing stiochiometric to rejection of contaminants and do not result in formation of sludge which requires periodical disposal.

2.6. They do not require sophisticated installations or additional equipment to usual RO system.

2.7. They have a partial sterilizing effect.

3. The main factors that enabled to control the membrane fouling which is the main application problem of NF & LERO are:

3.1. Periodical backwash of membranes coupled with intermittent choc chlorination.

3.2. Hindrance of the adsorption of organics and microorganisms on membrane surface through promoting its hydrophilicity.

3.3. High temperature NF or LERO which accelerated permeate rate at lower pressure and enabled control of organic and biofouling without significant loss of polyvalent ion rejection (HMCs, hardness components and radioactive isotopes).

4. High temperature NF or LERO suggest the direct treatment of hot wastewater streams or groundwater in order to save the installation and operation costs of the cooling facilities. It also enabled lower energy consumption and higher plant availability through lowered shutdown for cleaning.

Knowledge of the wastewater composition would enable the selection of the adequate temperature range of HTNF for the optimum product rate and rejection performance.

5. References

Albino, K. & Donald, B., 1988, *Brackish and Seawater Desalting, Reverse Osmosis Technology*, Parekh, B. S., ed, Marcel Dekker, p 271, New York.

Baker, J. S. & Dudley, L. Y., 1998, *Proceedings of Membranes in Drinking and Industrial Water Production*, p 237, Amesterdam.

Bartels, C, Wilf, M. Casey, W. & Campbell, J., 2008, *Desalination*, 221, p 158.

Childress, A.E & Elimelech, M., 2000, *Environ. Sci. Technol.*, 34, pp 3710-3716.

Cho, J., Amy, G. & Pellegrino, J., 2000, *J. Membrane Science*, 146, pp 89-110.

Clifford, D., 1991, *Radon, Radium, and Uranium in Drinking Water*, Cothern, C. R & Rebers, P. A., Eds., Lewis Publishers, p 225.

Clifferd, D., 1991, *Radon, Radium, and Uranium in Drinking Water*, Cothern, C.R & Reberes, P. A., Eds., Lewis Publishers, p 227.

Cort, S. L., 2005, *US Patent 6896815*, www.freepatentsonline.com.

Deraniyagala, Y & Hlophe, M., 2006, *Workshop on Nanotechnology, Water and Development*, Chennai, India.

Gabaldon, C. & Gonzalez, J., 2000, *Separation Science and Technology*, 35, p 1039.

Glater, J., Hong, S. K. & Elimelech, M., 1994, *Desalination*, 95(3), p.325.

Hahn, N., 1991, *Radon, Radium, and Uranium in Drinking Water*, Cothern, C. R. & Rebers, P. A., Eds., Lewis Publishers, pp 249-267.

Hassan, A. M, Al Sofi, M, Al Amoudi, A, Jamaludin, A., Farooque, A., Rowaili, A, Dalvi, A, Kither, N., Mustafa, G. & Al Tisan, I, 1998, *Desalination*, 118, pp 35-51.

Hassan, A. M., Farooque, A., Jamaludin, A, Al Amoudi, A., Al-Sofi, M, Al Rubaian, A., Kither, N, Al Tisan, I & Rowaili, A., 2000, *Desalination*, 131, pp157-171.

Hock, E. M. V., Kang, S. & Deshusses, M, 2003, Symposium on *"Molecular Level Aspects of Bacterial Adhesion Transport and Biofilm Formation"*, Div. Environmental Chemistry, Am. Chem. Soc., LA, USA.

Hofman, J. A., Beerendonk, E. F., Groennou, J. Th. & Taylor, J. S., 1995, *International Workshop on Membranes in Drinking Water Production*, 67, Paris.

Jansson, G. & Boesen, C.E., 1975, *Desalination*, 17, p 145.

Jensen, K & Thorsen, T., 1995, *International Workshop on Membrane in Drinking Water Production*, p 63, Paris.

Kaakinen, J. & Moody, C., 1984, *"Reverse Osmosis and Ultrafiltration"*, Sourirajan, S. & Matsuuri, T., Eds. 188th Meeting of the American Chemical Soc.

Khedr, M. G. & Varoqui, R., 1981, *Ber. Bunsenges. Phys. Chem.*, 85, p 116.

Khedr, M. G. & Lashien, A. M. S., 1989, *J. of the Electrochemical Soc.*, 136(4), p 968.

Khedr, M. G., Quadri, S. S., 1997, High *Temperature Reverse Osmosis and Control of Membrane Fouling", International Desalination Association World congress on Desalination & Water Reuse*, III, p 627, Madrid.

Khedr, M. G., 2000, *Desalination & Water Reuse*, 10(3), p 8.

Khedr, G, 2004, "Optimization of Reverse Osmosis Process Efficiency and Environmental Safety through Reject Processing", *Euromembarne International Conference*, p 600, Hambourg.

Khedr, M. G., 2008, *"Membrane Methods in Tailoring Simpler, More Efficient, and Cost Effective Wastewater Treatment Alternatives", Desalination*, 222, p 135.

Khedr, M. G., 2009, *"Nanofiltration Membrane Fouling, Definition, Mechanism, and Remedy Guidelines", International Desalination Association World Congress*, p 21, Dubai.

Lee, S. Y., Hall, S. K. & Bondietti, E. A., 1982, *Office of Drinking Water, US-Environmental protection Agency* 570/9-82/003.

Li, Q. & Elimelech, M., 2004, *Environ. Sci. Technol.*, 38, p 4683.

Mohammad, A. W., Othman, R. & Hilal N., 2004, *Desalination*, 168, p 241.

Nicolaisen, B., 2002, *Desalination*, 153, p.355.

Nita, K, Numata, K., Gotoh, S. & Matsui, Y., 1993, *Proceedings of IDA World Conference on Desalination and Water Treatment*, p 33, Yokohama.

Paul, D. & Abanmy, A., 1990, *Ultrapure Water*, 7(3), p 25.

Ridgway, H, 1988, *Reverse Osmosis Technology*, Marcel Dekker, NY, USA.

Rizwan, T. & Bhattacharjee, 2007, *Canadian J. of Chemical Engineering*, 85(5), p 570.

Seidel, A. & Elimelech, M., 2000, *J. Membrane Science*, 203, pp 245-255.

Sorg, T. J., 1988, *Journal of American Water Works Association*, 80(7), p 105.

Sorg, T., 1991, *Removal of Uranium from Drinking Water by Conventional Treament Methods, radon, Radium and Uranium in Drinking Water*, Cothern, C. R. & Rebers, P. A., eds, Lewis Publishers, p 173.

Thanuttamavong, K. Yamamoto, K., Ik Oh, J., Choo, K. H. and Choi, S. J., 2002, *Desalination*, 145, pp 257-264

Yuan, W. & Zydney, A. L., 1999, *Desalination*, 122, pp 63-76.

11

Assessment of UV Pre-Treatment to Reduce Fouling of NF Membranes

Di Martino Patrick[1], Houari Ahmed[1],
Heim Véronique[2] and Marconnet Cyril[3]
[1]*Laboratoire ERRMECe (EA1391), Institut des Matériaux, Université de Cergy-Pontoise*
[2]*Syndicat des Eaux d'Ile de France*
[3]*Veolia Water, Compagnie Générale des Eaux, France*

1. Introduction

Nanofiltration (NF) is an efficient technology to produce safe and biologically stable drinking water from surface water (Cyna et al. 2002). NF can respond to the increased demand of water with higher quality due to the evolution of the legislation and of the customer's expectation. NF treatment allows the reduction of the concentration of organic precursors to disinfection byproducts, and the reduction of the concentration of trace contaminants such as pesticides and pharmaceuticals. The introduction of NF in the drinking water production plant of Méry-sur-Oise, France, conducted to several changes in quality of the distributed water: reduction of total organic carbon (TOC) and Biodegradable dissolved organic carbon (BDOC) by a factor 3 to 5, reduction of THMs by a factor 2, reduction of viable bacteria population by a factor 10, reduction of chlorine demand of the distribution system by a factor 3, amount of pesticides below detection level.

Four membrane properties are important for the efficiency of a water treatment plant: high rejection of dissolved organics, low salt rejection, low energy consumption, and stable performance after repetitive cleanings. Membrane fouling generates flux decline leading to an increase in production cost due to increased energy demand and chemical cleaning. Moreover, fouling induces reduction in membrane life. Different types of NF fouling can be defined on the basis of fouling material: inorganic fouling due to deposition on membrane surface of inorganic scales; organic fouling due to humic acids, proteins and carbohydrates (natural organic material, NOM); biofouling due to biofilm formation at the membrane surface.

Flux decline associated with NOM fouling and with biofouling can be partially restored by chemical cleaning (Al-Amoudi et al. 2005, Di Martino et al. 2007, Houari et al. 2010, Roudman et al. 2000). Biofouling is distinct from NOM fouling caused by contaminated organic matter derived from biological systems (Flemming et al., 1997). Biofouling involves biologically active microorganisms which grow at the membrane surface as complex structures termed biofilms (Lappin-Scott et al. 1989). Biofilm formation precedes biofouling, which becomes an issue only when biofilms reach thickness and surface coverage that cause declined normalized flux and/or increase in normalized pressure drops during NF operation (Vrouwenvelder et al., 1998, Ridgway, et al. 1996).

Fouling and subsequent chemical cleaning of nanofiltration membranes used in water treatment are inevitable but can be decreased by adequate pretreatments (Al-Amoudi and Lovitt 2007, Di Martino et al. 2007, Flemming et al. 1997, Hilal et al., 2004, Houari et al. 2010, Speth et al. 2000). Pre-treatment must achieve a strong removal of microbial cells and growth-promoting compounds from the feed water (Vrouwenvelder et al. 1998, Wend et al. 2003, Vrouwenvelder and Van der Kooij 2001, Marconnet et al. 2009). Several pre-treatments schemes prior to NF have been evaluated including coagulation/precipitation, biofiltration, granular activated carbon (GAC) adsorption, microfiltration/ultrafiltration, ozonation, pre-chlorination, slow sand filtration, and UV irradiation (Cyna et al. 2002, Wend et al. 2003, Vrouwenvelder et al. 1998, 2008, Speth et al. 2000, Koyuncu et al. 2006). UV pre-treatment has several advantages: it has immediate germicidal effect; it is a closed and thus a safe system that requires only a small space for equipment; it does not generate byproducts; it acts only on water and cannot have deleterious effects on membranes. Nevertheless, since UV disinfection has no residual effect, the time between the UV irradiation and the membrane filtration has to be the shortest possible, so as to avoid bacterial regrowth (Salcedo et al. 2007). Few studies have evaluated the benefit of using UV as a pre-treatment upstream from NF and/or RO membranes (Conlon and Jhawar 1993, Koyunku et al. 2006, Munshi et al. 2005, Mofidi et al. 2000).
In the present study, the addition of a UV irradiation process upstream from NF modules was assessed to maintain membrane performances and to limit biofouling.

2. Materials and methods

2.1 Pilot units

Two identical nanofiltration pilot units (2 x 1 m^3/h) containing each a single 4-inch spiral wound NF element harbouring NF200B membranes (DOW Filmtec, Delft, The Netherlands) were used (Figure 1). A UV pilot including a low pressure mercury vapour lamp (monochromatic at 254 nm) was operated at a dose of 400 J/m^2 upstream from one of the pilots. The pilot units were fed by clarified river water pre-treated through sand filtration (sand-filtered water, SFW). Pilot 1 was fed SFW while pilot 2 was fed SFW irradiated by UV (SF + UVW), during a filtration run of 10 weeks (from February to May). Both NF pilots included an internal pre-treatment step made of pH neutralisation, 20 and 6 µm cartridge filtration and antiscalant injection, similar to what can be found in most of NF industrial units. The role of this pre-treatment was to reduce scaling (pH neutralisation and antiscalant addition) and particulate fouling (pre-filtration). A recycling loop enriches the module feed water with concentrate at a ratio concentrate/feed of 70%, in order to accelerate fouling.

2.2 Water quality parameters

Total direct counts (TDC) of microbial cells after 4',6-Diamino-2-phenyindole dihydrochloride (DAPI) staining, and active bacteria counts (ABC) of microbial cells after cyano 2-3 dytolyl – tetrazolium chloride (CTC) staining are determined by epifluorescence microscopy. Dissolved Organic Carbon (DOC) is measured with a Total Carbon Analyzer using persulfate oxidation of organic carbon, followed by infra-red non-dispersive detection of the CO_2 produced by the oxidation reaction. Biodegradable dissolved organic carbon (BDOC) is determined by batch incubation with sand-fixed bacteria according to the bioassay procedure developed by Joret and Lévi (1986) and the French standard XP T 90-319.

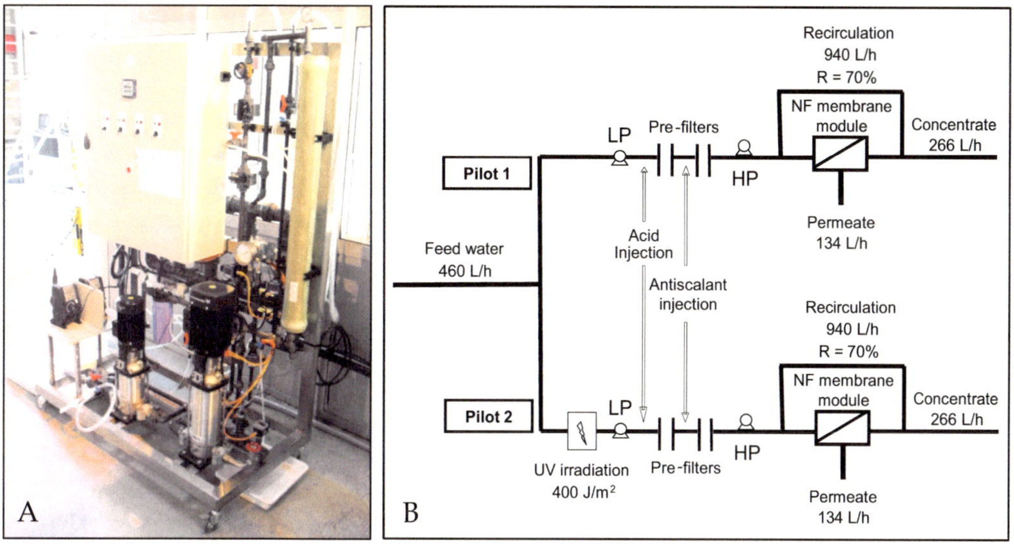

Fig. 1. A, photograph of the nanofiltration pilot showing the membrane module, the pumps and the electronic instrument panel. B, scheme of the experimental setup. Feed water was sand-filtered water. LP, low pressure pump. HP, high pressure pump.

2.3 Modules performance monitoring

The performances of the membrane modules, water permeability (L. h^{-1}. m^{-2}. bar^{-1}) normalized at 25°C and corrected of the osmotic pressure, and longitudinal pressure drop (LPD) also termed feed channel pressure drop (FCP) normalized by temperature and the flow rate along the module, were monitored on-line during both filtration tests.

2.4 Membrane autopsy

NF modules from the two pilot units were autopsied at the end of the filtration test. Membrane samples cut from randomly chosen areas of the module were air-dried and analysed by Attenuated Total Reflection Fourier Transform Infrared (ATR-FTIR) spectroscopy. Other membrane samples were fixed in paraformaldehyde before staining and fluorescence microscopy analysis, as described below.

2.4.1 Dry weight, protein content, microbial activity and wettability of the deposit

A membrane coupon of 50 x 90 cm was gently scraped to collect the wet deposit. The deposit was lyophilized, and the lyophilisate weight was measured. The lyophilisate was then dissolved in ultrapure water, vortexed and sonicated.

The protein concentration was determined after centrifugation by Bradford's colorimetric assay (Biorad protein assay, Biorad laboratories GmbH, München, Germany), bovine serum albumine (BSA, Sigma-Aldrich) being used as a standard (Bradford 1976).

For ATP assays, remnants of NF membrane (9.6 cm^2 for each one) were cut. Each sample was put in 3 mL of ultrapure water and sonicated (Branson sonifier 450). The concentration of ATP released by microbial cells after sonication was measured by the luminescence luciferine/luciferase test (ATP Determination Kit, Invitrogen). The intensity of light produced by the reaction was measured with a luminometer (Sirius Luminometer, Berthold

detection systems GmbH, Pforzheim, Germany). The ATP quantity was calculated by comparison with free ATP standards. The ATP concentration was an indicator of the microbial activity within the deposit.
The wettability of the membrane surface was determined by measuring the water contact angle of the foulant matter. A membrane coupon of 1 cm^2 was cut. An ultrapure water drop was deposited on the surface of the membrane, and the tangent angle between the drop and the surface was measured with a drop shape analysis-profile device (DSA-P, Kruss, Germany).

2.4.2 Analysis of membrane foulants by ATR-FTIR

Samples of air-dried fouled membranes were analyzed by ATR-FTIR spectroscopy. IR spectra were recorded using a Tensor 27 IR spectrophotometer with a 45° diamond/ZeSe flat plate crystal and an average depth penetration of 2 µm. Each spectrum presented was the result of 32 accumulations obtained with a resolution of 2 cm^{-1} with air as the background. IR bands indicative of biomass were detected near 1650 cm^{-1} (amide I, corresponding mainly to proteins), 1550 cm^{-1} (amide II, corresponding mainly to proteins), 1090 and 1040 cm^{-1} (corresponding mainly to polysaccharides) as previously defined (Doumèche et al. 2007). The IR signal near 700 cm^{-1} was used to calculate ratios corresponding to relative IR signals of biomass (amide I / membrane signal, amide II / membrane signal, and band near 1040 cm^{-1} / membrane signal) (Houari et al. 2009). Mean values ± standard deviation of relative IR signals of biomass are shown here.

2.4.3 DAPI and lectin staining of foulants

DAPI and lectin staining of foulants were done as previously described (Doumèche et al. 2007). Samples of the fouled membranes were treated with paraformaldehyde (4 %, v/v) prior to lectin application in order to fix the foulant matter. Double staining with a mixture of TRITC-labelled peanut agglutinin (PNA) and FITC-labelled *Lycopersicon esculentum* agglutinin (LEA) (Sigma, Saint Quentin Fallavier, France) was done. 100 µL of a mixture of two lectins (final concentration of 100 µg/mL each) were carefully applied directly on top of the membrane. After incubation during 30 min in the dark at room temperature, unbound lectins were removed by washing with filter-sterilized water. Following the lectin staining step, the fouled membranes were treated with a solution of DAPI (1 mg/L) (Sigma, Saint Quentin Fallavier, France) in filter-sterilized water. After incubation during 30 min in the dark at room temperature, unbound DAPI was removed by washing the membrane surface with filter-sterilized water. The stained preparations were then mounted with Mowiol (Calbiochem, Meudon, France) and stored at 4°C in the dark. For the negative control, the lectins and DAPI were replaced by filter-sterilized water.

2.4.4 Microscopy and image analysis

The fouled membranes stained with DAPI and fluorescently labelled lectins were examined with a Leica epifluorescence microscope (MPS 60) and with a Leica SP2 upright confocal laser scanning microscope (DM RAX-UV) equipped with the Acousto-Optical Beam Splitter (AOBS) system and using 63X, N.A. 1.32, oil immersion objective (Leica microsystems, Rueil-Malmaison, France). For epifluorescence images, DAPI was excited at 364 nm, CTC was excited at 450 nm, FITC was excited at 494 nm, and TRITC was excited at 550 nm. For confocal images, DAPI was excited at 405 nm and observed from 410 to 600 nm, FITC was excited at 488 nm and observed from 505 to 540 nm, TRITC was excited at 543 nm and observed from 560 to 600 nm. The selection of spectral emission window for each fluorophore has been determined through λ scan analysis on single stained membrane fragments. The gain and offset for each photomultiplier have been adjusted to optimize nucleic acid and lectins detection. Images of the Confocal Laser Scanning Microscopy (CLSM) observations (1024×1024 pixels) have been acquired through a sequential mode, between stacks, to exclude spectral crosstalk from the data. 400 Hz scan speed was used and signal/noise ratio has been increased through frame average. Overlay and maximum projection of the z-stacks files have been performed with post acquisition Leica confocal software (LCS) functions to obtain the presenting snapshots. Original z-stack Leica files have then been imported into Imaris 4.0 software (Bitplane AG, Zürich) to obtain snapshots illustrating xz and yz representations.

2.5 Statistical analysis

ATP concentrations, contact angles measurements and values of relative IR signals of biomass obtained with or without UV irradiation were compared by the equal-variance Student's *t* test, following the variance test with Fisher *F* statistics. *P* values below 0.05 are considered significant.

3. Results

3.1 Feed water characterizations

Mean values of total direct bacteria counts (TDC), active bacteria counts (ABC), Dissolved Organic Carbon (DOC) concentration and Biodegradable Dissolved Organic Carbon (BDOC) concentration of the different feed waters are presented on Figure 2.

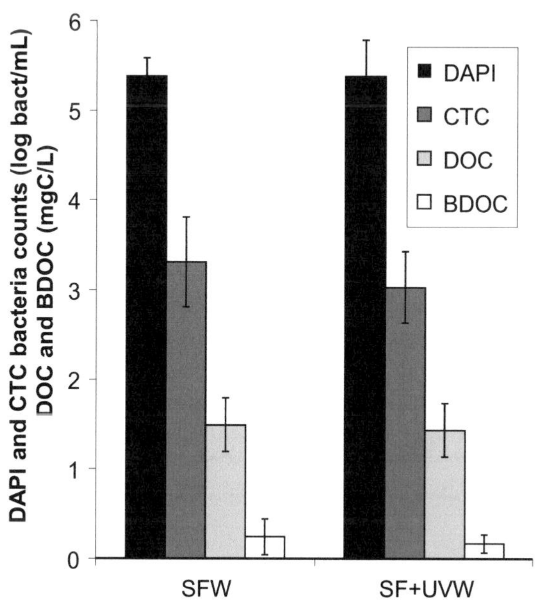

Fig. 2. Total direct Counts (DAPI) and active counts (CTC) of bacteria in feed waters, concentration in of dissolved organic carbon (DOC) and biodegradable dissolved organic carbon (BDOC) in the feed water with (SF + UVW) or without (SFW) UV irradiation.

UV pre-treatment did not affect the organic carbon concentration, as it was applied with a low pressure lamp at a reasonable dose of 400 J/m^2. The germicide effect of UV pre-treatment can be seen on the active bacteria counts (about 0.3 log removals).

3.2 Membrane performances

Water permeability and Longitudinal Pressure Drop (LPD) varied slightly all along the filtration period (Fig. 3). At the end of the test, water permeability decreased by 2.5 % with SFW as feed water and by 3.6 % with SF + UVW as feed water and the LPD value was 1.08 for SFW and 1.02 for SF+UVW. Both permeability and LPD evolutions were very low and the differences between the two conditions were not significant.

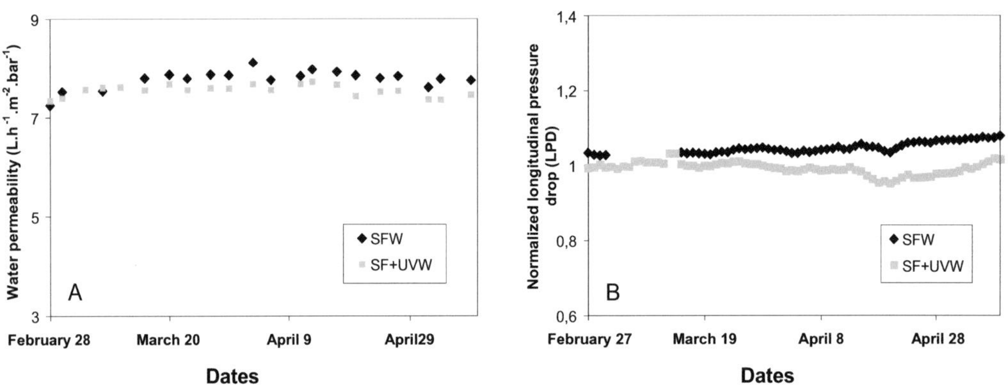

Fig. 3. NF module water permeability (A) and longitudinal pressure drop (B) evolution during filtration. Sand-filtered water with (SF + UVW) and without (SFW) UV pre-treatment.

3.3 Autopsy results: Characterization of the biofilm

3.3.1 Dry weight, protein content, microbial activity and wettability of the deposit

Table 1 gives mean values and standard deviations for the dry weight, protein content, microbial activity (ATP concentration) and wettability of the deposit. Dry weight, protein concentration and microbial activity of the membrane deposit were lowered after UV pre-treatment indicating that UV pre-treatment limited noticeably the formation of the deposit on NF membranes. The difference between the water contact angles of the module fed SFW and the module fed SF + UVW was in concordance with the other data: UV pre-treatment lowered the water contact angle; the membrane surface was more wettable. This may be linked to the decrease of the quantity of exopolymeric substances covering the surface of the membrane; therefore the water contact angle came closer to its initial low value (between 15 and 30°).

Parameter	Unit	SFW	SF + UVW
Dry weight	µg/cm^2	102.3	40.7
Protein concentration	µg/cm^2	1.15 ± 0.09	0.34 ± 0.03
ATP concentration	pmol/cm^2	0.40 ± 0.18	0.16 ± 0.12**
Water contact angle	degrees	83.4 ± 7.6	74.7 ± 4.3*

Table 1. Characterization of the foulant matter deposited on NF modules during filtration. Sand-filtered water with (SF + UVW) and without (SFW) UV pre-treatment. *, $P < 0.05$. **, $P < 0.01$.

3.3.2 Analysis of membrane foulants by ATR-FTIR

IR spectra acquired on the surface of NF modules are presented on Figure 4.

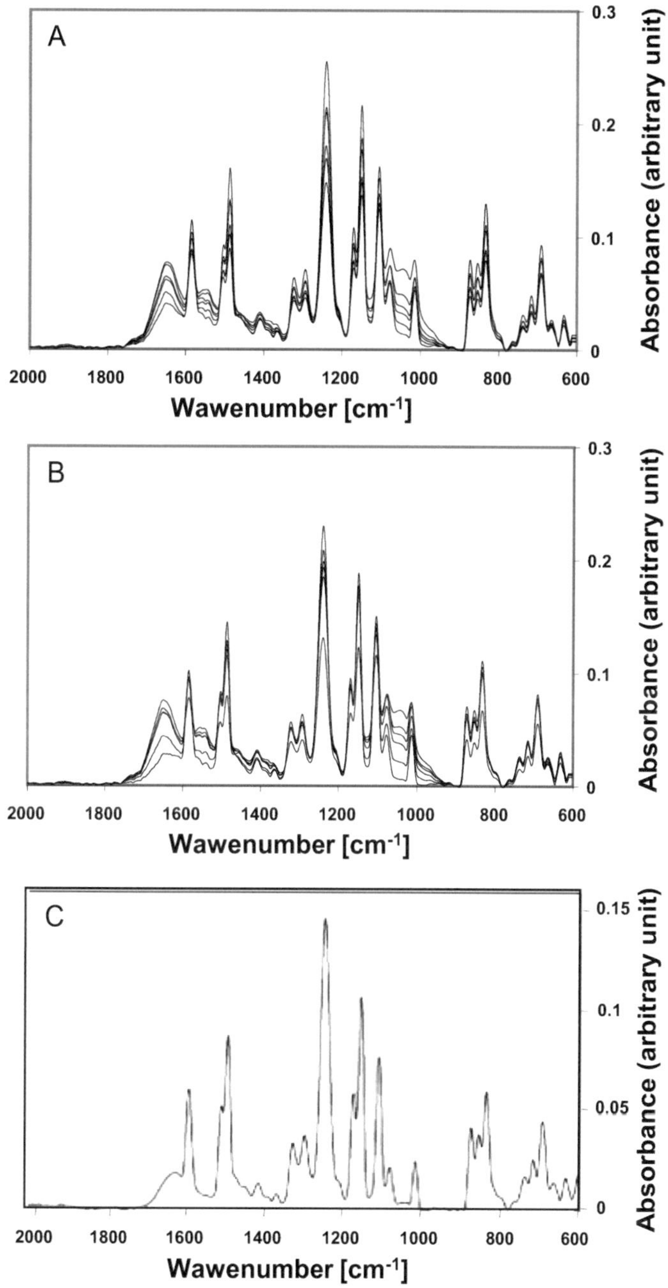

Fig. 4. IR spectra of the foulant matter deposited on NF modules. A, Sand-Filtered water (SFW) without UV pre-treatment. B, Sand-Filtered water with UV pre-treatment (SF + UVW). C, virgin membrane.

Different spectra illustrating heterogeneity among the foulant layer are presented for each fouled membrane. Signals corresponding to the presence of proteins (signal at 1650 cm^{-1}) and exopolysaccharides (signal at 1040 cm^{-1}) were observed on the two modules. Many IR bands corresponding to the membrane were detected for all the spectra, indicating only partial coverage of the membranes by the fouling layer. The membrane IR signal near 700 cm^{-1} was used to calculate relative ratios for the proteins and polysaccharides signals to compare semi-quantitatively the abundance of exopolymers on the different NF elements. Table 2 indicates the mean values and standard deviations of these ratios.

Relative IR signal ratios	SFW	SF + UVW
Amide (I) / membrane[a]	0.87 ± 0.22	0.74 ± 0.25**
Polysaccharides / membrane[b]	0.52 ± 0.22	0.44 ± 0.29*

Table 2. Relative IR signals measured at the surface of NF modules. Sand-filtered (SF)-water, with (SF + UVW) and without (SFW) UV pre-treatment. a, Amide (I) / membrane is the ratio of signal intensities at 1650 cm^{-1} and 700 cm^{-1}. b, Polysaccharides / membrane is the ratio of signal intensities at 1040 cm^{-1} and 700 cm^{-1}. *, $P < 0.05$. **, $P < 0.01$.

The relative signal of amide (I), characterizing the amount of proteins, and the signal near 1040 cm^{-1}, revealing the amount of polysaccharides, were significantly lowered for UV pre-treated waters. FTIR-ATR spectroscopy showed that the exopolymers amount on the membrane surface was decreased when the feed water was UV-irradiated.

3.3.3 Microscopy

Microscopic observations made on NF membrane surfaces are illustrated on Figures 5 and 6. Each micrograph presented corresponds to one area of the corresponding membrane after the observation of at least 5 distinct areas at a magnification of x400 and 10 distinct areas at a magnification of x630 randomly distributed at the membrane surface. The snapshots presented on Figure 6 illustrations are showing the tendency of the fouling extent.

The analysis of the foulant matter after DAPI staining showed that the membrane surface was colonized by many microorganisms organized as microcolonies (Figure 5). Lectin staining revealed the presence of exopolysaccharides on the surface, with a highly heterogeneous repartition. Some areas of the membrane surface were covered by important amounts of microorganisms and polysaccharides, while others were still virgin and did not present any microbial cells (Figure 5 and 6). Microorganisms and exopolysaccharides were organized as a biofilm highly structured in three dimensions clusters. The most intense signal for membranes fed SFW or SF + UVW was obtained with the LEA-FITC lectin.

The cell concentration on the surface of the membrane was decreased when the feed water was UV-irradiated (SF + UVW). DAPI total direct counts indicate a bacteria concentration of $1.07 \pm 0.4 \times 10^6$ cells/cm^2 with SF + UVW as feed water and $2.20 \pm 1.17 \times 10^6$ cells/cm^2 with SFW as feed water. As illustrated on Figures 5 and 6, the amount of polysaccharides on the surface appeared to be lower after UV irradiation. Moreover, UV irradiation seemed to increase the heterogeneity of the deposit: the distinction between colonized areas and virgin areas, already visible with SFW as feed water, is yet stronger with SF + UVW.

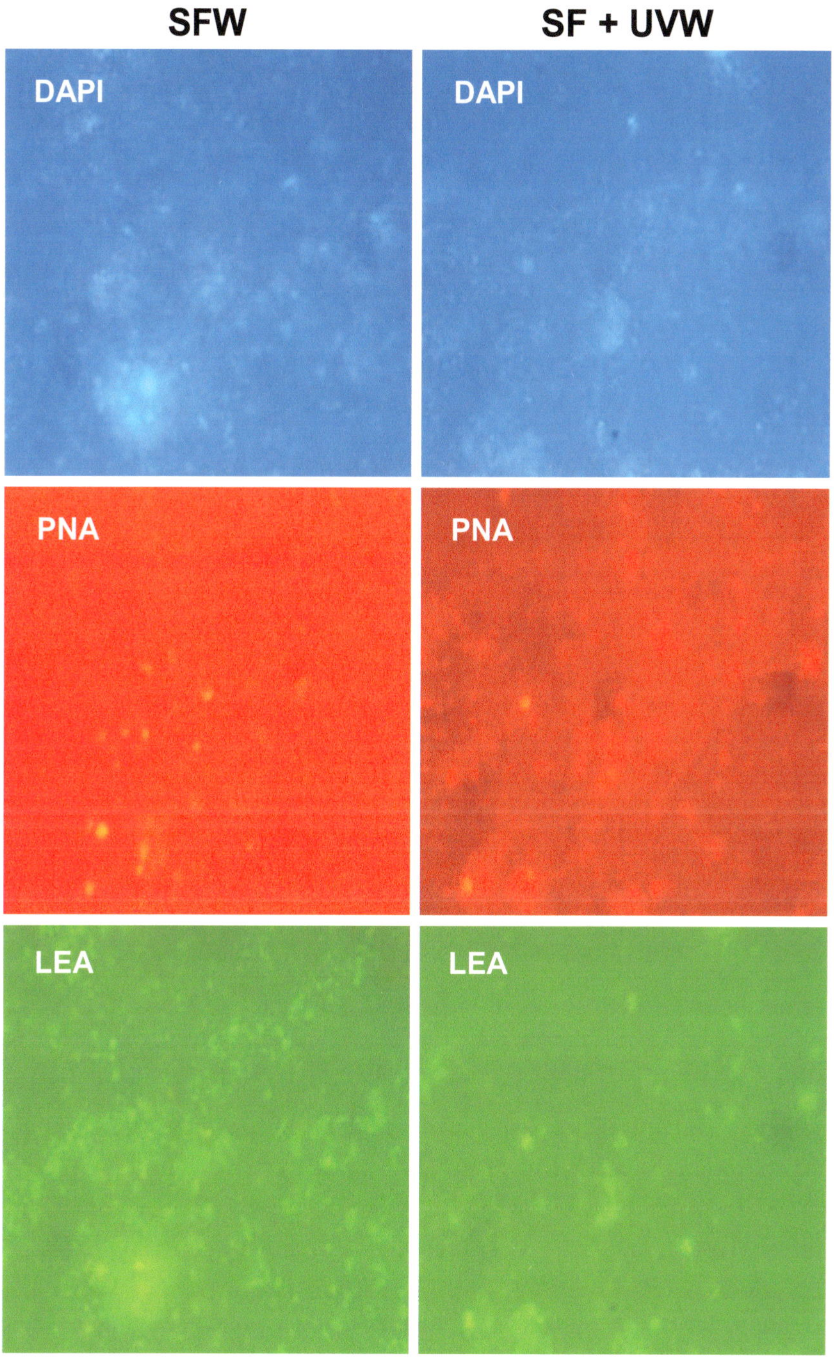

Fig. 5. Interactions between fluorescently-labelled lectins and fouled membranes, as visualized by epifluorescence microscopy. Sand-filtered (SF)- water, with (SF + UVW) and without (SFW) UV pre-treatment. PNA, TRITC-labelled peanut agglutinin. LEA, FITC-labelled *Lycopersicon esculentum* agglutinin. Magnification, x 630.

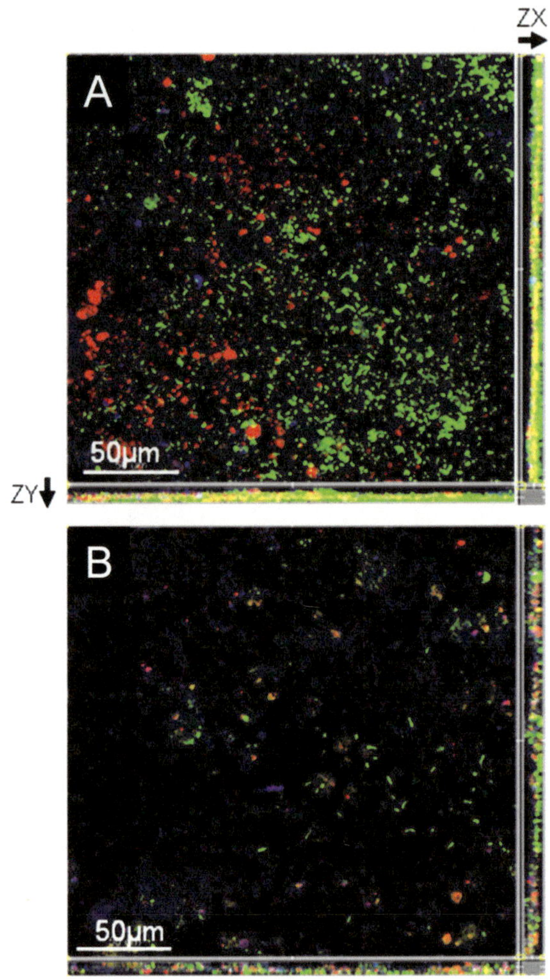

Fig. 6. Interactions between fluorescently-labelled lectins and fouled membranes, as visualized by CLSM. A, sand-filtered water without UV pre-treatment (SFW). B, sand-filtered water with UV pre-treatment (SF + UVW). Colour allocation: blue - DAPI, green - FITC-labelled *Lycopersicon esculentum* agglutinin (LEA), and red - TRITC-labelled peanut agglutinin (PNA). Magnification, x 630. Black arrows indicate the orientation of the foulant matter from the top to the bottom.

4. Discussion

The goal of this work was to study the influence of a UV irradiation pre-treatment located upstream from nanofiltration modules, on the biofouling of NF membranes at a pilot scale. The efficiency of UV pre-treatment to control biofouling was assessed with feed water obtained after coagulation/sedimentation/sand filtration.

We first measured the effects of UV irradiation on feed water parameters. UV pretreatment had no effect on the concentration of available dissolved carbonic microbial nutrients, since the DOC and BDOC concentrations were not modified by irradiation. UV treatment with a

low pressure mercury lamp (UV intensity varying from 10 mWs cm^{-2} to 240 mWs cm^{-2}) has been shown to be able to induce the decomposition of humic substances (Li et al. 1996, Lund et al. 1994). These effects of UV irradiation on humic acids have been observed for DOC concentrations of at least 15 mg C/L. The apparent discrepancy between the present results and the literature may be related to the relatively low concentrations of DOC and BDOC in the feed water in our experiments (less than 4 mg C/L). In the present study, UV irradiation decreased bacterial viable counts by 0.3 log. This may be related to the well-established mechanism of UV disinfection, i.e. bactericidal and bacteriostatic effect of UV mainly through genetic damages induction (Gaid et al. 2006, Al-Adhami et al. 2007). This effect of UV irradiation can be potentialized by other pretreatments decreasing BDOC like ozonation/GAC adsorption since dissolved organic matters are necessary to repair cell damages (Alkan et al. 2007, Camper 2004).
The second part of this work measured the effects of UV irradiation on the fouling of NF membranes, i.e. evolution of the filtration performances and formation of the deposit at the membrane surface. The analysis of the foulant matter by complementary technical tools showed that UV pre-treatment decreased the deposit formation. The global quantity of foulant matter and all the biofilm parameters measured were decreased when UV pre-treatment was applied: the total direct bacteria counts, viable bacteria counts, microbial activity (ATP concentration), and the concentration of exopolymeric substances (proteins and polysaccharides) were reduced at the membrane surface after UV irradiation. Nevertheless, UV pre-treatment did not modify the evolution of the permeability and of the longitudinal pressure drop. This may be linked to the short duration (10 weeks) of the test. Biofouling of nanofiltration membranes has been associated with a LPD-increase (Characklis and Marshall 1990, Vrouwenvelder et al., 2000, 2008) and/or a permeability decrease (Speth et al., 2000, Marconnet et al. 2009; Vrouwenvelder et al., 1998, 2009). Water permeability is a physical parameter which is related mainly to the material structure, to the porosity of the deposit, to its wettability, and to a lesser extent to its global quantity (Herzberg and Elimelech 2007). We did not observe any relation between membrane surface wettability and membrane permeability. Despite different evolutions of the membrane surface wettability of the two modules, the water permeability was the same.
On the whole, this work shows that controlling the concentration of active bacteria in the feed water is efficient to reduce deposit, biomass and biopolymer accumulation on the membrane surface. In this study, the 0.3 log removal of planktonic bacteria induced by UV irradiation is strong enough to decrease significantly the biofilm growth on the surface of the membrane but not fouling, i.e. filtration performances decrease. A longer duration test may be necessary to obtain higher biofilm development at the membrane surface with effects on membrane performances, i.e. permeability decrease and longitudinal pressure drop increase.

5. Conclusion

UV irradiation used as a pre-treatment upstream from nanofiltration is able to:

- lower the concentration of viable planktonic bacteria in the feed water.
- reduce the global quantity of deposit, the sessile bacteria concentration and the amount of extracellular polymeric substances present on the surface of the membrane.

UV pre-treatment, by limiting biofilm development at the membrane surface during nanofiltration of surface water, may be able to control the membrane performances decrease observed during long term operation. New experiments with longer filtration duration tests are needed to demonstrate that UV irradiation can help to maintain membrane performances as well as to limit biofilm growth at the NF membrane surface.

6. References

Al-Adhami, B.H., Nichols, R.A., Kusel, J.R., O'Grady, J., Smith, H.V. (2007) Detection of UV-induced thymine dimers in individual *Cryptosporidium parvum* and *Cryptosporidium hominis oocysts* by immunofluorescence microscopy. Applied and Environmental Microbiology 73(3): 947-955.

Al-Amoudi A., Farooque A.M. (2005) Performance, restoration and autopsy of NF membranes used in seawater pretreatment. Desalination 178: 261–271.Al-Amoudi, A., Lovitt, R.W. (2007) Fouling strategies and the cleaning system of NF membranes and factors affecting cleaning efficiency. Journal of Membrane Science, 303(1-2), 4-28.

Alkan, U., Teksoy, A., Atesli, A., Baskaya, H.S. (2007) Influence of humic substances on the ultraviolet disinfection of surface waters. Water and Environment Journal 21(1): 61-68.

Bradford, M.M. (1976) A rapid and sensitive for the quantitation of microgram quantities of protein utilizing the principle of protein-dye binding. Analytical Biochemistry 72, 248-254.

Camper, A.K. (2004) Involvement of humic substances in regrowth. International Journal of Food Microbiology 92(3): 355-364.

Characklis, W.G., Marshall, K.C. (1990) Biofilms. John Wiley & Sons, New York.

Conlon, W.J., Jhawar, M. (1993) Pretreatment of membrane processes with ultraviolet disinfection, Water Resources. Proceedings of AWWA Annual Conference and Exposition, 343-370.

Cyna, B., Chagneau, G., Bablon, G., Tanghe, N. (2002) 2 years of nanofiltration at the Méry-sur-Oise plant, France. Desalination 147, 69-75.

Di Martino P., Doumèche B., Galas L., Vaudry H., Heim V., Habarou H. (2007) Assessing chemical cleaning of nanofiltration membranes in a drinking water production plant: a combination of chemical composition analysis and fluorescence microscopy. Water Science & Technology, 55(8-9): 219-225.

Doumèche, B., Galas, L., Vaudry, H., Di Martino, P. (2007) Membrane foulants characterisation in a drinking water production unit. Food and bioproducts processing 85(C1), 42-48.

Flemming, H.C., Scaule, G., Griebe, T., Schmitt, J., Tamachkiarowa, A. (1997) Biofouling: the Achilles heel of membrane processes. Desalination 113, 215-225.

Gaid, K., Girodet, P., Bord, G., Féliers, C. (2006) La désinfection aux ultraviolets: une alternative de choix. L'eau l'industrie les nuisances 291: 37-44.

Herzberg, M., Elimelech, M. (2007) Biofouling of reverse osmosis membranes: Role of biofilm-enhanced osmotic pressure. Journal of Membrane Science 295, 11-20.

Hilal N. , Al-Zoubi H., Darwish N.A., Mohamma A.W., Abu Arabi M. (2004) A comprehensive review of nanofiltration membranes: treatment, pretreatment, modelling, and atomic force microscopy, Desalination 170 (3): 281–308.

Houari A., Seyer D., Couquard F., Kecili K., Démocrate C., Heim V., Di Martino P. (2010) Characterization of biofouling and cleaning efficiency of nanofiltration membranes. Biofouling 26(1), 15 - 21.

Houari, A., Habarou, H., Djafer, M., Heim, V., Di Martino, P. (2009) Effect of storage of NF membranes on fouling deposits and cleaning efficiency. Desalination and water treatment 1, 307-311.

Koyuncu I., Wiesner M.R., Bele C., Coriton G., Djafer M., Cavard J. (2006) Bench-scale assessment of pretreatment to reduce fouling of salt-rejecting membranes. Desalination 197, 94-105.

Lappin-Scott H.M., Costerton J.W. (1989) Bactierial biofilms and surface fouling, Biofouling 1: 323–342.

Li, J.W., Yu, Z., Gao, M., Zhang, L., Cai, X., and Chao, F. (1996) Effect of ultraviolet Irradiation on the characteristics and trihalomethanes formation potential of Humic acids. Water Research 30 (2), 347-350.

Lund, V., Hongre, D. (1994) Ultraviolet Irradiated Water Containing Humic Substances inhibit Baterial Metabolism. water Resarch 28 (5), 1111-1116.

Marconnet, C., Houari, A., Galas, L., Vaudry, H., Heim, V., Di Martino, P. (2009) Biodegradable Dissolved Organic Carbon concentration of feed water and NF membrane biofouling: a pilot train study. Desalination 242, 228-235.

Mofidi, A.A., Bartels, C.R., Coffey, B.M., Ridgway, H.F., Knoell, T., Safarik, J., Ishida, K., Bold, R. (2000) Ultraviolet Irradiation as a Membrane Biofouling Control Strategy. American Water Works Association Annual Conference and Exposition.

Munshi, H.A., Saeed, M.O., Green, T.N., Al-Hamza, A.A., Farooque, M.A., Ismail, A.R.A. (2005) Impact of UV irradiation on controlling biofouling problems in NF-SWRO desalination process. International Desalination Association (IDA) World Congress Conference.

Ridgway H.F., Flemming H.C. (1996) Membrane Biofouling in Water Treatment Membrane Processes, McGraw Hill, New York.Roudman A.R., DiGiano F.A. (2000) Surface energy of experimental and commercial nanofiltration membranes: effects of wetting and natural organic matter fouling, Journal of Membrane Science 175 (1): 61–73.

Salcedo, I., Andrade, J.A., Quiroga, J.M., Nebot, E. (2007) Photoreactivation and dark repair in UV-treated microorganisms: effect of temperature. Applied and Environmental Microbiology, 73(5), 1594-1600.

Speth, T.F., Gusses, A.M., Summers, R.S. (2000) Evaluation of nanofiltration pretreatments for flux loss control. Desalination 130, 31-44.

Vrouwenvelder, J.S., Buiter J., Riviere M., van der Meer W.G.J., van Loosdrecht M.C.M., Kruithof J.C. (2009) Impact of flow regime on pressure drop increase and biomass, accumulation and morphology in membrane systems. Water Research 44, 689-702.

Vrouwenvelder, H.S., Manolarakis, S.A, van der Hoek, J.P., van Paassen, J.A.M., van der Meer, W.G.J., van Agtmaal, J.M.C., Prummel, H.D.M., Kruithof, J.C., van Loosdrecht, MCM. (2008) Quantitative biofouling diagnosis in full scale nanofiltration and reverse osmosis installations. Water Research 42, 4856 - 4868.

Vrouwenvelder, H.S., Van der Kooij, D. (2001) Diagnosis, prediction and prevention of biofouling of NF and RO membranes. Desalination 139, 65-71.

Vrouwenvelder, J.S., Manolarakis, S.A., Veenendaal, H.R., Van der Kooij, D. (2000) Biofouling potential of chemicals used for scale control in RO and NF membranes. Desalination 132, 1–10.

Vrouwenvelder, H. S., van Paassen, J.A.M., Folmer, H.C., Hofman, Jan A.M.H., Nederlof, M. M., van der Kooij, D. (1998) Biofouling of membranes for drinking water production. Desalination 118, 157-166.

Wend, C.F., Stewart, P.S., Jones, W., Camper, A.K. (2003) Pre-treatment for membrane water treatment systems: a laboratory study. Water Research 37, 3367-3378.

12

PAC/UF for Removing Cyanobacterial Cells and Toxins from Drinking Water

Margarida Campinas and Maria João Rosa
Urban Water Division, Department of Hydraulics and Environment
LNEC - National Laboratory for Civil Engineering, Lisbon, Portugal

1. Introduction

Cyanobacterial blooms seasonally arise as a concern to some drinking water suppliers due to the massive input of cells, the release of algogenic organic matter (AOM) into the water and often the production of cyanotoxins as secondary metabolites. Cyanotoxins may cause a wide range of adverse health effects, including skin irritation, neurotoxic damage, gastroenteritis, liver damage, tumour promotion and ultimately death, and their removal is therefore a major goal in water treatment.

An optimal water treatment requires the removal of intact cyanobacterial cells (particulate matter) (Drikas et al., 2001), given that cyanotoxins are largely intracellular and cell-bound, as well as the removal of soluble compounds, since there is always a fraction of cyanotoxins that is released into the water by cell-ageing or induced cell damage (Sivonen & Jones, 1999; Pietsch et al., 2002). Conventional treatment is considered ineffective for the removal of dissolved cyanotoxins, making the study and implementation of alternative technologies crucial to minimise or eliminate their negative impact.

Hybrid membrane processes of PAC/UF (powdered activated carbon adsorption/ ultrafiltration) were recently developed for controlling pesticides, taste and odour compounds and disinfection-by-products in drinking water treatment. PAC/UF combines the PAC ability to adsorb organics, e.g. dissolved toxins, with the effective removal of particles by low-pressure UF membranes, including cyanobacterial cells and the associated intracellular and cell-bound toxins. Despite its high potential for controlling cyanobacteria and cyanotoxins in drinking water, there is a lack of knowledge regarding PAC/UF application on this environmental-health issue.

This chapter addresses the removal of cyanobacterial cells and toxins from drinking water by PAC/UF hybrid process, focusing on the main questions involving each technology and the singularities of these environmental-health hazards. The chapter summarises the results of a comprehensive study performed with a commercial fine-grade mesoporous PAC and a lab-scale UF apparatus. Laboratory cultured cells of *Microcystis aeruginosa* and the associated microcystins were studied.

M. aeruginosa is one of the most commonly occurring cyanobacteria and grows in laboratory as single cells, which is an advantage for this study. Due to its small size, *M.*

aeruginosa single cells represent the size of algae that is more prone to escape from a conventional water treatment plant. In addition, they may be used as surrogate to assess the removal efficiency of particles of problematic size range (3-10 μm), like *Giardia* cysts and *Cryptosporidium* oocysts. This cyanobacterium produces microcystins - cyclic peptides very relevant for drinking water supply as they are chemically stable, have both acute and chronic effects (hepatotoxic and tumour-promoting) and are the most widely spread cyanotoxins in freshwaters (Sivoven & Jones, 1999). Furthermore, microcystins are the cyanotoxins for which the World Health Organization (WHO) derived a drinking water provisional guideline value (1 μg/L for daily exposure to the microcystin-LR), adopted as national standard for drinking water quality in many countries, including Portugal.

One of the aims of this chapter is to evaluate the UF performance for removing laboratory grown *M. aeruginosa* cells under different growth ages, with special attention to cell damaging and subsequent degradation of the permeate quality. Cell lysis and AOM effects are actually the most controversial issues related with the removal of algal cells by conventional treatment. Some authors refer the occurrence of cell lysis and release of intracellular material into water (Himberg et al., 1989; Hrudey et al., 1999; Pietsch et al. 2002) while others report no effect on cell integrity (Chow et al., 1999; Drikas et al., 2001; Ribau Teixeira & Rosa, 2006, 2007). Membrane filtration is referred as an attractive technology to reach high removal efficiencies of microalgae (Chow et al., 1997; Mouchet & Bonnélye, 1998; Hudrey et al., 1999; Pietsch et al., 2002; Gijsbertsen-Abrahamse et al., 2006), however studies on its effect on cell integrity are rather scarce. A few studies pointed to a small portion of cells damaged by UF, with 0-2% increase of cell bound microcystin release detected in the permeate (Chow et al., 1997; Gijsbertsen-Abrahamse et al., 2006). Yet, the same studies recommend further investigation, particularly in what concerns the cell sensitivity to shear stress and the toxin release into water, phenomena that are both largely affected by the cyanobacterial cell age.

An additional purpose is to investigate the membrane foulant behaviour of the PAC itself and to analyse if different fractions of natural organic matter (NOM) interfere and change the overall PAC effect on the UF performance. PAC contribution to the fouling control by AOM in PAC/UF hybrid process is also assessed. The state-of-the-art concerning PAC effect on membrane fouling is contradictory, some authors reporting a positive impact on permeate flux, duration of filtration cycles or chemical washing frequency (Adham et al., 1991; Konieczny & Klomfas, 2002; Lee et al., 2007) and others presenting similar flux behaviour (Yiantsios & Karabelas, 2001; Tomaszewaska & Mozia, 2002; Matsui et al., 2006) or exacerbated flux decline (Lin et al., 1999; Li & Chen, 2004; Zhao et al., 2005; Zularisam et al., 2007). Besides membrane hydrophobicity, raw water diversity is often used to explain the differing PAC effects on membrane fouling, but a complete understanding is lacking.

The final objective of this chapter is to investigate the removal of *M. aeruginosa* cells and microcystins from natural waters by PAC/UF and to compare this process performance with the often used PAC application to conventional clarification by coagulation/flocculation/sedimentation (C/F/S). PAC addition is an attractive option to overcome the limited removal of dissolved microcystins by C/F/S and UF. Nevertheless, PAC adsorption is strongly influenced by NOM size and character (Cook et al., 2001; Li et al., 2003; Campinas & Rosa, 2006) and by NOM interaction with water background inorganics (Campinas & Rosa, 2006), which may significantly reduce the adsorption kinetics and capacity. As so, competitive adsorption is site-specific, and must therefore be studied with the natural water to be treated. In addition, cyanobacterial cells and toxins challenge the coagulation and UF performances. Both processes may lyse cyanobacterial cells, with a deleterious effect on treated water. Coagulation is particularly inhibited by high AOM concentration and protein-like compounds, whose concentration increases during cell lysis (Takaara et al., 2007). Hydrophilic AOM, such as polysaccharide-like compounds, may be responsible for high membrane fouling (Kimura et al., 2006; Lee et al., 2006; Yamamura et al., 2007). A comparative analysis of PAC/UF and PAC+C/F/S for removing cyanobacterial cells and toxins is therefore essential although ultimately missing.

2. Experimental

2.1 Cyanobacterial cells and toxins

M. aeruginosa culture (Pasteur Culture Collection, PCC 7820) was grown in the laboratory, in BG11 medium, at 23-24°C, under a light regimen of 12 h fluorescent light, 12 h dark. Depending on the experiment, cultures were harvested at different growth ages, namely 1, 2-3 and 4 months, corresponding to exponential phase, late exponential phase and stationary phase, respectively (Campinas & Rosa, 2010a), and used to simulate cyanobacterial blooms. Cyanotoxins are produced at all stages of cyanobacterial growth and usually stay inside the cell (intracellular) until age or stress-driven cell lysis causes their release into the water (extracellular) (Sivonen & Jones, 1999). Throughout the growth phases, there are dissimilarities in the cyanobacterial biomass, AOM production, total cyanotoxin content, and in the intra-/extracellular cyanotoxin ratio. Microcystins are cyclic heptapeptides that share a general structure containing five fixed amino acids and two variable L- amino acids, designated as X and Z. The most commonly occurring microcystin contains leucine in position X and arginine in position Z, and is therefore called MC-LR (Lawton & Edwards, 2001). The studied strain of *M. aeruginosa* produces four microcystin variants (MC-LR, -LY, -LW, -LF), with a dominance of MC-LR, which were used to supplement the assayed waters with the target contaminant. Microcystins are relatively hydrophobic hepatotoxins, with 985-1024 g/mol, depending on the variant, and a net negative charge at pH 6-9 (-1 for MC-LR, -2 for the other variants) (Newcombe et al., 2003).

2.2 Feed waters

The results presented in sections 3.1 and 3.2 are relative to *M. aeruginosa* suspensions prepared by spiking a background electrolyte with a predetermined volume of *M. aeruginosa* culture to obtain a chlorophyll-a concentration of ca. 20 μg/L (Table 1). This concentration aims to simulate a weak bloom or a strong bloom after preliminary treatment, i.e. it is less than half the guidance level 2 for recreational waters and ten times the WHO alert level 1 for drinking waters (Bartram et al., 1999). Cultures harvested at different growth ages (1, 2, 3 and 4 months old) were used to test different cell vulnerabilities. To avoid osmotic shock which may result in cyanobacterial cell lysis, the cells were spiked in a background electrolyte prepared with deionised water amended with potassium chloride until the water conductivity reached ca. 260 μS/cm. The pH was also corrected to 7.0 ± 0.3 with KOH and H_2SO_4.

The experiments discussed in section 3.3 used deionised water with 2.5 mM background ionic strength ($KCl+CaCl_2$). Results in section 3.4 were obtained with solutions of algal organic matter, both as extracellular organic matter excreted during cyanobacterial growth (EOM) and as AOM, which includes EOM and the intracellular organic matter released during cell lysis.

Suspension	pH	EC (μS/cm)	Turbidity (NTU)	Chl-a (μg/L)	TOC (mgC/L)	$MC\text{-}LR_{eq}$ (μg/L)	
						Extra	Intra
1 month	7.2	310	5.3	28.0	2.2	8.9	--
2 months	6.7	293	2.5	20.6	--	0.6	1.5
3 months	7.3	294	3.4	20.0	3.4	2.9	--
4 months	6.7	301	2.8	19.2	--	0.5	2.1

EC: electrical conductivity; Chl-a: chlorophyll-a

Table 1. Characteristics of *M. aeruginosa* suspensions used in the UF experiments discussed in sections 3.1 and 3.2.

EOM and AOM filtrates were prepared from a *M. aeruginosa* culture following Takaara's procedure with the adaptations described by Campinas and Rosa (2010b). Specific volumes of EOM and AOM filtrates were used to prepare the UF feed solutions with a total organic carbon (TOC) concentration of ca. 4 mgC/L (Table 2). Given the high content of salts in EOM solutions, no further salts were added, besides those involved in the pH adjustment to ca. 7. In AOM solutions, the conductivity was completed to ca. 300 μS/cm with KCl and the pH was adjusted to 7.

The low value of the specific UV absorbance (SUVA) indicates that algal derived organic matter has a dominance of the hydrophilic fraction, a result in agreement with those of Her et al. (2004) and Henderson et al. (2008). On the other hand, Henderson´s characterisation of *M. aeruginosa* EOM points to an AOM solution most likely composed of two fractions: a high molar mass fraction (> 100 kDa), dominated by hydrophilic polysaccharides and hydrophobic proteins, and a low molar mass fraction (< 1 kDa). Regarding the EOM solution, it should have an alike composition, but with a larger carbohydrates/DOC (dissolved organic carbon) ratio and with some minor part of hydrophobic proteins (Campinas & Rosa, 2010b).

NOM solution	pH	EC (μS/cm)	TOC (mgC/L)	SUVA (L/(mgC.m))
EOM	7.2	831	4.5	1.3
AOM	7.2	288	3.7	1.0

Table 2. Characteristics of the AOM solutions used in UF experiments discussed in section 3.4.

Ozonated water collected at Tavira's Water Treatment Plant (western Algarve, Portugal) was used in section 3.5 trials. This water had neutral pH (7.5), low concentration of organic matter (2.3 mgC/L TOC; 2.1 mgC/L DOC), low SUVA values (0.7 L/(m.mgC)), low alkalinity (40 mg $CaCO_3$/L) and hardness (51 mg $CaCO_3$/L). Given the low SUVA value, NOM was essentially hydrophilic and with low molar mass (Edzwald & Van Benschoten, 1990).

Ozonated water was spiked with a predefined volume of *M. aeruginosa* culture until a chlorophyll-a concentration of ca. 10 μg/L was achieved (Table 3), corresponding to guidance level 1 for recreational waters and ten times higher than WHO alert level 1 for drinking waters (Bartram et al., 1999). A specific volume of microcystins stock-solution was also added to adjust the concentration of dissolved cyanotoxins to ca. 5-8 μg/L $MC\text{-}LR_{eq}$.

Parameter	
pH	7.3-7.5
EC (μS/cm)	169-211
Turbidity (NTU)	2.1-3.0
UV_{254nm} (cm^{-1})	0.016-0.041
Chl-a (μg/L)	11-12
Extra $MC\text{-}LR_{eq}$ (μg/L)	5.6-7.1
Intra $MC\text{-}LR_{eq}$ (μg/L)	1.4-1.6

Table 3. Characteristics of the feed water used in section 3.5 trials (ozonated water spiked with *M. aeruginosa* culture and dissolved microcystins).

2.3 PAC

PAC/UF experiments used Norit SA-UF, a highly mesoporous PAC, with a point of zero charge of 9.6 (Campinas & Rosa, 2006). Norit SA-UF small average particle diameter (6 μm) makes it suitable for circulating within the narrow UF hollow-fibres. Its large pore size distribution (38% of primary micropore volume, 22% of secondary micropore volume and 40% of mesopore volume) (Li et al., 2002) has proved to be advantageous for the removal of microcystins (Donati et al., 1994; Huang et al., 2007). Norit SA-UF was not used in C/F/S experiments since its extremely fine grade makes it inadequate for efficient retention by a conventional separation process. C/F/S experiments were therefore performed with Filtracarb WP7, the PAC used in Tavira's WTP. Pore size distribution was not available, however WP7 has an iodine number of 850 g/g and a methylene blue adsorption capacity of 180 mg/g.

2.4 Membrane

An Aquasource cellulose acetate hollow-fibre (inside-out configuration) membrane was used in UF experiments. This UF hydrophilic membrane has a molar mass cut-off of 100 kDa and a hydraulic permeability of 250 L/(h.m^2.bar) (manufacturer data). The module (16 fibres, 1.1 m length and 0.93 mm internal diameter; 0.05 m^2 total membrane area) was operated in a cross-flow filtration mode using the inside-out configuration during the filtration cycles and under outside-in flow during backwashing. The manufacturer recommends a maximum UF pressure of 1.5 bar and a maximum backflushing pressure of 2.5 bar. The module was mounted in the lab system schematically illustrated in Figure 1.

2.5 UF and PAC/UF runs

The membranes were first washed, flushed and compacted with deionised water until achieving a steady permeate flux at the pressure and cross-flow velocity to be used in the experiments.

UF runs were performed at a constant permeate flow (3.5 L/h), an initial transmembrane pressure (TMP) of 0.65 bar and a cross-flow velocity of 0.5 m/s, corresponding to a Reynolds number of 463 at 20ºC (1 m/s was tested only once). A feed glass tank (FT) fed the raw water to the recirculating tank (RT) through a peristaltic pump at a constant flow of 3.5 L/h to balance the permeate outflow. The concentrate was conducted to RT and blended with the additional feed water from FT, whereas the permeate was continuously collected in a beaker until a 1 L sample was obtained. By that time, RT was sampled to characterise the UF feed water. Filtration cycles lasted 1 h (four samples of permeate were always collected).

Individual concentrations were determined and, whenever necessary for data interpretation, cycle-averaged values (0-60 min) were computed. After each filtration run, the membranes were washed with a 5 mg/L (as Cl_2) sodium hypochlorite solution to inhibit the biological activity on the UF system, and thoroughly flushed with deionised water.

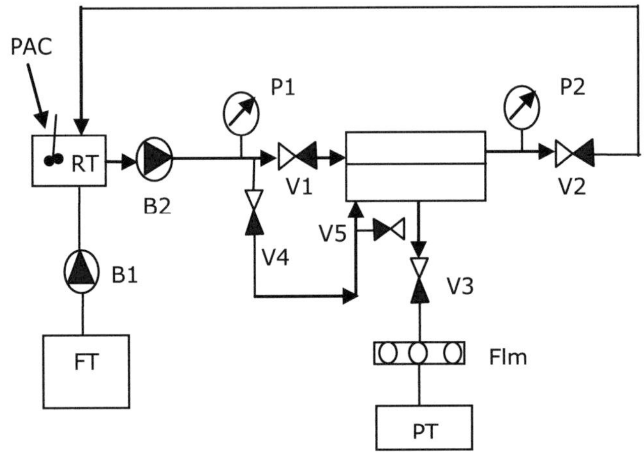

Fig. 1. Flow diagram of UF apparatus (FT - Feed tank; RT - Stirred recirculating tank; PT - Permeate tank; Flm - Flowmeter; P - Manometers; B1 - Peristaltic pump; B2 - Positive displacement pump; V1, V4, V5 - Valves for backwashing; V2 - Concentrate valve; V3 - Permeate valve).

In PAC/UF experiments, PAC was directly added to the feed tank that was continuously stirred at 150 rpm. The total mass of PAC was always added at the beginning of the filtration cycle, and was only discarded at the end of the run. PAC doses of 5 and 10 mg/L were tested. During time-depending fouling runs with AOM (section 3.4), the concentrate and permeate were both recycled back to the feed tank during two hours and permeate samples were collected at given time intervals (after 30 min., 1 and 2 hours of filtration).
Given the difficult accurate control of the permeate flow (since no automatic control was available), there were some minor oscillations of flow during the UF runs. Hence, membrane permeability rather than transmembrane pressure was represented as a function of time. Membrane permeability is given by $Q_p/(\Delta P.A_m)$, where Q_p is the permeate flow (L/h), ΔP is TMP (bar) and A_m is the membrane area (m^2).

2.6 PAC+C/F/S experiments

PAC+C/F/S experiments were performed in a laboratory jar test apparatus with four flat paddles (Flocumatic, Selecta). Operating conditions were as follows: a) rapid mixing at 104 rpm (196 s^{-1} velocity gradient) for 3 min; b) slow mixing at 20 rpm (17 s^{-1}) during 20 min; c) 1 h settling period. The aluminium polychloride coagulant PAX XL-14 (Quicom) was used at 50 mg/L. This dose is 2 to 2.5 times higher than the value used during Tavira's WTP regular operation (i.e. in the absence of cyanobacterial events), and was determined from earlier C/F/S lab tests designed to control the water turbidity during a cyanobacterial bloom with a chlorophyll-a concentration ten times the WHO alert level 1 for drinking water (Bartram et al., 1999). This scenario was simulated in the laboratory by supplementing *M. aeruginosa* culture to Tavira's ozonated water. Filtracarb WP7 PAC was also added during the rapid mixing, after

2 min of coagulation. An effective PAC contact time of 21 min was therefore ensured. Mixing conditions, coagulant type and conditions of PAC application were adjusted from Tavira's WTP. PAC+C/F/S trials were performed in duplicate. A PAC dose of 10 mg/L was tested.

2.7 Analytical methods

Samples were analysed for pH (at 20ºC, WTW 340 pH meter), electrical conductivity (EC at 25ºC, Crison GLP 32 conductimeter), turbidity (HACH 2100N turbidimeter of high resolution, 0.001 NTU), total and dissolved organic carbon (TOC, DOC) (measured as non-purgeable organic carbon by high temperature combustion method in a Shimadzu TOC 5000A), chlorophyll-a and UV_{254nm} (Beckman DU 640B UV/VIS spectrophotometer) using standard methods of water analysis. DOC and UV_{254nm} were determined on pre-filtered samples through 0.45 μm filters. For chlorophyll-a analysis, samples were filtered through GF/F filter paper and the chlorophylls were extracted using 10 mL acetone (90%). The optical densities of the extracts were measured at 665 nm and 750 nm using a Beckman DU 640B UV/VIS spectrophotometer and chlorophyll-a concentration was computed from Lorenzen equations (1967). Dissolved and intracellular microcystins were analysed by high performance liquid chromatography with photo-diode array detection (HPLC-PDA Dionex Summit System) following Meriluoto and Spoof's procedures (2005a, b, c) with the adaptations detailed by Ribau Teixeira and Rosa (2007). The overall concentration of microcystins was quantified as $MC\text{-}LR_{eq}$.

3. Results

3.1 Retention of cyanobacterial cells

A first set of trials was performed to evaluate the UF removal of *M. aeruginosa* cells under different growth ages. *M. aeruginosa* was obtained from laboratory cultures growing as single cells, whereas in natural environment they grow mainly as colonies. This different morphology was considered an advantage, since single cells are the smallest cyanobacteria and must therefore be the most difficult to retain by a solid-liquid separation. Furthermore, single cells may be used as surrogate to assess the removal of *Giardia* cysts and *Cryptosporidium* oocysts (Vlaski et al., 1996).

The removal efficiency of cyanobacterial cells was assessed through chlorophyll-a and turbidity measurements. Figure 2 depicts the cycle-averaged concentrations of chlorophyll-a and turbidity in the feed and permeate, during the UF cycles performed with *M. aeruginosa* cells 1, 2, 3 and 4 months old.

Results show the high quality of the permeate, i.e. turbidity was always below 0.1 NTU and chlorophyll-a was never detected in the permeate. Even though the feed turbidity varied between 2.8 ± 0.4 NTU and 5.3 ± 0.9 NTU, and chlorophyll-a between 28 ± 6 μg/L and 40 ± 14 μg/L during the UF filtration cycles, the rejections were always high, namely above 97% for turbidity and a complete rejection of chlorophyll-a. These results indicate an absolute removal of *M.aeruginosa* cells and confirm that UF is a safe barrier against cyanobacteria and protozoa, as reported in the literature (Ottoson et al., 2006). UF results of cell retention were already expected since cyanobacterial cells are 400-600 times larger than the membrane pores (4 - 6 μm vs. 0.01 μm; Chow et al., 1997), and are substantially higher than the algal removal efficiencies reported for conventional separation by C/F/S (69-99%) (Vlaski et al., 1996; Ribau Teixeira & Rosa, 2006, 2007) or even by C/F/dissolved air flotation (92-99%) (Ribau Teixeira & Rosa, 2006, 2007, 2010).

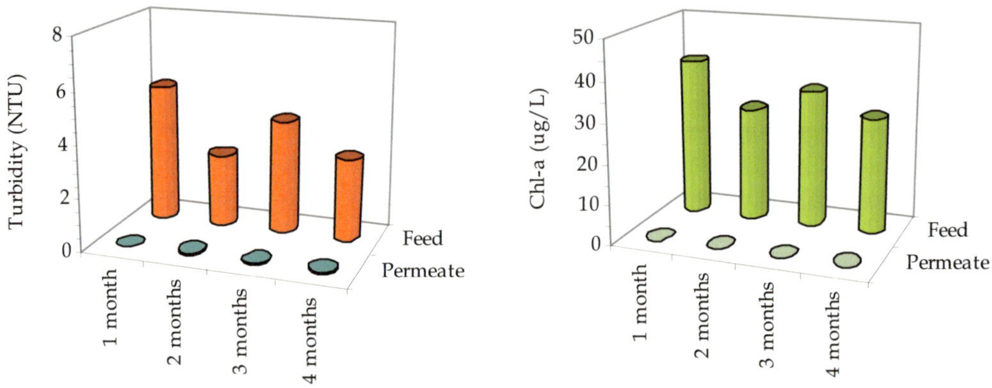

Fig. 2. Turbidity and chlorophyll-a cycle-averaged concentrations on the feed and permeate during constant flow runs with M. aeruginosa cells 1, 2, 3 and 4 months old.

3.2 Cell lysis

Experiments were performed to investigate cyanobacterial cell damaging during UF cycles (due to shear stresses developed at the membrane surface or from pumping) and subsequent release of intracellular cyanotoxins and AOM into the permeate.

Preliminary UF runs with model solutions of microcystins revealed a low adsorption of microcystins by the cellulose acetate membrane (4% average rejection (Campinas & Rosa, 2010a)). It was thus decided to indirectly assess cell lysis through the evolution of the dissolved microcystin concentration in the feed and permeate during the filtration cycle. A first set of experiments was performed with *M. aeruginosa* cultures 1 and 3 months old, and the dissolved microcystins concentration was monitored in the feed and in permeate. The resulting cycle-averaged rejections are presented in Figure 3.

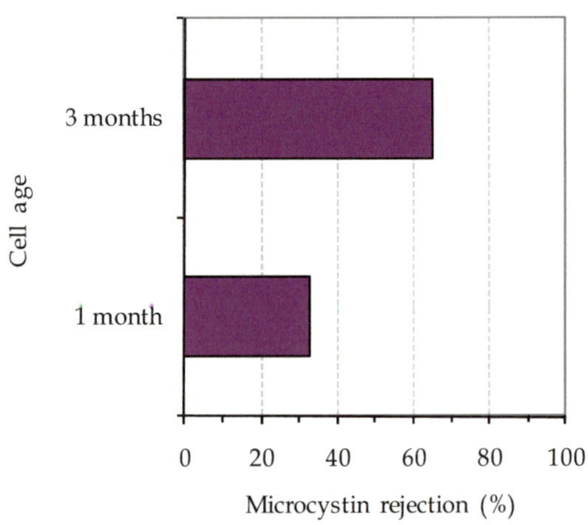

Fig. 3. Cycle-averaged rejection of dissolved microcystins during constant flow runs with *M. aeruginosa* cells 1 and 3 months old.

Figure 3 shows a cycle-averaged rejection of microcystins of 33% with 1 month old cells and 65% with 3 months old cells. The studied microcystins are relatively hydrophobic compounds with a molar mass (985–1024 Da) much below the cut-off of the hydrophilic membrane (100 kDa), making rejection through sieving not an option, but probably through adsorption. As single-solute solutions of microcystins revealed their low adsorption onto the studied UF membrane, a reasonable explanation for the microcystins rejection exhibited by both cultures is the AOM-microcystins interaction, which confers rejection properties to the hydrophilic membrane. Due to this AOM-driven microcystins adsorption onto the UF membrane, cell lysis assessment based only on the time progression of the dissolved microcystin concentration in the feed and permeate is not suitable and underestimates the cell lysis phenomenon (cell lysis and microcystins adsorption may happen simultaneously).

The first set of results prompted the replication of the experiments one month latter. This is, by the time of the second set of trials the cultures were 2 and 4 months old. This time, cell lysis was assessed through the evolution of intra-microcystin and chlorophyll-a contents in the feed during the filtration cycle. The results were analysed through concentration factors ($Cf = C/C_0$), comparing the experimental Cf values with the expected ones (Figure 4). The latter were computed by mass-balance equation assuming complete rejection, no adsorption and no cell lysis. Experimental Cf values lower than expected Cf values were therefore interpreted as cell lysis.

Figure 4 shows the occurrence of cell lysis in both cultures, with a stronger impact in the older culture and at the end of the filtration cycle. Chlorophyll-a concentration is not shown but was less sensitive to cell lysis than the intra-microcystins concentration (Campinas & Rosa, 2010a).

On the other hand, Figure 5 reveals a cycle-averaged rejection of total microcystins of 60% with the 2 months old culture and of 80% with the 4 months old cells, which indicates the importance of the adsorption phenomenon for this older culture. In parallel to cell damage, an enhancement of microcystin rejection by the UF cellulose acetate membrane must occur with cell ageing, probably associated with the greater content in segregated AOM (mucopolysaccharides) and/or protein lysed AOM of older cultures. Consequently, the permeate quality did not degrade at all with the oldest culture (39% of dissolved microcystin rejection) and suffered only a minor degradation with the youngest culture (-4% of dissolved microcystin rejection).

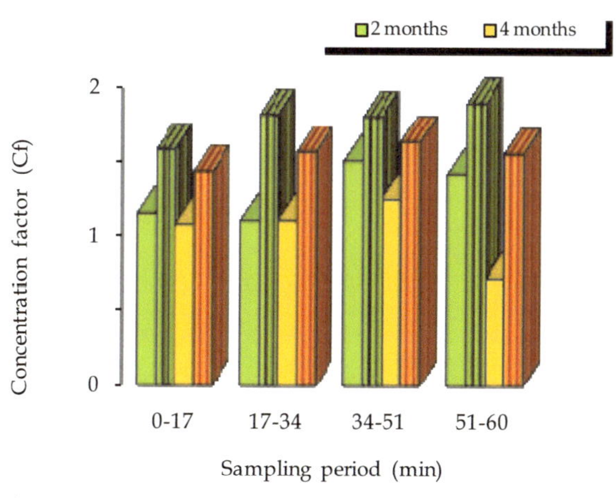

Fig. 4. Concentration factors for intracellular microcystins during filtration cycles with *M. aeruginosa* cells 2 and 4 months old: experimental (full) and expected (striped) values based on mass-balance equations.

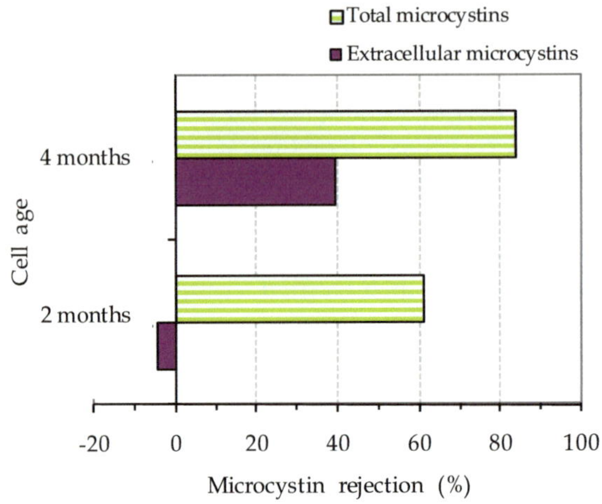

Fig. 5. Cycle-averaged rejection of total and dissolved microcystin during constant flow runs with *M. aeruginosa* cells 2 and 4 months old.

It can be therefore concluded that cell lysis occurs during the UF runs, particularly for older cultures, but this event does not necessarily degrade the permeate quality since in parallel to cell damage an enhancement of microcystin rejection by the UF membrane is observed with cell ageing, probably due to AOM-driven microcystin adsorption on the membrane. Pietsch et al. (2002) obtained also higher cell removals and cell lysis by flocculation of *M. aeruginosa* in the stationary growth phase than with cells in the exponential growth phase.

3.3 UF fouling potential of PAC

Time-depending fouling runs with PAC addition (5 mg/L) were performed with deionised water with 2.5 mM background IS (KCl+$CaCl_2$) and using two cross-flow velocities (CFV), 0.5 m/s (used in all previous runs) and 1.0 m/s. For comparison purposes, analogous UF runs were carried out with no PAC addition. Figure 6 presents the membrane permeability as a function of time during PAC/UF and UF runs.

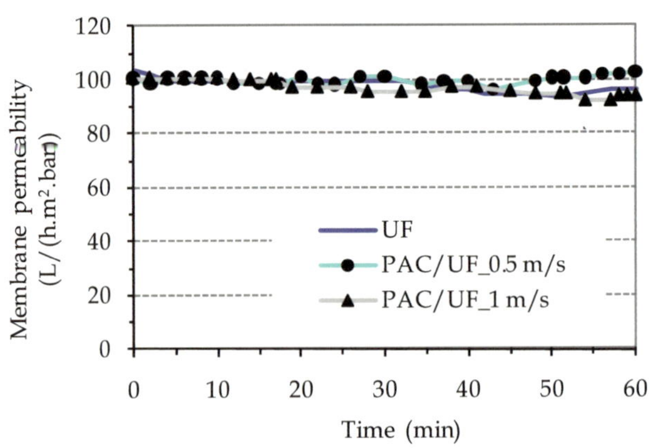

Fig. 6. Membrane permeability during UF and PAC/UF time-depending fouling runs at two cross-flow velocities (UF: no PAC addition; PAC/UF: 5 mg/L PAC).

As explained earlier in section 2.5, it was difficult to manually control the permeate flow, and there were in consequence some minor oscillations, especially at the end of the runs. Nevertheless, the curves with and without PAC almost overlap, meaning that PAC alone did not affect the membrane permeability of the hydrophilic UF membrane. Similar results were obtained during concentration-time depending fouling runs (Campinas & Rosa, 2010b) and by other authors such as Lin et al. (1999), Yantsios & Karabelas (2001), Li & Chen (2004) and Mozia et al. (2005). PAC should deposit on the membrane surface, however the PAC average diameter (6 μm) is probably large enough to avoid membrane pore blocking, forming a porous layer that allows the passage of water with no additional resistance.
Figure 7 displays the normalised turbidity of PAC/UF feed during the time-depending fouling runs at two CFV. With both velocities only a fraction of PAC was recycled as a suspension and just for a short period of time, and even the higher CFV did not yield a higher feed normalised turbidity. A similar PAC deposition rate was observed for both CFVs, a behaviour which is analogous to that found by Crozes et al. (1997) and Matsui et al. (2001). In addition, for the laminar conditions tested, a comparable membrane permeability decrease pattern was observed for CFV 0.5 m/s (Reynolds number of 463, at 20°C) and 1.0 m/s (Reynolds number of 926). As a result, the most cost-effective cross-flow velocity (0.5 m/s) was used in the following experiments.

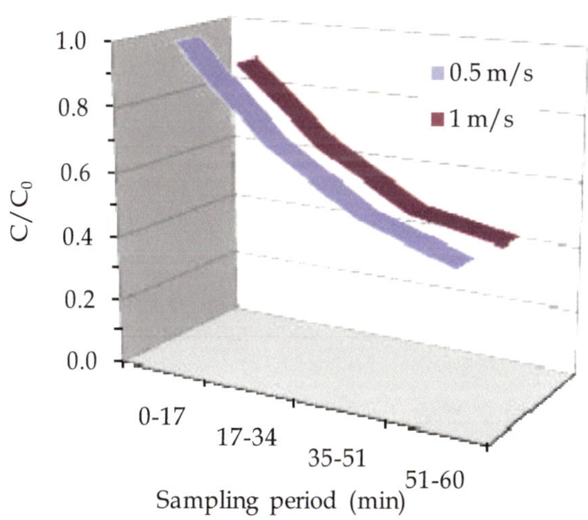

Fig. 7. Normalised turbidity of PAC/UF feed during the time-depending fouling runs at two cross flow velocities (5 mg/L PAC).

3.4 AOM fouling potential in PAC/UF

Some authors concluded that PAC itself does not impose considerable membrane fouling, yet the presence of NOM acts as glue that binds the PAC particles to one another and to the membrane surface, and increases the fouling resistance (Lin et al., 1999; Li and Chen, 2004). To investigate the combined effect of PAC and AOM on the membrane fouling, UF and PAC/UF (10 mg/L PAC) fouling runs were performed with algogenic organic fractions (AOM and EOM). The normalised flux as a function of time is depicted on Figure 8.

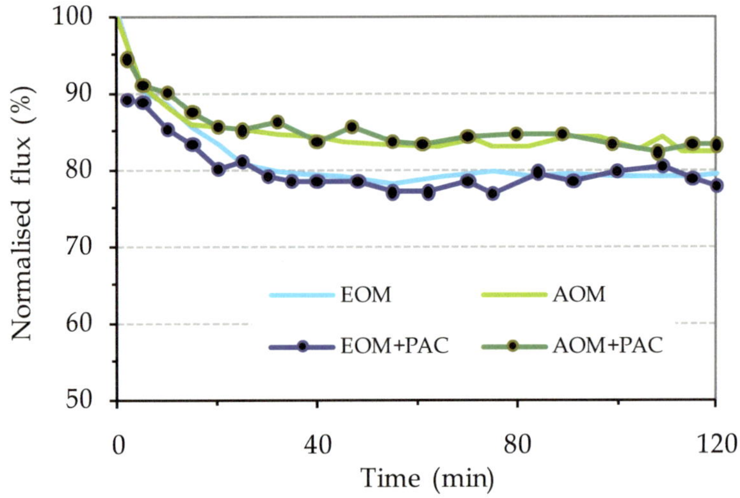

Fig. 8. Normalised flux during PAC/UF (10 mg/L PAC) and UF time-depending fouling runs with algogenic organic fractions (AOM and EOM).

Figure 8 shows no differences at all of the normalised flux with and without PAC addition, which indicates that PAC neither endorsed nor controlled the membrane reversible fouling. The same trend has already been obtained by others (Tomaszewska & Mozia, 2002; Mozia et al., 2005; Matsui et al., 2006;) and differ from the results of Jermann et al. (2008) with NOM-kaolinite, probably due to the larger particle size of Norit SA-UF PAC (6 μm) compared to kaolinite (0.1-2 μm). Figure 8 also shows that the different characteristics of the NOM used in the experiments, i.e. the higher or lower protein content or even the higher salt content in EOM solutions, did not seem to modify the PAC effect on the membrane reversible fouling. A similar behaviour was observed with NOM of different hydrophobicity, such as tannic and humic acids (Campinas & Rosa, 2010b).

EOM run resulted in a more severe membrane fouling than the AOM run, yielding a flux decline of 21% after 2 h of operation, somewhat higher than the AOM-driven flux decline (18%). Other authors have also obtained a high fouling potential of hydrophilic NOM, as that prevailing in AOM and EOM fractions (Table 2) (Her et al., 2004; Lee et al., 2006; Zularisam et al., 2007).

Figure 9 presents NOM rejections with and without PAC addition. PAC improved the TOC rejection of AOM (35% vs. 55%) and had no major impact on the UV_{254nm} absorbing substances of AOM and on TOC and UV_{254nm} rejections of EOM constituents. This dissimilar behaviour is most likely justified by the presence of more hydrophobic compounds in the AOM fraction, causing the PAC-driven enhancement of TOC rejection. It seems that the highly hydrophilic compounds of EOM (only expressed by the overall content as TOC, since UV_{254nm} is not a good indicator for polysaccharides and aromatic tryptophan-like proteins) were not easily adsorbed onto PAC, and as such PAC did not improve the membrane reversible fouling.

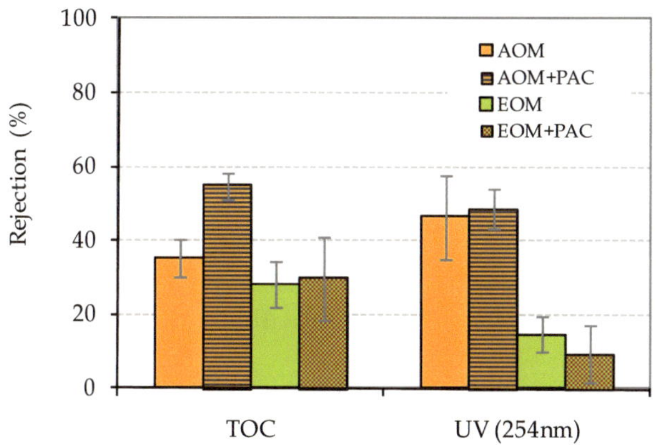

Fig. 9. TOC and UV_{254nm} rejections of algogenic organic fractions during PAC/UF (10 mg/L) and UF fouling runs (error bars represent standard deviations).

Given the higher salt content of EOM solution and the lower TOC and UV_{254nm} rejections, the higher fouling potential of EOM is attributed to interactions between polysaccharide-type compounds and multivalent cations. These interactions prompt a denser fouling layer which results in an augmented overall resistance to water permeation. These observations are in agreement with those found by Jermann et al. (2007) and Katsoufidou et al. (2007) in UF studies with alginate (a typical model polysaccharide used as EOM surrogate) in the presence of calcium.

3.5 Comparing UF, PAC/UF and PAC+C/F/S

The performance of UF, PAC+C/F/S and PAC/UF was compared for removing *M. aeruginosa* cells and microcystins from soft to moderately hard natural waters with hydrophilic low molar mass organics (Table 3). PAC/UF and PAC+C/F/S used the same PAC dose (10 mg/L) but different PAC type (Norit SA-UF and Filtracarb WP7, respectively) and effective PAC contact time (1 h and 21 min, respectively) representing the real full-scale scenario.

Figure 10 shows that all processes ensured an absolute removal of *M. aeruginosa* cells and chlorophyll-a was never detected in the treated water. Although a much lower size of the PAC carbon was used (6 µm), PAC/UF process attained a safer removal of particles, as turbidity was always below 0.1 NTU, much below the 0.48 NTU obtained with PAC+C/F/S (> 98% turbidity rejection by UF and PAC/UF vs. 84% by PAC+C/F/S). Between UF and PAC/UF there were some clear differences, in particular a negative rejection of dissolved microcystins and UV_{254nm} (-2% and -20%, respectively) was observed with UF, most likely due to cell lysis (Campinas & Rosa, 2010c). Comparing both processes with PAC addition, i.e. PAC/UF and PAC+C/F/S, greater removal of microcystins (intra and extracellular) was achieved by PAC/UF (90% vs. 36%). A reasonable explanation is the longer PAC effective contact time and the use of smaller PAC particles in PAC/UF process, thus providing longer adsorption time and faster kinetics. According to Cook et al. (2001) and Ho & Newcombe (2005), PAC application to a conventional clarification has the disadvantages of low contact time with suspended carbon (since during settling there is minimal or no adsorption) and floc interference with

PAC adsorption capacity, i.e. PAC is incorporated into the flocs and diffusion kinetics are reduced.

On the other hand, conventional clarification showed greater UV_{254nm} removals than PAC/UF (66% vs. 39% removal, Figure 10), which is thought to be connected with the cell lysis occurrence during UF cycles, causing a deleterious effect on the permeate quality (and responsible for the -20% UV_{254nm} rejection in UF runs). PAC/UF improved the UF permeate quality in terms of UV_{254nm} but was unable to reach the PAC + C/F/S quality. These results reflect the preferential coagulation of high molar mass compounds, a feature that benefits the C/F/S process, even if in the present study the conventional clarification was designed for particle removal by adsorption/neutralisation rather than for NOM enhanced coagulation.

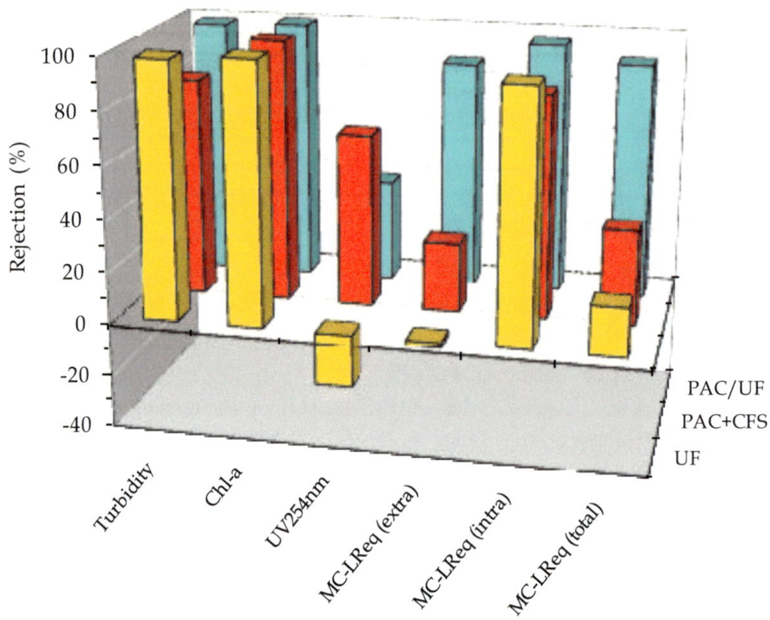

Fig. 10. UF, PAC+C/F/S and PAC/UF performances with ozonated water supplemented with *M. aeruginosa* culture and dissolved microcystins (7.2-8.5 μg/L $MC\text{-}LR_{eq}$, 4-5 extra/intracellular; 10 mg/L PAC).

Figure 11 displays the cycle-averaged concentration of microcystins (intra and extracellular) in UF, PAC+C/F/S and PAC/UF treated waters. The error bars depicted in the figure are standard deviations.

Both UF and conventional application of 10 mg/L PAC were unable to control the microcystins, attaining a permeate concentration of 6.2 ± 0.1 μg/L $MC\text{-}LR_{eq}$, and 5.4 ± 0.4 μg/L $MC\text{-}LR_{eq}$ respectively. By opposition, PAC/UF allowed a concentration of 0.72 ± 0.4 μg/L $MC\text{-}LR_{eq}$ with a PAC dose of 10 mg/L. Higher removal of dissolved microcystins would be expected if cell lysis could be avoided or PAC adsorption improved, e.g. by using higher PAC dosages, smaller PAC particles or longer PAC retention time.

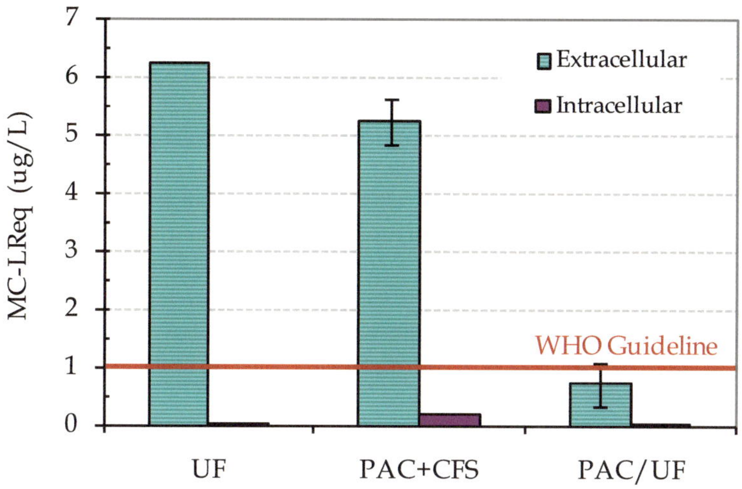

Fig. 11. Cycle-averaged microcystin concentration of treated waters produced by UF, PAC+C/F/S and PAC/UF application to ozonated water supplemented with *M. aeruginosa* culture and dissolved microcystins (7.2-8.5 μg/L MC-LR_{eq}, 4-5 extra/intracellular; 10 mg/L PAC).

4. Conclusions

This chapter summarises the results of a comprehensive study of the removal of cyanobacterial cells and toxins from drinking water by PAC/UF hybrid process, focusing on the main questions involving each technology and the singularities of these environmental-health hazards.

The first objective of the present research was to evaluate the UF performance for removing cyanobacterial cells under different growth ages, with special attention to cell damaging and subsequent degradation of permeate quality. The results obtained show UF as a safe barrier against cyanobacteria, ensuring an absolute removal of *M. aeruginosa* single cells, the smallest cyanobacterial cells and hence the most difficult to remove. Data showed an increased cell lysis with cell ageing, although it did not necessarily degrade permeate quality, as in parallel to cell damage an enhancement of microcystin rejection by the UF hydrophilic membrane (of cellulose acetate) was observed with cell ageing. This effect is most probably due to AOM-driven microcystin adsorption on the membrane, connected to the greater content of the older cultures in segregated AOM (mucopolysaccharides) and/or protein lysed AOM.

The second objective of this study was to investigate the foulant behaviour of the PAC itself and to analyse if different NOM fractions change the overall PAC effect on the UF performance. Data of UF time-depending fouling runs showed that UF membrane effectively retained the fine PAC particles and that PAC did not affect the hydrophilic membrane permeability. The different protein and salt contents of the algal organic fractions (AOM and EOM) did not change the PAC effect on the membrane reversible fouling as suggested by some authors.

The third objective was to assess PAC contribution to the fouling control by algogenic organic matter in PAC/UF hybrid process. Data showed that EOM had a greater impact on membrane fouling than AOM, probably due to polysaccharide-like substances interaction with multivalent ions, resulting in dense fouling layers. PAC improved the rejection of hydrophobic AOM but it was apparently ineffective for adsorbing the highly hydrophilic EOM, and as such did not improve the EOM-driven membrane reversible fouling.

This comprehensive study ultimately aimed to investigate the removal of *M. aeruginosa* cells and microcystins from natural waters by PAC/UF and to compare this process with PAC application to conventional clarification (PAC+C/F/S). Data showed that the membrane-based processes (PAC/UF and UF alone) ensured an absolute removal of *M. aeruginosa* cells, and chlorophyll-a was never detected in treated waters. In addition, both UF processes achieved an excellent overall control of particles (turbidity below 0.1 NTU), but PAC/UF remarkably improved the rejection of dissolved microcystins and UV_{254nm} absorbing substances, from negative values observed with UF due to cell lysis, to 87% and 30%, respectively.

Comparing PAC/UF and PAC+C/F/S, the former ensured a significant improvement in turbidity rejection (99% vs. 84%) and especially in total microcystins rejection (90% vs. 36%) and concentration in the treated water (0.72 vs. 5.4 μg/L $MC\text{-}LR_{eq}$), as a result of enhanced adsorption kinetics in PAC/UF (due to longer PAC effective contact time, smaller PAC particles and PAC not incorporated in the flocs). Nevertheless, PAC application to C/F/S achieved higher removal of UV_{254nm} absorbing substances than PAC/UF (66% vs. 39%), justified by privileged coagulation of large compounds and detrimental UF effect on cell integrity. The fouling behaviour of algal organic matter and the cell lysis occurrence during UF, with subsequent release of dissolved microcystins and AOM to water, indicate that UF with no PAC addition is inadequate to treat cyanobacterial-rich waters and highlight the relevance of a roughing clarification step prior to PAC/UF.

5. Acknowledgments

The authors would like to thank the Portuguese Foundation for Science and Technology for providing a Ph.D scholarship to Margarida Campinas (BD/10356/2002). Special thanks are also addressed to Águas do Algarve, SA (Portugal) for partially funding this project.

6. References

Adham, S., Snoeyink, V., Clark, M. & Bersillon, J.-L. (1991) Predicting and verifying organics removal by PAC in an ultrafiltration system. *Journal of American Water Works Association*, 12, 81-91.

Bartram, J., Burch, M., Falconer, I., Jones, G. & Kuiper-Goodman, T. (1999). Situation assessment, planning and management. *In Toxic Cyanobacteria in Water: a Guide to Their Public Health Consequences, Monitoring and Management*. World Health Organization (Eds.), E&SPON, London.

Campinas, M. & Rosa, M.J. (2006). The ionic strength effect on microcystin and natural organic matter surrogate adsorption onto PAC. *Journal of Colloid and Interface Science*, 299, 520-529.

Campinas, M. & Rosa, M.J. (2010a). Evaluation of cyanobacterial cells removal and lysis by ultrafiltration. *Separation and Purification Technology*, 70, 345-353.

Campinas, M. & Rosa, M.J. (2010b). Assessing PAC contribution to the NOM fouling control in PAC/UF systems. *Water Research*, 44, 1636-1644.

Campinas, M. & Rosa, M.J. (2010c). Comparing PAC/UF and conventional clarification with PAC for removing microcystins from natural waters. *Desalination and Water Treatment*, 16, 120-128.

Chow, C., Panglish, S., House, J., Drikas, M., Burch, M.& Gimbel, R. (1997). A study of membrane filtration for the removal of cyanobacteria. *Journal of Water Supply: Research and Technology - Aqua*, 46(6), 324-334.

Chow, C., House, J., Drikas, M., House, J., Burch, M. & Velzeboer, R. (1999). The impact of conventional water treatment processes on cells of the cyanobacterium *Microcystis aeruginosa*. *Water Research*, 33(15), 3253-3262.

Cook, D., Newcombe, G. & Sztajnbok, P. (2001). The application of powdered activated carbon for MIB and geosmin removal: predicting PAC doses in four raw waters. *Water Research*, 35(5), 1325-1333.

Crozes, G., Jacangelo, J., Anselme, C. & Laîné, J. (1997) Impact of ultrafiltration operating conditions on membrane irreversible fouling. *Journal of Membrane Science*, 124(1), 63-76.

Donati, C., Drikas, M., Hayes, R. & Newcombe, G. (1994). Microcystin-LR adsorption by powdered activated carbon. *Water Research*, 28(8), 1735-1742.

Edzwald, J. & Van Benschoten, J. (1990). Aluminium coagulation of natural organic material. *In Chemical Water and Wastewater Treatment*. Hahn H.H. and Klute K. (Eds). Berlin: Springer-Verlag.

Gijsbertsen-Abrahamse, A., Schmidt, W., Chorus, I. & Heijman, S. (2006). Removal of cyanotoxins by ultrafiltration and nanofiltration. *Journal of Membrane Science*, 276, 252-259.

Henderson, R., Baker, A., Parsons, S. & Jefferson, B. (2008) Characterisation of algogenic organic matter extracted from cyanobacteria, green algae and diatoms. *Water Research*, 42(13), 3435-3445.

Her, N., Amy, G., Park, H.-R. & Song, M. (2004). Characterizing algogenic organic matter (AOM) and evaluating associated NF membrane fouling. *Water Research*, 38, 1427-1438.

Ho, L. & Newcombe, G. (2005). Effect of NOM, turbidity and floc size on the PAC adsorption of MIB during alum coagulation. *Water Research*, 39, 3668-3674.

Hrudey, S., Burch, M., Drikas, M. & Gregory, R. (1999). Remedial measures. *In Toxic Cyanobacteria in Water: a Guide to Their Public Health Consequences, Monitoring and Management*. World Health Organization (Eds.), E&SPON, London.

Huang, W.-J, Cheng, B.-L. & Vheng, Y.-L. (2007). Adsorption of microcystin-LR by three types of activated carbon. *Journal of Hazardous Materials*, 141, 115-122.

Jermann, D., Pronk, W., Meylan, S. & Boller, M. (2007) Interplay of different NOM fouling mechanisms during ultrafiltration for drinking water production. *Water Research*, 41(8), 1713-1722.

Jermann, D., Pronk, W., Kägi, R., Halbeisen, M. & Boiler, M. (2008) Influence of interactions between NOM and particles on UF fouling mechanisms. *Water Research,* 42(14), 3870-3878.

Katsoufidou, K., Yiantsios, S. & Karabelas, A. (2007) Experimental study of ultrafiltration membrane fouling by sodium alginate and flux recovery by backwashing. *Journal of Membrane Science,* 300, 137-146.

Kimura, K., Yamamura, H. & Watanabe, Y. (2006). Irreversible fouling in MF/UF membranes caused by natural organic matters (NOMs) isolated from different origins. *Separation Science Technology,* 41, 1331-1344.

Konieczny, K. & Klomfas, G. (2002) Using activated carbon to improve natural water treatment by porous membranes. *Desalination,* 147, 109-116.

Lawton, L. & Edwards, C. (2001). Purification of microcystins. *Journal of Chromatography A,* 912, 191-209.

Lee, N., Amy, G. & Croué, J-P. (2006). Low-pressure membrane (MF/UF) fouling associated with allochthonous versus autochthonous natural organic matter. *Water Research,* 40, 2357-2368.

Lee, C., Bae, S., Han, S. & Kang, L. (2007) Application of ultrafiltration hybrid membrane processes for reuse of secondary effluent. *Desalination,* 202, 239-246.

Li, F., Yuasa, A., Ebie, K., Azuma, Y., Hagishita, T. & Matsui, Y. (2002) Factors affecting the adsorption capacity of dissolved organic matter onto activated carbon: modified isotherm analysis. *Water Research,* 36, 4592

Li, Q., Snoeyink, V., Mariñas, B. & Campos, C. (2003). Pore blockage effect of NOM on atrazine adsorption kinetics of PAC: the roles of PAC pore size distribution and NOM molecular weight. *Water Research,* 37, 4863

Li, C.-W. & Chen, Y.-S. (2004) Fouling of UF membrane by humic substances: effects of molecular weight and powder-activated carbon (PAC) pre-treatment. *Desalination,* 170, 59-67.

Lin, C.-F., Huang, Y.-J. & Hao, O. (1999) Ultrafiltration processes for removing humic substances: effect of molecular weigth fractions and PAC treatment. *Water Research,* 33(5), 1252-1264.

Lorenzen, C. J. (1967) Determination of chlorophyll and pheo-pigments: spectrophotometric equations. *Limnology and Oceanography,* 12(2), 343-346.

Matsui, Y., Yuasa, A. & Colas, F. (2001) Effect of operational modes on the removal of a synthetic organic chemical by powdered activated carbon during ultrafiltration. *Water Science and Technology: Water Supply,* 1(5/6), 39-47.

Matsui, Y., Sanogawa, T., Aoki, N., Mima, S. & Matsushita, T. (2006) Evaluating submicron-sized activated carbon adsorption for microfiltration pretreatment. *Water Science and Technology,* 6(1), 149-155.

Meriluoto, J. & Spoof, L. (2005a). SOP: Solid phase extraction of microcystin in water samples. *In Toxic Cyanobacterial Monitoring and Cyanotoxin Analysis.* J. Meriluoto and G. Codd (Eds.), Abo Akademi University Press, Finland.

Meriluoto, J. & Spoof, L. (2005b). SOP: Extraction of microcystins in biomass filtered on glass-fibre filters or in freeze-dried cyanobacterial biomass. *In Toxic Cyanobacterial Monitoring and Cyanotoxin Analysis.* J. Meriluoto and G. Codd (Eds), Abo Akademi University Press, Finland.

Meriluoto, J. & Spoof, L. (2005c). SOP: Analysis of microcystins by high-performance liquid chromatography with photodiode-array detection. *In Toxic Cyanobacterial Monitoring and Cyanotoxin Analysis.* J. Meriluoto and G. Codd (Eds.), Abo Akademi University Press, Finland.

Mouchet, P. & Bonnélye, V. (1998). Solving algae problems: French expertise and world-wide application. *Journal of Water Science Research and Technology - Aqua,* 47(3), 125-141.

Mozia, S., Tomaszewaska, M. & Morawski, A. (2005). Studies on the effect of humic acids and phenol on adsorption-ultrafiltration process performance. *Water Research,* 39, 501-509.

Newcombe, G., Cook, D., Brooke, S., Ho, L. & Slyman, N. (2003). Treatment options for microcystins toxins: similarities and differences between variants. *Environmental Technology,* 24(3), 299-308.

Ottoson, J., Hansen, A., Bjorlenius, B., Norder, H. & Stenstrom, T. (2006). Removal of viruses, parasitic protozoa and microbial indicators in conventional and membrane processes in a wastewater pilot plant. *Water Research,* 40, 1449-1457

Pietsch, J., Bornmann, K. & Schmidt, W. (2002). Relevance of intra and extracellular cyanotoxins for drinking water treatment. *Acta Hydrochimica et Hydrobiologica,* 30(1), 7-15.

Ribau Teixeira, M. & Rosa, M.J. (2006). Comparing dissolved air flotation and conventional sedimentation to remove cyanobacterial cells of *Microcystis aeruginosa.* Part I: The key operating conditions. *Separation and Purification Technology,* 52, 84-94.

Ribau Teixeira, M. & Rosa, M.J. (2007). Comparing dissolved air flotation and conventional sedimentation to remove cyanobacterial cells of *Microcystis aeruginosa.* Part II: The effect of water background organics. *Separation and Purification Technology,* 53, 126-134.

Ribau Teixeira, M. & Rosa, M.J. (2010). Investigating dissolved air flotation performance with cyanobacterial cells and filaments. *Water Research,* 44, 3337-3344.

Sivonen, K. & Jones, G. (1999). Cyanobacterial toxins. *In Toxic Cyanobacteria in Water: a Guide to Their Public Health Consequences, Monitoring and Management.* World Health Organization (Eds.), E& SPON, London.

Takaara, T., Sano, D., Konno, H. & Omura, T. (2007). Cellular proteins of *Microcystis aeruginosa* inhibiting coagulation with polyaluminum chloride. *Water Research,* 41, 1653-1658.

Tomaszewska, M. & Mozia, S. (2002) Removal of organic matter from water by PAC/UF system. *Water Research,* 36, 4137-4143.

Vlaski, A., Breemen, A. & Alaerts, G. (1996). Optimisation of coagulation conditions for the removal of cyanobacteria by dissolved air flotation or sedimentation. *Journal of Water Science Research and Technology - Aqua,* 45(5), 253-261.

Yamamura, H., Kimura, K. & Watanabe, Y. (2007). Mechanism involved in the evolution of physically irreversible fouling in microfiltration and ultrafiltration membranes used for drinking water treatment. *Environmental Science and Technology,* 41, 6789-9794.

Yiantsios, S. & Karabelas, A. (2001) An experimental study of humic acid and powdered activated carbon deposition on UF membranes and their removal by backwashing. *Desalination*, 140, 195-209.

Zhao, P., Takizawa, S., Katayama, H. & Ohgaki, S. (2005) Factors causing PAC cake fouling in PAC-MF (powdered activated carbon-microfiltration) water treatment systems. *Water Science and Technology*, 51(6-7), 231-240.

Zularisam, A., Ismail, A., Salim, M., Sakinah, M. & Ozaki, H. (2007) The effects of natural organic matter (NOM) fractions on fouling characteristics and flux recovery of ultrafiltration membranes. *Desalination*, 212, 191-208.

13

Fabrication of Tubular Membrane Supports from Low Price Raw Materials, Using Both Centrifugal Casting and/or Extrusion Methods

Abdelhamid Harabi and Ferhat Bouzerara
Ceramics Laboratory, Mentouri University - Constantine, Constantine, Algeria

1. Introduction

Porous ceramics supports are, generally, needed for membranes manufacturing. For the development of high-quality supports, the following properties are of major importance: pore size distribution, total porosity ratio, surface quality with the absence of large defects or large pores, good mechanical properties and chemical stability (Gestel et al., 2002). In fact, the top layer is closely related to its support. In addition, the quality of the support is of crucial importance to the integrity of the membrane layers that are applied in the subsequent preparation steps. The required thickness of the membrane is further limited by the smoothness of the support because the membrane material must cover all irregularities of the support to form a continuous, defects free layer (Biesheuvel et al., 2001).
The conventional method of preparing ceramic tubes is extrusion. Nevertheless, a problem of extruded ceramic tubes may be encountered such as low surface smoothness (Nijmeijer et al., 1998) and larger average pore sizes. Consequently, an alternative method for such a supports preparation has been proposed (centrifugal casting) [2-9] (Biesheuvel et al., 2001; Nijmeijer et al., 1998; Chen et al., 2005; Steenkamp et al., 2002; Steenkamp et al., 2002; Falamaki & Veysizadeh, 2008; Steenkamp et al., 2001; Kim et al., 2002; Pinggen et al., 2003). Although this method is rather more expensive than extrusion, it is very suitable for manufacturing high-quality tubes (Steenkamp et al., 2002). A centrifuged tube when compared to an extruded tube shows an extremely good roundness, smooth inside surface and a narrow pore size distribution are obtained, which are essential for the quality of membranes that are, afterwards, deposited on this surface (Nijmeijer et al., 1998).
This is the main objective of this work which consists of the preparation of adequate and less expensive membrane supports using centrifugal casting process and to investigate the effect of organic content (such as starch) on porosity. For comparison, the raw materials with and without starch addition were prepared.
These tubular supports are destined to be used as supports of Micro-Filtration (MF) or Ultra-Filtration (UF) membranes. They permit to provide mechanical strength to a membrane top layer to withstand the stress induced by the pressure difference applied over the entire membrane and must simultaneously have a low resistance to the filtrate flow. The usual starting materials (Al_2O_3, Mullite, ZrO_2, …) are replaced in this work by a native raw material (Tamazert Kaolin: TK, Djebel Debbag kaolins: DD2 and DD3, dolomite, calcite), in order to reduce the cost of supports fabrication.

It should be noticed that TK type has been selected in this work (to be used) on the basis of a preliminary study. In fact, the other kaolin types did not behave likely when they were used individually. This result might be due to their differences in chemical compositions and constituting phases.

Because of their application in the treatment of big amounts of wastewaters (Benito et al., 2005), there is much current interest in the application of membranes in separation procedures. The use of ceramic membranes has many advantages such as high thermal and chemical stability, pressure resistance, long lifetime, and good defouling properties (Lee et al., 2002; Ding et al., 2006). MF and UF are often used to remove particles, microorganisms, and colloidal materials from suspensions (Gaucher et al., 2002). Asymmetric membrane usually consists of a thin top-layer responsible for separating components, and a porous ceramic support with single or multiple intermediate layers imparting the required mechanical strength to the membrane composite (Mori et al., 1998). In fact, the commercial support made of artificial material is an important part of the high price of the membranes, which is why some authors have focused their researches to develop the preparation of low coast supports made of natural raw materials such as clays (Saffaj et al., 2004). The industrial membrane production uses a limited choice of materials. As a consequence, ceramic membranes have a high price. A significant effort was then provided these last years in membrane technology field in order to find new porous ceramics materials at low price (Masmoudi et al., 2005; Rakib et al., 2000). In order to decrease this cost and to evaluate our natural resources, the supports have been manufactured (Elmoudden et al., 2001; Bouzerara et al., 2006), in this work, from kaolin ($Al_2O_3 \cdot 2SiO_2 \cdot 2H_2O$) and the calcium carbonates ($CaCO_3$) local raw materials. The choice of these materials has been dictated by their natural abundance (low price) and their thermal stability (Khemakhem et al., 2004).

In this chapter, the manufacture of a tubular supports ceramics membrane, using both centrifugal casting and extrusion methods, is described in more details. The main objective of this work consists of the preparation and characterization of adequate and less expensive membrane supports, using abundant local raw materials. A porous raw materials tube of 20 mm in diameters and 170 mm in length were successfully fabricated by centrifugal casting technique. Moreover, the obtained samples were characterized, using different techniques. The structure was analyzed by X-ray diffraction (XRD) and mercury porosimetry techniques. The pore size and the presence of possible defects in the supports were determined by Scanning Electron Microscopy (SEM). It has been found that tubular ceramic membrane supports had a highly homogeneous product and a smooth inside-surface. The influence of the sintering temperature on the total porosity, average pore size and pore size distribution of supports is taken into account. The obtained results enabled to conclude that clay supports can be used alone (without any additions), successfully, in tangential microfiltration or as a support for UF membranes. Finally, this investigation demonstrates that centrifugal casting may be also considered as a promising technique in order to fabricate tubes for membrane application. Moreover, obtained tubes were characterized in terms of porosity; these supports were extremely homogeneous as can be seen from the very sharp pore size distribution. Moreover, the raw materials employed were easily obtainable at low costs. Membrane supports manufactured from raw materials and starch mixtures presented features of porosity (porous volume and average pore size) more important than those elaborated from TK; the manufactured membrane supports are mainly constituted of mullite and quartz phases. The presence of these phases may also extend further their use, even under severe atmosphere conditions.

For comparison, other low cost raw materials (DD2 kaolin and $CaCO_3$) have been chosen in order to fabricate membrane supports but using extrusion technique. Finally, a correlation between microstructure and properties of all the prepared membrane supports was also taken into account.

Since the ceramic filters are generally constituted of a thick support (2000 µm) and one or multi thin membranes (from 10 to 40 µm for each one). That is why this work or chapter is mainly focussed on ceramic supports rather than its deposed membranes. Therefore, replacing the more expansive starting materials, mentioned above, by other low cost raw materials for supports fabrication (which constitutes about 99% of the filter mass) is significantly important. So, what low cost raw materials does mean? It is incomparable; the alumina price is at least about 100 times more expensive than that of kaolin.

Another important advantage is the substantial gain in energy by decreasing the sintering temperature from about 1600°C (Harabi & DAVIS, 1995a) to about 1250°C, when alumina supports were replaced by the proposed supports. Besides this, about 50% the prepared supports is pores (porosity) which may also be considered as a gain in its mass. The relatively lower theoretical density of the prepared supports (2.8 g/cm^3) when compared to that of alumina (3.98 g/cm^3) (Harabi & DAVIS, 1995a) is also a further exiting advantage. The authors ask readers to do calculation by themselves about the advantages mentioned above.

One can claim that there is no need to do this calculation if these low price manufactured supports do not obey to the main internationally required support characteristics. Consequently, this chapter is devoted to go through these characteristics one by one in more details.

2. Important definitions

For reader whom not well familiarized with this kind of technology the definition of certain technology words such as inorganic membrane, extrusion and centrifugal methods is of great importance.

2.1 Inorganic membranes

A membrane is a physical barrier allowing selective transport of mass species, widely used for separation and purification in many industries. Membranes can be classified into organic, inorganic and hybrids of organic/inorganic systems.

Inorganic membranes can be divided firstly into dense and porous structures, and for porous membranes into asymmetric and symmetric ones. Symmetric membranes exhibit homogeneous pore size throughout the membrane. Asymmetric membranes present a change of structure through the membrane. An inorganic membrane can be described as an asymmetric porous ceramic formed by a macroporous support with successive thin layers deposited on it. The support provides mechanical resistance to the medium. The successive layers are active in MF, UF or Nano-Filtration (NF), depending on their pore diameters. Most commercial ceramic membranes are in plate or tubular configuration in order to increase the surface area to volume ratio, which gives more separation area per unit volume of membrane element.

The most common fabrication processes are extrusion, pressing, tape casting, slip casting, centrifugal casting, Sol-Gel process, dip and spin coating. From these, extrusion, pressing, slip casting, centrifugal casting and tape casting are used for support systems, tape casting and slip casting are used for microfiltration membranes, Sol-Gel process, dip and spin coating are used for UF and NF membranes.

2.2 Extrusion method

Extrusion is a process technology for the production of ceramic tubes. In extrusion: a stiff paste is compacted and shaped by forcing it through a nozzle. In general the manufacturing process of tubular ceramic supports using extrusion method includes the following steps:
-Steps of mixing various materials, including raw material, organic additive, and other extrusion-aid materials to form a paste.
-Passing the paste through an extruder to form a cylindrical, tubular supports.
-Placing the tubular supports on rotating rollers to cause the support to rotate and to dry.
-Firing the tubular ceramic support. The firing will remove all organic binders and chemicals utilized in the ceramic manufacturing process, such as the starch, amijel and methocel. It will also remove any residual water not removed.

2.3 Centrifugal method

Centrifugal casting is a new process technology for the production of ceramic tubes. In centrifugal casting a cylindrical mold is filled with suspension and rotated rapidly around its axis. This results in the movement of the particulate phase towards the cylinder wall and the formation of a tubular cast.

3. Experimental procedure

3.1 Analysis of the raw materials

The chemical compositions of the 3 clays (KT, DD2 and DD3) used in the present work given in weight percentages of oxides are given in table 1.

Clay	SiO_2	Al_2O_3	TiO_2	Fe_2O_3	K_2O	Na_2O	CaO	MgO	MnO	I. L. at 1100°C
KT	50.56	34.15	0.28	1.15	7.18	-	0.31	0.28	-	06.4
DD2	45.29	38.36	0.44	0.09	0.33	0.07	0.23	0.03	0.19	15.8
DD3	45.90	37.49	0.44	0.40	0.41	0.07	0.41	0.01	1.52	16.5

Table 1. Chemical composition of the three clay types expressed as wt% of equivalent oxides, used in this chapter.

When extrusion technique is used, the supports were prepared from domestic kaolin (DD2) and calcium carbonates derived from Guelma and Constantine regions (Algeria), respectively. The purity of CaO obtained from calcined calcite was about 99.7%. The other starting raw materials were domestic kaolin (DD3) and dolomite derived from Guelma and Batna regions (Algeria), respectively. The chemical composition of this kaolin is given in table 1, where the main impurities are CaO, MnO and Fe_2O_3. The doloma (CaO.MgO) has been obtained from dolomite after calcination at 950°C for 4 hours. The purity of the added doloma is about 99,6 wt%. It contains, mainly, 0.27 wt% Fe_2O_3, 0.07 wt% Al_2O_3 and 0.02 wt% Na_2O as impurities.

3.2 Characterization techniques

The structure was analyzed by X-ray diffraction (XRD) and Hg-porosimetry techniques. The presence of possible defects in the prepared supports was checked by using Scanning Electron Microscopy (SEM). The tensile strength testing of sintered samples at room temperature was carried out using the diametral compression test, at a constant displacement rate of 0.2 mm/min., using hard metal plattens. Generally, three samples of each composition sintered under the same conditions were tested and an average value was taken. Following previous strength testing procedures, packing strips (Manilla office file) of 0.30 mm thickness were used. This technique is well detailed elsewhere (Harabi & DAVIS, 1995b).

3.3 Supports preparation

There are three different procedures (1, 2 and 3) which were followed in this chapter for supports preparation (1 by centrifugation and 2 by extrusion).
Firstly, the main preparation steps, used in the first one, are shown in Fig. 1.

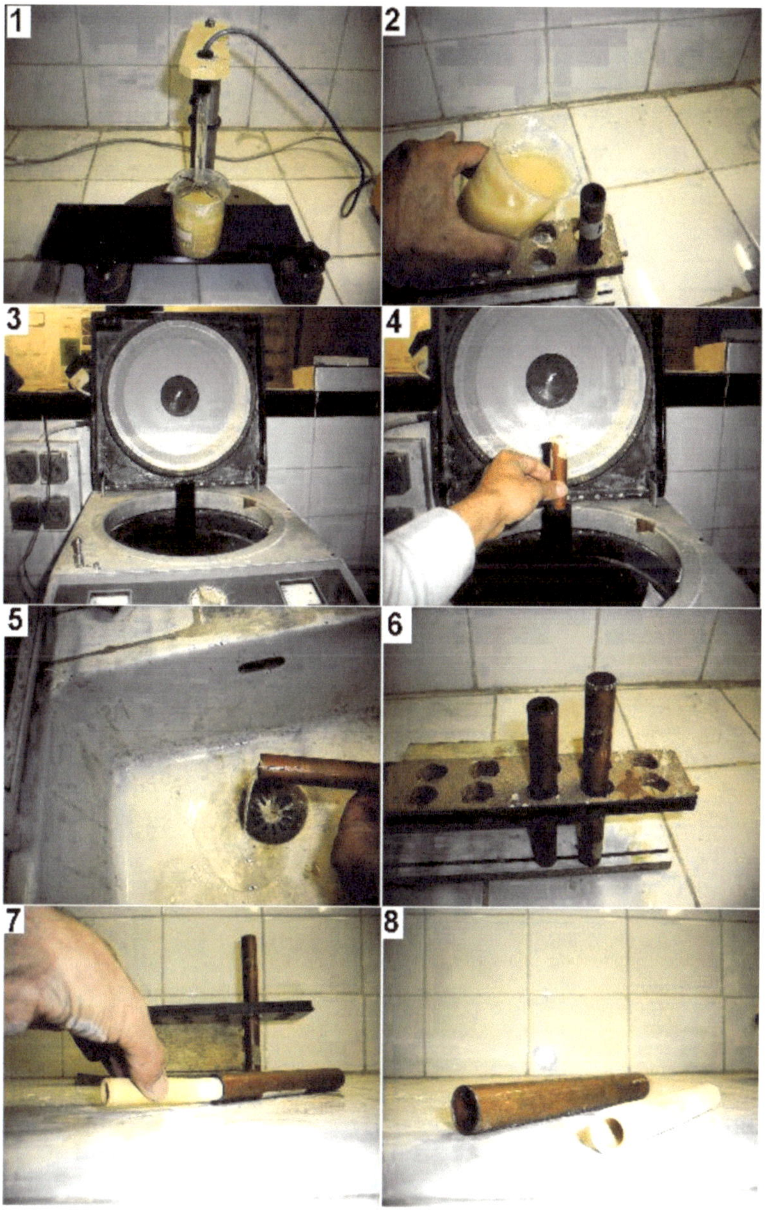

Fig. 1. Steps order of membrane supports fabrication using centrifugal casting technique.

50 g (33.3 wt%) of the raw material powders were mixed with 100 ml (66.7 wt%) distilled water by magnetic stirring. Then, the prepared suspensions were ultrasonically treated during 10 minutes at 35 kHz to break down particle agglomerates. Afterwards, the obtained mixtures were poured into cylindrical metal moulds 20 mm in inner diameter and 170 mm in length. Then rotated rapidly around its central axis for 8 minutes at 6000 rpm and the remaining liquid was poured out of the moulds. The green tubes were vertically dried inside the moulds during 3 days at 22°C and 50 % relative humidity. After drying, the green tubes were removed out from the moulds and sintered horizontally at 1000, 1150, 1200 and 1250°C during 1 hour. In order to study the influence of the starch addition on the porosity and pore size distribution of the tubes, a series of different starch concentrations in the suspension has been prepared. In the stepwise method, the same procedure was repeated to form the next laminate. Suspensions, identified as S2 were prepared by adding the starch to raw materials.

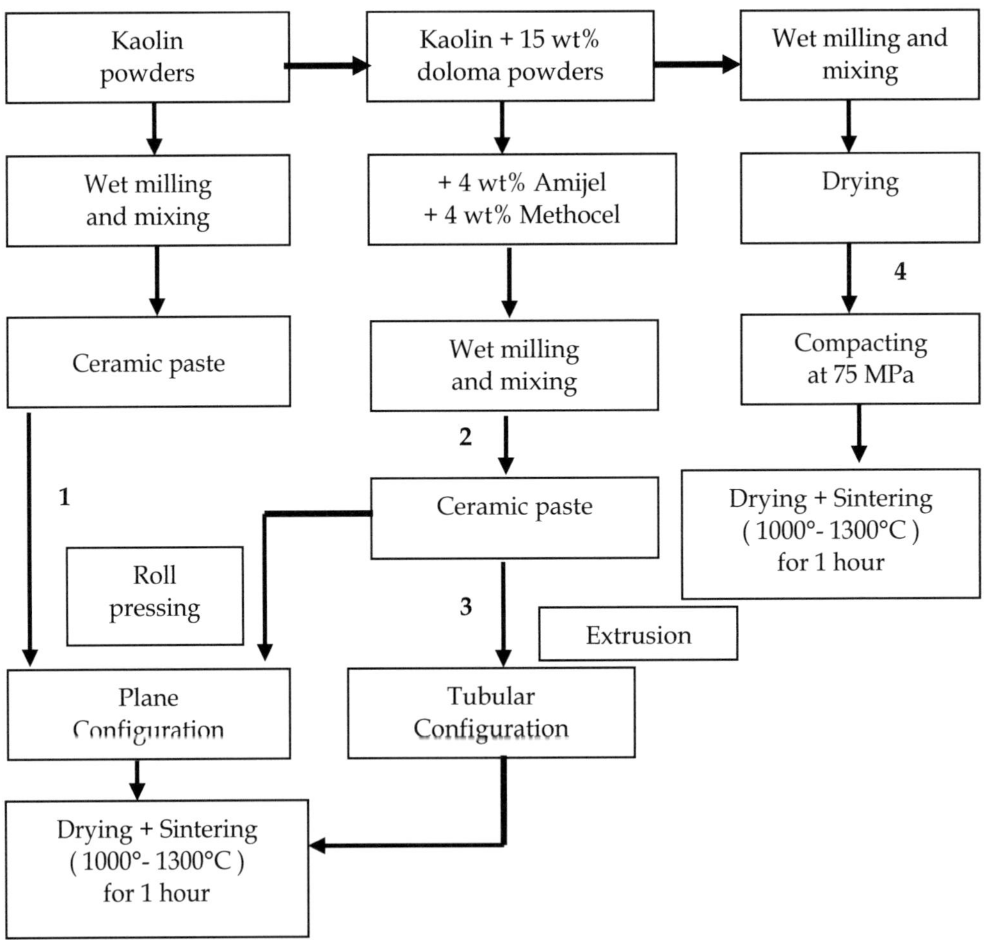

Fig. 2. A schematic diagram showing the main processes (1, 2, 3 and 4), used for membrane supports preparation, in this work.

Secondly, the kaolin (DD2) is properly crushed, then calcinated at 540°C for 1 hour to be later on sieved at 150 μm. After that, a quantity of 28 wt% of calcium carbonate powder is

added. In order to improve the properties that facilitate the forming, some organic materials, such as 3 wt% methocel, as a plasticizer and 3 wt% amijel, as a binder have been added. This mixture should be continuously mixed up with water so as to get the plastic paste. For a good diffusion of the water in the paste, this latter should be properly covered in a plastic case for at least 12 hours. After that, extrusion technique takes place to make some tubular samples. For a good drying of these tubular samples, they should be placed at room temperature on rotating aluminum roll. In order to eliminate organic materials added and avoid the microcrackes in the samples, the rate of sintering chosen is 2°C/min.
Thirdly, the main steps of the 4 processing routes (noted 1, 2, 3 and 4) for samples preparation, used in this work, are described in Fig. 2. One of the main differences between these processes is the shape of the product. The flat configuration is obtained when processes 1, 2 and 4 were applied while a tubular one was achieved when process 3 was used. Another important difference is the doloma addition (process 4) and doloma coupled with methocel and amijel additions (processes 2 and 3). After doloma addition to kaolin, the mixture loses its plastic properties. To improve these properties and facilitate the shaping of products, organic materials have been added (processes 2 and 3). The organic additions used are: 4 wt% of methocel, derived from methylcellulose (The Dow Chemical Company) as a plasticizer and 4 wt% of amijel, derived from starch (Cplus 12072, Cerestar), as a binder.

3.4 Membranes preparation

Due to the chemical and thermal properties of zirconium oxide, it is widely used for membranes preparation. This powder has a specific surface area which is about 43.5 m^2/g and average particle size of about 0.22 μm. This material has been prepared by Cezus Chimie campany.
For preparing a microfiltration layer with zirconia powder, a deflocculated slip was obtained by mixing 10 wt% zirconia powder, 25 wt% PVA (12 wt% aqueous solution) and water (65 wt%). The deposition of the slip on the support was performed by the slip casting method (Khemakhem et al., 2004). In this case of the tubular membranes, the tube was closed at one end and filled with the solution. The coating was carried out by capillary suction. The thin layer thickness was determined by the capillary pressure and is depended on the support porosity, on the coating time and on the suspension viscosity (Masmoudi et al., 2005). The deposition time was between 5 and 10 min. After drying at room temperature for 24 hours, the microfiltration layer was sintered at 1150°C for 2 hours. A temperature plate at 250°C for 15 minutes is necessary in order to eliminate completely the PVA, which is in great quantities in the slip. A relatively slow temperature rate (2°C/min) was needed in order to avoid the formation of cracks on the layer (Masmoudi et al., 2006). The distribution of the pore diameters of the membrane was determined by mercury porosimetry.

4. Results and discussion

4.1 Phases identification

Phases identification is of great importance before any supports and membranes manufactures. Because of the presence of certain phases may limit their application for water filtration only, rather than acids filtration or gases separation.

For example, Fig. 3 shows the X-ray diffraction (XRD) spectra of samples sintered at 1000°, 1100° and 1200°C during 1 hour. The main observed phases are: mullite ($3Al_2O_3.2SiO_2$) and quartz (Harabi et al., 2009). At 1000°C, the main dominant phase appearing at this temperature is quartz while at higher temperatures the main dominant phase is mullite. Mullite becomes more crystallized at 1100°C (Fig. 3).

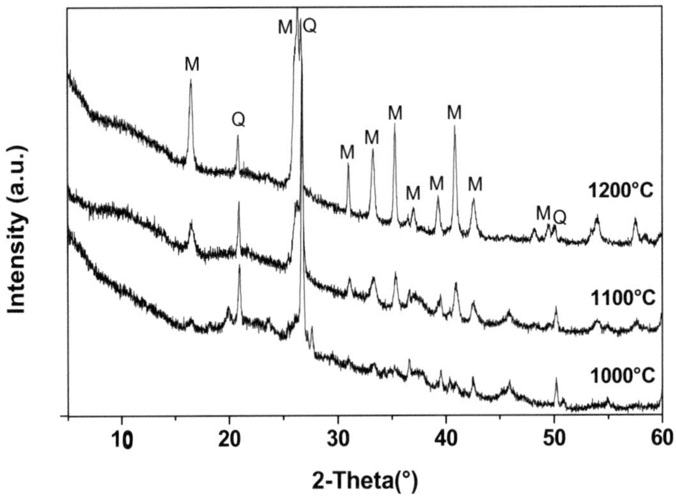

Fig. 3. XRD spectra of samples sintered at different temperatures during 1 hour. M: Mullite; Q: quartz.

Its content increased, thereafter, with increasing temperature. These identified phases are of great importance because of their promising physical and mechanical properties. For example, mullite is a useful refractory for high-temperature ceramics applications, because of its low thermal expansion and high creep resistance. In addition to this, it has a high load bearing capacity, abrasion and corrosion resistance.

Moreover, in the case of supports prepared by extrusion method from kaolin DD2 type ($Al_2O_3 \cdot 2SiO_2 \cdot 2H_2O$) and the calcium carbonates ($CaCO_3$) local raw materials the main observed phases are: mullite and anorthite (Boudaira et al., 2009).

By contrast, when supports were prepared only from kaolin DD3 type the main observed phases are: mullite and cristoballite ((Bouzerara et al., 2009). Apart from cistoballite all the above formed phases are chemically stable in acids. In order to transform this udiserible phase into another stable phase, about 37 wt% of ZO_2 was added into kaolin. It has been found that the main formed phases were mullite and zicon ($ZrSiO_4$) (Mecif et al., 2010).

4.2 Pore characterization of supports prepared by extrusion technique

For the development of high-quality supports, the following properties are of major importance: pore size distribution, total porosity ratio, surface quality with the absence of large defects or large pores, mechanical properties and chemical stability. Structural characteristics of the final membrane supports, prepared from kaolin and kaolin + doloma, according to the four different processing routes are summarized in Table 2.

Used processes	Sintering temperature (°C)	Total Porosity (%)	A.P.S. (μm)	PSD modal	Tensile Strength (MPa)
Process 1	1000	37.26	0.02	BMPSD	
	1200	15.68	3.49	MMPSD	
	1250	13.50	3.32	SMPSD	
	1300	15.94	4.28	SMPSD	
Process 2	1000	56	7.68	MMPSD	
	1200	53.31	12.72	MMPSD	
	1250	51.23	22.31	SMPSD	
	1300	51.9	39.40	SMPSD	
Process 3	1150	39.89	1.65	MMPSD	
	1200	38.29	2.89	BMPSD	
	1250	43.18	27.72	SMPSD	
	1300	40.53	48.93	SMPSD	
Process 4	1000	41.94	0.69	MMPSD	6
	1100	-	-	-	9
	1200	40.99	3.77	SMPSD	15
	1250	40.81	11.09	BMPSD	15
	1300	37.63	23.48	SMPSD	9

Table 2. Some properties of samples prepared from kaolin and kaolin + doloma according to the four different processes.

The total porosity, average pore size and pore size distribution have been determined by mercury intrusion porosimetry (Micromeritics, Model Autopore 9220) for supports sintered at different temperatures for 60 min. The obtained results are illustrated in Fig. 4. As would be expected, this figure shows, generally that there is an increase in average pore size and a decrease in total porosity in samples, when the sintering temperature is increased. Moreover, it can be said that both the average pore size and porous volume are closely related to the preparation method. The obtained results (Table 2 and Fig. 4) show that the doloma addition to kaolin has a positive effect on the porosity ratio of supports compared to those prepared from kaolin alone. For example, the kaolin support (process 1) had a porosity ratio of ≈13% and an average pore size around 3 μm, whereas the kaolin-doloma supports (process 2) had a porosity ratio of ≈ 51% and an average pore size around 22 μm, sintered under the same conditions (1250°C for 1 hour).

The pores characterisation may be divided into three main features (categories). These consist of total porosity, average pore size and the modal distribution of pore size. The pore size distribution modal may also be classified into three distinct modals; single or Gaussian distribution, bi-modal and multi-modal pore size distributions. The Single (mono) Modal of Pore Size Distribution (SMPSD) is generally obtained for samples having a uniform pore size distribution. When pore volume (%) is plotted against pore size, the curve is characterised by a single peak (Fig. 5a).

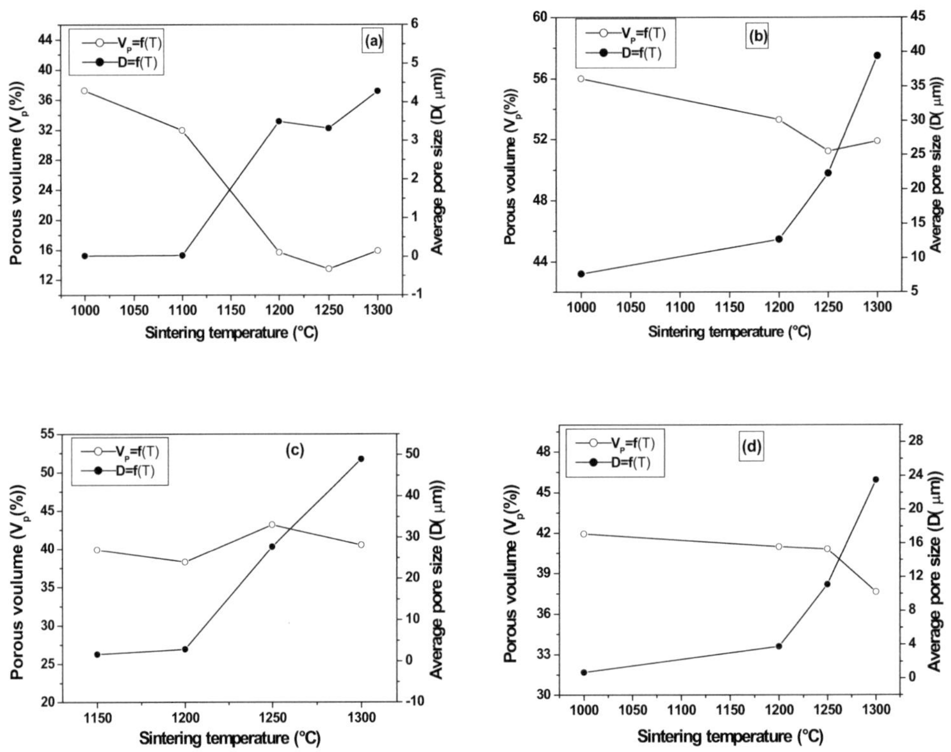

Fig. 4. Porous volume (%) and average pore size versus sintering temperature for kaolin samples, using process 1(a) and for kaolin + 15 wt% doloma samples, using processes 2(b), 3(c), 4(d).

However, the Bi-Modal of Pore Size Distribution (BMPSD) is characterised by two different or overlapping peaks. This means that there are two classes of pore size distribution (Fig. 5b). Finally, the Multi Modal of Pore Size Distribution (MMPSD) is characterized by the presence of more than two distinct (Fig. 5c) or overlapping peaks (Fig. 5d). The BMPSD observed in Fig. 5b, using process 1, consists of two different pore origins, existing at this lower sintering temperature (1000°C). The origin of finer pore sizes may be attributed to the H_2O escaping, while the coarser ones are due to the voids existing between kaolin starting particles. When the sintering temperature is increased (1300°C), the coalescence of finer pores leads to the disappearance of the second peak (Table 2) and a SMPSD is obtained, within an average pore size of about 4 μm. The BMPSD (mentioned above) becomes MMPSD (Figs. 5c and 6a) when

further starting materials have been added such as $Mg(OH)_2$, $Ca(OH)_2$, amijel and methocel, for samples sintered at a lower temperature (1000°C).
Fig. 5c is a good example for that it illustrates well that the number of peaks is about five which corresponds to the number of starting materials (kaolin, $Mg(OH)_2$, $Ca(OH)_2$, amijel and methocel). A uniform pore size distribution (SMPSD) is also obtained when samples were sintered at higher temperature (1300°C), as shown in Figs. 5a and 6c. Therefore it can be said that these modal distributions are closely related to the processing routes used in this work. Moreover, the compacting pressure is a major factor controlling the modal distribution of pore size. For example, a uniform pore size distribution is already obtained for samples sintered at 1200°C (Fig. 6b), using process 4 (75 MPa), while the appearance of this modal is delayed to 1300°C, using process 2 (hand roll pressing). It should be mentioned that the pore size distribution is almost uniform for samples sintered at 1250°C for 1 h, using process 4 (Fig. 6c).

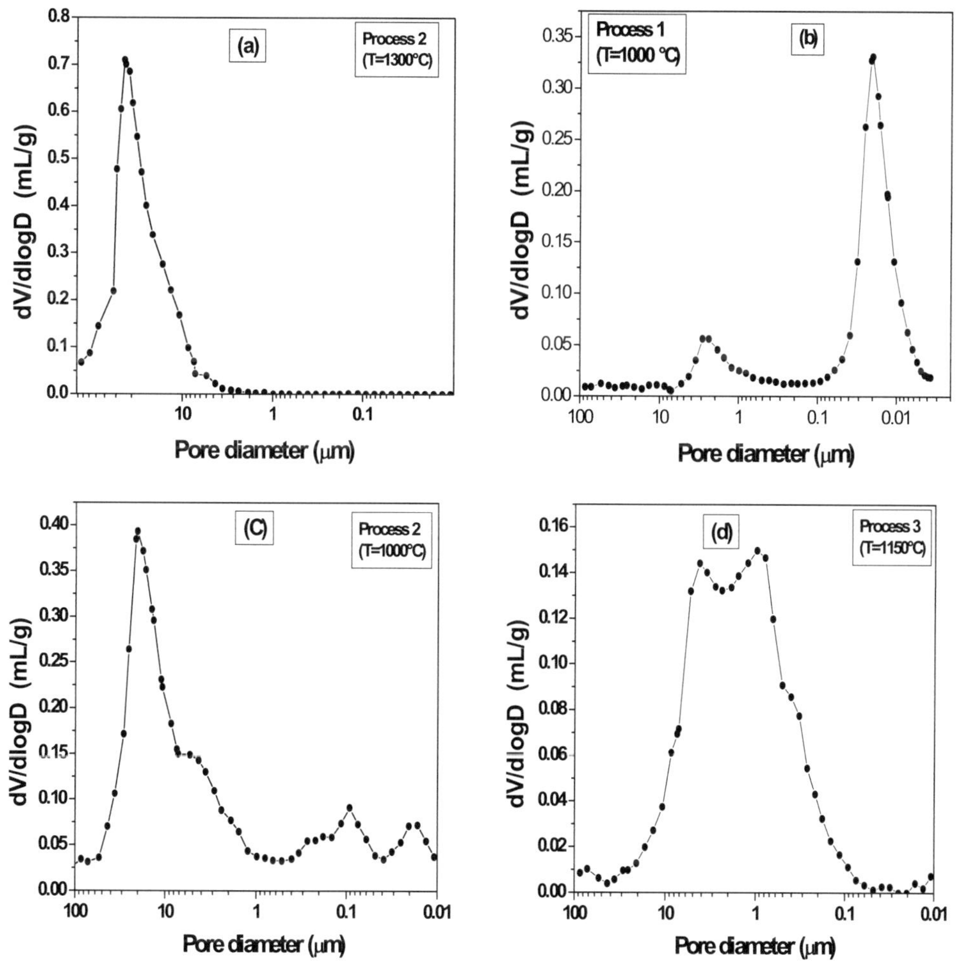

Fig. 5. Pore size distribution in kaolin using process 2 (a) and in kaolin + 15 wt% doloma samples using processes 1 (b), 2(c) and 3(d).

Fig. 6 displays the modal distributions of pore size, for samples sintered at different temperatures for 1 h, using process 4. It is clear that the pore size distribution modal is temperature dependent. It is almost mono-modal distribution (uniform distribution) in compacts sintered at 1200°C (Fig. 6b) and 1300°C (Fig. 6d), whereas the bi-modal and multi-modal pore size distributions were obtained in those sintered at 1250°C (Fig. 6c) and 1000°C (Fig. 6a), respectively. It should be mentioned that even though the modal of pore size distribution is the same (homogeneous) for samples sintered at 1200 and 1300°C, they may behave differently. In fact, the large interval of pore size distribution (0.1–10 μm) in samples sintered at 1200°C is shifted towards a narrow interval of pore size distribution (8–50 μm). On the basis of the above results, it can be said that the increase in sintering temperature encourages the coalescence of pores, which in turn, leads to a larger average pore size (Table 2). These kinds of pore size distribution curves do not tell us more information about both cumulative porous volume (%) and average pore size (μm). Consequently, curves of cumulative porous volume (%) versus pore diameter were replotted in Figs. 7a–d, using four different processing routes. It should be mentioned that the average pore size is directly given by mercury intrusion porosimetry. Moreover, these values are in good agreement with those determined also from the pore size corresponding to 50% of cumulative volume (Figs. 7a–d).

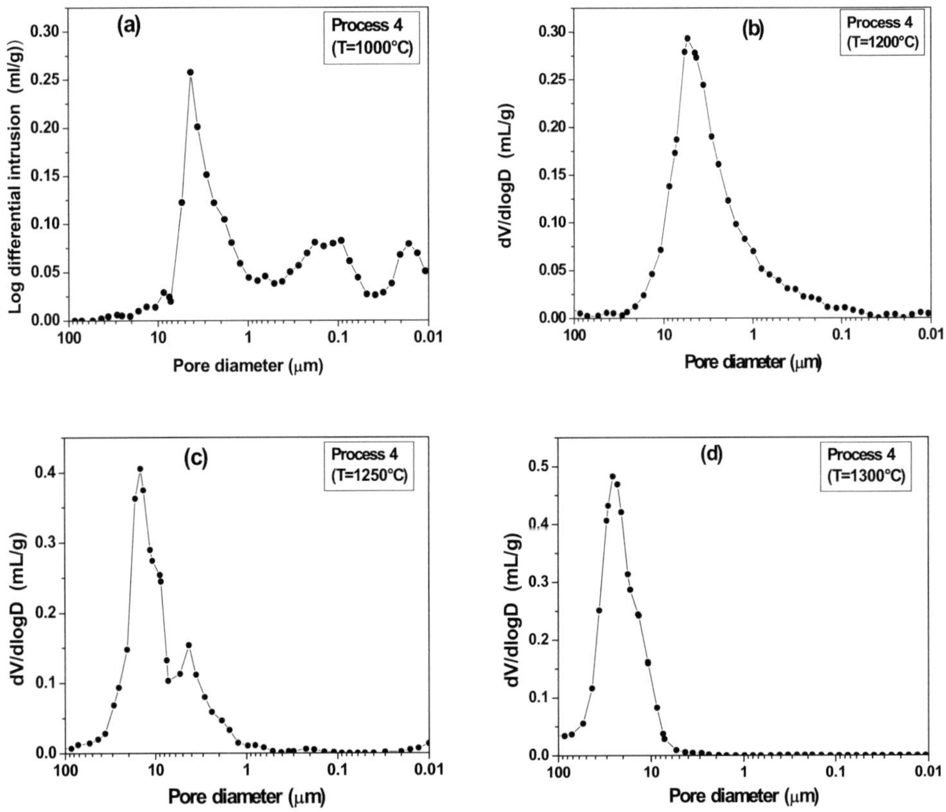

Fig. 6. Pore size distribution in kaolin + 15 wt% doloma samples, using process 4 sintered at different temperatures for 1 hour.

These figures show mainly that the average pore size is shifted towards higher values when sintering temperature is increased. For example, the average pore size values in samples sintered at 1150, 1200, 1250 and 1300°C, using process 3 were 2, 3, 27 and 50 µm, respectively. These values correspond to a cumulative porous volume = 50% (in Fig. 7c). Furthermore, these curves may inform us about the percentage (%) of any pore size interval, at a given sintering temperature and processing route. In some applications this information is required. All in all, it can be said that each processing route has its advantages and potential uses.

One can notice that a uniform pore size distribution is already obtained for samples sintered at 1200°C for 1 h (Fig. 6b), using process 4 (75 MPa). Its average pore size value is about 4 µm within a cumulative porous volume = 41%. These characteristics may be considered (amongst others) as a candidate support for MF and UF membranes. A carful exam of this paragraph and table 2 may allow without any hesitation that these parameters or characteristics of these proposed supports are under control.

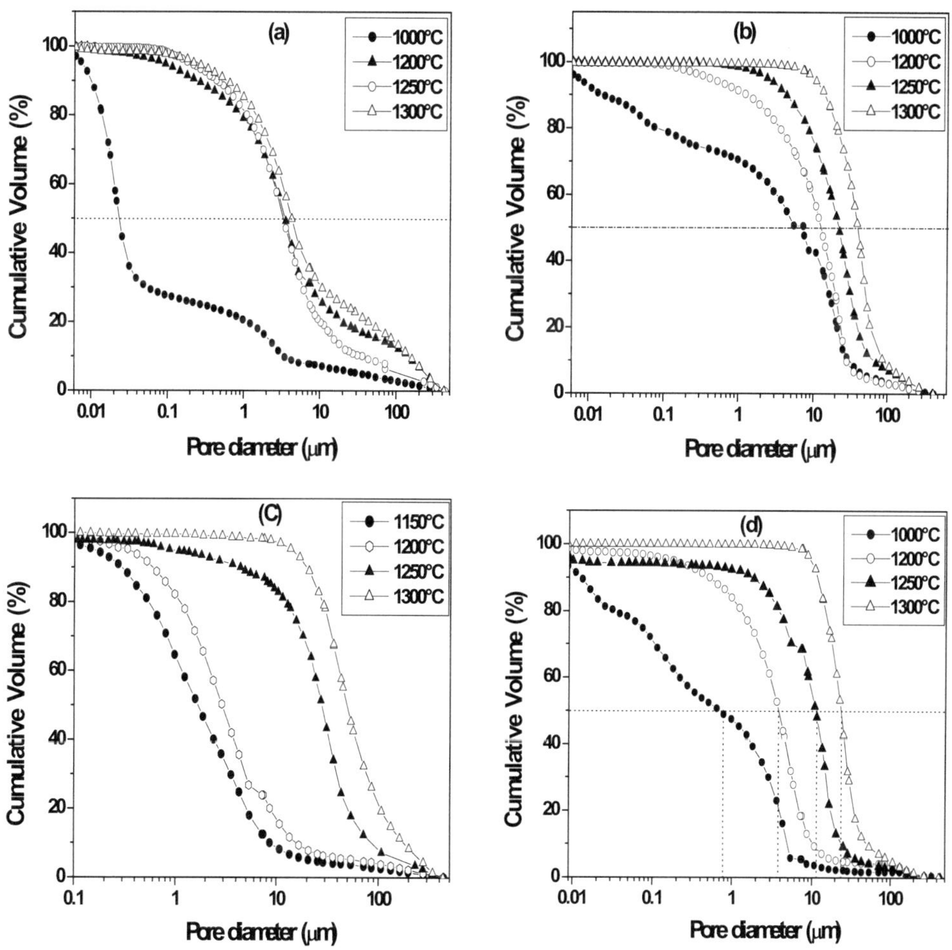

Fig. 7. Cumulative porous volume (%) as a function of pore size for kaolin using process 1 (a) and in kaolin + 15 wt% doloma samples using processes 2 (b), 3(c) and 4(d).

4.3 Pore characterization of supports prepared by centrifugation technique

In order to select the appropriate starch amount that should be added into the studied kaolin, different percentages have been taken into account, as shown in Fig. 8. This figure shows clearly that both average pore size and porous volume (%) values increased with increasing starch percentages. For many applications, higher values of these 2 parameters were recommended; the 20 wt% starch addition has been selected.

The porosity measurement and the average pore size have been carried out for supports sintered at different temperatures during 60 minutes. The obtained results are illustrated in Fig. 9. As would be expected, these figures show, generally, that there is an increase in average pore size and a decrease in total porosity in the samples, when the sintering temperature is increased. On the basis of the above results, it can be said that the increase in sintering temperature encourages the coalescence of pores which, in turns, leads to a larger average pore size.

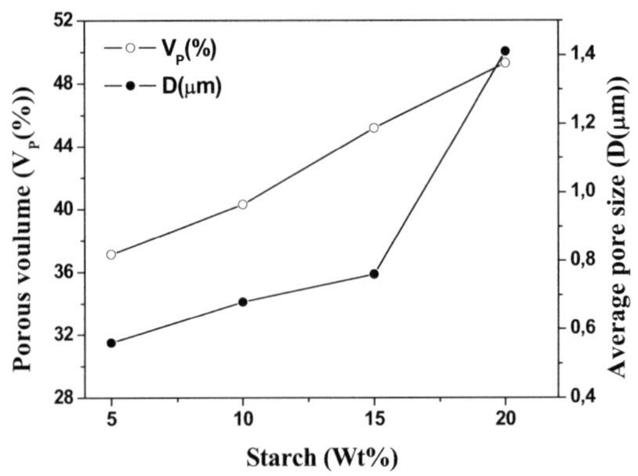

Fig. 8. Porous volume (%) and average pore size versus sintering temperature for kaolin with different wt% starch samples.

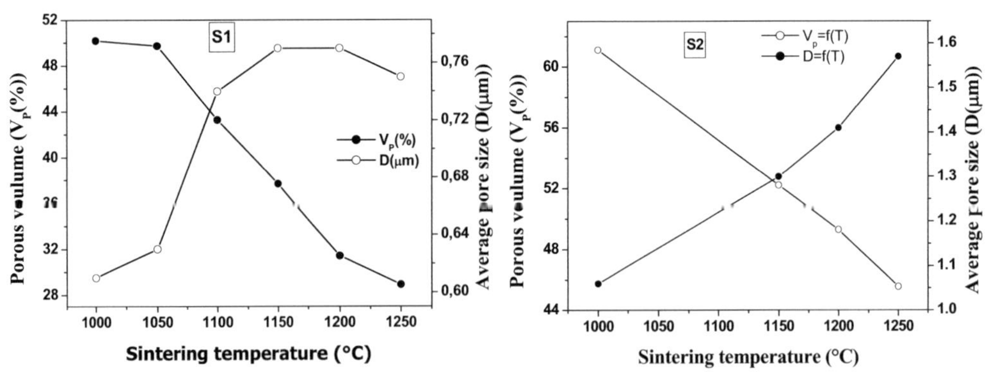

Fig. 9. Porous volume (%) and average pore size versus sintering temperature for kaolin (S1) and for kaolin with 20 wt% starch (S2) samples.

Moreover, it can be said that both the average pore size and porous volume are closely related to the preparation method. The obtained results show that the starch addition to raw materials has a positive effect on the porosity ratio of supports compared to those

prepared from raw materials alone. For example, the raw materials supports had a porosity ratio around 31% and an average pore size around 0.77 µm, whereas the raw materials with 20 wt% starch supports had a porosity ratio around 49% and an average pore size around 1.41 µm, sintered under the same conditions (1200°C during 1 hour).
Fig. 10 shows typical pore size distributions of samples (kaolin samples, S1 and kaolin with 20% wt% starch samples, S2) composed from raw materials and mixtures of 20 wt% starch. The mixtures of raw materials and starch samples have higher median pore diameters than the samples prepared from raw materials alone, proving the presence of larger pores in the mixtures. Apparently, the removed starch particles cause an increase in median pore diameter.

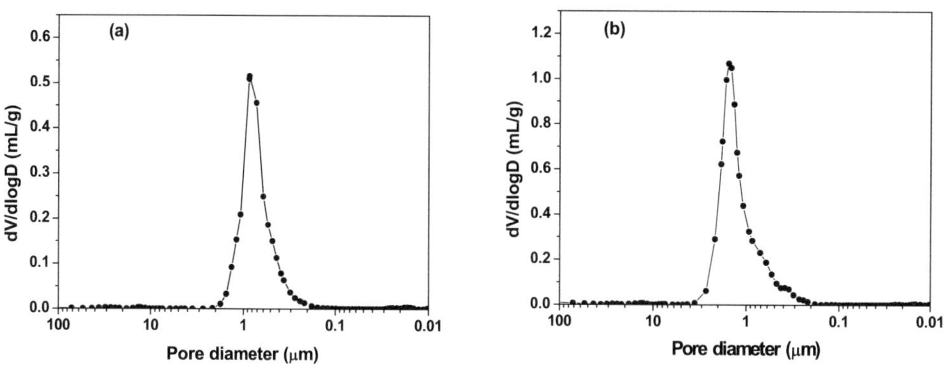

Fig. 10. Pore size distribution in samples sintered at 1200°C during 1 hour. (a) kaolin and (b) kaolin with 20 wt% starch samples.

The pore size distribution modal may also be classified into 1 modal; single or Gaussian modal, The Single Modal of Pore Size Distribution is generally obtained for samples having homogeneous pore size distribution. When pore volume (%) is plotted versus pore size, the curve is characterized by a single peak. Moreover centrifugal casting showed to be a very convenient way of preparing high quality tubular membrane supports. These supports have smooth inside surface and a narrow pore size distribution. This is necessary for a good integrity of the membrane.
Finally, tangential filtration experiments were performed on a typical prepared membranes support, using a home-made pilot plant at room temperature. The working pressure was obtained using a nitrogen gas source. The prepared supports (S1) sintered at 1050°C was characterized by their water permeability.
Fig. 11a shows that the water permeability through the support measured as a function of time depends on the applied pressure. A stable flux is obtained after few minutes. The relatively low water permeate flux value is as would be expected, because of its lower average pore size (0.6 µm) and its thickness (2 mm). Additionally, the effect of the applied pressure (bar) on water permeate flux has been taken into account. The flux increases linearly with the applied pressure and the average permeability is about 107 L/h.m^2.bar as we can see in Fig. 11b . This indicates that the pressure difference is the only driving force for permeation (Sekulie et al., 2004). For transport driven only by convection, the volume flow rate is proportional to the pressure difference, following the Darcy law.
It should be mentioned that when the average pore size is around 0.77 or 1.41 µm, this support may be directly used for UF membranes deposition without any intermediate MF membranes.

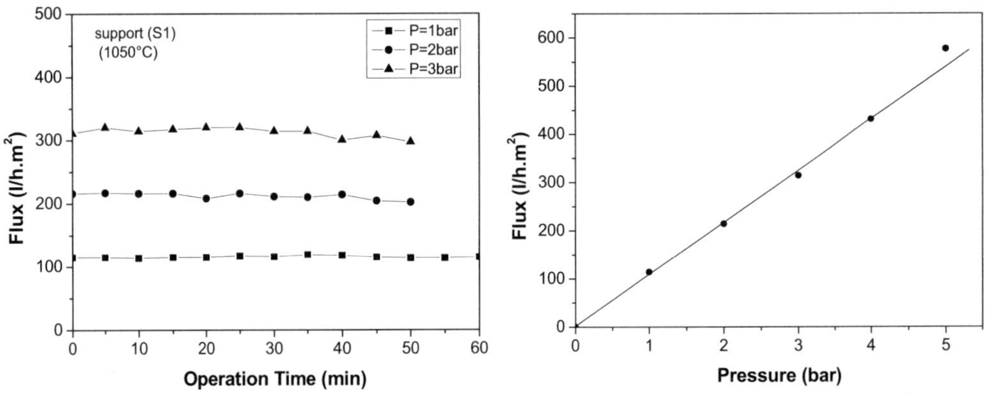

Fig. 11. Water permeability versus time, at 3 working pressure values (on the left) and water permeability versus pressure (on the right).

4.4 Tensile strength

After their preparation and pore characterization, the supports should mechanically be tested, before any membranes deposition. For example, the effect of sintering temperature on tensile strength of supports (using extrusion technique) is illustrated in Fig. 12a. This figure shows that there are three different stages. A state of constant tensile strength is reached for supports sintered in the range (1150 and 1200°C) and this is due to stability in

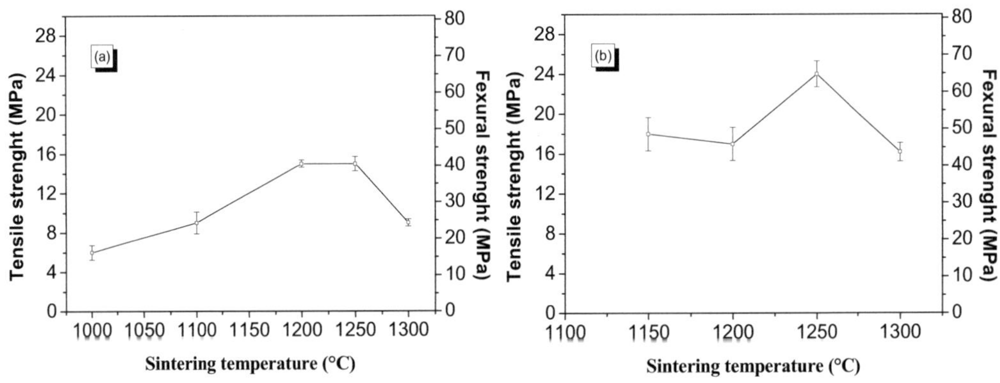

Fig. 12. Tensile strength as a function of sintering temperature (a) (Bouzerara et al., 2006) and (b) (Boudaira et al. 2009).

the pore size and porosity ratio. However, in the range (1200 and 1250°C) the tensile strength increased rapidly, the mechanical properties have been improved, this improvement is due to the effect of sintering in samples where materials grains have been densified; and this is well shown in Fig. 12a (samples sintered at 1250°C). Afterwards, in the range (1250 and 1300°C) there is a sharp decrease in tensile strength which is due to the considerable increase in pores volume. A maximum equivalent flexural strength of about 40 MPa for samples sintered at 1250°C for 1 h.

The same behavior was observed for other supports but with enhanced values (Fig. 12b). In fact, this improvement is well justified by the SEM micrographs shown in Fig. 13. Moreover, the reader may readily notice the significant effect of both process and starting materials on flexural strength. Indeed, the flexural strength value (65 MPa) was nearly doubled (within the error bars) for samples sintered under the same conditions, as shown in Fig. 12.

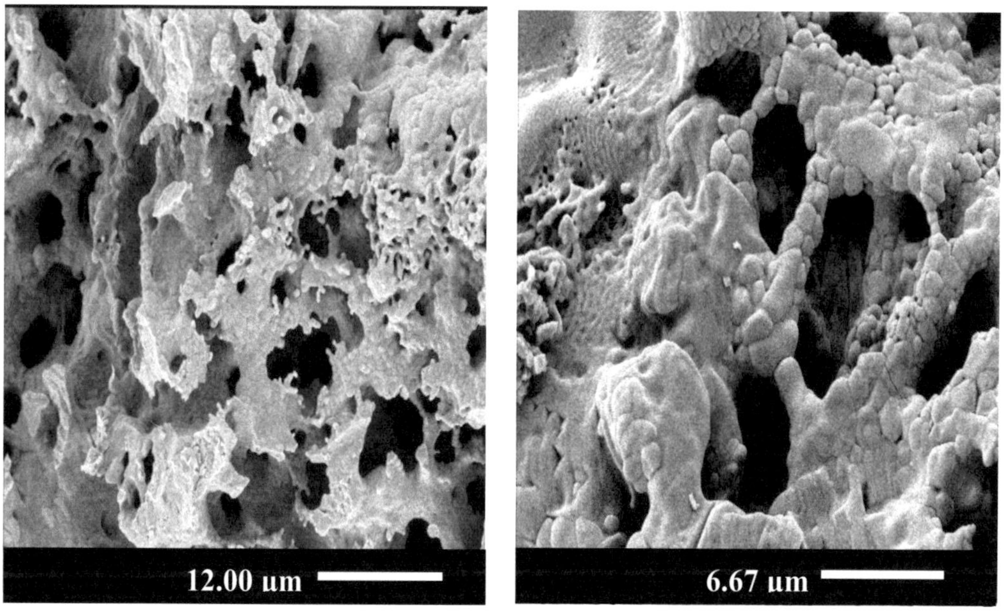

Fig. 13. SEM micrographs of cross-section, left, and the surface, right, of support, sintered at 1250°C for 2 h

4.5 Microfiltration membrane characterization

After their preparation and characterisation these low price supports, they should be used for membranes deposition in order to check out their adherence. Fig. 14 is a good example.
The average pore diameters and the porous volume of the active layer are around 0.35 μm (Fig. 14) and 52%, respectively. It is known that the particle size distribution (PSD) has an influence on the pore size distribution, as a narrow PSD results in a narrow pore size distribution, while a wide PSD would result in a wider pore size distribution (Bissett et al., 2008). The pore size indicates that this kind of membranes can be utilized in the microfiltration range (Masmoudi et al., 2005). SEM images of the prepared membranes are shown in Fig. 15. This figure gives information on the texture of the elaborated membrane surface (Khemakhem et al., 2006). It is noticed that there is no cracks and the pore distribution of the membrane is uniform.
The thickness of the microfiltration layer is about 24 μm (Fig. 15), it can be controlled by the percentage of the mineral powder added in the suspension and the period of the deposited time. Tangential filtration tests were carried out at room temperature. The membrane is immersed in distilled water for 24 hours. The water flux through the membrane was measured as a function of time at diffrent transmembrane pressure values, where the flux are stables after period of 15-40 min of filtration depending on the working pressure (Fig. 16).

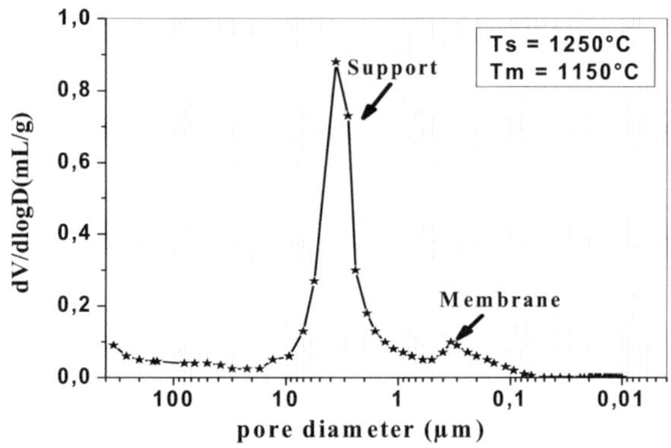

Fig. 14. Pore size distribution of the support and its membrane.

Fig. 15. SEM micrographs of cross-section, left, and the surface, right, of membranes, sintred at 1150°C for 2 h.

The permeability was determined from the different flux values for each working pressure. The obtained curve is a straight line with a slope equal to around 1440 l/h.m^2.bar. It is amongst the best permeability values when compared to those reported by other authors [11] (Khemakhem et al., 2004).

Moreover, these ceramics supports were also used for other MF membranes such as ZnO, anorthite, within a good adherence.

More recently, UF membranes were successfully deposited on this low price fabricated supports (Harabi et al., 2010).

Since the average pore size ranges of these deposited MF and/or UF membranes were situated between 350- 500 nm and 30-50 nm respectively, a huge number of applications are expected. It is well known that these MF and /or UF membrane domains may be used from used water purification to water sterilization.

This work is in progress for the deposition of NF membranes on these proposed low price supports for further applications.

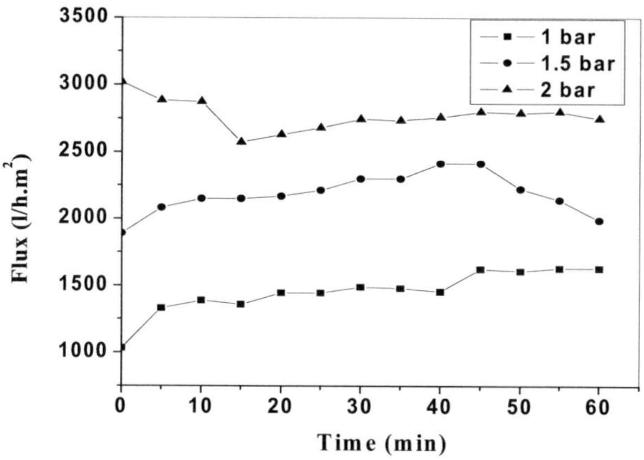

Fig. 16. Water flux as a function of time for 3 working pressure values.

On the basis of the whole chapter body from phase identification to the membranes deposition step, the following remarks and/or conclusions may be drawn and commented.

- All the supports were mainly constituted of one or more of mullite, cistoballite, anorthite and quartz phases. Apart from cristoballite, all the manufactured supports containing these phases may be used even for acids filtration. Besides this, there are no signs that these phases or raw materials may be considered as against environment.
- As far as pore characterization of manufactured supports, using both centrifugal casting and extrusion methods, it can be said that homogeneous pore distributions (Gaussian distribution) were successfully obtained for all followed procedures (under given conditions of sintering). Moreover, the average pore size was also under control, within a volume pore size generally ranging between 42 and 50%.
- The measured flexural strength (ranged between 40 and 65 MPa) is significantly acceptable for highly porous supports (about 50%).
- The selected supports (according to international standards) were afterwards used for different MF and NF membranes deposition. The adherence between these membranes and their supports as well as abscence of large defects were carefully visually checked and observed by SEM.

Consequently, it can be said that these characteristics may modestly confirm that the commercial alumina supports may be replaced by these low expansive supports when commercialized.

5. Conclusions

As far as centrifugal casting technique is concerned, this chapter allowed fabricating aluminosilicate tubes for membrane applications using this technique (procedure 1). Obtained tubes were characterized in terms of porosity; these supports were extremely homogeneous as can be seen from the very sharp pore size distribution. Moreover, the raw materials employed were easily obtainable at low costs. Membrane supports manufactured from raw materials and starch mixtures presented features of porosity (porous volume and average pore size) more important than those elaborated from TK kaolin alone; the manufactured membrane supports are mainly constituted of mullite and quartz phases. The presence of these phases may also extend further their use, even under severe atmosphere conditions.

Even though, these interesting characteristics of the manufactured supports encounter some difficulties such as their shaping limitations and their limited production at the commercial scale.
However, when the extrusion technique was used 2 procedures (2 and 3) have been followed. The attractiveness in the present work (using procedure 2) is the development of membrane supports manufactured from native kaolin and doloma (obtained from dolomite) mixtures, available in our country. This kind of supports presents features of porosity (porous volume and average pore size) more important than those elaborated from kaolin as a raw material. They can be used as supports of membranes of MF and UF. Moreover, these supports are characterised by a reduced manufacture cost since the used raw materials are very abundant (in Algeria) and their mechanical properties seem to be acceptable. The manufactured membrane supports are mainly constituted of mullite, cordierite and anorthite phases. The presence of these phases may also extend further their use, even under severe atmosphere conditions. Finally, it has been found that the pore structures (modal distributions of pore size, total porosity and average pore size) may be controlled by the sintering temperature, additions and processing routes. For examples, a uniform pore size distribution within a total porosity ratio of 43% and 28 μm as an average pore size, were obtained, for samples prepared according to process 3 and sintered at 1250°C for 1 hour. However, a relatively higher porosity ratio (51%) was reached when process 2 was applied, under the same sintering conditions.
Using procedure 3, other kind of membrane supports was manufactured from local kaolin (DD2) and Calcium carbonate mixtures, available in our country. The ceramic support was formed by extrusion of a ceramic paste from kaolin and calcium carbonats mixtures. The microfiltration layer, deposited on the supports, was obtained by the slip casting technique using suspensions of zirconia powder. This membrane can be used for MF and also used as supports for UF.
Since the ceramic filters are generally constituted of a thick support (2000 μm) and one or multi thin membranes (from 10 to 40 μm for each one). That is why this work or chapter is mainly focussed on ceramic supports rather than its deposed membranes. Therefore, replacing the more expansive starting materials, mentioned above, by other low cost raw materials for supports fabrication (which constitutes about 99% of the filter mass) is significantly important. So, what low cost raw materials does mean? It is incomparable; the alumina price is at least about 100 times more expensive than that of kaolin.
Another important advantage is the substantial gain in energy by decreasing the sintering temperature from about 1600°C to about 1250°C, when alumina supports were replaced by the proposed supports. Besides this, about 50% the prepared supports is pores (porosity) which may also be considered as a gain in its mass. The relatively lower theoretical density of the prepared supports (2.8 g/cm^3) when compared to that of alumina (3.98 g/cm^3) is also a further exiting advantage.
Finally, this work is in progress for the deposition of NF membranes on these proposed low price supports for further applications.

6. Acknowledgment

The authors would like to thank both Prof. A. Larbot and Dr. S. Condom - Institut Européen des Membranes, UMR 5635 CNRS ENSCM UMII, Montpellier, France – for their helpful aids and pores characterization of the manufactured supports and membranes.

7. References

[1] Gestel, T.V. ; Vandecasteele, C. ; Buekenhoudt, A. ; Dotremont, C. ; Luyten, J. ; Leysen, R. ; Van der Bruggen, B. & Maes, G. (2002). Alumina and Titania Multilayer Membranes for Nanofiltration: Preparation, Characterization and Chemical Stability. *J. Membr. Sci.*, Vol.207, pp. 73–89.

[2] Biesheuvel, P. M.; Breeveld, V.; Higler, A. & Verweij, H. (2001). Graded Membrane Supports Produced by Centrifugal Casting of a Slightly Polydisperse Suspension. *Chemical Engineering Science*, Vol. 56, pp. 3517–3525.

[3] Nijmeijer, A.; Huiskes, C.; Sibelt, N.G.M.; Kruidhof, H. & Verweij, H. (1998). Tubular Membrane Supports by Centrifugal Casting. *Am. Ceram. Soc. Bull., Vol.* 77, N 4, pp. 95-98.

[4] Chen, C.H. ; Takita, K. ; Ishiguro, S. ; Honda, S. & Awaji, H. (2005). Fabrication on Porous Alumina Tube by Centrifugal Molding. *J. Eur. Ceram. Soc.*, Vol. 25, pp. 3257–3264.

[5] Steenkamp, G.C. ; Keizer, K. ; Neomagus, H.W.J.P. & Krieg, H.M. (2002). Copper(II) Removal From Polluted Water With Aalumina/Chitosan Composite Membranes. *J. Membr. Sci.*, Vol.197, pp. 147–156.

[6] Falamaki, C. & Veysizadeh, J. (2008). Taguchi Design of Experiments Approach to the Manufacture of One-Step Alumina Microfilter/Membrane Supports by the Centrifugal Casting Technique. *Ceramics International*, Vol. 34, pp. 1653–1659.

[7] Steenkamp, G.C. ; Neomagus, H.W.J.P.; Krieg, H.M. & Keizer, K. (2001) . Centrifugal Casting of Ceramic Membrane Tubes and the Coating With Chitosan. *Sepration and Purification Technology*, Vol. 25, pp. 407–413.

[8] Kim, K.H. ; Choa, S.J. ; Yoon, K.J. ; Kim, J.J. ; Ha, J. & Chun. D. (2002). Centrifugal Casting of Alumina Tube for Membrane Application. *J. Membr. Sci.*, Vol. 199, pp. 69–74.

[9] Pinggen, R. ; Mikio, I.; Takahiro, T. & Isao. K. (2003). Centrifugal Casting of Al_2O_3–15 wt % ZrO_2 Ceramic Composites. *Ceramics International*, Vol. 29, pp. 209–212.

[10] Benito, J.M.; Conesa, A.; Rubio, F. & Rodriguez, M.A. (2005). Preparation and Characterization of Tubular Ceramic Membranes for Treatment of Oil Emulsions. *J. Eur. Ceram. Soc.*,Vol. 25, pp. 1895–1903.

[11] Lee, S. H.; Chung, K. C.; Shin, M. C.; Dong, J. I.; Lee, H. S. & Auh., K. H.(2002). Preparation of Ceramic Membrane and Application to the Cross flow Microfiltration of Soluble Waste Oil. *Mater. Lett.*, Vol. 52, pp. 266–271.

[12] Ding, X.; Fan, Y. & Xu, N. (2006). A new Route for the Fabrication of TiO_2 Ultrafiltration Membranes With Suspension Derived From a Wet Chemical Synthesis. *J. Membr. Sci.*, Vol. 270, pp. 179–186.

[13] Gaucher, C., Jaouen, P., Comiti, J. & Legentilhomme, P. (2002). Determination of Cake thickness and Porosity during Cross-flow Ultrafiltration on a Plane Ceramic Membrane Surface Using an Electrochemical Method. *J. Membr. Sci.*, Vol. 210, pp. 245–258..

[14] Mori, H.; Mase, S.; Yoshimura, N.; Hotta, T.; Ayama, K.; Tsubaki, J. I. (1998).Fabrication of Supported Si3N4 Membranes Using the Pyrolysis of Liquid Polysilazane Precursor. *J. Membr. Sci.*, Vol.147, pp. 23-33.

[15] Saffaj, N.; Younssi, S. A.; Albizane, A.; Messouadi , A.; Bouhria, M.; Persin, M.; Cretin, M. & Larbot, A.(2004). Preparation and characterization of ultrafiltration membranes for toxic removal from wastewater. *Desalination*, Vol.168, pp. 259-263.

[16] Masmoudi, S.; Ben Amar, R.; Larbot, A.; El Feki, H.; Ben Salah, A. & Cot, L. (2005). Elaboration of Inorganic Microfiltration Membranes With Hydroxyapatite Applied to the Treatment of Wastewater From Sea Product Industry. *J. Membr. Sci.*, Vol.247, pp. 1-9.

[17] Rakib, S. ; Sghiar, M.; Rafik, M.; Cot, D.; Larbot, A. & Cot, L. (2000). Elaboration Et Caractérisation d'une Céramique Macroporeuse à Base d'Arène Granitique. *Ann. Chim. Sci. Mat.*, 25 567–576.

[18] Elmoudden, N., Elghazouali, A., Rakib, S., Sghyar, M., Rafiq, M., Larbot, A. & Cot, L. (2001). Nouveaux Supports Membranaires à Base de Chamotte d'Argile. *Ann. Chim. Sci. Mat.*, Vol.26 N°(2), pp. 5-11.

[19] Bouzerara, F.; Harabi, A. ; Achour, S. & Larbot, A. (2006). Porous Ceramic Supports for Membraes Prepared From Kaolin and Doloma Mixtures. *J. Eur. Ceram. Soc.*, Vol. 26, pp. 1663-1671.

[20] Khemakhem, S.; Ben Amar, R.; Ben Hassen, R.; Larbot, A.; Medhioub, M.; Ben Salah, A. & Cot, L. (2004). New Ceramic Membranes for Tangential Waste-Water Filtration. *Desalination*, Vol.167, pp. 19-22.

[21] (Harabi & DAVIS, 1995a) Harabi, A. & Davies, T.J. (1995a). Densification and Grain Growth in Sintered Alumina- Chromia Powder Mixtures. *Br. Ceram. Trans. J.*, Vol. 94 N° (3) pp. 97-102.

[22] Masmoudi, S.; Larbot, A.; El Feki, H. & Ben Amar, R.(2006). Elaboration and Properties of New Ceramic Microfiltration Membranes from Natural and Synthesised Apatite. *Desalination*, Vol.190, pp. 89-103.

[23] Harabi, A.; Bouzerara, F. & Condom, S. (2009). Preparation and Characterization of Tubular Membrane Supports Using Centrifugal Casting, *Desalination and Water Treatment*, Vol.6, pp. 222-226.

[24] Boudaira, B.; Harabi, A.; Bouzerara, F. & Condom, S. (2009). Preparation and Characterization of Microfiltration Membranes and Their Supports Using Kaolin (DD2) and $CaCO_3$, *Desalination and Water Treatment*, Vol.9, pp. 142-148.

[25] Bouzerara, F.; Harabi, A. & Condom, S. (2009). Porous Ceramic Membranes Prepared From Kaolin, *Desalination and Water Treatment*, Vol.12, pp. 415-419.

[26] Mecif, A.; Soro, J.; A. Harabi, & Bonnet, J.P. (2010). Preparation of Mullite and Zircon Based Ceramics Using Kaolinite and Zirconium Oxide: a Sintering Study, *J. Am. Ceram. Soc.*, Vol. 93 [5] 1306–1312.

[27] Bouzerara, F.; Harabi, A. ; Achour, S. & Larbot, A. (2006). Porous Ceramic Supports for Membranes Prepared From Kaolin and Doloma Mixtures. *J. Eur. Ceram. Soc.*, Vol. 26, pp. 1663-1671.

[28] Sekulie, J.; Magraso, A.; Ten Elshof, J.E. & Blank, D.H.A. (2004). Influence of ZrO_2 Addition on Microstructure and Liquid Permeability of Mesoporous TiO_2 Membranes. *Microporous and Mesoporous Materials*, Vol. 72, pp. 49–57.

[29] Bissett, H.; Zah, J. & Krieg, H.M.(2008). Manufacture and Optimization of Tubular Ceramic Membrane Supports. *Powder Technology*, Vol.181, pp. 57–66.

[30] Khemakhem, S.; Larbot, A. & Ben Amar, R. (2006). Study of Performances of Ceramic Microfiltration Membrane From Tunisian Clay Applied to Cuttlefish Effluents Treatment. *Desalination*, Vol.200, pp. 307–309.

[31] Harabi, A.; Bouzerara, F. & Condom, S. (2010). Fabrication of tubular membranes for MF an UF filtration using Kaolin and ZrO_2, Patent, *INAPI*, N° 100019, Algeria.

[32] (Harabi & DAVIS, 1995b) Harabi, A. & Davies, T.J. (1995b). Mechanical Properties of Sintered Alumina - Chromia Refractories, *Br. Ceram. Trans. J.*, Vol. 94 N° (2) pp. 79-84.

Part 4

Special Applications

14

Determination of Optimal Conditions for Separation of Metal Ions Through Membrane Dialysis/Electrodialysis Using Statistical Experimental Methods

Jau-Kai Wang, Jir-Ming Char and Ting-Chia Huang
Fooyin University / School of Environmental and Life Sciences, Kaohsiung City, Taiwan, R.O.C.

1. Introduction

The membrane process for industrial application is characterized by a large market profile, such as brackish and sea water desalination, the production of ultrapure water, or hemodialysis and a large number of small market segments in the food, chemical and pharmaceutical industries, analytical laboratories and especially in the treatment and purification of industrial waste water streams (Lonsdale, 1982; Mears, 1976). It is seemed to be difficult to make a reasonably accurate forecast about the future developments of the market for membranes (Strathmann, 2004). However, due to the fact that the environmental problem of Global Warming is increasing drastically resulting in that the demand of industrial water and fresh water with the required quality are steadily decreasing worldwide(Wang, & Hsieh, 2009). There should be a need for energy efficient and affordable processes for the production of high quality water from sea and brackish water sources as well as from waste or polluted surface waters. Since membrane processes have proven to be among the most energy efficient and economic means for this purpose it is quite likely that for the foreseeable future the membrane water purification industry will continue to grow (Hwang, & Kammermeyer, 1975). The growth will also depend on further developments of membranes with improved selectivity and higher fluxes as well as better chemical, thermal and mechanical stability. Long-term experience with their application in large plants will also contribute to increase the useful life of the membranes thus making the processes more reliable and economical (Huang, 1991).

The separation of a metal ion from a multi-ionic mixture by selective transport through membrane processes was of importance from both fundamental and practical viewpoints owing to the serious lack of earth resource. For the selective permeation, the desired metal ion in a mixture has to be preferentially absorbed into the membrane phase and then be transported across the membrane phase. Subsequently, such membranes were used for desalination purposes (Helfferich, 1962). An appropriate driving force was essential for an uphill transport resulting in building up of the concentration of the permeating ion across the concentration gradient (Nobel, 1987). The driving force could be due to difference in pH of the solutions, concentration difference of the permeating species or an electrical potential. The pH difference was generally responsible for the transport in liquid membrane systems, whereas the electrical potential difference was the driving force for the transport across ion exchange membranes. Donnan dialysis was essentially the transport of the charged species across a membrane due to concentration difference. The combined effect of electrical potential difference and concentration gradient results in the process of electro-dialysis. Ion exchange membranes, used in electro-dialysis permit the separation of the species, without any change in the phase. Depending on the type of the driving ion used, Donnan dialysis can be applied to the recovery of metal ions from wastewater. There were many studies on the use of ion exchange membranes for Donnan dialysis (Nobel, 1987; Mizutani, 1990). Electrodialysis was principally used for the production of brine from seawater. For this specific application, special cation exchange membranes were often used to reduce the permeability of alkaline earth metal ions with respect to sodium ions (Gering, & Scamehorn, 1988). Since ion exchange membranes were used for various purposes, requirements for membranes with specific properties have increased (Bungay, 1986). Ion exchange membranes find many industrial applications primarily due to their property of perm-selectivity.

The aim of this chapter is to introduce systematically the preferential transport behavior of a mutli-ion system across the ion exchange membrane in dialysis and electrodialysis with combined use a complex agent. Then the mathematical description of the mass transport in membranes in general and in specific membrane processes is described. Finally, determination of optimal conditions for separation of metal Ions through membrane dialysis/electrodialysis using statistical experimental methods are discussed in order to create an advantage position for industrial application in the future.

2. General considerations of dialysis process

Ion exchange membrane dialysis is a useful membrane process in which ions move across an ion exchange membrane based on the Donnan membrane equilibrium principle (Donnan, 1924). Application of this process can simultaneously solve two important industrial problems, for the recovery and concentration of valuable as well as for the removal of undesired ions from waste streams (Dibenedetto & Lightfoot, 1962). Many investigations of ion transfer across the ion exchange membrane have been reported (Brajter, & Slonawaska,1988; Costa, et al.,1984; Cox & DiNunzio, 1977; Coz et al., 1984; Huang & Wang, 1992a). One of the prominent features in the development of modern technology is the cross-fertilization of ideas among different disciplines (Sata, 1991). Therefore, separation processes combined use of complex agents with cation exchange membrane for the recovery and the enrichment of heavy metal ions have received much attention in recent years (Hopfenberg et al., 1978; Huang & Wang, 1993a; Kojima et al., 1982; Kojima et al.,1991).

Dialysis is a membrane based separation process in which the concentration gradient across the membrane is the driving force resulting in a flow of material from one side of the membrane to the other. The transport is due to the solution-diffusion mechanism (Bohere, 1983). Most researchers have employed an ion exchange membrane indicating the process as diffusion or Donnan dialysis (Davis, et. al, 1971; Mizutani,1990; Wen,& Hamil, 1981). The process is based on the Donnan equilibrium principle (Donnan, 1924) with the usual aim being the separation of ions or charged species. The transport in the Donnan dialysis is a function of two phenomena, i.e. diffusion due to the concentration gradient, and migration due to the potential gradient arising from the difference in ionic concentrations across the membrane. Another important parameter affecting the permeation of ions is the membrane charge, which depends on the pH, and the electrolytes in the bulk solution. Dialysis is able to solve two important industrial problems, i.e. recovery of valuable materials and removal of undesired ions from waste streams (Spencer & Ellison, 1953; Wallace, 1967). The process has been used to recover various acids including organic acids (Boudet,1991) from waste streams generated in steel, metal-refining, and electro-plating industries (Wen, & Hamil, 1981), and to remove Na^+ (Takahashi, 1989), Cu^{+2}(Sudoh,1987), Ni^{+2} (Tombalakian,1967) ternary (Cu^{+2}-Ni^{+2}-Zn^{+2})ion system (Wang, & Chang, 2004), cyanides (Kim, 1979) and fluoride (Rush, & Baker, 1980).In dialysis, the improvement of permeation and selectivity is the key point for industrial applications. For this purpose, ion-exchange membranes are combined with complex agents (such as EDTA or citric acid) for the enrichment of metal ions (Ciric, & Graydon, 1962; Wang, 2004).

3. Theoretical consideration for dialysis process

The dialysis process of mass transport was indicated in Figure 1, which show schematically the mass transport through membranes separating two homogeneous phases. We consider an ion exchange system in which a cation exchange membrane separates two electrolyte solutions i.e., one is the feed phase consisting of three bivalent cations species i with the common anion Y, and the other is the stripping phase composed of the N-Y electrolyte solution. The mechanism of the transport of counterion across the cation exchange membrane is postulated to consist of three rate processe (Huang, & Wang, 1992b; Saito, et al.,1982): (1) diffusion through the two liquid films adjacent to the membrane surface, (2) the ion exchange reaction occurring at two membrane-solution interfaces, and (3) diffusion through the membrane.It is assumed that the concentration of N-Y electrolyte solution and the agitation speed in the stripping side are sufficiently high to ignore the mass-transfer resistance of the stripping phase as compared to the other ones. Furthermore, the film-membrane system has been allowed to adjust to a steady state during the dialysis operation, so that the ionic flux of the membrane is equal to that of the liquid phase.

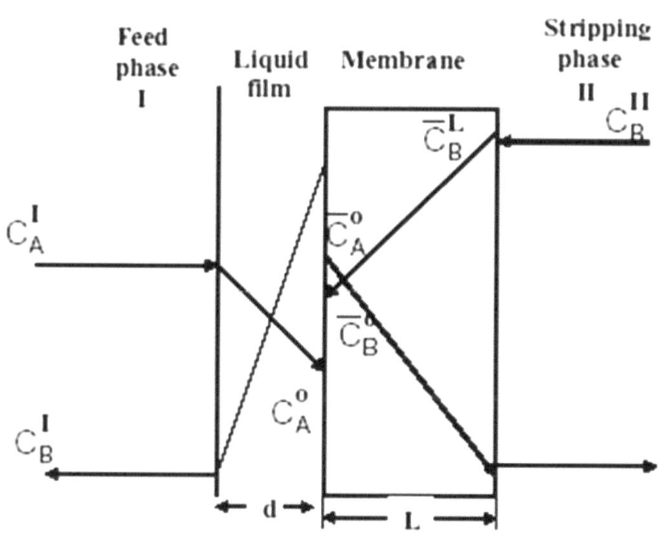

Fig. 1. Schematic diagram for theoretical consideration

First, let us consider the bulk solution-phase mass transport. If metal ion transport in one dimension, and mass flux of each ion in the solutiom can be represented on the basis of the Nernst-Planck equation as follows (Lightfoot & Scattergood, 1965; Saito, et al.,1991; Sherwood, 1975):
For the metal ions and counter ion i, the transport flux is

$$J_i = -D_i\left(dC_i/dx + C_i z_i F/RT \times dE/dx\right) \quad \text{i =A, B, C, N} \tag{1}$$

For the common anion Y, the transport flux is

$$J_Y = -D_Y\left(dC_Y/dx + C_Y z_Y F/RT \times dE/dx\right) = 0 \tag{2}$$

where A, B and C denote the coded metal ions, N denotes the counter ion. E, D_i and C_i denote the diffusion potential, the diffusion coefficient and the concentration of ion i in the liquid phase, respectively. F, x and z_i denote the Faraday constant, the distance normal to the membrane and the valence of ion i, respectively.
On the basis of the mass balance in the steady state, the following equation can be obtained.

$$z_A J_A + z_B J_B + z_C J_C + z_N J_N = 0 \tag{3}$$

The electroneutrality condition must be satisfied at any location of the electrolyte solution. i.e.

$$z_A C_A + z_B C_B + z_C C_C + z_N C_N + z_Y C_Y = 0 \tag{4}$$

The potential gradient in eqn.1 and 2 can be eliminated by using eqn.3 and 4. Then, the concentration distribution of ion i in liquid film can be obtained by integrating these equations with the boundary condition, $C = C_i^0$ at x=0, as follows:

$$C_i = \left\{C_i^o\left[Q^o\right]^{z_i} - \left[J_i/D_i Q^{z_i}\right]\right\}/Q^{z_i} \quad \text{i =A, B, C, N} \tag{5}$$

where

$$Q^o = z_A C_A^o + z_B C_B^o + z_C C_C^o + z_N C_N^o \tag{6.1}$$

$$Q = Q^o - \left(\frac{J_A}{D_A} + \frac{J_B}{D_B} + \frac{J_C}{D_C} + \frac{J_N}{D_N}\right)x \tag{6.2}$$

Next, let us consider the membrane-phase mass transport. If common ion, do not exists in the membrane and metal ions transport in one dimension, then the mass flux of each ion in the membrane can be expressed as follows (Wang, & Hsieh, 2008; Wills, & Lighfoot, 1961):

$$J_i = -\overline{D}_i\left(d\overline{C}_i/dx + \overline{C}_i z_i F\Big/ RT \times \frac{d\overline{E}}{dx}\right) \quad \text{i=A, B, C, N} \tag{7}$$

where $\overline{E}$, $\overline{D}_i$ and $\overline{C}_i$ denote the diffusion potential, the diffusion coefficient and the concentration of ion i in the membrane, respectively.

Assume that the experiments are operated at steady state; the effect of ionic strength of solution on the mass transport is negligible. This is consistent with the Goldman approximation (Tomicki, 1991), in which the constancy of the electric field existed in the membrane, can be adopted in this study. Eqn. 7 can be reduced to the following:

$$J_i/\overline{D_i} = -\left(d\overline{C_i}/dx + \overline{C_i}\, z_i F \Big/ RT \times \frac{\overline{E}}{L} \right) \quad \text{i=A, B, C, N} \tag{8}$$

where L denotes the thickness of the membrane and $\overline{E}$ denotes the membrane potential that can be estimated by using the multi-ionic potential theory as follow (Lake& Melsheimer, 1978):

$$\overline{E} = \frac{RT}{z_i F} \ln \left\{ \frac{\sum_j \left(\overline{D}_j a_j^I / \overline{f}_j \right)}{\sum_k \left(\overline{D}_k a_k^{II} / \overline{f}_k \right)} \right\} \tag{9}$$

where a and f denote the activity and the molar activity coefficient; the subscripts j refer to ions A, B, C and N in compartment I and the subscript k refers to counter ion N in compartment II.

Based on the boundary conditions $\overline{C_i} = \overline{C_i^o}$ at $x = 0$ and $\overline{C_i} = 0$ at $x = L$, eqn. 8 can be integrated to obtain the concentration profile in membrane. The concentration profiles in membrane are given as:

$$\overline{C_i} = \frac{J_i}{q\overline{D_i}} \left[e^{qL}/e^{qx} - 1 \right] \quad \text{i=A, B, C, N} \tag{10}$$

where

$$q = \frac{z_i F}{RT} \left(\overline{E}/L \right)$$

Based on the Teorell-Meyer-Sievers hypothesis (Dranoff & Lapidus, 1957; Tombalakian, et al.,1967), a pseudo- equilibrium state between each metal ion and counter ion Na^+ is assumed to exist at the film-membrane interface, respectively. An apparent selectivity coefficient, K_{Na}^i, can be defined as

$$K_{Na}^i = \frac{C_i}{\overline{C_i}} \times \frac{\overline{C}_{Na}^2}{C_{Na}^2} \tag{11}$$

Substituting eqn.5 and 10 into eqn. 11, we obtained

$$\frac{\left[J_i/q\overline{D_i}\left(e^{qL} - 1\right) \right]}{\left[J_{Na}/q\overline{D_{Na}}\left(e^{qL} - 1\right) \right]^2} = K_{Na}^i \frac{\left[C_i^o \left(Q^o\right)^2 - \left(J_i/D_i Q_d^2\right) \right]}{\left[C_{Na}^o \left(Q^o\right) - \left(J_{Na}/D_{Na} Q_d\right) \right]^2} \tag{12}$$

where

$$Q = Q^o - \left(\frac{J_A}{D_A} + \frac{J_B}{D_B} + \frac{J_C}{D_C} + \frac{J_N}{D_N} \right) d$$

and d denotes the thickness of stagnant liquid film.
Eqn.(12) is a general equation describing the preferential transport behaviors of ternary bivalent ions sharing with the same counter-ion across the cation-exchange membrane. The permeation flux ratio of the metal ions, J_i/J_{Na}, can be calculated numerically from eqn.12 by the iterative technique under the restrictions of eqn.(3) and (4) as the apparent selective coefficient K_{Na}^{i}, obtained from equilibrium measurement, respectively.The dimensionless permeation transport of metal ion A relative to sodium ion, $P_{i\text{-}Na}$, in ion exchange membrane dialysis can be defined by

$$P_{i-Na} = (J_i/J_{Na})/(C_i^o/C_{Na}^o) \tag{13}$$

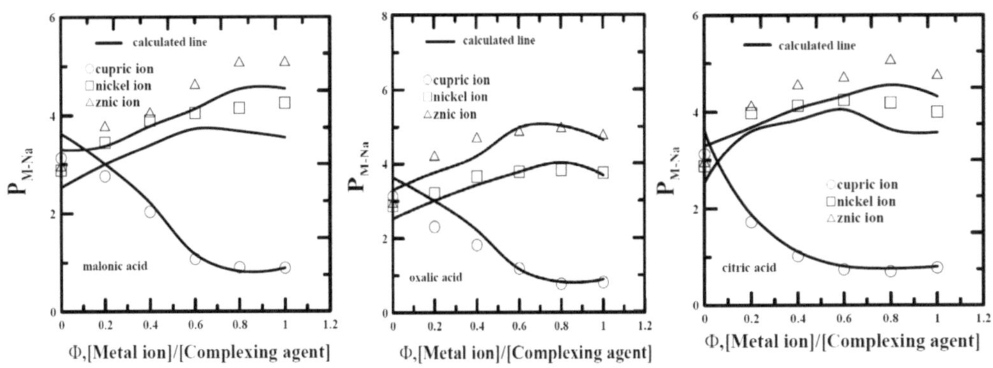

Fig. 2. Effect of the stoichiometric ratio of complex agent to metal ions on the dimensionless permeation fluxes of metal ions in the feed solution at pH 4.0 and $[CuCl_2] = [NiCl_2] = [ZnCl_2]$=0.001M.

As can be seen from Figures 2 (a), (b) and (c) the experimental results show that the preferential transport behaviors of ternary system(Cu^{+2}-Ni^{+2}-Zn^{+2}) through dialysis process combined using the respective complex agent, malonic acid, citric acid and oxalic acid, respectively (Wang, & Hsieh, 2008). In the experiment with solutions of investigated metal ions and a complexing agent with a total sodium ion concentration of 0.1M in the feed compartment, the dimensionless permeation fluxes of metal ions were measured as the stoichiometric ratio of the complexing agent to metal ions Φ= [complex agent]/[metal ions], varied from 0 to 1.0 at pH 4.0 for the investigated system.The experimental results show that the uptake of metal ions in the cation exchange membrane is changed by adding a complex agent in the solution phase, and both the anion ligands and the kinds of metal ions can differentiate the permeation fluxes of metal ions so as to increase the membrane selectivity of the metal ions. The optimal dimensionless permeation flux ratio of $P_{Zn\text{-}Na}/P_{Ni\text{-}Na}/P_{Cu\text{-}Na}$ is obtained about 5.3/3.6/0.7. As a result, the combined use of a cation exchange membrane and a complex agent seem to be a feasible process for the simultaneous separation and concentration of metal ions. However, more experimental work on multi-ions system with the dialysis method seems to be needed in order to develop an ion fractionation process for the enrichment of metal ions as well as the separation of metal ions from waste water in industrial practicc.

The experimental results in Figures 3, (a) (b) and (c) show that the values of P_{Cu-Na} decreases with an increasing pH value of feed solution, but a peak of P_{Ni-Na} and P_{Zn-Na} arise with the change of the pH value of solution under the conditions of various complex agents. This can be explained from the fact that the stability constants of metal ion complexes have the following order zinc ion complex < nickel ion complex < cupric ion complex (Zeleznik & Gordon, 1968). The higher pH value leads to the complex agent releasing more complex ligands to react with cupric ion as well as makes it in an advantage position to react with nickel ions and zinc ions to form complexes when the values of pH is larger than 3.0. Thus, the values of P_{Ni-Na} and P_{Zn-Na} increase sharply as the pH value varies from 2.0 to 3.0, resulting from the formation of cupric ion complexes which decrease the cupric ion flux. There is a decrease the values of P_{Ni-Na} and P_{Zn-Na} at pH values ranging from 4.0 to 5.0 because the formation of nickel ion complexes and zinc ion complexes decreases the nickel ion flux and zinc ion flux. The optimal values of preferential transport behavior can be obtained while the pH value of solution ranges from 3 to 4 . The optimal values of pH are found to be 3.0 and 4.0 for various kinds of complex agents. The maximum values of P_{Cu-Na}/ P_{Ni-Na}/ P_{Zn-Na} is found to be 0.78/4.30/4.85 in case of using the citric acid as the complex agent at pH 3.0. The optimal value of P_{Cu-Na}/ P_{Ni-Na}/ P_{Zn-Na} is found to be 0.70/3.60/5.3 and 0.70/3.0/4.6 for the respective complex agent being malonic acid and oxalic acid at pH 4.0.

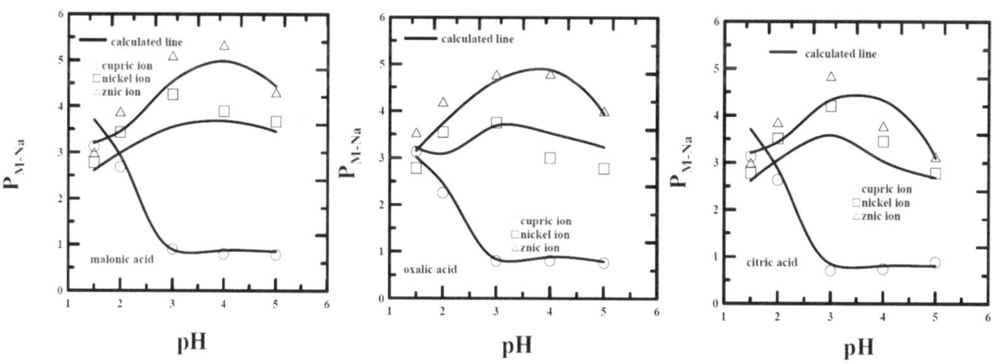

Fig. 3. Effect of the pH of solution on the dimensionless permeation fluxes of metal ions at $[CuCl_2] = [NiCl_2] = [ZnCl_2]$=0.001M.

It can be concluded that preferential transport of metal ions with ternary system across the cation- exchange membrane can be effectively enhanced by the addition of a complex agent. The suitable complex agents are found to be malonic acid for the investigated system Cu^{2+}-Ni^{2+}-Zn^{2+}. The kinds of complex agents, the stoichiometric ratio of complex agent to metal ions, the pH value of solution are the primary factors on the preferential transport behavior of metal ions across the cation-exchange membrane. On the basis of the Nernst-Planck equation and interfacial equilibrium hypothesis, a theoretical approach calculated from Eqn. 12 with numerical calculation is in agreement with the experimental data. Data from the present investigation have shown that the cation-exchange membrane with use of a suitable complex agent is found to be a feasible process for the simultaneous separation and concentration of metal ions. It is also highlighted that an effective multi-ion fractionation process could be build from cation-exchange membrane being the separation interface; the driving force being the concentration gradient of counter ion; the suitable complex agent being the separation agent and the pH value is the separation index.

4. General considerations of electrodialysis process

Ion exchange membrane electrodialysis is a membrane separation process (Leitz, F. B. & Eisenamann, 1981;) in which ions transport across ion-permselectivity membranes from one fluid to another under an applied direct electric field. As shown in Figure 4, there are at least five element parts for electrodialysis applications: (1) direct current supply, which proves effective to initializing ion migration; (2) electrodes, where the oxidation/reduction reactions occur to realize the transformation from ionic conduction to electron conduction and thus provide the original driving force for ion migration; (3) ion exchange membranes, the key components which permit the transport of counter ions and block the passage of co-ions; (4) solvents, which make a continuum for ion transport by filling the space between electrodes and membranes; (5) electrolytes, the current carriers between cathode and anode. In such a way, it should be able to deionize one stream so as to produce a purified stream that can be either discharged or reused and a concentrated electrolyte stream that can be disposed of or processed for recovery of dissolved salts. The cations and anions permeate through cation selective and anion selective membranes, respectively as a result of applied electrical energy. In electrodialysis, electrolytes permeate through the membranes and, the solvent and nonelectrolytes generally do not permeate through the membranes. The solution from which electrolytes are removed gets depleted of salt and the solution which receives the solute gets enriched with salt. Owning to its distinguished functions, applications of ion exchange membrane electrodialysis have been widely conducted on desalination of brackish water (Farrell & Smith,1962; Korngold et al., 1987; Seto et al., 1978), recovery of metal ions from waste water (Gering. & Scamehorn, 1988; Itoi et al.,1980), purification of protein solutions (Lounis & Gavach, 1990), recovery of acids (Graillon et al., 1996), and removal of inorganic impurities from sugar solutions (Sandeaux et al.,1991).

One of the prominent features in the development of modern science and technology is the cross-fertilization of ideas among different disciplines (King, 1980). The combined use of complex agents with ion exchange membranes has received much attention. It has been found that different kinds of complex agents can differentiate the permeation fluxes of metal ions across the membrane. Therefore, the permselectivity of metal ions through membrane dialysis can be effectively increased resembling a support liquid membrane characterized by its high selectivity of metal ions. Some fundamental work on the preferential transport of ions through an ion exchange membrane with electrodialysis in the absence of complex agents was done by many researchers (Huang & Wang, 1993b; Kitamoto & Takashima, 1964; Margerum, 1965). It was proved that the applications of ion exchange membrane electrodialysis with complex agents to be an effective technique not only for concentration of metal ions but also for their separation.

In this section, preferential transport of ions through a cation exchange membrane electrodialysis has been discussed combined using complex agents. Based on the Nernst-Planck equation (Wang, 2004; Keister & Kasting, 1986)and interfacial quasi-equilibrium hypothesis (Rosenberg.& Tirrell, 1957), a theoretical approach was obtained with numerical calculation and compared with the experimental results. Then the mathematical description of the mass transport in membranes in general and in specific membrane processes is described. Finally, determination of optimal conditions for separation of metal Ions through membrane electrodialysis using statistical experimental methods are discussed in order to create an advantage position for industrial application in the future.

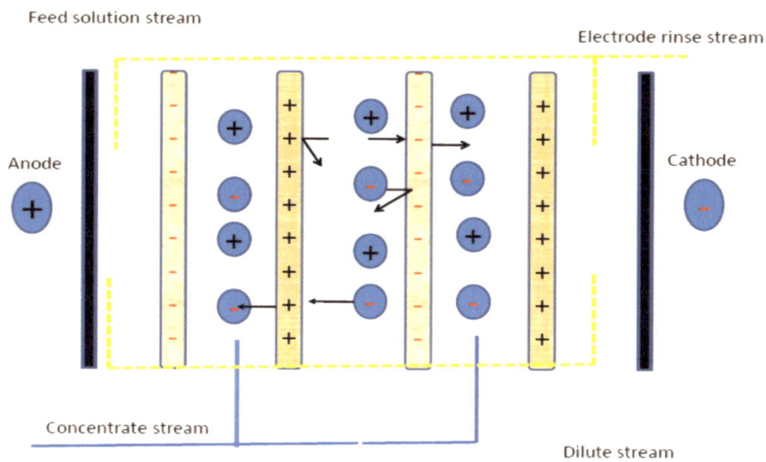

Fig. 4. Schematic drawing illustrating the principle of an electrodialysis process;
- and, + refer anion and cation selective membranes, respectively.

5. Theorectical considerations of electrodialysis process

Consider a cation exchange membrane in an electrolyte solution which consists of cations *A* and *B* with a common anion *C*(-1). When an electric field is applied to this electrodialysis system, the cations pass through the diffusion liquid film in the desalinating compartment from the bulk dilute solution to the cation exchange membrane surface, then transfer through the membrane into the concentrate compartment.

First, let us consider the mass transport in the diffusion liquid film adjacent to the membrane surface. According to experimental conditions, the following assumptions can be made(1)Anions cannot pass through the cation exchange membrane (2) Electroneutrality holds in all parts of the membrane-solution system. (3) There are no pressure and temperature gradients across the system. (4) Solvent transport can be neglected (Belfort & Guter, 1972). Otherwise, assumed that the electrodialysis cell has been operated at a constant current density, and during this operation the membrane-film system has been allowed to adjust to a steady state. At this point the Nernst-Planck flux equation can be written for the ionic fluxes, Ji, in each liquid film and in the membrane (Huang & Wang, 1992a; Keister & Kasting, 1986; Kuroda et al., 1983).

$$J_i = -D_i \left(dC_i/dx + C_i z_i F/RT \times dE/dx \right) \quad \text{i =A, B, C, N} \tag{14}$$

where E, D_i and C_i denote the electrical field, the diffusion coefficient and the concentration of ion i in the liquid phase, respectively. F, x and z_i denote the Faraday constant, the distance normal to the membrane and the valence of ion i, respectively.

On the basis of electroneutrality condition is assumed to be existed in all part of systeme, the following equation can be obtained.

$$z_A C_A + z_B C_B + zyCy = 0 \tag{15}$$

By means of the simultaneous application of the one flux equation of anion and two flux equations of cations with using the boundary condition: C= C° at X=0, the concentration profiles of cations in the film can be derived,

$$C_Y = C_Y^0 \exp[F/RT\Psi X] \quad \text{i=A, B, C, N} \tag{16}$$

where

$$\Psi = (dE / dX) = cons\tan t$$

The electric flux transport by each ionic species can be expressed as

$$I = Z_A F J_A + Z_B F J_B = t_A I \ + t_B I \tag{17}$$

Substituting Eqns. (16) and (17) into Eqns. (14) and (15) , the concentration profile of cation can be formulated by integrating method to give:

$$C_i = \left\{ C_i^o \left[C_Y{}^o \right]^{Z_i} - P \left[C_Y{}^o \right]^{Z_i} \left[1 - \exp[Z_i F \Psi X / RT] \right] \right\} \Big/ C_Y^{Zi} \quad \text{i =A, B} \tag{18}$$

where

$$P = t_i IRT / (Z_i F)^2 D_i \Psi$$

Next, consider the mass transport in the membrane phase. The driving force of permeation flux is mainly due to the electric migration, and the normal diffusion as well as convective transport can be neglected. Therefore the ionic concentration in the membrane can be derived from Nernst-Planck equation as following (Yamabe & Seno, 1967):

$$\overline{C_i} = \frac{t_i I}{Z_i F D_i} \frac{RT}{Z_i F \Psi_m} \tag{19}$$

Here $\overline{C_i}, \overline{D_i} and \Psi_m$ repent the ionic concentration, ionic diffusivity and membrane potential in the membrane phase; the subscripts i and k refer to ions and to all components in the system, respectively.

Based on the Teorell-Meyer-Sievers hypothesis (Tombalakian, et al.,1967), a pseudo-equilibrium state between each metal ion is assumed to exist at the film-membrane interface, respectively. An apparent selectivity coefficient, K_B^A , can be defined as

$$K_B^A = \frac{\overline{C}_A^{Z_B}}{C_A^{Z_B}} \times \frac{C_B^{Z_A}}{\overline{C}_B^{Z_A}} \tag{20}$$

Substituting eqn 16 and 19 into eqn. 20, we obtained

$$\frac{\left\{C_A^o - P\left(1-\exp[Z_A F\Psi\delta / RT]\right)\right\}^{Z_B}}{\left\{t_A IRT / (Z_A F)^2 \overline{D}_A \Psi_m\right\}^{Z_B}} = K_B^A \frac{\left\{C_B^o - P\left(1-\exp[Z_B F\Psi\delta / RT]\right)\right\}^{Z_A}}{\left\{t_B IRT / (Z_B F)^2 \overline{D}_B \Psi_m\right\}^{Z_A}} \qquad (21)$$

where δ denotes the thickness of stagnant liquid film, $t_A + t_B = 1$.
Eqn.21 is a general equation describing the preferential transport behaviors of metal ions sharing with the same counter-ion across the cation-exchange membrane in electrodialysis process. In the above equation, Ci, Di, ti, Ui and Zi refer to the concentration, diffusivity, transport number, mobility and charge of ion I, respectively. F, I, L and $K A B$ are the Faraday constant, current density, film thickness and selectivity coefficient, respectively. The superscripts o and – represent the bulk solution phase and the membrane phase. Since the selectivity coefficient $K A B$ can be obtained from equilibrium measurement and film thickness L estimated from I-V curves (Mandersloot et al., 1965), the transport number t_A and t_B can then be calculated by numerical method at each operating condition.
In general, the permselectivity of ion A relative ion B, T_B^A in ion exchange membrane electrodialysis can be defined by (Cowan & Brown, 1959):

$$T_B^A = \left(J_A / J_B\right) / \left(C_A^o / C_B^o\right) = \left(Z_B t_A / Z_A t_A\right) / \left(C_A^o / C_B^o\right) \qquad (22)$$

The permelectivity coefficient can also be determined experimentally from the following equation derived from a shell mass balance [Huang & Wang, 1992b].

$$T_B^A = \ln\left(C_{A,o}^o / C_{A,t}^o\right) / \ln\left(C_{B,o}^o / C_{B,t}^o\right) \qquad (23)$$

where $C_{i,o}^o$ is the initial concentration of ionic species i in the bulk solution, and $C_{i,t}^o$ is that of the bulk solution at tme t.

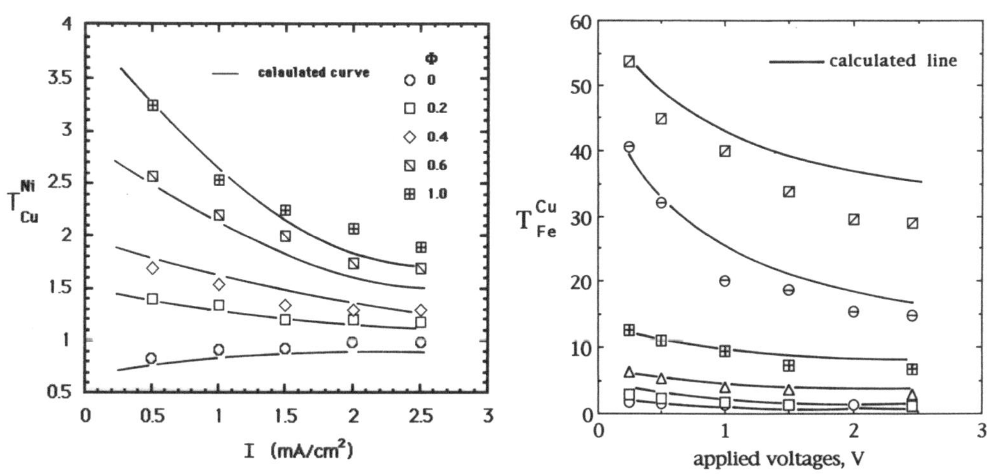

Fig. 5. Preferential transport behaviors of metal ions in a five compartment electrodialyzer with using citric acid as a complex agent at $[CuCl_2] = [NiCl_2] = [FeCl_3] = 0.015M$.

The preferential transport of nickel and cupric ions as well as ferric and cupric ions through cation exchange membranes was studied with electrodialysis in chloride solutions containing a complex agent (Huang & Wang, 1993b). Citric acid was were employed as complex agents in order to increase the permselectivity of metal ions. As can be seen from Figures 5 (a) and (b), citric acid was found to be an effective complex agents to increase the permselectivity coefficient T_{Cu}^{Ni} from 0.83 to 3.32 and T_{Cu}^{Fe} from 0.85 to 58.78. The values of T_{Cu}^{Ni} and T_{Cu}^{Fe} increase as the stoichiometric ratio of complex agent to metal ions and pH of complex solutions increase, but decrease with an increased current density and applied voltages. Based on the Nernst-Planck equation and interfacial quasi-equilibrium hypothesis, a theoretical approach obtained from numerical calculation is consistent with the experimental data. The experimental results show that electrodialysis combined with the use of complex agents has been demonstrated to be a useful technique not only for the enrichment of metal ions but also for their separation.

6. Optimization method of dialysis/electrodialysis process

As a result, the combined use of a cation exchange membrane and a complex agent seem to be a feasible process for the simultaneous separation and concentration of metal ions. However, many factors such as membrane and complex agent sources, pH of solution, concentrations of complex agent and metal ions are important variables affecting permselectivity of metal ions. It is seemed to need a simple model to optimize such an ion fractionation process that the enrichment of metal ions as well as the separation of metal ions from waste water in industrial practice because of economic consideration.

Conventional practice of single variable optimization by maintaining other variables involved at a specified constant level does not depict the combined effect of all the variables involved. It was reported that the complexities and uncertainties associated with industry separation usually come from lack of knowledge of the sophisticated interactions among various variables. Unlike traditional optimization, statistical optimization methods can take into account the interactions of variables in generating the process response. These limitations of a single variable optimization can be eliminated by using of experimental design and response surface methodology (Dean & Voss, 1999). A combination of variables generating a certain optimum response can be identified through Box-Behnken design and the use of response surface methodology. This pattern is designed by using statistical methods to yield the most information by a minimum number of experiments and had been successfully applied to the optimization of various engineer process (Wang & Hsieh, 2009).

In general usage, statistics experimental design is the design of any information-gathering exercises where variation is present, whether under the full control of the experimenter or not. However, in statistics, these terms are usually used for controlled experiments. A methodology for designing experiments was proposed by Ronald A. Fisher, (Montgomery, 1991). As an example, he described how to test the hypothesis that a certain lady could distinguish by flavour alone whether the milk or the tea was first placed in the cup. While this sounds like a frivolous application, it allowed him to illustrate the most important ideas of experimental design. In statistics, Box–Behnken designs are experimental designs for response surface methodology, devised by George E. P. Box and Donald Behnken (Khuri & Cornell, 1987), to achieve the following goals:

- Each factor, or independent variable, is placed at one of three equally spaced values. (At least three levels are needed for the following goal.)

- The design should be sufficient to fit a quadratic model, that is, one containing squared terms and products of two factors.
- The ratio of the number of experimental points to the number of coefficients in the quadratic model should be reasonable .
- The estimation variance should more or less depend only on the distance from the centre (this is achieved exactly for the designs with 4 and 7 factors), and should not vary too much inside the smallest (hyper)cube containing the experimental points.

Each design can be thought of as a combination of a two-level (full or fractional) factorial design with an incomplete block design. In each block, a certain number of factors are put through all combinations for the factorial design, while the other factors are kept at the central values. For instance, the Box–Behnken design for 3 factors involves three blocks, in each of which 2 factors are varied through the 4 possible combinations of high and low. It is necessary to include centre points as well (in which all factors are at their central values).In statistics, response surface methodology (RSM) explores the relationships between several explanatory variables and one or more response variables. The method was introduced by G. E. P. Box and K. B. Wilson in 1951 (Dean, & Voss, 1999). The main idea of RSM is to use a sequence of designed experiments to obtain an optimal response. Box and Wilson suggest using a second-degree polynomial model to do this. They acknowledge that this model is only an approximation, but use it because such a model is easy to estimate and apply, even when little is known about the process.

A prior knowledge and understanding of the process and process variables under investigation are necessary for achieving a more realistic model. Based on the results of preliminary experiments, we selected the significant variables from sreening experimental variables in first. Then the Box-Behnken design and response surface methodology (RSM) were employed to obtain the optimal variable conditions for higher permselectivity of metal ions. Each variable had three levels to be examined at high level (+1), medium level (0) and low level (-1). The high and low levels we selected for this study represented the extremes of normal operating ranges. The range and the levels of the experimental variables investigated in this study are given in the Table 1. Based on Box-Behnken design, a set of 17 experiments was carried out for for each complex agent, respectively. In developing regression model, the experimental variables were coded according to the following equation:

$$x_i = \frac{X_i - X_0}{\Delta X} \tag{24}$$

where x_i is the coded value of the variable X_i, X_0 is the value of X_i at the center point and ΔX is the step change value. Once the experiments were performed, the regression model could be constructed by fitting the experimental results. According to the RSM methodology, a second-order polynomial model was used to fit the experimental variables using the following equation :

$$Y = b_0 + b_1 x_1 + b_2 x_2 + b_3 x_3 + b_4 x_1^2 + b_5 x_2^2 + b_6 x_3^2 + b_7 x_1 x_2 + b_8 x_1 x_3 + b_9 x_2 x_3 \tag{25}$$

where Y is the permselectivity of metal ions; b_0, …, b_9 are coefficients of the model; x_1, x_2, x_3 are the coded variables.

In this study, Design-Expert package (version 6.0.10, Stat-ease Inc., Minneapolis, USA) was used for regression analysis of the data obtained, to estimate the coefficients of regression model and to search the optimal conditions. The fit of regression model attained was checked by the multiple correlation coefficients (R^2). The optimization process was depicted in Figure 6.

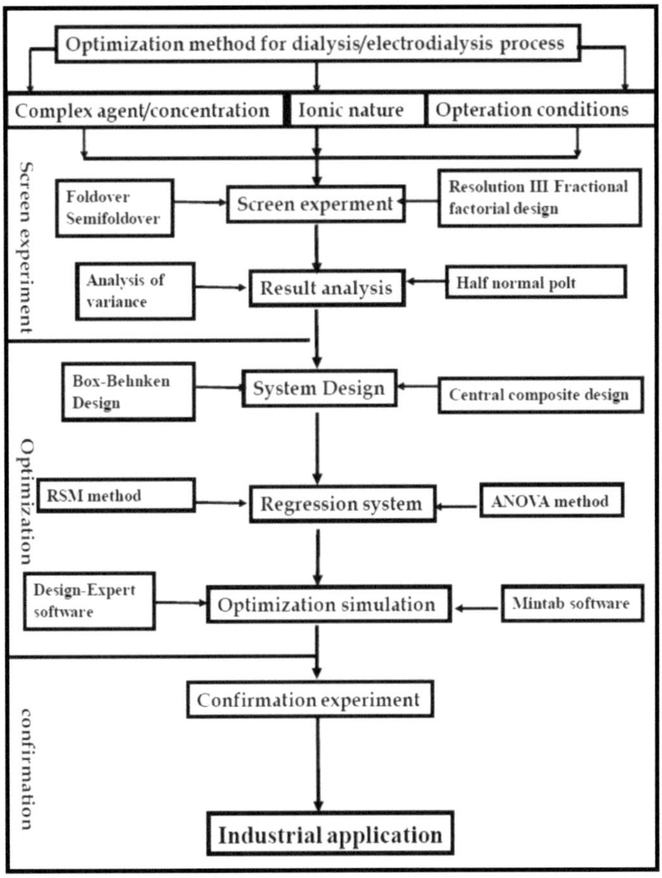

Fig. 6. The flowchart of optimization process

Statistical experimental methods were used to optimize the variable conditions of membrane dialysis process with a complex agent. Permselectivity of metal ions has been investigated in a two-compartment dialyzer with reflux flow. Cu^{+2}-Ni^{+2} ion system was chosen for the experimental investigation. Oxalic, malonic acid, acid and citric acid were selected as a complex agent used in the investigated solution in order to increase the discrepancy between transport fluxes of metal ions. The concentration ratio of complex agent to metal ions (X_1), pH value of investigated solution (X_2) and concentration of metal ions (X_3) were conducted to be optimized. The Box-Behnken design and Response surface methodology were employed in the design of experiments and the analysis of results. The experimental results indicated that oxalic acid is a better complex agent in this study.

The Box-Behnken design and experimental results are shown in the Table 2. The experimental data indicates that permselectivity of metal ions can be enhanced with use of a complex agent. This is due to the fact that metal ions will compete with each other to react with complex ligands to form complexes which are hardly permeated through the membrane, while the complex agent is added in feed. The difference in the concentration gradients of free metal ions across the membrane can be obtained because of the difference in stability and quantity of complexes (Ng & Snyder, 1981; Kuroda, 1983). Thus, the various permselectivity of metal ions can be observed in Table 2 for various kinds of complex agent included oxalic acid, malonic acid and citric acid. From the experimental results in Table 2, the means of permselectivity of metal ions are 2.31 for malonic acid as complex agent; 4.24 for oxalic acid as complex agent; 1.95 for citric acid as complex agent, respectively. It shows that the effect of complex agent on permselectivity of metal ions for Cu+2-Ni+2 ion system has the following order: oxalic acid >malonic acid >citric acid. The permselectivity of metal ions can be enhanced with a complex agent under suitable conditions. The difference in the concentration gradients of free metal ions across the membrane can be obtained because of the difference in stability and quantity of complexes. The larger the difference that exists between stability constants of metal ion complexes, the higher difference that the permselectivity of metal ions can be obtained.Theses experimental results were analyzed using statistical software package Design-Expert 6.0.

Trial no.	Variables			Permselectivity (Y)		
	x_1	x_2	x_3	oxalic	malonic	citric
1	-1	-1	0	0.91	0.91	0.91
2	1	-1	0	2.20	1.30	1.25
3	-1	1	0	0.95	0.95	0.95
4	1	1	0	10.88	5.25	4.13
5	-1	0	-1	0.87	0.87	0.89
6	1	0	-1	6.00	3.86	3.03
7	-1	0	1	0.93	0.93	0.93
8	1	0	1	5.78	3.53	2.40
9	0	-1	-1	2.43	1.53	1.20
10	0	1	-1	5.50	2.34	2.78
11	0	-1	1	1.13	1.03	1.03
12	0	1	1	9.43	4.30	3.78
13	0	0	0	5.03	2.53	1.88
14	0	0	0	4.95	2.45	1.98
15	0	0	0	5.20	2.40	2.05
16	0	0	0	4.95	2.53	1.89
17	0	0	0	5.20	2.45	1.99

Table 1. The Box-Behnken design and results for dialysis with using a complex agent

1. *For oxalic acid as complex agent*

According to the RSM methodology, a second-order polynomial model, Eqn.(25), was used to fit the experimental variables. Multiple regression analysis of the experimental results gave the following equation:

$$Y = 5.07 + 2.65x_1 + 2.65x_2 + 0.42x_3 - 1.14x_1^2 - 0.19x_2^2 - 0.53x_3^2 + 2.16x_1x_2 - 0.07x_1x_3 + 1.08x_2x_3 \quad (26)$$

The results of analysis of variable (ANOVA) for regression model how that x_1 , x_2 , x_1^2, x_1 x_2 and x_2 x_3 have significant effect on permselectivity of metal ions because the value of Pob>F is less than 0.05 (Montgomery, (1991). The R^2 (multiple correlation coefficient) of the regression equation obtained from analysis of variance is always larger than 0.9685 (a value >0.75 indicates aptness of the model), which means that the model can explain 96.85% variation in the response (Wang & Hiesh, 2009).

2. *For malonic acid as complex agent*

From the analysis of sequential model and lack of fit, a second-order regression model was suggested. Multiple regression analysis of the experimental results gave the following equation:

$$Y = 2.47 + 1.28x_1 + 1.02x_2 + 0.16x_3 - 0.17x_1^2 - 0.20x_2^2 + 0.98x_1x_2 - 0.098x_1x_3 + 0.59x_2x_3 \quad (27)$$

The results of analysis of variable (ANOVA) for regression model are given that x_1, x_2, x_1 x_2 and x_2x_3 have significant ($P<0.05$) effect on permselectivity of metal ions. The results showed the R^2 (multiple correlation coefficient) of the regression equation obtained from analysis of variance is 0.9807 (a value >0.75 indicates aptness of the model), which means that the model can explain 98.07% variation in the response.

3. *For citric acid as complex agent*

From the analysis of sequential model and lack of fit, a second-order regression model was suggested. Multiple regression analysis of the experimental results gave the following equation:

$$Y = 1.96 + 0.89x_1 + 0.91x_2 + 0.03x_3 - 0.27x_1^2 + 0.12x_2^2 + 0.12x_3^2 + 0.71x_1x_2 - 0.17x_1x_3 + 0.29x_2x_3 \quad (28)$$

The results of analysis of variable (ANOVA) for regression model are show that x_1, x_2 and x_1 x_2 have significant ($P<0.05$) effect on permselectivity of metal ions. The R^2 (multiple correlation coefficient) of the regression equation obtained from analysis of variance is always larger than 0.9680 (a value >0.75 indicates aptness of the model), which means that the model can explain 96.80% variation in the response.

Trial no.	Permselectivity					
	Oxalic acid		Malonic acid		Citric acid	
	Actual	**Predicted**	**Actual**	**Predicted**	**Actual**	**Predicted**
1	0.91	0.79	0.91	0.77	0.91	0.72
2	2.20	1.58	1.30	1.39	1.25	1.09
3	0.95	1.57	0.95	0.86	0.95	1.12
4	10.88	11.19	5.25	5.39	4.13	4.32
5	0.87	0.25	0.87	0.75	0.89	0.72
6	6.00	5.69	3.86	3.52	3.03	2.84
7	0.93	1.24	0.93	1.27	0.93	1.12
8	5.78	6.40	3.53	3.65	2.40	2.57
9	2.43	2.36	1.53	1.68	1.43	1.55
10	5.50	5.49	2.34	2.55	2.78	2.78
11	1.13	1.04	1.03	0.82	1.03	1.03
12	9.43	8.95	4.30	4.05	3.78	3.43
13	5.03	5.07	2.53	2.47	1.88	1.96
14	4.95	5.07	2.45	2.47	1.98	1.96
15	5.20	5.07	2.40	2.47	2.05	1.96
16	4.95	5.07	2.52	2.47	1.89	1.96
17	5.20	5.07	2.45	2.47	1.99	1.96

Table 2. Comparisons of the predicted values and actual values for dialysis process

Once the optimal level of the process variables is identified for each complex agent, the confirmation experiment was conducted to validate the optimization model. As shown in Table 2, comparisons of experimental value and predicted value of regression model, it was observed that agreement was satisfactory.The results also indicates that the model could be an simple method to stimulate the permselectivity behaviors of metal ions through membrane dialysis with a complex agent.

In the prceeding section there is a demonstration case that statistics based experimental designs were used to optimize preferential behavior for the electrodialysis process combined use a complex agent. Cu^{+2}-Ni^{+2} and Cu^{+2}-Fe^{+3} ions system were chosen for the experimental investigation. Citric acid was employed as a complex agent used in the investigated solution in order to increase the discrepancy between transport fluxes of metal ions. The Box-Behnken design and experimental results are tabulated in the Table 3. Once the experiments were performed, the regression model could be constructed by fitting the experimental results with a second-order polynomial. The optimal conditions of selected variables were searched using the regression model and also by analyzing the response surface method. Based on Box-Behnken design, a set of 17 experiments was carried out. According to the RSM methodology, a second-order polynomial model was used to fit the experimental variables. The statistical software package Design-Expert 6.0 was used to analyze the experimental results.

Trial no.	Variables			Permselectivity	
	X_1	X_2	X_3	Cu^{+2}-Ni^{+2}	Cu^{+2}-Fe^{+3}
1	-1	-1	0	0.91	0.85
2	1	-1	0	1.30	12.20
3	-1	1	0	0.95	1.03
4	1	1	0	4.25	42.78
5	-1	0	-1	0.87	2.43
6	1	0	-1	3.86	36.00
7	-1	0	1	0.93	0.88
8	1	0	1	3.53	25.78
9	0	-1	-1	1.43	9.43
10	0	1	-1	2.34	25.50
11	0	-1	1	1.23	11.23
12	0	1	1	4.50	23.50
13	0	0	0	2.53	15.03
14	0	0	0	2.45	14.95
15	0	0	0	2.40	15.20
16	0	0	0	2.53	14.95
17	0	0	0	2.45	15.20

Table 3. The Box-Behnken design and results for electrodialysis process

The results of analysis of variable (ANOVA) for regression model are listed in Table 4 for Cu^{+2}-Ni^{+2} system. It is indicated that the main effect X_1, X_2 and two-variable interaction X_1X_2 had significant effect ($P < 0.05$) on permselectivity of metal ions. The R^2 (multiple correlation coefficient) of the regression equation obtained from analysis of variance is always larger than 0.9436 (a value >0.75 indicates aptness of the model), which means that the model can explain 94.36% variation in the response. By using the regression analysis method, the regression model for Cu^{+2}-Ni^{+2} system in the coded form was established as follows:

$$Y = 5.06 + 2.65x^1 + 2.70x^2 + 0.52x^3 - 1.24x_1^2 - 0.092x_2^2$$
$$-0.43x_3^2 + 2.16x^1x^2 - 0.07x^1x^3 + 1.18x^2x^3 \quad (29)$$

Effect	Sum of Squars	DF	Mean Square	F- Value	Prob>F
X_1	6.35	1	6.35	68.98	< 0.0001
X_2	6.57	1	6.57	71.32	< 0.0001
X_3	7.2x10-3	1	7.2x10-3	0.078	0.7855
X_1 X_2	2.02	1	2.02	21.89	0.0009
X_1 X_3	0.11	1	0.11	1.22	0.2956
X_3	0.34	1	0.34	3.71	0.0828

Table 4. Analysis of variance for electrodialysis of Cu^{+2}-Ni^{+2} system Table 5 Analysis of variance for electrodialysis of Cu^{+2}-Fe^{+3} system

Effect	Sum of Squares	DF	Mean Square	F- Value	Prob>F
X_1	56.18	1	56.18	81.12	< 0.0001
X_2	58.10	1	58.10	83.90	< 0.0001
X_3	2.14	1	2.14	3.09	0.1220
X_1 X_2	6.34	1	6.34	9.15	0.0192
X_1 X_3	0.028	1	0.028	0.041	0.8455
X_2 X_3	0.75	1	0.75	1.08	0.3327

Table 5. Analysis of variance for electrodialysis of Cu^{+2}-Fe^{+3} system

The results of analysis of variable (ANOVA) for regression model for Cu^{+2}-Fe^{+3} system are tabulated in Table 5. It is indicated that the main effect X_1, X_2 and two-variable interaction X_1X_2 had significant effect ($P < 0.05$) on permselectivity of metal ions. The R^2 (multiple correlation coefficient) of the regression equation obtained from analysis of variance is always larger than 0.9678 (a value >0.75 indicates aptness of the model), which means that the model can explain 96.78% variation in the response. By using the regression analysis method, the regression model for Cu^{+2}-Fe^{+3} system in the coded form was established as follows:

$$Y = 3.90 + 88.61x_1 - 28.87x_2 - 79.94x_3 - 150.52{x_1}^2 + 47.05{x_2}^2$$
$$+189.83{x_3}^2 - 24.95x_1x_2 - 32.42x_1x_3 - 275.7x_2x_3 \quad (30)$$

Based on the regression model, the predicted maximum permselectivity was calculated to be 5.25 from Eq (29) Cu^{+2}-Ni^{+2} system. The corresponding optimal variable conditions were 0.85 for Φ, 4.83 for pH, and current density 12.43A/dm². The predicted maximum permselectivity was calculated to be 43.5 from Eq (30) for Cu^{+2}-Fe^{+3} system. The corresponding optimal variable conditions were 1.00 for Φ, 3.78 for pH, and current density 17.43A/dm².Comparisons of experimental value and predicted value of regression model), it was observed that agreement was satisfactory.

Data from the present investigation have shown that statistics based experimental designs is an effective tool to find the optimal separation conditions for the ion membrane dialysis process combined use a complex agent. It is also highlighted that an effective multi-ion fractionation process (as shown in Figure 7)could be build from cation-exchange membrane being the separation interface; the driving force being the applied electric field; the suitable complex agent being the separation agent and the pH value is the separation index.

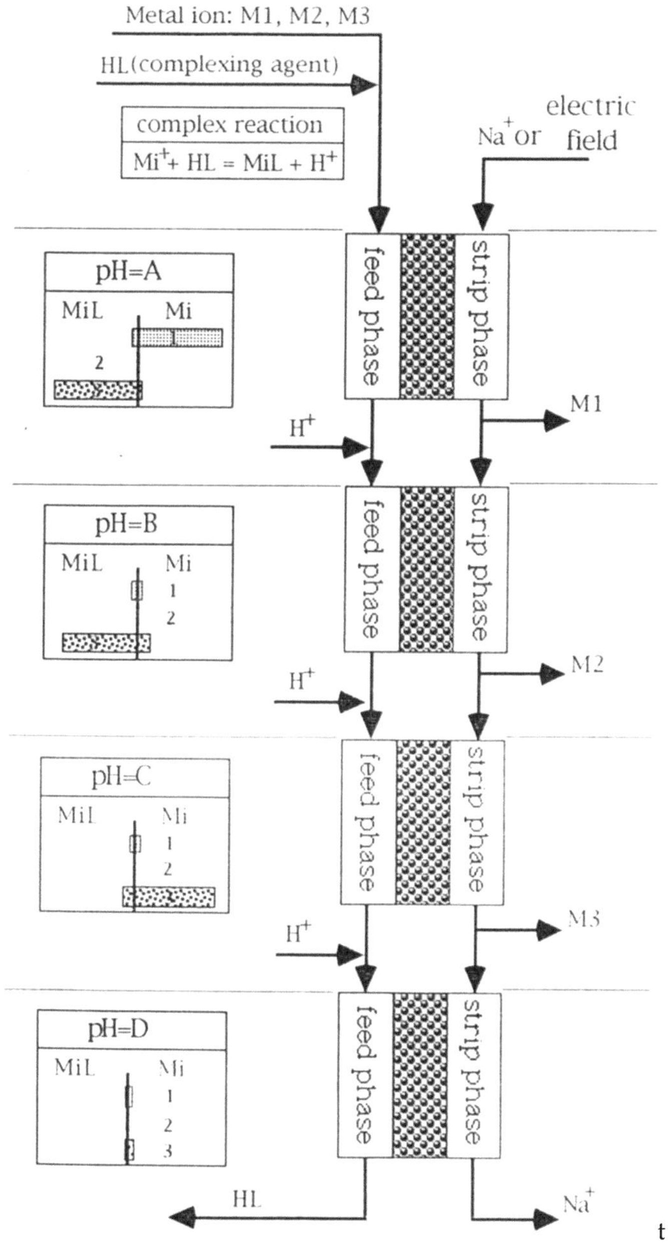

Fig. 7. Schematic mechanism of multi-ion fractionation process

7. Conclusion and outlook

Ion exchange membrane processes included dialysis and electrodialysis are used today on a large industrial scale in the production of potable and industrial process water as well as in the treatment of industrial effluents. However, today's membrane processes and components used in these processes have still severe technical and commercial limitations and there is a substantial need for basic as well as applied research to improve the processes. More recently developed processes will be discussed in more detail and their technical as well as economic advantages are described, and present and potential future applications indicated. Problems and technical limitations of these processes as well as commercial opportunities are indicated. They will be brought to industrial practice as long as the developers or users recognize that these separation technologies are versatile tools to deal with specific treatment requirements. It should also be noted that dialysis/electrodialysis separations are promising technologies for clean production and separation but have a long way to go before industrialization in most industries. The obstacles include the limited choice of membranes and the extremely high equipment cost of electrodialyzer. The increase in environmental awareness and raw material cost have accelerated the application of dialysis/electrodialysis integrations, especially in highly industrialized and densely populated countries. Accordingly, researchers should give attention to interdisciplinary knowledge and solve some interesting problems, such as membrane exploration, processes simulation, and operation optimization in dialysis/electrodialysis separation processes.

8. References

Belfort, G. & Guter, G. A. (1972). An Experimental Study of Electrodialysis Hydrodynamics, *Desalination*, 10, 221

Bohere, M. P. (1983). Diffusional Boundary Layer Resistance for Membrane Transport, *Ind. Eng. Chem. Fundam.*, 22, 27

Brajter, K. & Slonawaska K. (1988). Separation of Noble Metals by Donnan Dialysis, *Sep. Sci. Technol.*, 23, 321

Boudet, D.; Lindheimer, M.; Gavach, A. & C. (1991). Transport Properties of Anion Exchange in Contact with Hydrochloric Acid Solutions--Membranes for Acid Recovery by Electrodialysis, *J. Membr. Sci.*, 57, 57

Bungay, P. M.; Lonsdale H. K. & DePinhp M. N. (1986). Synthetic Membrane: Science, Engineering and Application. *ASISeries*, Vol. 181

Ciric, J. & Graydon, W. F. (1962). Ion-Exchange Membranes (IV)--Univalent Cation Transfer Rates, *J. Phys. Chem.*, 66, 1549

Costa, E.; Lucas, A. D. & Gonzalez, M. E. (1984). Ion Exchange Determination of Interdiffusion Coefficients, *Ind. Eng. Chem. Flundam.*, 23, 400

Cox, J. A. & DiNunzio, J. E. (1977). Dialysis Enrichment of Cations, *Anal. Chem.* 49, 1272

Coz, J. A., Slonawska, J.; Gatchell, D. K. & Hiebert, A. G. (1984). Metal Speciation by Donnan Dialysis, *Anal. Chem.*, 56, 650

Cowan, D. A. & Brown, J. H. (1959). Effect of Turbulence on Limiting Current in Electrodialysis Cells, *Ind. Eng. Chem.*, 51(12), 1445

Davis, T. A.; Wu, J. S. & Baker, B. L. (1971). Use of the Donnan Equilibrium Principle to Concentrate Uranyl Ions by an Ion-Exchange Membrane Process, *AIChE J.*, 17 (4), 1006

Dean, A. & Voss, D. (1999). *Response surface methodology. In: Design and analysis of experiments,* Springer-Verlag, New York, pp. 547–86.

Dibenedetto, A. T. & Lightfoot, E. N. (1962). The Separation of Ions With Permselectivity Membranes, *AIChE J.*, 79

Donnan, F. G. (1924). Theory of Membrane Equilibria, *Chem. Rev.*, 1, 73

Dranoff, J. & Lapidus, L. (1957). Equilibria Ternary Ion-Exchange System", *Ind. Eng. Chem.*, 49, 1297

Farrell, J. B. & Smith, R. N. (1962). Process Application of Electrodialysis, *Ind. Eng. Chem.*, 54(6), 29

Gering, K. L. & Scamehorn, J. F. (1988). Use of Electrodialysis to Remove Heavy Metals form Wast, *Sep. Sci. Technol.*, 23 ,2231

Graillon, S.; Persin, F.; Pourcelly, G.& Gavach, C.(1996). Development of electrodialysis with bipolar membrane for the treatment of concentrated nitrate effluents, *Desalination*, 107, 159-169.

Helfferich, F. (1962). *Ion Exchange,*McGraw-Hill Book Company, Inc., New York

Hopfenberg, H. B.; Lee, K. L. & Wen, C. P. (1978). Improved Membrane Separations by Selective Chelation of Metal Ions in Aqueous Feeds, *Desalination*, 24, 175

Huang, R. Y. M. (1991). *Prevaporation Membrane Separation Process,* Elsevier, Amsterdam

Huang, T. -C. & Wang, J. -K. (1992a). Separation of Cupric and Ferric Ions with a Cation-exchange Membrane in the Presence of Complexing Agents, *Sep. Sci. Technol.*, 27, 1137

Huang, T. -C. & Wang, J. -K. (1992b). Preferential Transport of Nickel and Cupric Ions through Cation Exchange Membrane in Electrodialysis with a Complex Agent, *Desalination*, 86, 257

Huang, T. -C. & Wang, J. -K. (1993a). Selective Transport of Metal Ions through Cation Exchange Menbrane in the Presence of Complexing Agent, *Ind. Eng. Chem. Res.*, 32(1), 133-139

Huang, T. -C. & Wang, J. -K. (1993b). Preferential Transport of Cupric and Ferric Ions through Ion Exchange Membrane with Electrodialysis in the Presence of Citric Acid, *Chem. Eng. Commun.*, 122, 213-225

Hwang, S. -T. & Kammermeyer, K. (1975). *Membrane in Separations,* John Wiely & Son, New York

Itoi, S.; Nakamura, I. & Kawahara, T. (1980). Electrodialysis Recovery Process of Metal Finish ing Waste Water, *Desalination*, 32, 393

Keister, J. C. & Kasting, G. B. (1986). Ionic Mass Transport through a Homogeneous Membrane in the Presence of a Uniform Electric Field, *J. Membr. Sci.*, 29, 155

Khuri , A. I.& Cornell, J. A.(1987). *Response Surfaces Design and Analysis*, Dekker, New York

Kim, B. M. (1979). Donnan Dialysis for Removal of Chromates and Cyanides. *AIChE Symb. Ser.*, 76(197), 184

King, C. J. (1980). *Separation Process 2th ed.*, McGraw-Hill Inc., New York

Kitamoto, A. & Takashima, Y. (1964). Preferential Ion Transport through Ion Exchange Membrane by Electrodiaysis, *J. Chem. Eng. Japan*, 37, 1443

Kojima, T.; Furusaki, S.; Takao, K. (1982). A Fundamental Study on Recovery of Copper with a Cation Exchange Membrane Part 2-- Transfer Rate of Copper and Hydrogen Ion through Cation Exchange Membrane, *Candian J. Chem. Eng.*, 60, 650

Korngold, E.; Kock, K. & Strathmann, H. (1987). Electrodialysis in Advanced Waste Water Treament, *Desalination*, 24, 129

Kuroda, O.; Takahashi, S. & Nomura, M. (1983). Characteristics of Flow and Mass Transfer Rate in an Electrodialyzer Compartment Including Spacer, *Desalination*, 46, 225

Lake, M. A. & Melsheimer, S. S. (1978). Mass Transfer Characterization of Donan Dialysis, *AIChE J.*, 24(1), 130

Leitz, F. B. & Eisenamann, J. L. (1981). Electrodialysis as a Separation Process, *AIChE Sym. Ser.*, 77(204), 204

Lightfoot, E. N. & Scattergood, E. M. (1965). Suitability of the Nernst-Planck Equation for Describing Electrokinetic Phenomena. *AIChE J.*, 11, 175

Lonsdale, H. K. (1982). The Growth of Membrane Technology, *J. Membr. Sci.*, 10, 81

Lounis, A. & Gavach, C. (1990). Electrotransport of Carbonate and Hydrogen Carbonate Ions in Anion Exchange Membranes. *J. Membr. Sci.*, 54, 63

Mandersloot, W. G. B. & Hicks, R. E. (1965). Concentration Polarization on Ion-exchange Resin Membrane in Electrodialytic Demineralization, *Ind. Eng. Chem. Process Des. Dev.*, 4, 304

Margerum, D. W.; Janes, D. L.; Rosen, H. M. (1965). Multidentate Ligand Kinetics. VII. The Stepwise Nature of the Unwrapping and Transfer of Ethylenediamine-tetraacetatonickelate from Nickel (II) to Copper (II), *J. Am. Chem. Soc.*, 87, 4463

Matsuyama, H.; Fuji, K. &Teramoto, M. (1991). Selective Separation of Rare Earth Metals by Donnan Dialysis in the Presence of Water-Soluble Complexing Agent, *J. Chem. Eng. Japan*, 24, 253

Mears, P. (1976). *Membrane Separation Process,* Elsevier, Amsterdam

Mizutani, Y.(1990). Ion Exchange Membrane with Preferential Permselectivity for Monovalent Ions, *J. Membr. Sci.*, 54 233

Montgomery, D. C. (1991). *Design and Analysis of Experiments*, Wiley, New York, 3rd edn.

Ng, P. K. & Snyder, D. D. (1981). Mass Transport Characterization of Donnan Dialysis: The Nickel Sulfate System, J. *Electrochem. Soc.*, 128(8), 1717

Nobel, R. D. (1987). A Overview of Membrane Separations, *Sep. Sci. Technol.*, 22(2&3), 731

Nott, B. R. (1981). Electrodialysis for Recovering Acid and Caustic from Ion-Exchange Regeneration Wastes, *Ind. Eng. Chem. Process Res. Dev.*, 20, 170

Pyrzynska, K. (1991). Membrane method for preconcentrating and separating gold complexes from aqueous solutions containing other platinum group metals, *Anal. Chim. Acta*, 255, 169-175.

Rosenberg, N. W. & Tirrell, C. E. (1957). Limiting Currents in Membrane Cells, *Ind. Eng. Chem.*, 49, 780

Rush, W. E. & Baker, B. L. (1980). Quantitative Description of Ion Transport in Donnan Ion Exchange Membrane Systems, *Sep. Sci. Technol.*, 15(4), 1153

Saito, K., S. Furusaki & Kojima (1982). A Method for Determination of Integral Interdiffusion Coefficient in a Cation Exchange Membrane, *J. Cham. Eng. Japan*, 15(b), 487

Sandeaux J.; Sandeaux, R. & Gavach, C. (1991). Competition between the Electrotransport of Acetate and Chloride Ions through a Polymeric Anion Exchange Membrane, *J. Membr. Sci.*, 59, 265

Sata, T. (1991). Ion Exchange Membranes and Separation Process with Chemical Reaction, *J. App. Electrochem.*, 21 ,283

Saito, K.; Sakairi, T.; Yonemoto, T. &T. Tadaki, (1991). Ionic Transport in a Continuous Donnan Dialyzer with Two Parallel-plate Channels, *J.Chem. Eng. Japan*, 24(5), 575

Seto, T.; Ehara, M. L.; Komori, R.; Yamaguchi, A. & Miwa, T. (1978). "Seawater Desalination by Electrodialysis", *Desalination*, 25,

Sherwood, T. K.; Pigford, R. L. & Wilke, C. R. (1975). *Mass Transfer*, McGraw-Hill, Book Co. Inc., New York

Spencer, H. G. & Ellison, T. M. (1953). Effects of pH and Divalent Cations on the Transport of Univalent Cation Across a Weak Acid Membrane, *J. phys. Chem.*, 57, 687

Stratmanm, H. (2004). *Ion-exchange Membrane Separation Process, Elsevier*, Amsterdam

Sudoh, M.; Kamei , H. & Nakamura, S. (1987). Donan Dialysis Concentration of Cupric Ions, *J. Chem. Eng. Japan*, 20(1), 34

Takahashi, K.; Tsuboi, K. & Takeuchi, H. (1989). Mass Transfer across Cation Exchange Membrane, *J. Chem. Eng. Japan*, 22(4), 352

Tombalakian, A. S.; Yeh, C. Y. & Graydon, W. F. (1967). Cation Exchange across Ion Exchange Membrane, *J. Phys. Chem.*, 71, 435

Tomicki, B. (1991). A Thermodynamic Description of Membrane Equilibria, Dissipation and Transport, *J. Membr. Sci.*, 60, 169

Wang, J.-K. & Hsieh, C.-S.(2009a). Determination of optimal conditions for separation of metal ions through membrane dialysis using statistical experimental methods, *Desalination and Water Treatment*, 2, 246–253

Wang, J.-K. & Hsieh, C.-S. (2009). Optimization of the Medium Components by Statistical Experimental Methods to Enhance Nattokinase Activity" , Fooyin J. Health Sci., 1(1), 21-27

Wang, J.-K. & Hsieh, C.-S. (2008). Preferential transport behaviors of ternary system Cupric-Nickel-Zinc ions through cation- exchange membrane with a complex agent by dialysis, *Chem. Eng. J.*, Vol 138/1-3, 103-110

Wang, J.-K. (2004). Preferential transport behaviors of ternary system ferric-cupric-nickel ions through cation ion exchange membrane with a complex agent by dialysis, *Desalination*, 161 , 277-285

Wang, J.-K. & Chang, C.-P. (2004). Selective transport behaviors of ternary system Cupric-Nickel-Cobalt ions through cation ion exchange membrane with a complex agent by dialysis, *Desalination*, 164,269-280

Wallace, R. M. (1967). Concentration and Separation of Ions by Donan Membrane Equilibrium, *Ind. Eng. Shem. Process Des. Dev.*, 6, 423

Wen, C. P. & Hamil, H. F. (1981). Metal Counterion Transport in Donan Dialysis, *J. Membr. Sci.*, 8, 51

Wills, G. B. & Lighfoot, E. N. (1961). Membrane Selectivity, *AIChE J.*, 7(2), 273

Yamabe, T. & Seno, M. (1967). The Concentration Polarization Effect in Ion-Exchange Membrane Electrodialysis, *Desalination*, 2(2), 148

Zeleznik, F. J. & Gordon, S. (1968). Calculation of Complex Chemical Equilibria, *Ind. Eng . Chem.*, 60, 27

15

Developing Nano-Structured Carbon Electrodes for Capacitive Brackish Water Desalination

Linda Zou
University of South Australia, Adelaide
Australia

1. Introduction

Capacitive deionisation (CDI) is a promising alternative technology in desalination. It targets to remove the salt ions which are only a small percentage of the feed solution, as compared to most other technologies that aim to shift water which accounts of 90% of the feed solution. As a result, CDI requires less energy to operate and the electrodes are easily regenerated. The basic concept of CDI is electrical potential induced surface adsorption of ions on to electrodes. An electrical field forces charged sodium (Na^+) and chloride (Cl^-) ions in the brackish water to move towards oppositely charged electrodes, by forming electrical double layers, and remove them from the water. It needs low voltage to operate, and it does not require harsh chemical cleaning process. The most critical component of CDI is the carbon electrode materials, as reported that the electrosorptive capacity strongly depends on the physical properties such as surface area and conductivity of the electrode. This chapter reports the current research efforts and progress in the development of porous carbon electrodes conducted in our research group, the commercial activated carbons (ACs), carbon nanotubes (CNTs) including single walled CNTs (SWCNTs) and double walled CNTs (DWCNTs), ordered mesoporous carbons (OMCs), and more recently, graphene nano-flakes were prepared and used as the electrodes of capacitive deionization device. Their salt removal performances were investigated. Their morphology and specific surface area were characterized. The electrosorption experiment results showed that their electrosorptive capacities are in the order of OMCs >SWCNTs>DWCNTs>ACs, ie. 0.93, 0.81, 0.80, 0.62 mg/g respectively. This is attributed to their differences in pore size distribution, pore pattern arrangement and specific surface area. In addition, the ion sorption onto these mentioned materials follows a Langmuir isotherm, indicating monolayer adsorption. Finally, we have investigated the regeneration property of both CNTs and AC through charge-discharge experiment and found that their regeneration was effective.

Chemically modified graphene has been studied in the context of many applications due to its excellent electrical, mechanical and thermal properties. Chemical modification of graphene oxide, which is generated from graphite powder as start material, has been a promising route to achieve mass production of graphene platelets. Hummers and Offeman (1958) developed a powerful oxidation method involving reacting graphite with a mixture of potassium permanganate ($KMnO_4$) and concentrated sulfuric acid (H_2SO_4) to achieve an efficient level of oxidation. It is believed that the desalting capacity can be greatly improved by increasing the surface area of the graphene. Recently, graphene-like nano-flakes (GNFs) with relatively high specific surface area have been prepared in our research group and used as electrodes for capacitive deionization. The TEM image of GNFs indicates that GNFs are showing interlayer pattern which is beneficial to ions accessing and adsorbing on the surface of the flakes. Its electrosorptive performance was much better than commercial activated carbon (AC), suggesting a great potential in capacitive deionization application. Further, the electrosorptive performance of GNFs electrodes with different electrical potentials, flow rates and ionic strengths were measured and the electrosorptive isotherm and kinetics were investigated as well. The results showed that GNFs prepared by this process had the specific surface area of 222.01 m^2/g. The specific electrosorptive capacity of the GNFs was 1.35mg/g for Na^+ when the initial concentration was at 25 mg/l, which was higher than that of previously reported data using graphene and AC under the same experimental condition. In addition, the equilibrium electrosorptive capacity was determined as 73.47 μmol/g at 2.0 V by fitting data through the Langmuir isotherm, and the rate constant was found to be 1.01 min^{-1} by fitting data through pseudo-first-order adsorption. At last, the conductivity of the synthesized graphene materials and activated carbons are determined and then compared by using a four-pin resistivity meter. The results suggested that the chemically synthesized GNFs can be a promising candidate of electrode material in CDI process for brackish water desalination.

2. Current desalination technologies for water challenge

In recent decades, the most widely used processes for desalination have included two membrane separation ones. The list is reverse osmosis (RO), electrodialysis (ED), thermal separations: multistage flash distillation (MSF), multi-effect distillation (MED) and mechanical vapor compression (MVC). Among these processes, RO or MSF methods are employed in the bulk of the seawater desalination plants (90%) worldwide [1]. At membrane based desalination plants, RO occupies 86% while ED only represents 14%. Compared to the thermal systems, membrane processes, such as RO and ED, are widely adopted for desalination from the point view of energy costs [2].

A recent review written by Anderson [2] who has systematically analysed all the information available in the literature regarding energy costs of water desalination. He hs pointed out that the ED process is not suitable for water with a concentration of dissolved solids below 400 ppm due to low conductivity. Furthermore, both RO, ED systems require feed pre-treatment to remove species that may, precipitate onto, or otherwise "foul" the surface of the ion exchange membranes and thus decrease the desalination performance of ED as well as RO. Hence, fouling of membranes is still a major challenge for RO and ED technologies.

2.1 Capacitive desalination

Capacitive deionization (CDI) is an electrochemically induced alternative approach for removing salt ions from concentrated aqueous solutions by means of forcing charged ions into the electrical double layer at an electrode-solution interface when the electrode is connected to an external power supply. The schematic mechanism of CDI is shown in Fig.1. The water containing salt ions flows through a pair of charged electrodes and anions and cations migrate to separately anode and cathode, respectively. As a consequence, the ions are ideally held onto the double layer. No redox process occurs and the process can easily be regenerated when the electrical power is short circuited or even reversed.

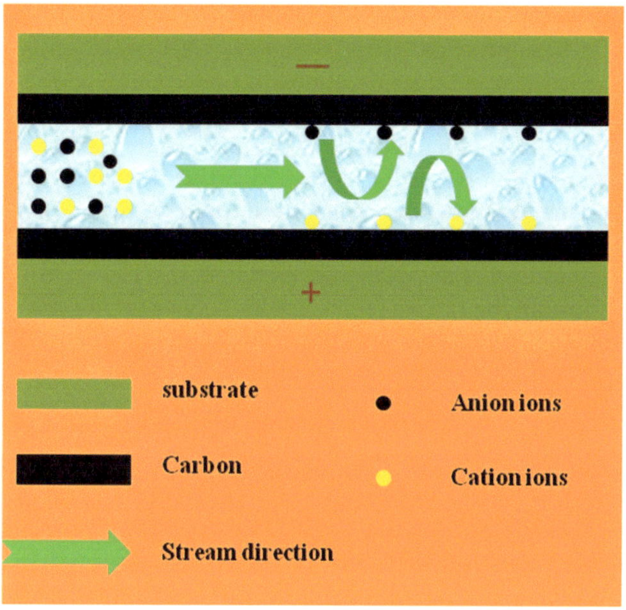

Fig. 1. Schematic diagram of capacitive deionization process

Historically, the double layer was firstly proposed by Helmholtz in 1883 [3], who described the charge distribution between two different phases as a capacitor. Gouy and Chapman then developed the model by considering the gradient of electron density at the charged interface. Later in 1924, the model was completed by Stern by assuming that the double layer can be divided into an "inner" region where the ion distribution followed Langmuir adsorption, while the region further from the surface could be roughly described with the Gouy-Chapman model. Thus, the electrochemical capacitance could be obtained like a series union of inner double layer and diffuse layer contributions.

From the perspective of double layer capacitance, the electrode materials play a crucial role in CDI and therefore to desalination performance. Generally, the electrode is made of porous materials with high specific surface area as well as very good conductivity, both of which will be of great advantage in absorbing large quantities of salt ions for desalination. Regarding the history of CDI, the pioneering work was conducted by the Caudle and Johnson groups in the mid-1960s and the early 1970s [4,5].

More detailed information could be found in a recent review of CDI progress by Oren [6]. Oren has named his version of CDI electrochemical parametric pumping [7,8]. He has published several comprehensive papers analysising the use of new carbon materials in the CDI process as well as fundamental electrochemical double layer issues of the electrodes [9-13]. In the mid-1990s, Farmer and co-workers [14-17] from the Lawrence Livermore National Labs (LLNL), developed a new kind of high specific surface area material which was called carbon aerogel for their CDI device and at the same time rekindled interest in CDI within the scientific community.

In recent years, most reports in terms of CDI typically utilized various carbons, such as carbon aerogels [14-16], carbon cloth [18-20], carbon sheets [21], carbon nanotubes [22-30]and meso-porous carbon [31-34] etc as electrode materials because these materials have the advantage of high specific surface area, easy large scale production and environmental-friendly.
In addition to CDI, membrane capacitive deionization technology (MCDI) as an modification of CDI has attracted great attentions. CDI which combined with an ion exchange membrane has shown great desalination performance compared to CDI without the membrane [35-38]. In a standard CDI device, the salt removal efficiency would be slightly reduced because of co-ion effects (ions of equal polarity as the electrode). These co-ions are near the electrode but cannot be electro-adsorbed efficiently. To avoid this negative effect, ion-exchange membranes have been introduced in front of the electrodes [38]. Specifically, a cation-exchange membrane is placed in front of the electrode that is negatively polarized, and an anion-exchange membrane is placed in front of the positive electrode. Counter-ions can then move freely into and out electrode while co-ion transport is restricted. In this chapter, we will introduce some of typical carbon materials, such as activate carbon, carbon nanotube (single walled and double walled) as well as a new rising star in carbon community "graphene", for application of CDI electrode base on a lab-scale CDI unit. Further, the optimized electrodes' materials were employed to carry out the desalination experiment with MCDI unit. We believe that these would be valuable for industrializing CDI technology in near future.

3. Lab scale CDI Experiments

The typical bench scale CDI unit cell consists of a retaining plate, carbon electrode and spacer which are shown schematically in Fig. 2. To assemble the MCDI unit, the ion-exchange membranes were soaked in distilled water for 2-3 days to make them sufficiently saturate before use and then inserted into either side of the insulated spacer but in front of carbon electrodes. In the adsorption experiments, the solution was pumped into the unit cell by means of peristaltic pump and the effluent returned to the unit cell. The solution volume was maintained at 50 ml and the ambient temperature was kept at 298 K, respectively. Meanwhile, the applied voltage was adjusted from 1.0 V to 2.0 V to determine the optimum working voltage. The above mentioned experiments were performed using a synthetic NaCl solution which had an initial conductivity of around 55 μS/cm. The relationship between conductivity and concentration was established according to a calibration table made prior to the experiment. The concentration variation of NaCl solution was continuously monitored at the outlet of the unit cell by using conductivity meter.
In these experiments, the salt removal capacity and salt removal efficiency were defined as follow, respectively:

$$Salt\ removal\ capacity\left(\mu mol\, g^{-1}\right) = \frac{(c - c_0)V}{58.44 \times M} \tag{1}$$

$$Salt\ removal\ efficiency(\%) = \frac{(c - c_0)}{c_0}\% \tag{2}$$

Where C and C_0 (mg/l) represent the final and initial concentration, respectively, and V is the volume of the container (ml), and M is the mass of carbon (g).

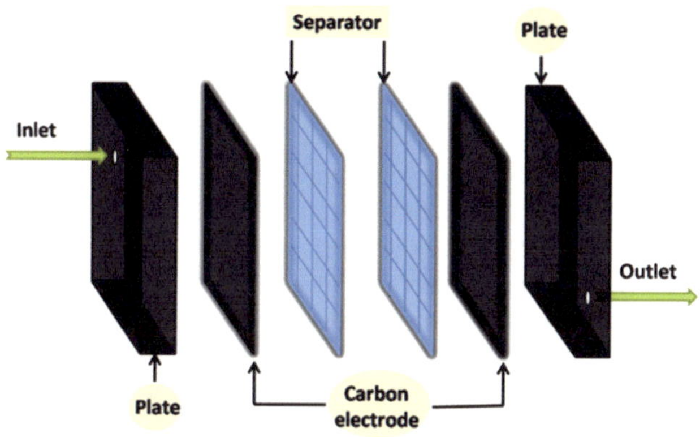

Fig. 2. Schematic diagram of CDI unit

3.1 Fabrication of carbon electrode

In this experiment, commercial ACs and CNTs were employed to fabricate the CDI electrodes. The specifications of as-purchased CNTs are listed in Table 1. The AC/CNTs, graphite as conductive material and PTFE as binder were used to fabricate the CDI electrodes. Their respective percentages in the final electrodes are 72%, 20% and 8%. Each electrode was 70 mm wide × 140 mm long × 0.3 mm thick, and had a flow-through hole with a diameter of 4 mm. To achieve adhesion between the carbon mixture and graphite layer, the raw mixture of powders (carbon material, graphite powder and PTFE powder) should be grind for several hours. Ethanol (10-20 ml) was added dropwise to the mixture to make it moist and it was then pressed onto graphite sheet.

Parameters	Double walled carbon nanotube	Single walled carbon nanotube
Diameter	< 5nm	< 2nm
Length	5 ~ 15μm	5 ~ 15μm
Purity of CNTs	≥ 90%	≥ 90%
Ash	≤ 2wt%	≤ 2wt%
BET	> 400m²/g	> 400m²/g
Amphorous Carbon	< 5%	< 5%

Note: the actual specific surface areas of the AC, DWCNTs and SWCNTs characterized by means of a N_2 adsorption system were 999, 415 and 455 m^2/g, respectively. The dominant peak present in the pore size distribution was below 1 nm for AC and around 2 nm for CNTs.

Table 1. Specifications of as-purchased CNTs

3.2 Characterization of carbon electrode materials

The structures of the electrode materials were characterized by transmission electron microscopy (TEM, CM200) imaging. The N_2 adsorption-desorption isotherms of the carbon materials were performed at -196°C on a Belsorp system (BEL JAPAN, INC). The pore size distribution curve was calculated by the Barrett-Joyner-Halenda (BJH) method from the desorption branch. The specific surface area was calculated from the adsorption data in the relative pressure interval from 0.04-0.2 using the BET method. The total volume (V) was estimated from the amount adsorbed at a relative pressure of 0.98. Dubinin-Radushkevich

(DR) theory was employed for estimating the micropore volume (V_{mi}), and the as-plot method was used for the external surface area (S_e) and the micropore surface area (S_{mi}). The mesopore fraction was obtained from ($V-V_{mi}$).

3.3 Desalination performance of various carbon electrodes

Fig. 3 presents the conductivity trends at various electrical voltages in AC, DWCNTs and SWCNTs, respectively. As expected, the high voltage resulted in high salt removal efficiency because of high electrostatic force between the relevant ions and the polarized electrode. In all cases, the conductivity reduction becomes much smaller after 30 minutes during adsorption process, implying that the electrodes are close to saturation. It is also observed that the slopes of the adsorption curves in (b) and (c) are higher than that in (a) at corresponding working voltages, indicating that the ions can be transferred easily in CNTs (both in DWCNTs and SWCNTs) due to the superior conductivity of CNTs. This leads to a higher ion adsorption rate. Furthermore, the regeneration curve shows that all electrodes could be regenerated efficiently without electrodialysis occurring at the voltage of 1.2 V. Fig.3 (d) illustrates the salt removal capacity with respect to electrical voltage by using AC, DWCNTs and SWCNTs based CDI. At each given voltage, the salt removal capacity follows the order of SWCNTs > DWCNTs > AC. This can be explained in terms of their respective

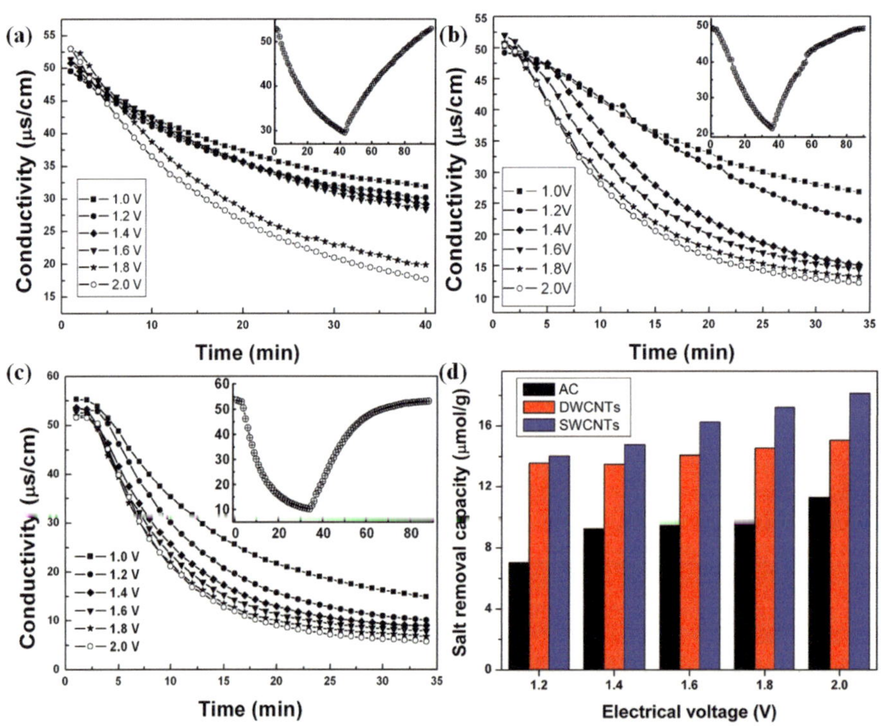

Fig. 3. The conductivity transient in (a) AC (b) DWCNTs and (c) SWCNTs based CDI, respectively. Inset is the corresponding adsorption-desorption curves at an applied voltage of 1.2 V. (d) presents the corelation of salt removal capacity over electrical voltage with an initial conductivity of 80 µS/cm.

pore structures. Although AC has the biggest specific surface area (999 m^2/g) among the three targeted materials (compared to 455 m^2/g for SWCNTs and 415 m^2/g for DWCNTs), but a large fraction of its pores are ascribed to the micropore range according to the IUPAC classification and these are not accessible to ions. It even has a negative effect on salt removal due to the cut-off pore effect. In case of CNTs, the dominant pore sizes are concentrated on 2-3 nm which are ascribed to the mesopore range (2-50 nm) and thereby it is beneficial to electrosorption. In addition, the conductivity at the electrode materials also plays an important role in CDI. Thus, the salt removal strongly depends on many factors, and the most important influence factors are pore size distribution as well as conductivity.

Regeneration is another important issue that needs to be considered in evaluating CDI electrode materials. Fig.4 presents the regeneration properties of AC, DWCNTs and SWCNTs. It is noted that all carbon electrodes can be regenerated very well by shorting the circuit. SWCNTs in particular can be regenerated at the level of initial concentration in a very short time. It takes much longer in case of AC electrodes. This can probably be attributed to the weak conductivity and large fraction of micropores. The SWCNTs would be a good candidate electrodes material for CDI. It will also be employed to fabricate MCDI electrode in the following study on MCDI.

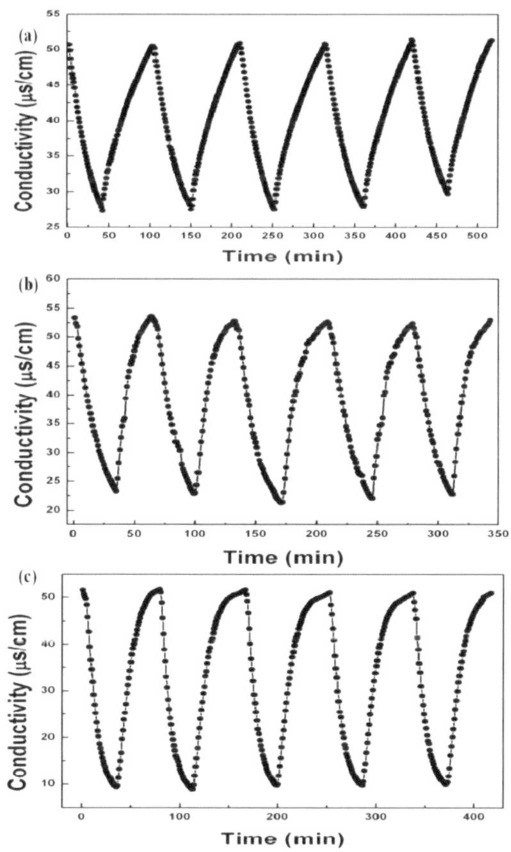

Fig. 4. Regeneration properties of (a) AC, (b) DWCNTs and (c) SWCNTs at an initial salt conductivity of 55 µS/cm

3.4 Comparison

The TEM images in the Fig. 5 show the ordered porous texture of ordered mesoporous carbon (OMCs, the corresponding CDI data from ref.31), the single and double wall pore structure of CNTs and the random irregular pore arrangement in AC.

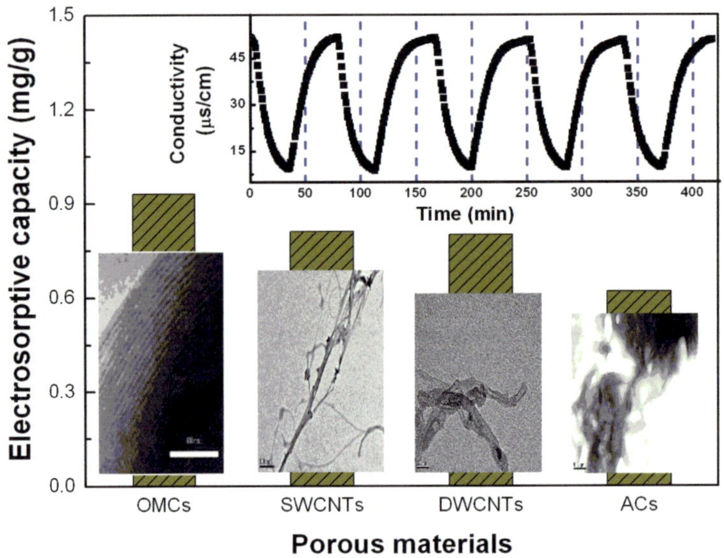

Fig. 5. Comparison of porous carbon materials

Table 2 summarises the characteristics of these four types of porous carbon materials. OMCs have the highest BET surface area of 1491 m^2/g, followed by Activated carbon of 845 m^2/g, DWCNTs of 413m^2/g and SWCNTs of 454m^2/g. Although ACs has a higher BET than CNTs, its mean pore size is smaller (2.1nm) than that of both SWCNTs (5.1nm) and DWCNTs (4.8nm). It suggests that ACs have a large amount of micropores whereas OMCs and CNTs both have more mesopores. These differences in pore characteristics have influence on the electrosorptive removal of salt ions.

Sample	S_{BET} (m^2/g)	S_{mi} (m^2/g)	V_t (cm^3/g)	D_{me} (nm)
OMC-S	1491	206	1.589	3.7
AC	845	579	0.605	2.1
DWCNTs	413	307	0.523	5.1
SWCNTs	454	293	0.543	4.8

Table 2. Characteristics of porous carbon materials

During the CDI test, the brackish water flowed through the cell by a peristaltic pump. The experimental results showed that their electrosorptive capacities are in the order of OMCs >SWCNTs>DWCNTs>ACs, ie. 0.93, 0.81, 0.80, 0.62 mg/g respectively (Fig. 5). This is attributed to their differences in pore size distribution, pore pattern arrangement and specific surface areas. OMCs and CNTs have pore size that are more suitable for electrosorption, and more ordered structure that facilitate the easy movement for adsorption and desorption of Na^+ and Cl^-. In contrast, some surface areas of ACs appear to be unavailable for ion adsorption as the pores are too small to form electrical double layer. In addition, the randomly arranged pore networks of ACs may contribute to the difficulty in the movement of the ions.

3.5 Graphene nanoflakes as electrodes in CDI

3.5.1 Preparation of graphene nanoflakes

Graphite oxide (GO) powders were synthesized by a modified Hummers method [39]. 2.5 g of graphite powder (<20µg, Sigma–Aldrich, Inc.) was added into an 353 K solution of a concentrated nitric acid and sulphuric acid mixture with volume ratio of 1:1 (120 ml each), the suspension was mixed for 4.5 h. After that, the mixture was then cooled down to the room temperature and diluted with 0.5 l of deionized (DI) water and left overnight. Later, the mixture was filtered and dried in ambient condition for 24 h. The reaction vessel contained 120 ml of concentrated H_2SO_4 and pre-treated graphite powder was immersed in an ice bath, and potassium permanganate was added slowly. Successively, the mixture was stirred for 2 h, then, after the dilution with DI water, 20 ml 30% H_2O_2 was added to the mixture. The colour of mixture changed into brilliant yellow along with bubbling. Finally, the mixture was filtered and washed with HCl aqueous solution (1:10 in volume) to remove residual metal ions, and washed with DI water until pH equalled 7. Exfoliation was carried out by adding hydrazine and mixing for 24 h at a temperature of 353 K. Successively, the as-prepared graphene suspension was washed several times and filtered carefully, followed by air dry at 333 K, the GNFs obtained were ready to use.

The fabrication of the electrode involved using 72% of the as-synthesized GNFs as electrode material, 20% of graphite as conductive material, and 8% of PTFE powder as binder. Each electrode was pressed into the dimension of 70 mm wide × 140 mm long × 0.3 mm thick, and had a flow-through hole with a diameter of 4 mm.

3.5.2 Effect of working conditions on electrsorptive performance of GNFs.

The key factors that affect the electrosorptive performance of Na^+ on the surface of GNFs include the flow rate and bias potential. As expected, higher voltage leads to higher electrosorptive capacity because of stronger Coulombic interaction. Fig.6 (a) shows the electrosorptive performance of GNFs at different bias potentials, and the inset represents the electrosorptive capacity as a function of voltage. With the voltage increased from 0.8 to 2.0 V, the electrosorptive capacity is gradually increased from 4.44 µmol/g to 23.18 µmol/g, which is nearly a 5-fold increase. When the applied voltage is at 2.0 V, no visible gas bubbles were observed, indicating that no water electrolysis was taking place. When taking both electrosorptive capacity and energy consumption into account, the optimum working voltage for GNFs based CDI was 2.0 V. Fig.6 (b) depicts the variation of conductivity in solution with flow rate, and the inset demonstrates the correlation between the electrosorptive capacity and flow rate. It is clearly observed from the figures that at a lower flow rate, for example, 15 mL/min, GNFs has a lower electrosorptive capacity of 9.03 µmol/g. When the flow rate was increased from 15 mL/min to 25 mL/min, the electrosorptive capacity reached the maximum value of 13.9 µmol/g. However, the electrosorptive capacity of GNFs electrode was reduced from 13.9 µmol/g to 8.49 µmol/g when the flow rate was further increased. This was due to the equilibrium between electrostatic force and the driving force in the flow rate of 25 mL/min.

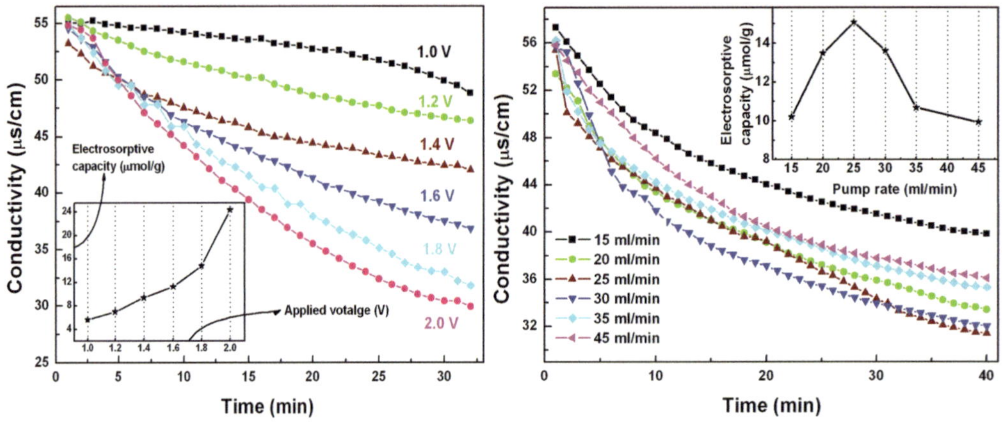

(a) (b)

Fig. 6. The electrosorption of Na^+ onto GNFs electrode at (a) different bias potentials, (b) different flow rate, respectively, 295 K [40].

3.5.3 Effect of initial concentration on electrsorptive performance of GNFs

To investigate the electrosorption behaviour of GNFs, the experiment was carried out at different initial concentrations, and thereby the electrosorptive isotherm was obtained. The electrosorption of Na^+ onto the GNFs electrode was evaluated at a constant temperature of 298 K for the isotherm as well as the kinetic models. The initial concentrations of the NaCl solutions were 25, 50, 100, 250, 400 and 500 mg/l, respectively, as shown in Fig.7. Langmuir Equation (3) and Freundlich Equation (4) were used to validate the experimental data for electrosorption of Na^+ onto GNFs, respectively.

$$q = \frac{q_m K_L C}{1 + K_L C} \tag{3}$$

$$q = K_F C^{1/n} \tag{4}$$

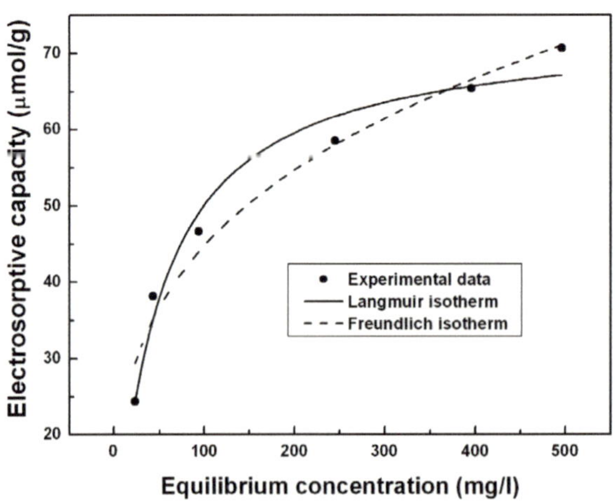

Fig. 7. The electrosorption isotherm of Na^+ onto GNFs electrode at 2.0 V [40]

Where C is the equilibrium concentration (mg/L), q is the amount of adsorbed Na^+ (in micromoles per gram of GNFs), q_m is the maximum adsorption capacity corresponding to complete monolayer coverage. It was found that the Langmuir isotherm correlated better with the experimental data according to the r^2. This phenomenon suggested that the monolayer adsorption was dominant during the electrosorption process. The parameter q_m in the Langmuir isotherm model is considered as the maximum adsorption capacity, as mentioned above. Thus the equilibrium electrosorption capacity, at polarization of 2 V and the flow rate of 25 ml/min, was 73.47 μmol/g, which is higher than previously reported data.

3.5.4 Comparison of GNFs and AC electrodes

At the applied voltage of 2.0 V, comparative results of the electrosorptive capacity as well as BET surface area between the GNFs and AC based CDI in the same experimental conditions are shown in Fig.8 (a). Although having the larger surface area (989.54 m^2/g) than GNFs

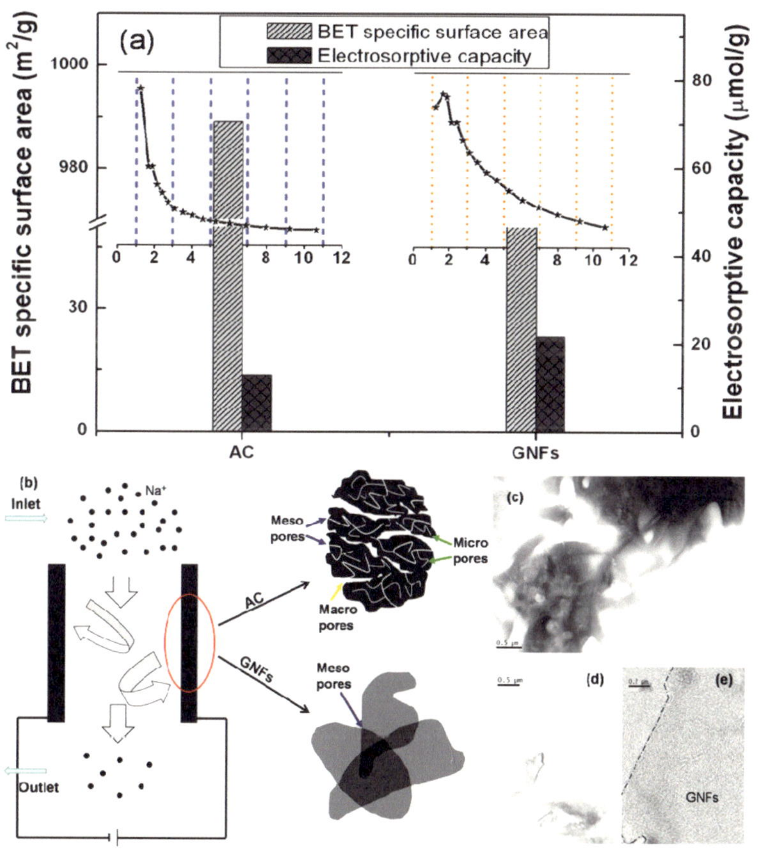

Fig. 8. (a) Comparison of electrosorptive performance by employing GNFs and AC at the same experimental condition, the pictures at top-left and top-right depict the pore size distribution of AC and GNFs below 10 nm, respectively. (b) Mechanism of CDI employing AC and GNFs electrode. TEM observation images of AC (c) and GNFs in low (d) and high (e) magnification, respectively [40].

(222.01 m^2/g), AC has an electrosorptive capacity of only 13.73 μmol/g, which is much lower than that of GNFs (whose electrosorptive capacity is 23.18 μmol/g). This can be attributed to the fact that GNFs have an interlayer structure which is more accessible for ions, while AC has a large fraction of inaccessible small micropores. As a result, the effective surface area of GNFs is higher than that of AC. Fig.8 (b) illustrates the principle of CDI and the mechanism of how ions are electrosorbed onto AC and GNFs electrodes, respectively. TEM images of AC and CNFs are used to confirm the hypothesis. The TEM image of GNFs indicates that GNFs are aggregated together, showing a semi-transparent flower shape interlayer pattern. It also shows that GNFs are homogenous flakes with micro-size that are beneficial to ions accessing and are adsorbed on the surface of the flakes. In contrast, the structure of AC on the TEM image shows that it presents a beehive-type pore structure so that the ions cannot access to the inner pores and therefore a high electrosorptive capacity is difficult to achieve. In addition, it is believed that the conductivity of carbon materials also plays a vital role in the electrosorptive process. Having higher bulk conductivity is equivalent to having a higher applied voltage between the two electrodes. It is reported that the conductivity of graphene prepared via chemical approaches is normally above 200 S/m. It is much higher than that of commercial AC whose electrical conductivity is normally between the orders of 10^{-6} and 10^{-3} S/m. Further, several experiments regarding conductivity measurement were also performed by measuring GNFs and AC film, respectively and found that the conductivity of GNFs was much better than that of AC. Another possible reason for the high conductivity for GNFs is the presence of conductive graphitized chunks in the GNFs, which was caused by the incomplete grapheme preparation from the graphite precursor. Thus, considering both effective specific surface area and electrical conductivity, it is believed that the GNFs with high specific surface area have the potential as an excellent candidate electrode material for the CDI.

3.6 Removal efficiency of MCDI for the removal of salt from synthetic brackish water

The following experiments were conducted to investigate the salt removal efficiencies of the MCDI device. The initial salt concentration for MCDI testing was 110 μS/cm, as the MCDI is more efficient in reducing the conductivity.

Fig.9 (a) depicts the typical SWCNTs based MCDI process at different working voltages and the corresponding salt removal efficiencies. The salt removal efficiency increased from 91.3 to 98.1% by varying the polarized bias from 0.8 to 1.6 V. However, by taking into account both energy consumption and salt removal efficiency, the optimum working voltage for SWCNTs based MCDI is recommend as 1.2 V. Another useful piece of information relates to MCDI is shown in Fig.9 (a). It is observed that the slops of the adsorption curve are very steep which indicates a high adsorption rate constant and therefore confirms the theory that the co-ion effects can be significantly reduced by incorporation of ion-exchange membrane. Furthermore, the high adsorption rate constant also indicates that the ions access the electrode faster and thus the desalination process can be completed in a very short time as well as the corresponding regeneration process. In addition, several experiments have also been performed to check the regeneration property of SWCNTs based MCDI. Fig.8 implies that the salt ions in the testing solution can be nearly depleted during the adsorption process and the electrodes are easily regenerated in the desorption process for three adsorption-desorption cycles.

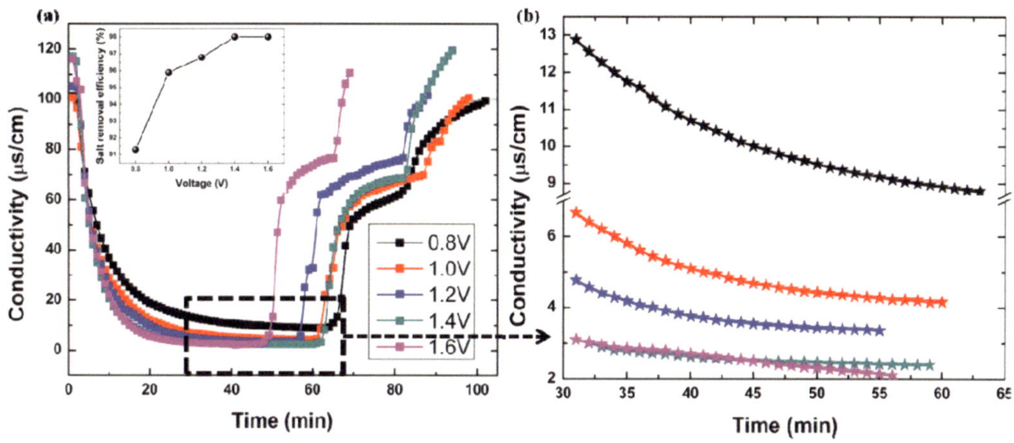

Fig. 9. SWCNTs based MCDI processes at different working voltages (a) (inset shows the corresponding salt removal efficiency)and (b) soprtion performance from 30 to 60 minutes. The initial salt concentration was 110 μS/cm [41]

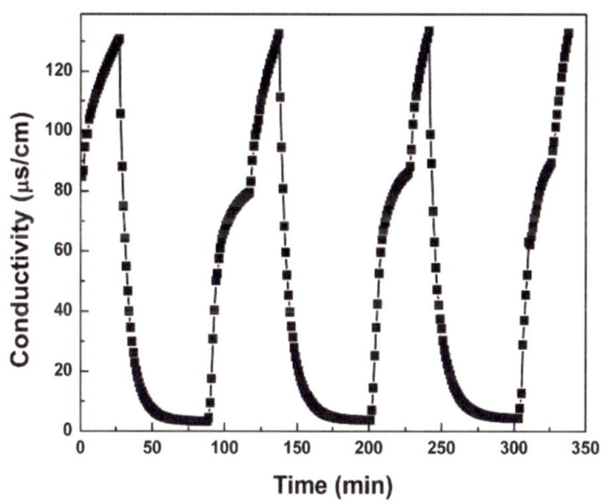

Fig. 10. Regeneration perperty of SWCNTs based MCDI process at an initial salt concentration of 130 μS/cm [41]

To further investigate the desalination performance of SWCNTs based MCDI at much higher salt concentrations, several experiments were conducted at an initial conductivity of 500 μS/cm and 1000 μS/cm respectively.

Fig.11 (a) shows the conductivity reduction in term of adsorption time at an initial salt conductivity of 500 μS/cm. It is noted that the salt removal efficiency increased from 55.4 to 88.2% with the increase of applied voltage from 0.8 to 1.6 V but was still much lower than corresponding salt removal efficiency with lower initial conductivity, for instance, 100 μS/cm. Particularly, when the initial conductivity was further increased to 1000 μS/cm, the salt removal efficiency continued to decrease as shown in Fig.11 (b). However, the regeneration curve keeps very well by compared with results at low conductivity.

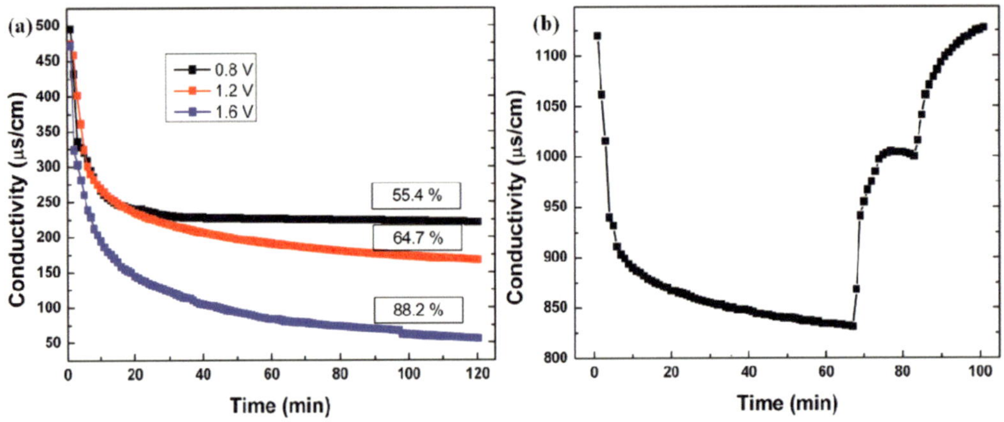

Fig. 11. Desalination performances of SWCNTs based MCDI process at (a) an initial salt conductivity of 500 μS/cm by changing working voltage from 0.8 to 1.6 V (b) an initial conductivity of 1000 μS/cm with applied voltage of 1.2 V [41].

3.7 Comparing desalting results between CDI and MCDI processes

In this section, the SWCNTs based CDI and corresponding MCDI are compared on their desalination performance at the same experimental conditions. The result is showed in Fig.12 which draws the typical adsorption-desorption process based on two devices.

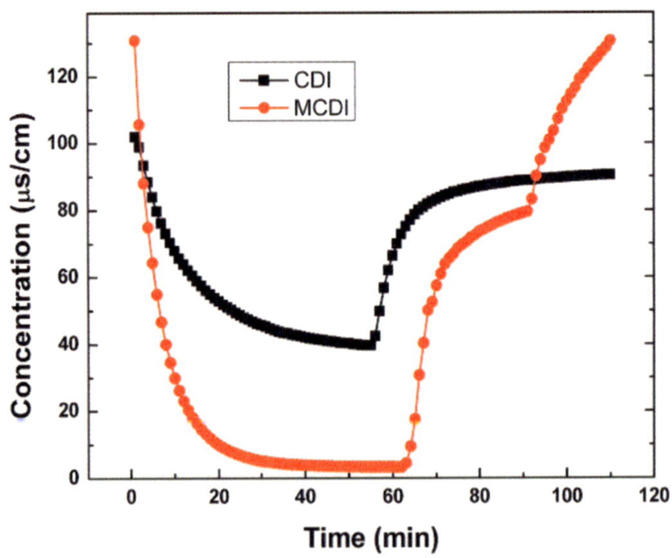

Fig. 12. Comparative desalination result by SWCNTs based CDI and MCDI. (the initial conductivity is around 100 μS/cm, the electrical voltage is of 1.2 V, the other experimental conditions are according to the experimetal section) [41]

In Fig.12, the salt conductivity dramatically decreased within first 30 minutes and gradually approached to saturation and then increased to the initial level once the electricity was cut off. Apparently, the salt removal efficiency of CDI (about 60%) is much less than that of

MCDI (about 98%). As discussed in the preceding section, the salt removal efficiency would be affected by co-ions effect. On the other hand, the MCDI is much more beneficial for selective ion transfer as a result of the presence of the ion-exchange membrane and this can be confirmed from the slope of the adsorption curve as well.

4. Conclusions

The electrosorption experiment results of different porous carbon electrodes showed that their electrosorptive capacities are in the order of OMCs >SWCNTs>DWCNTs>ACs, ie. 0.93, 0.81, 0.80, 0.62 mg/g respectively. This is attributed to their differences in pore size distribution, pore pattern arrangement and specific surface areas. It was found that the ion sorption onto these mentioned materials follows a Langmuir isotherm, indicating monolayer adsorption. Finally, the regeneration property of both CNTs and AC were investigated through charge-discharge experiment and it was found that their electrode regenerations were effective. The MCDI is much more beneficial for selective ion transfer as a result of the presence of the ion-exchange membranes and this is confirmed by the more efficient ion adsorption by MCDI than CDI.

Graphene nanoflakes (GNFS) were synthesised by modified Hummer's method. The obtained GNFs were made into electrodes for CDI. Its electrosorptive performance was much better than commercial activated carbon (AC), suggesting to be a good candidate of electrode material in capacitive deionisation application.

5. Acknowledgement

The authors acknowledge the research grants received from Australian Research Council Linkage Grant (LP0883282), Water Quality Research Australia and National Centre of Excellence in Desalination.

6. References

[1]Borsani, R.; & Rebagliati, S. (2005). Fundamentals and costing of MSF desalination plants andcomparison with other technologies. *Desalination,* Vol.182, No.1-3, pp. 29-37, ISSN 0011-9164

[2] Anderson, M.A.; Cudero, A.L.; & Palma, J. (2010). Capacitive deionization as an electrochemical means of saving energy and delivering clean water. Comparison to present desalination practices: Will it compete? *Electrochimica Acta* Vol.55, No.12, pp. 3845-3856, ISSN 0013-4686

[3] Conway, B.E. (1999). *Electrochemical supercapacitors,* Kluwer, ISBN 978-030-6457-36-4, New York, American.

[4] Caudle, D.D.; Tucker, T.H.; Cooper, J.L.; Arnold, B.B.; & Papastamataki, A. (1966). Research report oklahoma university research institute

[5] Johnson, A.M.; & Newman, J. (1971). Desalting by means of porous carbon electrodes. *Journal of the electrochemical society,* Vol.118, No.3, pp. 510-517, ISSN 0013-4651

[6] Oren, Y. (2008). Capacitive deionization (CDI) for desalination and water treatment – past, present and future (a review). *Desalination,* Vol. 228, No.1-3, pp. 10-29, ISSN 0011-9164

[7] Oren, Y. & Soffer, A. (1983). Water desalination by means of electrochemical parametric pumping, separation properties of a multistage column. *Journal of Applied Electrochemistry,* Vol.13, No.4, pp. 489-505, ISSN 0021-891X
[8] Oren, Y. & Soffer, A. (1978). Electrochemical parametric pumping. *Journal of the Electrochemical Society,* Vol.125, No.6, pp. 869-875, ISSN 0013-4651.
[9] Oren, Y. & Soffer, A. (1986). The electrical double layer of carbon and graphite electrodes Part III. Charge and dimensional changes at wide potential range. *Journal of Electroanalytical Chemistry,* Vol.206, No.1-2, pp. 101-114, ISSN 1572-6657.
[10] Cohen, H.; Soffer, A. & Oren, Y. (1987). Adsorption of suspended alumina particles on fibrous carbon and graphite electrodes, *Journal of Colloid and Interface Science,* Vol.120, No.1, pp. 272–280, ISSN 0021-9797
[11] Oren, Y.; Tobias, H. & Soffer, A. (1983). Removal of bacteria from water by Electroadsorption on porous carbon electrodes. *Bioelectrochemistry and Bionergetics,* Vol.11, No.4-6, pp. 347-354, ISSN 1567-5394
[12] Oren, Y.; Tobias, H. & Soffer, A. (1984). The electrical double layer of carbon and graphite electrodes, part I: Dependence on electrolyte type and concentration. *Journal of Electroanalytical Chemistry,* Vol.162, No.1-2, pp. 87-99, ISSN 1572-6657
[13] Oren, Y. & Soffer, A. (1985). The electrical double layer of carbon and graphite electrodes, part II: Fast and slow charging processes, *Journal of Electroanalytical Chemistry,* Vol.186, No.1-2, pp. 63-77, ISSN 1572-6657
[14] Farmer, J.C.; Fix, D.; Mack, G.; Pekala, R. & Poco, J. (1996). Capacitive deionization of NH_4ClO_4 solutions with carbon aerogel electrodes. *Journal of Applied Electrochemistry,* Vol.26, No.10, pp. 1007-1018, ISSN 0021-891X
[15] Farmer, J.C.; Fix, D.; Mack, G.; Pekala, R. & Poco, J. (1996). Capacitive Deionization of NaCl and NaNO3 Solutions with Carbon Aerogel Electrodes. *Journal of the Electrochemical Society,* Vol.143, No.1, pp. 159-169, ISSN 0013-4651
[16] Farmer, J.C.; Bahowick, S.M.; Harrar, J.E.; Fix, D.V.; Martinelli, R.E.; Vu, A.K. & Carroll, K.L. (1997). Electrosorption of chromium ions on carbon aerogel electrodes as a means of remediating ground water. *Energy and Fuels,* Vol.11, No.2, pp. 337-347, ISSN 0887-0624
[17] Mayer, S.T.; Pekala, R.W. & Kaschmitter, J.L. (1993). The aerocapacitor: an electrochemical double-Layer energy-storage device. *Journal of the Electrochemical Society,* Vol.140, No.2, pp. 446-451, ISSN 0013-4651
[18] Ryoo, M.W.; Kim, J.H. & Seo, G. (2003). Role of titania incorporated on activated carbon cloth for capacitive deionization of NaCl solution. *Journal of Colloid and Interface Science,* Vol.264, No.2, pp. 414–419, ISSN 0021-9797
[19]Oh, H.J.; Lee, J.H.; Ahn, H.J.; Jeong, Y.; Kim, Y.J. & Chi, C.S. (2006). Nanoporous activated carbon cloth for capacitive deionization of aqueous solution. *Thin Solid Films,* Vol.515, No.1, pp. 220 – 225, ISSN 0040-6090
[20] Ahn, H.J.; Lee, J.H.; Jeong, Y.; Lee, J.H.; Chi, C.S. & Oh, H.J. (2007). Nanostructured carbon cloth electrode for desalination from aqueous solutions. *Materials Science and Engineering A,* Vol.449–451, No.25, pp. 841–845, ISSN 0921-5093
[21] Park, K.K.; Lee, J.B.; Park, P.Y.; Yoon, S.W.; Moon, J.S.; Eum, H.M. & Lee, C.W. (2007). Development of a carbon sheet electrode for electrosorption desalination. *Desalination,* Vol.206, No.1-3, pp. 86–91, ISSN 0011-9164

[22] Dai, K.; Shi, L.Y.; Zhang, D.S. & Fang, J.H. (2006). NaCl adsorption in multi-walled carbon nanotube/active carbon combination electrode. *Chemical Engineering Science,* Vol.61, No.2, pp. 428 - 433, ISSN 0009-2509

[23] Zhang, D.S.; Shi, L.Y.; Fang, J.H.; Dai, K. & Li, X.K. (2006). Preparation and desalination performance of multiwall carbon nanotubes. *Materials Chemistry and Physics,* Vol.97, No.2-3, pp. 415–419, ISSN 0254-0584

[24] Wang, X.Z.; Li, M.G.; Chen, Y.W.; Cheng, R.M.; Huang, S.M.; Pan, L.K. & Sun, Z. (2006). Electrosorption of ions from aqueous solution with carbon nanotubes and nanofibers composite film electrodes. *Applied Physics Letters,* Vol.89, No.5, PP. 053127-3, ISSN 0003-6951

[25] Wang, X.Z.; Li, M.G.; Chen, Y.W.; Cheng, R.M.; Huang, S.M.; Pan, L.K. & Sun, Z. (2006). Electrosorption of NaCl solutions with carbon nanotubes and nanofibers composite film electrodes. *Electrochemical Solid-State Letters,* Vol.9 No.9, pp. E23-26, ISSN 1099-0062

[26] Zhang, D.S.; Shi, L.Y.; Fang, J.H.; Dai, K. & Li, X.K. (2006). Influence of carbonization of hot-pressed carbon nanotube electrodes on removal of NaCl from saltwater solution. *Materials Chemistry and Physics,* Vol.96, No.1, pp. 140–144, ISSN 0254-0584

[27] Pan, L.K.; Wang, X.Z.; Gao, Y.; Zhang, Y.P.; Chen, Y.W. & Sun, Z. (2009). Electrosorption of anions with carbon nanotube and nanofibre composite film electrodes. *Desalination,* Vol.244, No.1-3, pp. 139-143, ISSN 0011-9164

[28] Gao, Y.; Li, H.B.; Pan, L.K.; Zhang, Y.P.; Chen, Y.W. & Sun, Z. (2009). Electrosorption behavior of cations with carbon nanotubes and carbon nanofibres composite film electrodes. *Thin Solid Fims,* Vol.517, No.5, pp. 1616-1619, ISSN 0040-6090

[29] Li, H.B.; Pan, L.K.; Zhang, Y.P. & Sun, Z. (2009). Ferric ion adsorption and electrosorption by carbon nanotubes and nanofibres films. *Water Science and Technology,* Vol.59, No.8, pp. 1657-1665, ISSN 0273-1223

[30] Li, H.B.; Pan, L.K.; Zhang, Y.P.; Zou, L.D.; Sun, C.Q.; Zhan, Y.K. & Sun, Z. (2010). Kinetics and thermodynamics study for electrosorption of NaCl onto carbon nanotubes and carbon nanofibres electrodes. *Chemical Physics Letters,* Vol.485, No.1-3, pp. 161-166, ISSN 0009-2614

[31] Li, L.X.; Zou, L.D.; Song, H.H. & Morris, G. (2009). Ordered mesoporous carbons synthesized by a modified sol–gel process for electrosorptive removal of sodium chloride. *Carbon,* Vol.47, No.3, pp. 775 -781, ISSN 0008-6223

[32] Zou, L.D.; Li, L.X.; Song, H.H. & Morris, G. (2008). Using mesoporous carbon electrodes for brackish water desalination. *Water Research,* Vol.42, No.8-9, pp. 2340 - 2348, ISSN 0043-1354

[33] Lim, J.A.; Park, N.S.; Park, J.S. & Choi, J.H. (2009). Fabrication and characterization of a porous carbon electrode for desalination of brackish water. Desalination, Vol.238, No.1-3, pp. 37–42, ISSN 0011-9164

[34] Richard, T. M.; Costas, T.; James O. Kiggans Jr.; Shannon M.M.; David W.D. & Dai, S. (2010). Hierarchical ordered mesoporous carbon from phloroglucinol-glyoxal and its application in capacitive deionization of brackish water. *Journal of Materials Chemistry,* Vol.20, No.39, pp. 8674-8678, ISSN 0959-9428

[35] Lee, J.B.; Park, K.K.; Eum, H.M. & Lee, C.W. (2006). Desalination of a thermal power plant wastewater by membrane capacitive deionization. *Desalination,* Vol.196, No.1-3, pp. 125–134, ISSN 0011-9164

[36] Li, H.B.; Gao, Y.; Pan, L.K.; Zhang, Y.P.; Chen, Y.W. & Sun, Z. (2008). Electrosorptive desalination by carbon nanotubes and nanofibres electrodes and ion-exchange membranes. *Water Research*, Vol.42, No.20, pp. 4923 – 4928, ISSN 0043-1354

[37] Kim, Y.J. & Choi, J.H. (2010). Improvement of desalination efficiency in capacitive deionization using a carbon electrode coated with an ion-exchange polymer. *Water Research*, Vol.44, No.3, pp. 990 – 996, ISSN 0043-1354

[38] Biesheuvel, P.M. & Van der Wal, A. (2010). Membrane capacitive deionization. *Journal of Membrane Science*, Vol.346, No.2, pp. 256-262, ISSN 0376-7388

[39] William, S.; Hummers, J. & Offeman, R.E. (1958). Preparation of graphitic oxide. *Journal of the American Chemical Society*, Vol.80, No.6, pp. 1339–1341, ISSN 0002-7863

[40] Li, H.B.; Zou, L.D.; Pan, L.K. & Sun, Z. (2010). Novel Graphene-Like Electrodes for Capacitive Deionization. *Environnemental Science and Technology*, Vol.44, No.22, pp. 8692–8697, ISSN 0013-936X

[41] Li, H.B. & Zou, L.D. (2011). Ion-exchange Membrane Capacitive Deionization: A New Strategy for Brackish Water Desalination, *Desalination*, Accepted, ISSN 0011-9164

16

Copper Ions Biosorption Properties of Biomass Derived from Algerian Sahara Plants

Abdelkrim Cheriti[1], Mohamed Fouzi Talhi[1], Nasser Belboukhari[1] and Safia Taleb[2]

[1]*Phytochemistry & Organic Synthesis Laboratory, University of Bechar, Bechar*
[2]*Catalysis & Materials Laboratory, University D. Liabes, Sidi Bel Abbes, Algeria*

1. Introduction

Biosorption of heavy metals from aqueous solutions is a relatively new technology for the treatment of industrial wastewater, which utilized naturally occurring waste materials derived from biomass. Generaly Heavy metals eliminated by biosorption are usually classified on three categories: toxic metals (such as Hg, Cr, Pb, Zn, Cu, Ni, Cd, As, Co, Sn, etc.), precious metals (such as Pd, Pt, Ag, Au, Ru etc.) and radionuclides (such as U, Th, Ra, Am, etc.), whose specific weight is usually more than 5.0 g/cm^3 [Bishop, 2002; Volesky, 1990; Wang & Chen, 2006].

The heavy metal ions are stable and persistent environmental contaminants since they cannot be degraded and destroyed. These metal ions can be harmful to aquatic life and water contaminated by toxic metal ions remains a serious public health problem for human health. Numerous methods exist to remove heavy metals ions from aqueous solutions by chemical precipitation or by activated carbon as the most used adsorbent, nevertheless it is relatively expensive [Demirbas, 2008; Gabaldon et al, 1996].

The toxicity of heavy metals is apparent in reducing growth and development in microorganisms and plants, and seriously harming the health of animals and humans. In particular, heavy metals may disrupt the normal function of the central nervous system and cause changes in the blood content, and adversely affect the function of lungs, kidneys, liver and other organs. The long-term action of heavy metals may cause the development of cancer, allergy, dystrophy, physical and neurological degenerative processes, Alzheimer's and Parkinson's diseases. However, in small amounts, heavy metals are indispensable for many organisms, but their enhanced doses induce acute or chronic poisoning [Kvesitadze et al, 2006]

Adsorbent materials derived from Biomass can be used for the effective removal and recovery of heavy metal ions from wastewater: from algae [Farooq et al, 2010; Hamdy, 2000], fungi [Kapoor et al., 1999], bacteria [Ozturk, 2007], sea-weeds [Elangovan et al., 2008], some higher plants (Rahman et al., 2005), and agricultural wastes[Demirbas, 2008; Park et al, 2006].

Several reviews articles have been published about the use of certain cellulosic agricultural waste materials for the removal of heavy metal ions [Sud et al., 2008], the use of microbial and certain plants derived biomass types [Ahluwalia & Goyal, 2007], use of Saccharomyces cerevisiae [Wang & Chen, 2006], the adsorption capacity of low cost adsorbents derived from agricultural waste for heavy metals removal [Febrianto et al, 2009; Kurniawan et al, 2006]. A number of other reviews are available in the literature [Davis et al., 2003; Lodiero et al., 2006; Nurchi & Villaescusa, 2008].

The major advantages of biosorption technology are its effectiveness in reducing the concentration of heavy metal ions to very low levels and the use of inexpensive biosorbent. Among these, is a rapid, reversible, independent of physiological constraints of living cells, no aseptic conditions required, chemical or biological sludge is minimized, economical and eco-friendly technology in contrast to traditional methods used for removal of heavy metals from wastewaters [Cheriti et al, 2010; Talhi et al, 2009]. However, there are certain disadvantages as irrespective of the value of the metal, it needs to be desorbed from the material to be further re-employed, the characteristics of the biosorbents can not be biologically controlled [Farooq et al, 2010].

Wastes containing metals are directly or indirectly discharged into the environment increasingly, especially in developing countries, having brought serious environmental pollution, and threatened biolife [Bishop, 2002; Wang, 2002; Wang et al, 2006]. Algeria, with more than 34 million inhabitants, is located in a semi arid region, and is increasingly confronted with the problem of the scarcity of water. The country is characterized by irregular rainfall with a generally variable pattern of distribution and the major environmental issues include: soil erosion from overgrazing and other poor farming practices, desertification, dumping of raw sewage, petroleum refining wastes, and other industrial effluents is leading to the pollution of rivers and coastal waters.

Continuing our effort on the valorization of Algerian Sahara plants as biomaterials for the biosorption of toxic heavy metals from water, We present in this chapter a brief review on the use of the locally available Sahara medicinal plants *Acacia raddiana, Anabasis aretioides, Retama raetam and Tamarix gallica* as an adsorbent for removal of copper ion from aqueous solution. Effect of operating conditions like Temperature, pH and initial metal concentration, on copper biosorption were investigated. The relation between the phytochemical composition (Polyphenol, alkaloids, terpenoids, carbohydrates...) of the plants and the percent of adsorption for copper ion was examined. Thus, biosorption of copper ion occurs as a result of ion exchange or complex formation between metal ions and functional groups (hydroxyl, amine, carboxyl...) on the cell surface of the biomass derived from Algerian Sahara plants

2. Characteristics and taxonomie of Sahara plants

Algeria is a large North African country located on the Mediterranean Sea. Three-quarters of the country is part of the Sahara Desert lying beyond the broad ranges of the Atlas Mountains which parallel the coast. In Sahara, the climate is hot and dry in the summer, and cooler with average annual rainfall of 50 mm in the winter, although there may be years where there is no rain at all. The highest temperature ranges between 40 and 50 °C, creating evaporation that far exceeds the amount that falls as rain and water is a serious constraint.

The flora of the central Sahara is very poor and estimated to include only 500 species [Le Houérou, 1990]. Vegetation is very contracted along the oued and the dayas with *Acacia sp, Anastatica hyrochuntina, Astragalus gombo, Calligonum sp., Calotropis procera., Luanaea*

arborescens, *Neurada procumbens, Panicum turgidum, Pistacia atlantica, Pithuranthos sp. and Tamarix sp.*, where there is sufficient ground water. Hammadas are covered by *Anabasis aretioides, Anrthirrnum ramosissimuma, Arthrophytum schmittianum, Gymnocarpos decander, Helianthemum sp., Ononis angustissima* and *Pithuranthos chloranrhus.* On sandy there are several plants, including *Aristida acutiflora, Calligonum comosum, Ephedra alata ssp alenda, Genista saharae, Oudneya africana, Panicum turgidim, Retama raetam, Rhanterium suaveolens and Zilla spinosa.* There is great seasonal diversity in the Saharan plant communities. Some evergreen perennials, such as *Luanaea arborescens, Pityranthus tortuosus* and *Stachys aegyptiaca*, may flower all year, whereas others have restricted flowering during the summer (*Zygophyllum coccineum*) or autumn (*Haloxylon salicornicum*), or other germinate in January during the wet season [Chehma, 2006; Ozenda, 1991; Quezel, 1965].
Plants used in this study (Figure 1) are common in Sahara and used by local population in traditional pharmacopeia.

(a) *Acacia raddiana* (Saoura region)

(b) *Anabasis aretioides* (Saoura region)

(c) *Retama raetam* (El Bayadh region)

(d) *Tamarix gallica* (Saoura region)

Fig. 1. Saharian Plants in their natural habitat

2.1 *Acacia raddiana* savi (Vernacular name: Talh, Mimosacea family) (Fig 1 a)

Acacia is a cosmopolitan genus containing in excess of 1350 species that together with the African and the Middle Eastern monotypic genus. In Algerian Sahara, *Acacia raddiana* is a medium umbrella-shaped tree 4-15 m tall, with round irregular crown and reddish bark. The flowers are arranged in heads. The legume is flattened and spirally twisted. Tree provide food and shelter for many desert animals and is a major source of livestock feed and firewood for the native people [Cheriti, 2000; Cheriti et al, 2006]. This specie is reported to tolerate annual precipitation of 10-100 mm, arid climates with temperatures as high as 50 °C and alkaline soils. The tree has been recommended for reclaiming dunes.

Gum from the tree is dispensed in water and used to treat ocular affections, jaundice, anthelmintic, antidiarrhoea, asthma and pulmonary diseases. Seeds, entire or powdered, are taken as antidiarrhoeic, vermifuge and dusted onto skin. Dried powdered bark used as disinfectant, for healing wounds and is a good source of tannin and used in tanning [Belakhdar, 1997; Cheriti et al, 2004; Cheriti, 2000; Ibn Baytar, 1992].

2.2 *Anabasis aretioides* Coss. & Moq. (*Syn. Fredolia aretioides* Coss. & Dur., Vernacular name: Dgaa, Chenopodaceae family) (Fig 1 b)

Anabasis aretioides is an endemic specie of South-west Algeria and South-east Morocco, grows on rocky and stony plateaux (reg and hamada). It's a vigorous cylindrical shrub, it looks like a huge cauliflower. The branches are very compact with sand in the interstices. The small fleshy leaves, not exceeding 5 mm, are opposite, very tightly attached to the stems, and blue-green. Each leaf ends in a small spine. The long roots that creep through vertical crevices enable the plant to thrive in severe climatic conditions where the annual rainfall does not exceed 100 mm. per year. The fruit is a small achene surrounded by transparent wings of the evergreen perianth [Ozenda, 1991].

In local traditional medicine, *A. aretioides* is used for rheumatism, as a diuretic, as an antidote and in treatment of gynaecological disorders [Belakhdar, 1997; Cheriti et al, 2004; Cheriti, 2000].

2.3 *Retama raetam* Forssk. (Vernacular name: R'tem, Fabaceae family) (Fig 1 c)

Retama raetam Forssk is an saharo-mediterranean species; it's a glabrescent large shrub that grows to about 3 m on sandy soils and in dry conditions (rainfall around 100 mm. per year), common in the Northern Africa and East Mediterranean region [Chehma, 2006; Quezel, 1965]. The leaves, which are very small (about 6-7 mm. long), simple, subsessile and narrow (only 1 mm. wide), drop quickly and the plant remains leafless for most of the year. The flowers are 8-10 mm. long, white and pea-like, appearing close to the stem in clusters of 3-15. The hairless grape-shaped seed pod (10-15 mm. diameters) contains one or two kidney-shaped seeds, which are about 6.5 mm. long and may be yellow, green, brown or black. The plant flowers from March to April and plays an important ecological role, it is widely used in dune stabilisation and soil fixation and as a good fuel wood.

In Algeria, *R. raetam* it is largely located in Sahara and arid steppe regions, provide food for many desert animals and is largely recommended by traditional herbal healers for treatment of various illness as diabetes, hypertension, gastro-intestinal disorders and as an antiseptic for wounds, skin rash and pruritus. Powdered branches mixed with honey are emetic, given as a purgative and vermifuge. The flowers are an important source of fodder for dromedaries [Belakhdar, 1997; Cheriti et al, 2006; Cheriti, 2000].

2.4 *Tamarix gallica L.* (Vernacular name: L'ariche, Tamaricaceae family) (Fig 1 d)

Tamarix gallica L. is a perennial small tree, densely ramified, abundant along streams and oued with a high salt content, rich clay soil and with a moderate sand accumulation. The purple-brown bark of this desert tree is initially smooth with numerous elongated lenticels, later developing shallow splits and becoming rough when mature. The leaves are tiny, grey-green, 1-3 mm. Flowering starts around March and lasts until May. The small flowers have 5 lavender pink or white petals 1.5-2 mm. long, numerous on long, very slender, spike-like racemes in terminal panicles. The fruit is a small dry capsule containing small cottony seeds[Quezel, 1965]. Its wood is used in local construction, for dyeing and as fuel.
The decoction of leaves and young branches are used in Saharan traditional medicine as a vermifuge, antihaemorrhoid and in treatment of diarrhoea, rheumatism and as an antialgetic for toothache and to cure dromedary galls [Belakhdar, 1997; Cheriti et al, 2006; Cheriti, 2004].

3. Biosorption of copper ion by different biomass

Recently, the surge of industrial activities has intensified more environmental problems as seen for example in the deterioration of several ecosystems due to the accumulation of dangerous polluants such as heavy metals [Park et al, 2006]. The heavy metal pollution has naturally become one of the most serious environmental problems today. Due to their persistence in nature, increased susceptibility to disease in man and animal (hepatic, kidney, nerves and the immune system damage and block functional vital groups.....) and toxic characteristics of heavy metals displayed as follows: The toxicity of heavy metals occurs even in low concentration of about 1.0–10 mg/L. Some strong toxic metal ions, such as Hg and Cd, are very toxic even in lower concentration of 0.001–0.1 mg/L; the toxicity can last for a long time in nature; some heavy metals even could be transformed from relevant low toxic species in to more toxic forms in a certain environment; the bioaccumulation by food chain could damage normal physiological activity and endanger human life; cannot be degraded by any methods including biotreatment [Wang and Chen, 2006; Wang, 2002].
Knowledge about toxicological effects of heavy metals on the environment and in drinking water is well recognized and therefore, it is inevitable to search for different methods to reduce water pollution. Conventional methods for removing heavy metal ions from aqueous solution have been studied, such as chemical precipitation, ion exchange, electrochemical treatment, membrane technologies and adsorption. Than, Copper (Cu) is one of the most important heavy metals often found in effluents discharged from industries and not biodegradable and travels through the food chain via bioaccumulation, which is potentially toxic. The excessive intake of copper results in its accumulation in the liver causing haemolysis, liver and kidney damage, irritation of the upper respiratory tract, chronic asthma, gastrointestinal disturbance, anemia and diarrhoea, Wilson desease and dermatitis [Cheriti et al, 2009; Kurniawan et al, 2006]. Thus, it becomes essential to search for different methods to reduce water pollution and to remove these heavy metals from wastewaters.
Biosorption is again promoted as a potential biotechnology for removal of heavy metals and related organic substances from waste streams and effluents. A wide range of microbial biomass types have been investigated in biosorption studies, including many kinds of macroalgae (seaweeds), plant materials (leaves, bark, sawdust), animal materials (hair, crustaceans) have also been studied [Gadd, 2009; Niua et al 2007; Zhang & Banks, 2006]. A common rationale is that 'waste' biomass will provide an economic advantage with low costs and availability in large quantities, easily regenerable, and cheap [Hashem, 2007].

Chemical modification of biomass may create derivatives with altered metal binding abilities and affinities [Bae et al, 2000; Gadd, 2009]. *Aspergillus niger* mycelium was modified by introducing additional carboxy or ethyldiamino groups which increased metal biosorption [Kramer & Meish, 1999] (Table 1).

BioSorbent	q_m (mg/g)
Acacia raddiana (Talhi et al, 2010)	82.63
Apple wastes	10.8
Bacillus subtilis	146.09
Banana pith	13.5
Cellulose	7.06
Chitin	29.04
Coirpith carbon	39.7
Cork	20
Cork bark	2.96
Cottonseed hulls	19.1
Dehydrated wheat bran (DWB)	51.5
Ganoderma	23.83
Grape stalks	10.10
Lignite	6.35
Mentha spicatum	10.37
Oil-palm fibre	1.98
Padina sp.	50.8
Peanut hull carbon	65.57
Penicillium ochrochloron	80.70
Plant root tissues	0.50
Pollen	6.23
Rice bran	33.58
Saccharomyces cerevisiae	30.50
Saragassum fluitans	108.03
Sawdust	1.79
Scenedesmus obliqus	10.04
Soybean hulls	38.7
Sphaerotilus natans	60
Sphagnum moss peat	16.4
sugar beet pulp	17.1
Sunflower	89.37
Tree fern	11.7
Washed sugar beet pulp	21.1
Yohimbe bark	9.42

Table 1. Maximum capacities of copper(II) ions biosorption by various biosorbents (compiled from [Benaissa & Elouchdi, 2007])

Biomass used for biosorption may be living or dead. While the use of dead biomass or derived products may be easier by reducing complexity, the influence of metabolic processes on sorption is often unappreciated, particularly where there is scant biological input to the problem [Gadd, 2009]. Removal of heavy metals using agricultural waste and its industrial by products has been massively investigated due to the abundance of agricultural materials in nature and its low cost [Febrianto et al, 2009].

It was noted that the pH value of the solution is of great importance in both cation and anion biosorption. However, the optimum pH for anion biosorption is opposite to that of cation biosorption. While cation biosorption is favored at increased pH > 4.5 [Kratochvil 1997], anion adsorption is preferred in a lower pH range of pH 1.5–4 [Gaff, 1992]. This was determined based on the characteristics of the biomass as well as considering the speciation of metals in the solution. The pH value of solution strongly influences not only the site dissociation of the biomass surface, but also the solution chemistry of the heavy metals: hydrolysis, complexation by organic and/or inorganic ligands, redox reactions, precipitation, the speciation and the biosorption availability of the heavy metals [Wang, 2002].

The temperature can influence the sorption process [Naja et al. 2008] whereas temperatures of 60°C or more caused a change in the texture of the sorbent and a loss in the sorption capacity due to the material deterioration. Biomass usually contains more than one type of sites for metal binding. This was confirmed in Cu adsorption by potassium-saturated microbial biomass. For most metals, the heat of reaction was constant, independent of the degree of site occupation. For Cu, however, the heat of reaction decreased with increasing degree of site occupation from 27 to 14 kJ/mol, indicating the involvement of different binding sites or the formation of different types of Cu complexes with the biomass [Naja & Volesky, 2011; Weppen & Hornburg, 1995].

Biosorption is rather difficult to define because many mechanisms may contribute to the overall process depending on the substance to be sorbed, the biosorbent used, environmental factors and the presence or absence of metabolic processes in the case of living organisms [Gadd, 2009].

4. Mechanisms of biosorption

Biosorptive capacity of heavy metals is influenced by many factors, including the status of biosorbent, properties of metal ions (radius of ion, valence, etc.) in aqueous solution, experimental conditions (such as pH, temperature, contact time, co-ions in solution, initial concentration of metal and biomass.

Biosorption is a physico-chemical process and includes such mechanisms as absorption, adsorption, ion exchange, surface complexation and precipitation [Gadd, 2009]. A variety of mechanisms are involved in the removal of such diverse substances from solution. A cation can associate with a surface as an inner-sphere or outer-sphere complex depending whether a chemical bond is formed between the metal and the electron donating oxygen ion in this case (inner-sphere complex) or if a cation approaches the surface negative groups to a critical distance. It is necessary to distinguish between *chemisorption*, which involves chemical binding, and *physisorption* that depends strictly on the surface-based physical forces of interfacial imbalance and attraction (e.g. van der Waals). Also there is deference between active metabolically mediated metal uptake by living cells as opposed to passive metal sequestering by dead biomass. In general, biosorption can be defined as the passive sequestering of metal ions by metabolically inactive biomass and is considered to be a fast physical/chemical process [Febrianto et al, 2009; Naja & Volesky, 2011; Wang & Chen, 2006].

In the literature another models have been reported with different schematizations of the interfacial structure in terms of supposed reactions, description of the electric double layer introduction of cooperative phenomena and continuous distribution for the equilibrium constants among active sites and ionic species in solution [Pagnanalli, 2011].

4.1 Biosorption models

Biosorption isotherms are used to describe the experimental adsorbed amount q_e *(mg/g)* as a function of the equilibrium concentration Ce (mg/L) of the adsorbing molecule at constant temperature. Than, Equilibrium isotherm models are usually classified into the empirical equations and mechanistic models, based on the mechanism of metal ion biosorption. Mechanistic models can be used not only to represent, but also to explain and predict the experimental behavior [Liu et al., 2002; Pagnanelli et al., 2002). The empirical models for single solute systems used to describe the biosorption equilibrium are Langmuir (based on monolayer adsorption of solute), Freundlich (developed for heterogeneous surfaces), Brunauer–Emmett–Teller (BET), Skips, Henry and Temkin models. However, Langmuir and Freundlich are the most widely accepted and used in a number of references. These models can provide information of metal-uptake capacities and differences in metal uptake between various species and have been widely applied since they are simple, give a good description of experimental behaviour in a large range of fixed operating conditions, and are characterised by a limited number of adjustable parameters [Kapoor & Viraraghavan, 1995; Pagnanelli et al., 2002]. Several biosorption isotherms originally used for gaz phase adsorption are vailable readily adopted to correlate adsorption equilibria in heavy metals biosorption.

4.1.1 Freundlich isotherm

Freundlich isotherm is an empirical equation (Eq. 1) most widely used for the description of adsorption equilibrium. Freundlich isotherm is capable of describing the adsorption of organic and inorganic compounds on a wide variety of adsorbents including biosorbent.

$$\log q_e = \log K_F + \frac{1}{n}\log C_e \qquad (1)$$

Where q_e is the amount of metal ion sorbed at equilibrium per unit weight of sorbent (mg/g), C_e the equilibrium concentration of metal ion in the solution (mg/L), The Freundlich model constants K_F and n indicate respectively the relative sorption capacity of the sorbent and the intensity of the sorption. The plot log q_e versus log C_e, gives a straight line of slope, which will permit us to obtain the constants from their intercepts and slopes [Cheriti et al, 2009].
log K_F is equivalent to log qe when Ce equals unity. However, in other case when $1/n \neq 1$, the K_F value depends on the units upon which qe and C_e are expressed. On average, a favorable adsorption tends to have Freundlich constant n between 1 and 10. Larger value of n (smaller value of $1/n$) implies stronger interaction between biosorbent and heavy metal while $1/n$ equal to 1 indicates linear adsorption leading to identical adsorption energies for all sites [Delle Site, 2001].
Freundlich isotherm has the ability to fit nearly all experimental adsorption–desorption data, and is especially excellent for fitting data from highly heterogeneous sorbent systems but does not predict an adsorption maximum and implies that the adsorption energy of a surface varies with surface coverage [Pagnanelli, 2001]. Table 2 abridges a number of studies that draw upon Freundlich isotherm to interpret biosorption equilibrium data of Copper ions.

Biosorbent	Operational condition		Freundlich parameters		
	pH	T ◦ C	K_F	*n*	R^2
Acacia raddiana [Talhi et al, 2010]	5.0	25	8.75	1.54	0.971
Aspergillus niger	–	20	1.77	2.05	0.886
Crab shell particles	3.5	–	2.05	1.68	0.979
Tea waste	5–6	22	0.70	1.35	0.984
Tea industry waste	5.5	25	0.45	1.18	0.992
Atriplex canescens	5.0	24 ± 2	4.02	1.04	0.979
Caulerpa lentillifera	5.0	-	10.44	1.27	0.971
green algae *Cladophora fascicularis*	5.0	25	84.38	2.45	0.998
Sour Orange Residue	4.5	28	9.52	2.90	0.980
Sour Orange Residue (NaOH treated)			5.52	3.74	0.980
Brown seaweed	5	25	5.5.70	2.30	0.908
		40	5.20	2.10	0.953
		55	5.40	2.00	0.917
Litter of poplar forests	4.5	25	6.14	3.89	0.966
		45	8.32	3.31	0.875
		60	14.18	5.57	1.000
Pretreated *Aspergillus niger*	5.5 ± 0.5	–	5.35	2.38	0.890
Lentil shell	6	20	0.65	2.77	0.993
		40	1.20	2.58	0.964
		60	1.14	2.87	0.901
Wheat shell		20	0.019	10.91	0.910
		40	0.034	9.59	0.955
		60	0.022	13.43	0.945
Rice shell		20	0.108	2.79	0.996
		40	0.148	2.82	0.929
		60	0.130	3.19	0.938
Myriophyllum spicatum L			1.60	1.20	0.950
Cedar sawdust	5–6	25	0.59	1.02	0.938
		35	1.04	1.13	0.969
		45	1.19	1.25	0.986
Crushed brick		25	0.75	1.19	0.957
		35	1.04	1.23	0.968
		45	1.54	1.49	0.964
PVA-*Sargassum*	5	22 ± 1	0.19	7.143	0.83
Free suspended *Sargassum*			0.56	3.226	0.95
Rhyzopus oryzae (viable)	4–6	21	8.65	6.67	0.921
Rhizopus oryzae (NaOH treated)			11.07	11.07	0.836

Table 2. Freundlich parameters and conditions for biosorption of copper ion by various biosorbents (compiled from [Febrianto et al 2009])

Sorption of copper on any sorbent can occur either by physical bonding, ion exchange, complexation, chelation or through a combination of these interactions. In the first case of physical bonding. Miscellaneous functional groups such as hydroxyl, carbonyl, carboxyl, sulfhydryl, thioether, sulfonate, amine, imine, amide, imidazole, phosphonate, and phosphodiester groups, can present within the structure of biosorbent, the mechanism of adsorption will not be restricted to physical bonding [Cay et al, 2004; Febrianto et al, 2009; Popuri et al, 2007]. Different mechanisms can be involved as the interaction between sorbent and solute molecules is expected to be strong. The parameter n of Freundlich equation in Table 2 expresses these phenomena.

4.1.2 Langmuir isotherm

The Langmuir equation (Eq. 2) was derived using the following assumptions [Pagnanelli, 2011]:

- The surface of the adsorbent has a fixed number of identical sites which can each retain one molecule, resulting in adsorption monolayer coverage.
- Reversible Adsorption process..
- Adsorption energy for all sites is identical, regardless of surface coverage

$$\frac{C_e}{q_e} = \frac{1}{bq_m} + \frac{C_e}{q_m} \tag{1}$$

Where q_e is the amount of metal ion sorbed at equilibrium per unit weight of sorbent (mg/g), C_e the equilibrium concentration of metal in the solution (mg/L), q_m and b the Langmuir model constants, indicate respectively the monolayer sorption capacity and the constant related to the free energy of sorption. The plot of C_e / q_e versus Ce give a straight line of slope, which will permit us to obtain the constants from their intercepts and slopes. Example of Langmuir parameters and conditions for biosorption of copper ion by various biosorbents are tabulated in Table 3.

In biosorption process, saturation limit of certain biomass is affected by several factors such as the number of sites in the biosorbent material, the accessibility of the sites, the chemical state of the sites and the affinity between the site and the metal.

The easy interpretation of Langmuir parameters made the fortune of this model: in fact q_m is often used to compare biosorbent performances in terms of maximum capacity, while b, which characterise the initial slope of the isotherm, is taken as a measure of the biosorbent affinity for a metal [Pagnanelli, 2011].

Within the Langmuir model, the saturation capacity $q$max is supposed to coincide with saturation of a fixed number of identical surface sites and as such, it should logically be independent of temperature. The decrease of b value with temperature rise signifies the exothermicity of the adsorption process (physical adsorption) [Shaker, 2007], while the opposite trend indicates that the process needs thermal energy (endothermic), leading to chemisorption [Dundar et al, 2008; Vilar et al, 2008]. In physical adsorption, the bonding between heavy metals and active sites of the biosorbent weakens at higher temperature in contrast with chemisorption bonding which becomes stronger.

4.2 Biosorption kinetic

Several biosorption kinetic models have been established to understand the kinetics and rate limiting step. These include pseudo-first and second-order rate model [Febrianto et al,2009], Weber and Morris sorption kinetic model, Adam–Bohart–Thomas relation, first-order reversible reaction model, external mass transfer model, first-order equation of Bhattacharya and Venkobachar, Elovich's model and Ritchie's equation. The pseudo-first and second-order kinetic models are the most used to study the biosorption kinetics of heavy metals and quantify the extent of uptake in biosorption kinetics.

Biosorbent	Operational condition		Langmuir parameters		
	pH	T◦ C	*b*	q_m	R^2
Acacia raddiana [Talhi et al, 2010]	5.0	25	0.073	82.63	0.985
Aspergillus niger	–	20	0.0120	33.11	0.944
		25	0.0150	33.23	0.941
		35	0.0260	33.54	0.939
Green alga *Cladophora fascicularis*	5.0	15	0.1406	47.019	
		25	0.1607	70.53	
		35	0.1649	92.77	
		45	0.1882	110.56	
Tea waste	5–6	22	0.0076	48.00	0.994
Spent grain	4.2	–	0.0800	10.47	0.995
Pretreated *Aspergillus niger*	5.5	–	0.1888	2.61	0.990
Lentil shell	6	20	0.0190	8.98	0.980
		40	0.0220	9.51	0.981
		60	0.0410	9.59	0.999
Wheat shell		20	0.0020	7.39	0.925
		40	0.0040	16.08	0.791
		60	0.0050	17.42	0.883
Rice shell		20	0.0110	1.85	0.991
		40	0.0160	2.31	0.982
		60	0.0180	2.95	0.987
Brown seaweed	5	25	0.0260	82.60	0.998
		40	0.0240	88.00	0.997
		55	0.0254	93.90	0.996
Caulerpa lentillifera	5.0	–	0.0760	5.57	0.968
	4.0		0.0618	42.37	0.995
	3.0		0.0290	2.64	0.995
Atriplex canescens	5.0	24	9.0900	590.92	0.0001
Non-living green algae *Cladophora fascicularis*	5.0	25	0.0652	102.55	0.997
		35	0.1241	106.81	0.998
		45	0.1750	112.97	1.000
Sour orange residue	4.5	28	0.0660	52.08	0.990
Sour orange residue (NaOH treated)			0.0540	23.47	0.980
Litter of poplar forests	4.5	25	1.66 × 10−4	19.53	0.991
		60	3.574 × 10−4	29.33	0.972
Pre-treated arca shell biomass		25	0.0590	26.88	0.979
Myriophyllum spicatum L.		20	0.0120	113.00	0.110
Algal waste		20	0.0028	16.7	0.950
		35	0.0053	17.0	0.950
Cedar sawdust	5–6	25	3.584 × 10−3	294.12	0.977
		35	6.290 × 10−3	144.93	0.977
Crushed brick		25	5.274 × 10−3	153.85	0.973
PVA-*Sargassum*	5	22 ± 1	28.0700	0.21	0.960
Freely suspended *Sargassum*			2.8600	0.96	0.980
Rhyzopus oryzae (viable)	4 – 6	21	0.1290	19.40	0.999
Rhyzopus oryzae (NaOH treated)			0.0280	43.70	0.999
Lignin	5.5	20	0.4309	22.87	0.978
Non-living lichen biomass of *Cladonia rangiformis* hoffm		15	0.1051	7.69	0.998
Lignocellulosic substrate (Wheat bran extract)	4.5		2.9600	12.58	0.980

Biosorbent	Operational condition		Langmuir parameters		
	pH	T◦ C	*b*	q_m	R^2
Phosphorylated orange waste	4.5	30	0.0452	67.35	0.990
Birch wood Betula sp.	5.5	22	0.2000	4.90	>0.997
Brown alga *Fucus vesiculosus*			1.1000	23.40	>0.997
Terrestrial moss *Pleurozium schreberi*			1.1500	11.10	>0.997
Cupriavidus taiwanensis TJ208	5	37	0.0363	19.00	0.887
Mimosa pudica inoculated with TJ208	5		0.0990	25.40	0.966
Mimosa pudica without inoculation	5		0.0614	22.70	0.923

Table 3. Langmuir parameters and conditions for biosorption of copper ion by various biosorbents (compiled from[Febrianto et al 2009])

A comprehensive reviews is available and can be referred upon by interested readers [Febrianto et al, 2009]. In biosorption of copper ion using cedar sawdust , two control mechanisms were observed, the film diffusion process controlled the early stage of biosorption process while in the later stage, chemical reaction become the limiting mechanism. The diffusion coefficient value of copper ions was 6.31×10^{-11} cm^2/s. Similar mechanism was also observed [Febrianto et al, 2009].

The exothermicity or endothermicity of the biosorption process can be determined via heat of adsorption. This thermodynamic property is commonly obtained through integrated Van't Hoff equation, which relates the Langmuir constant, *b* to the temperature. Heats of adsorptions *Ea* (kJ/mol) calculated for copper-biosorbent systems: Cedar sawdust [Djeribi & Hamdaoui, 2008], and *Cladophora fascicularis* [Deng et al, 2006], Shells of lentil and Shells of rice[Aydin et al, 2008] are respectively as: −9.70, 39.00, 15.37 and 18.79.

Unfortunately, knowledge of the biosorption mechanism is not easily obtained since we are not dealing with simple, clearly defined chemical compounds. Biosorbents comprise different types of cells with a highly complex structure whose various building blocks consist of a multitude of different molecules which in turn can display several binding sites [Naja & Volesky, 2011]. Numerous chemical groups have been proposed to contribute to biosorption metal binding by e.g. algae, bacteria or biopolymers, including hydroxyl, carbonyl, carboxyl, sulfhydryl, thioether, sulfonate, amine, imine, amide, imidazole, phosphonate, phosphodiester groups. In another hand biosorption depends on factors as: quantity of sites in the biosorbent material, accessibility of the sites, chemical state of the site and affinity between site and metal [Naja & Volesky, 2011]. Competition will also occur between cations and such an effect can also depress biosorption of the metal of interest. Various selectivity series have been published which reflect such competition, e.g. $Cu^{2+} > Cd^{2+} > Ni^{2+} > Pb^{2+} > Zn^{2+}$, $Co^{2+} > Cr^{3+}$ for *Chlorella vulgaris*, and $Cu^{2+} > Sr^{2+} > Zn^{2+} > Mg^{2+} > Na^{+}$ for *Vaucheria* sp.[Gadd, 2009]. In some cases, cations may increase biosorption of anionic species by enhancing binding of the negatively-charged anions. In some cases, cation loading of biomass may enhance biosorption of another cation because of pH buffering effects[Gadd, 2009].

5. Biosorption of copper ion by Sahara plants

Conventional wastewater treatment including sludge separation, chemical precipitation, electrochemical process, membrane separation, reverse osmotic treatment, ion-exchange and solvent extraction are often expensive, require high energy, have low selectivity and are impractical when they are used to treat the wastewaters with heavy metal ions amount lower than 100 mg l^{-1} [Cheriti et al, 2009; Talhi et al, 2010].

Screening of vegetable biomass types for heavy metals biosorption constitutes an important way of identifying the most promising types of biomass. Considerable efforts go into testing many different materials in order to assess their metal-sorbing potential. This is done mainly based on simple batch *equilibrium* sorption tests [Naja & Volesky, 2011]. Thus, several approaches have been studied and developed for the effective removal of heavy metals using biosorbents like peat, fly ash, algae, soya bean, hulls leaf mould, sea weeds, coconut husk, sago waste, peanut hull, hazelnut, bagasse, rice hull, sugar beet pulp, plants biomass and bituminous coal. It has also been observed that these biosorbents need further modifications to increase the active binding sites and also made them readily available for sorption [Talhi et al, 2010]
In the program of the valorization of Algerian Sahara plants initiated by our laboratory, we attempt to use some Saharan plants as low-cost biomaterials for the biosorption of toxic heavy metals from water and a suitable for application in small industries and for sustainable development. So, the objective of our studies was to utilize the locally available plants: *Acacia raddiana, Anabasis aretioides, Retama raetam and Tamarix gallica* as an adsorbent for removal of copper ion from aqueous solution. Effects of operating conditions like initial metal concentration, pH and temperature, on copper biosorption were investigated. The relation between the phytochemical composition and the percent of adsorption for copper ion was examined.

5.1 Materials and methods

5.1.1 Biosorbent preparation

All plants were collected from their natural habitat in El Bayadh region (south Algeria) for *Retama raetam* and from Saoura desert (south Algeria) for *Acacia raddiana, Anabasis aretioides, and Tamarix gallica*. We used the bark of tree *Acacia raddiana and Tamarix gallica* and the aerial parts of *Anabasis aretioides* and *Retama raetam.* The biomass was washed with distilled water several times to remove soil-associated particles and water soluble materials. The dried biomass, ground in a mortar to powder and sieved into a size ranging from 125 to 250 µm, was stored in a desiccator until use for the biosorption process.

5.1.2 Biosorption studies

All chemicals used in this study were of analytical grade and solutions were prepared using double distilled water. Cu^{2+} solution was prepared by dissolving copper sulfate in distilled water. Hydrochloric acid and sodium hydroxide solutions were used to adjust the solution pH. The pH was measured using Hanna pH meter at the beginning and at the end of the experiments.
Batch experiments were carried out by shaking 100 mg of biosorbent mixed with 100 ml of copper sulfate solution of known concentration in 200 ml Erlenmeyer flasks stirred at constant speed in a magnetic shaker in a thermostatic bath. After, the solid was removed by filtration through a filter paper (Whatman GF/A). Blank runs, with only the sorbent in 100mL of double distilled water, were conducted at similar conditions. The equilibrium metal concentration in the filtrates as well as in the initial solution was analyzed using atomic adsorption spectrophotometer (Perkin Elmer Analyst 700). Each experiment was carried out in duplicate.

5.2 Quantifying copper metal–biomass interactions

The identification of the physico-chemical interactions between active sites on the adsorbent and metallic species, and the quantification of metal removed according to different mechanisms (physical adsorption, ion exchange, complexation, chelation, surface micropreciptation) is an intriguing target that would allow the precise description of the effect of the environmental factors. Than, the biosrption capacity is governed by a series of properties, such as pore and particle size distribution, specific surface area, cation exchange capacity, pH, surface functional groups, and temperature [Naja & Volesky, 2011].

5.2.1 Effect of initial copper ions concentration

According to our recent works [Cheriti et al, 2009; Talhi et al, 2010], we have determined the optimum time of biosorption for different initial concentrations (varied between 50 and 300 mg l^{-1}). Than, biosorbed copper ion concentrations increased with time and reached equilibrium after 04 h for all initial copper ion concentrations tested, an increase of removal time to 24 h does not show notable effects.

An important parameters in biosorption of heavy metal ions is the equilibrium solid phase metal ion concentration. We noted that q_e, the Cu^{2+} equilibrium concentration in solid phase (mg Cu g biomass^{-1}), increased rising initial Cu^{2+} concentration . In the case of *Acacia raddiana*, Copper ion percent removal decreased from 94% to 39% (Fig. 2) and the final Cu^{2+} concentration increased from 3 mg l^{-1} to 183 mg l^{-1} (Fig. 3) when the initial Cu^{2+} concentration was raised from 50 mg l^{-1} to 300 mg l^{-1} at constant pH 5.0. At low initial Cu^{2+} concentrations, such as 50 mg l^{-1} and 100 mg l^{-1}, all copper ions were biosorbed onto binding sites on *Acacia raddiana* bark surfaces.

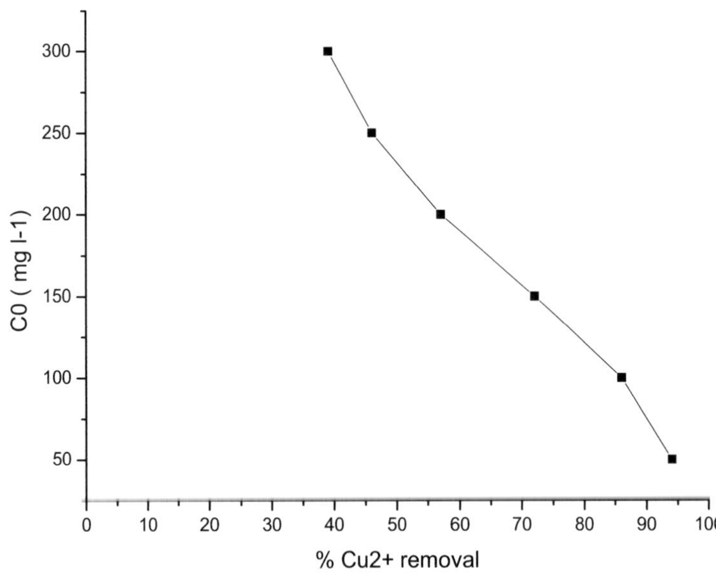

Fig. 2. Variation of percent copper removal with the initial copper ion concentration at the end of 4h, pH 5, T = 25 ± 0.1 °C for *Acacia raddiana*

For *Retama reatem*, at pH 5. the percent copper ion removal decreased from 88% to 38% and the final Cu^{2+} concentration increased from 6 mg l^{-1} to 186 mg l^{-1} when the initial Cu^{2+} concentration was raised from 50 mg l^{-1} to 300 mg l^{-1}. At low initial Cu^{2+} concentrations, such as 50 mg l^{-1} and 100 mg l^{-1} (Fig. 4), all copper ions were biosorbed onto binding sites on *R. raetam* plant surfaces. However, at high initial copper ion concentrations, such as 250 mg l^{-1} and 300 mg l^{-1}, a large fraction of binding sites on biomass surfaces were occupied by copper ions.

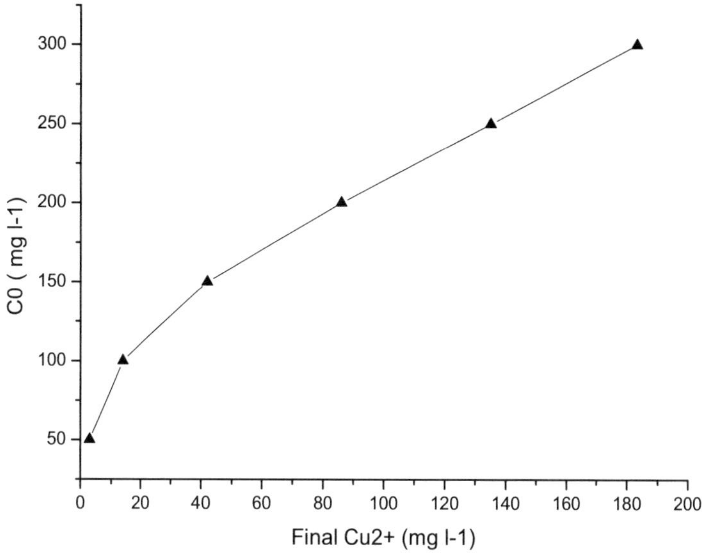

Fig. 3. Variation of final Cu^{2+} concentration with the initial copper ion concentration at the end of 4h, pH 5, T = 25 ± 0.1 °C – *Acacia raddiana*

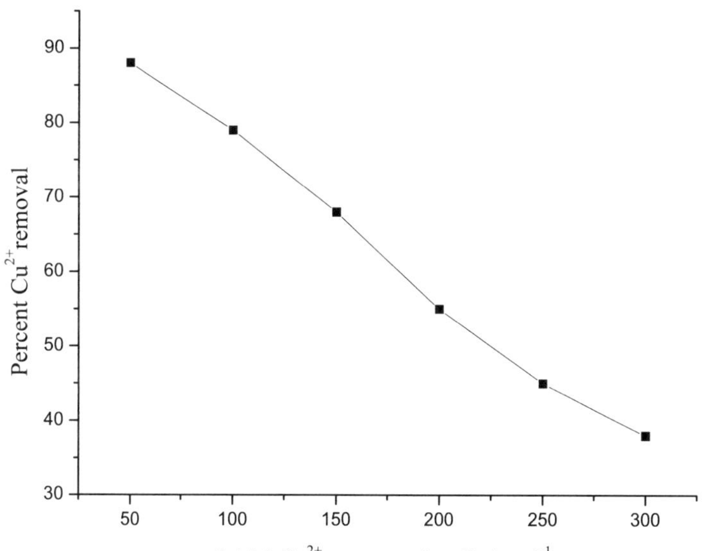

Fig. 4. Effects of initial copper ion concentration on the percent Cu^{2+} removal by *Retama raetam* plant

5.2.2 Effect of pH

The pH of the solution is the most important parameter that in the biosorption process can significantly influence the removal of heavy metals . The effect of pH on the sorption of copper (II) have been studied using different sorbent types and optimum sorption capacities have been reported at pH values of 5, 6 and 7 [Cheriti et al, 2009; Talhi et al 2010].

Keeping the same experiment conditions with 100 mg/L of copper ions solution as mentioned previously (times = 04h, T = 25 °C), pH of solution was varied from 3 to 7
Data on variations of percent copper removal and biosorbed copper concentration by the four saharan plants used under different pH values were summarizes in Table 4

pH	Percent Cu^{2+} removal			
	Acacia raddiana	*Retama raetem*	*Anabasis aretioides*	*Tamarix gallica*
3	11	09	05	10
4	64	58	14	58
5	86	79	35	81
6	85	70	33	76
7	83	67	25	71

Table 4. Effects of pH on percent copper removal and biosorbed copper concentration. Initial Cu^{2+}concentration C_0 = 100 mg l^{-1}, T = 25 °C, 04h

The greatest increase in the biosorption rate of Cu^{2+} ions on the plant biomass was observed at pH 5. At lower pH, H^+ ions compete with copper cation for the exchange sites in the system. The heavy metal cations are completely released under circumstances of extreme acidic conditions. At pH (3 - 7) there are three species present in solution, Cu^{2+}, $CuOH^+$ and $Cu(OH)_2$. These species are adsorbed an electrostatically interaction at the surface of the biomass. As the pH decrease, the surface of the *A. raddiana, T. gallica* and *R. raetam* plants exhibits an increasing positive characteristic. H^+ ions present at a higher concentration in the reaction mixture and compete with Cu^{2+} ions for the biosorption sites resulting in the reduced uptake of Copper cation. At higher pH values, precipitation of $Cu(OH)_2$ occurred and both sorption and precipitation would be the effective mechanisms to remove the copper ions in aqueous solution. At around pH 5, copper cations, mainly Cu^{2+}, would be expected to interact more strongly with the negatively charged binding sites in the sorbent. We obtained poor biosorption for the *Anabasis aretiode,* we think its due to the site saturation in this plants due to its ecological grows.

5.2.3 Effect of temperature solution

According to our previous sutdies on desert plant [Cheriti et al, 2009], we have found the same effect of the temperature on the biosorption of Copper cation. The sorption of copper cations increased slightly with the increase in temperature up to 50°C and then started decreasing. The temperature higher than 50°C caused a change in the texture of the biomass and thus reduced its sorption capacity (Table 5). Usually the physical sorption reaction is exothermic and preferred at lower temperature.

T °C	Percent Cu^{2+} removal			
	Acacia raddiana	*Retama raetem*	*Anabasis aretioides*	*Tamarix gallica*
25	86	79	35	81
30	84	79	30	81
40	75	61	23	72
50	49	42	11	49
60	36	21	11	30

Table 5. The biosorption quantity of Cu^{2+} at different temperature. Initial Cu^{2+}concentration C_0 = 100 mg l^{-1}, pH 5, 04 h

For the excellent source of biosorbent *Acacia raddiana,* we have determined the biosorption isotherms. Than, the isotherms were found to be linear over the entire concentration range studied with a good linear regression coefficient for Langmuir and Freundlich models respectively (R^2=0.985) and (R^2=0.917). The parameters of both models determined by least squares fit of the experimental biosorption data are presented in Table 6.

Langmuir			Freundlich		
q_m (mg/g)	b	R^2	K_F	n	R^2
82.63	0.073	0.985	8.75	1.54	0.917

Table 6. Parameters of the Langmuir and Freundlich isotherms for the biosorption of Copper ions onto *Acacia raddiana* bark

Fig. 5 shows the comparison of the experimental equilibrium data and the theoretical isotherm predicted by the Langmuir and Freundlich models. It was observed that Langmuir acceptably fit the experimental results with an acceptable coefficient. The maximum sorption capacity q_m of the *Acacia raddiana* bark was 82.63 mg/g which indicates that this biosobent is a quite good and attractive in treatment of diluted ions copper solutions.

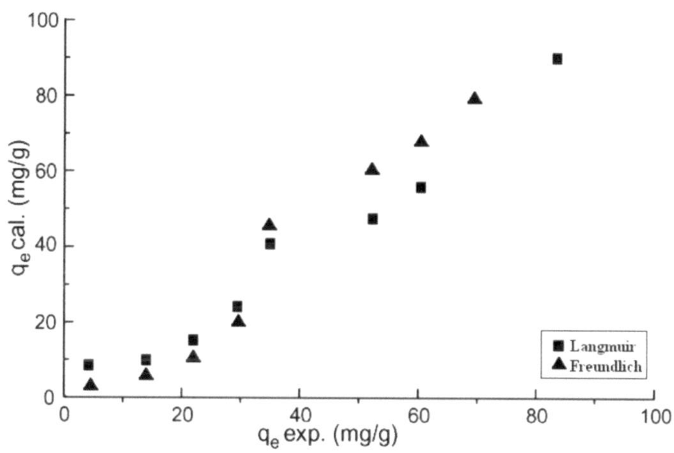

Fig. 5. Comparison of Langmuir and Freundlich models with experiment data of copper sorption by *Acacia raddiana* bark

5.3 Relation biosorption - phytochemical composition

Acording to our previous phytochemical screening of the Saharan plants, we reported the presence an important quantities in all plants of Flavonoids, Tritepenoids saponin, Alkaloids, Phytosterol, Cellulose, Hemicellulose and Lignin. Thus the substances detected are known to contain the functions (hydroxyl OH, carboxyl COOH, in Flavonoids, Tritepenoids saponin, Phytosterol, Cellulose, Hemicellulose, Lignin and amine function NHR, NR_2 in Alkaloids). It has been reported that the *Acacia raddiana, Retama reatem* and *Tamarix gallica* contain various polyphenols compunds and are rich on flavones and glycosides flavonoids [Cheriti, 2000; Cheriti et al, 2004; Cheriti et al, 2006]. It's known that these natural compounds present in the cell wall are the most important sorption sites. Therefore, the important of these natural compounds is that they contain functional groups as hydroxyl, carboxylic, carbonyl and amine groups which are important sorption sites for the binding of metal ions. Thus, biosorption of copper ion occurs as a result of ion exchange or complex formation between metal ions and functional groups (hydroxyl, amine, carboxyl...) on the cell surface of the biomass. In another hand, all active sites involved in heavy metal biosorption can be extremely heterogeneous according to the wide variety of biosorbents used for this application. Direct comparison of the used sorbent materials is difficult, since vegetable materials applied are different. The sorption capacities of *Acaciai raddiana , Tamarix gallica and Retama raetam* are relatively similar

6. Conclusion

Biosorption is an important process in the environment, and in several conventional waste treatment processes. By using Saharn plants, biosorpion of copper ions from aqueous solution was investigated as functions of important parameters, such as contact time, concentrations of adsorbate (Cu^{2+} ions), solution pH and temperature. The process of biosorption has nearly reached equilibrium in 04 hours and the biosorption of metals was pH and temperature dependent, respectively optimal pH was 5 and temperature was 25-30°C. We obtained better results with *Acacia raddiana* bark, *Retama reatam* and *Tamarix galliga* with percent of copper cations removal for the same initial concentration (100 mgl^{-1}) respectively: 86%, 79% and 81%. Therefore, the present study demonstrates the possibility of usage of inexpensive biosorbent, as suitable alternatives for the removal of copper ions from wastewater

Finally, additional work will be required in order to determine the biosorption of other heavy metals ions, and to determine the mechanism of copper sorption by these sorbents biomaterials.

7. References

Ahluwalia, S.S., Goyal, D., (2007). Microbial and plant derived biomass for removal of heavy metals from wastewater. Biores. Technol. 98, 2243–2257.

Aydin H., Bulut Y., Yerlikaya C. (2008), Removal of copper (II) fromaqueous solution by adsorption onto low-cost adsorbents, Journal of Environmental Management 87 37–45.

Bae W, ChenW, Mulchandani R and Mehra RK, (2000), Enhanced bioaccumulation of heavy metals by bacterial cells displaying synthetic phytochelatins. *Biotechnol Bioeng* 70:518–24.

Bellakhdar, J., (1997), La pharmacopée marocaine traditionnelle. Médecine arabe ancienne et savoirs populaires. IBIS Press.

Bishop PL., (2002), Pollution prevention: fundamentals and practice. Beijing Tsinghua University Press

Benaıssa H., Elouchdi M.A. (2007), Removal of copper ions from aqueous solutions by dried sunflower leaves, Chemical Eng. Nad processing, 46, 614-622

Chehma A. , (2006), *Catalogue des plantes spontanées du sahara septentrional algérien*, Ain Mila, Ed. Dar el Houda.

Cay S., Uyanik A., Ozasik A., (2004), Single and binary component adsorption of copper(II) and cadmium(II) from aqueous solutions using tea-industry waste, Separation and Purification Technology 38 273–280.

Cheriti A. (2000), *Plantes médicinales du sud ouest algérien : Ethnopharmacologie* , CRSTRA, Alger

Cheriti A. Belboukhari , N. &Hacini S., (2004) *Ir. J. Pharm. Res.*, 3(2), 51

Cheriti A., Belboukhari N., Sekkoum K. & Hacini S. (2006), *J. Algerien des regions arides,* 5, 07-10.

Cheriti A., Talhi M.F., Belboukharia N., Taleb S., Roussel C. (2009), Removal of copper from aqueous solution by *Retama raetam* Forssk. growing in Algerian Sahara , Desalination and Water Treatment, 10, 317–320

Delle Site A., (2001), Factors affecting sorption of organic compounds in natural sorbent/water systemsand sorption coefficients for selected pollutants.Areview, Journal of Physical and Chemical Reference Data 30, 187–439.

Davis, T.A., Volesky, B., Mucci, A., (2003). A review of the biochemistry of heavy metal biosorption by brown algae. Water Res. 37, 4311–4330.

Demirbas A., (2008), Heavy metal adsorption onto agro-based waste materials: A review , Journal of Hazardous Materials 157 220–229

Deng L., Su Y., Su, H. Wang X., Zhu X., (2006) , Biosorption of copper (II) and lead (II) fromaqueous solutions by nonliving green algae *Cladophora fascicularis*: equilibrium, kinetics and environmental effects, Adsorption 12 267–277

Djeribi R., Hamdaoui O. (2008), Sorption of copper(II) from aqueous solutions by cedar sawdust and crushed brick, Desalination 225(1-3), 95–112.

Dundar M., Nuhoglu C., Nuhoglu Y., (2008), Biosorption of Cu(II) ions onto the litter of natural trembling poplar forest, Journal of Hazardous Materials 151 86–95.

Farooq U., Kozinski JA., Khan MA.,Athar M., (2010), Biosorption of heavy metal ions using wheat based biosorbents - A review of the recent literature Bioresource Technology 101 5043–5053

Febrianto J., Kosasih AN., Sunarso J., Ju YH. , Indraswati N., Ismadji S., (2009), Equilibrium and kinetic studies in adsorption of heavy metals using biosorbent: A summary of recent studies, Journal of Hazardous Materials 162, 616–645

Gadd GM., (2009), Biosorption: critical review of scientific rationale, environmental importance and significance for pollution treatment, *J Chem Technol Biotechnol*; 84: 13–28

Gaff R., (1992) Chitin chemistry. Macmillan, London

Hamdy, A.A., (2000). Biosorption of heavy metals by marine algae. Curr. Microbiol. 41, 232–238.

Hashem M.A. (2007), Adsorption of lead ions from aqueous solution by okra wastes, Int. J. Phys. Sci. 2178–184.

Ibn Baytar (1992), *El djamia limoufradet el Adouia*, Ed Dar el koutob Elmia, Beirut.

Kapoor A, Viraraghavan T. (1995), Fungi biosorption,—an alternative treatment option for heavy metal bearing wastewaters: a review. Bioresour Technol;53:195–206.

Kapoor, A., Viraraghavan, T., Cullimore, D.R., (1999). Removal of heavy metals using the fungus Aspergillus niger. Biores. Technol. 70, 95–104.

Kramer M and Meisch H-U, (1999), New metal-binding ethyldiamino and dicarboxy products from *Aspergillus niger* industrial wastes. *Biometals* 12:241–246

Kratochvil D (1997) Biosorption of heavy metals by *Sargassum* seaweed biomass. Ph.D. thesis, chemical engineering, McGill University, Montreal, Canada

Kurniawan T.A., Chan G.Y.S., Lo W., BabelS., (2006), Comparisons of low-cost adsorbents for treating wastewaters laden with heavy metals, Science of The Total Environment 366, 409–426.

Kvesitadze G, Khatisashvili G., , Sadunishvili T., Ramsden JJ., (2006), Biochemical Mechanisms of Detoxification in Higher Plants Basis of Phytoremediation, Springer-Verlag Berlin Heidelberg

Le Houérou, H.N. (1990), Recherches écoclimatique et biogéographique sur les zones arides de L'Afrique du Nord. CEPE/CNRS, Montpellier

Liu N, Luo SZ, Yang YY, Zhang TM, Jin JN, Liao JL. (2002), Biosorption of americium-241 by Saccharomyces cerevisiae. J Radioanal Nucl Chem ;252: 187–91.

Lodiero, P., Herrero, R., Sastre de Vicente, M.E., (2006). Thermodynamic and kinetic aspects on the biosorption of cadmium by low cost materials: a review. Environ. Chem. 3, 400–418.

Luo F., Liu Y., Li X., Xuan Z. Ma, J. (2006), Biosorption of lead ion by chemicallymodified biomass of marine brown algae *Laminaria japonica*, Chemosphere 64, 1122–1127.

Naja G., Volesky B., (2011), The mechanism of metal cation biosorption, Chapitre 3,pp 19 in P. Kotrba et al. (eds.), *Microbial Biosorption of Metals,* Springer Science+Business Media B.V.

Naja G, Mustin C, Volesky B, Berthelin J (2008) Biosorption study in a mining wastewater reservoir., Int J Environ Poll 34:14–27

Niua CH, Volesky B and Cleiman D. (2007), Biosorption of arsenic (V) with acid-washed crab shells. *Water Res* 41:2473–2478

Nurchi, V.M., Villaescusa, I., (2008). Agricultural biomasses as sorbents of some trace metals. Coord. Chem. Rev. 252, 1178–1188.

Ozenda, P.,(1991), Flore et végétation du Sahara. Ed. CNRS, Paris. 662 p.

Ozturk, A., (2007). Removal of nickel from aqueous solution by the bacterium Bacillus thuringiensis. J. Hazard. Mater. 147, 518–523.

Padmavathy V., (2008), Biosorption of nickel(II) ions by baker's yeast: kinetic, thermodynamic and desorption studies, Bioresource Technology 99 3100 - 3109.

Pagnanelli F., (2011), Equilibrium, kinetic and dynamic modelling of biosorption processes, chapiter 4, p. 59 in in P. Kotrba et al. (eds.), *Microbial Biosorption of Metals,* Springer Science+Business Media B.V

Pagnanelli F, Esposito A, Veglio F. (2002), Multi-metallic modelling for biosorption of binary systems. Water Res;36:4095–105.

Park D., Yun Y., Jo J.H., Park J.M., (2006) , Biosorption process for treatment of electroplating wastewater containing Cr(VI): Laboratory-Scale Feasibility Test, Industrial and Engineering Chemistry Research 45, 5059–5065.

Popuri S.R., Jammala A., Reddy, K.V.N.S. Abburi K. (2007), Biosorption of hexavalent chromium using tamarind (*Tamarindus indica*) fruit shell—a comparative study, Electronic Journal of Biotechnology 10 ,358–367.

Quézel, P. (1965), La vegetation du Sahara, du Tchad a la Mauritanie. Fisher Verlag, Stuttgart. 333pp.

Rahman, H.U., Shakirullah, M., Ahmad, I., Shah, S., Shah, A.A., (2005). Removal of copper (II) ions from aqueous medium by sawdust of wood. J. Chem. Soc. Pak. 27, 233–238.

Shaker M.A., (2007) , Thermodynamic profile of some heavy metal ions adsorption onto biomaterial surfaces, American Journal of Applied Sciences 4 605–612.

Sud, D., Mahajan, G., Kaur, M.P., (2008). Agricultural waste material as potential adsorbent for sequestering heavy metal ions from aqueous solutions - a review. Biores. Technol. 99, 6017–6027.

Talhi M.F., Cheriti A., Belboukharia N., Agha L., Roussel C. (2010), Biosorption of copper ions from aqueous solutions using the desert tree *Acacia raddiana,* Desalination and Water Treatment, 21 323-327

Vilar V.J.P., Botelho C.M.S., Boaventura R.A.R., (2008) Copper removal by algae *Gelidium,* agar extraction algal waste and granulated algal waste: kinetics and equilibrium, Bioresource Technology 99 750–762.

Wang J., Chen C. (2006), Biosorption of heavy metals by Saccharomyces cerevisiae: A review, Biotechnology Advances 24 427–451

Wang JL. (2002), Biosorption of copper (II) by chemically modified biomass of Saccharomyces cerevisiae. Process Biochem;37: 847–50.

Weppen P, Hornburg A (1995) Calorimetric studies on interactions of divalent-cations and microorganisms or microbial envelopes. Thermochim Acta 269/270:393–404

Zhang Y, Banks C, (2006), Acomparisonof theproperties of polyurethane immobilised *Sphagnum* moss, seaweed, sunflower waste and maize for the biosorption of Cu, Pb, Zn and Ni in continuous flow packed columns. *Water Res* 40:788–798.

17

Chelating Agents of a New Generation as an Alternative to Conventional Chelators for Heavy Metal Ions Removal from Different Waste Waters

Dorota Kołodyńska
Maria Curie-Skłodowska University, Poland

1. Introduction

In the aqueous media free metal ions catalyze many reactions disadvantageous for man. Therefore it is necessary to control accessibility and a number of metal ions both during technological processes and in their final products. This aim could be achieved owing to metal ions complexation by the compounds characterized by complex formation ability. Therefore, the chelating agents belonging predominantly to two different groups i.e. aminopolycarboxylates (APCAs) and polyphosphonates are commonly used (Nowack, 2007; Knepper, 2003). APCAs containing carboxylic groups connected to one or a few atoms of nitrogen are able to complex metal ions by formation around them one or a few stable heteroatom rings. This phenomenon is called chelating. Forming of stable complexes with metal ions is the base of their application for analytical and industrial purposes. The complexes formed due to chelation are dissolved in water and metal ions found in them do not exhibit such chemical activity as uncomplexed ions. Because of the importance of chelating agents they are produced and used in large quantities and their behaviour as well as their effect on the environment have received considerable attention.

2.Traditional chelating agents and their complexes with heavy metal ions

2.1 EDTA and NTA

The first synthesis of a compound from the group APCAs i.e. NTA (nitrilotriacetic acid) was described by Heintz in 1862 (Heintz, 1862). Much later in 1935 I.G. Farbenindustrie carried out the synthesis of EDTA (ethylenediaminetetraacetic acid) which consisted in the reaction of monochloroacetic acid with ethylenediamine in the presence of sodium hydroxide. Another way to obtain EDTA is the reaction of ethylenediamine with sodium cyanide and formaldehyde in the presence of sodium hydroxide. Depending on the amine used also other APCAs can be obtained using this method. Since that time, on a worldwide scale over 100,000 tons of aminopolycarboxylic acids have been produced annually. It was estimated that 65,000 tons of EDTA and DTPA would be used up a year by the pulp and paper industry in the USA (Tucker, et al. 1999). Its annual consumption in West Europe was about

30,000 tons in 1987. However, in 1998 this amount was doubled and reached 64,000 tons. In 1996 the amount of EDTA in Germany was 3,686 tons and in 1999 – 3,894 tons. The pulp and paper industry is a major user of EDTA. It was estimated that about 15,000-20,000 tons of EDTA are used in pulp processing each year (Jones & Williams, 2002). From the late 1960s, NTA generally replaced phosphates in commercial detergents due to the increasing surface water eutrophication and its disastrous effect on natural environment (Anderson, et al. 1985). The Environmental Protection Agency (EPA) estimated that 35,000 to 38,000 tons of NTA were produced in the United States, and 27,000 to 30,000 tons were exported. However, in March 2006, the Commission's Working Group on the Classification and Labelling of Dangerous Substances decided to classify NTA, trisodium salt as a Category 3 Carcinogen with an R40 label (limited evidence of a carcinogenic effect) with a specific concentration limit of 5%.

2.2 Phosphonates

As for the second group of chelating agents, according to (Nowack, 1998, 2003) in 1998 the worldwide consumption of phosphonates was 40,000 - 56,000 tons in the USA, 15,000 tons in Europe and less than 800 tons in Japan. The demand for phosphonates grows steadily at 3% annually. Besides DTPPH (diethylenetriaminepenta(methylenephosphonic acid)) and NTMP (nitrilotris(methylenephosphonic acid)), HEDP (1-hydroxyethane-1,1-diphosphonic acid) is the most important in this group. HEDP is widely used in a broad variety of applications, among others, as ingredients of detergents. Bonding Ca(II) ions, which deactivate the surfactants improves their cleaning process. Its ability to prevent precipitation of calcium salts (called the threshold effect) finds also wide application in water treatment for scale inhibition in circulating cool water system, oil field and low-pressure boilers in such fields as electric power, chemical industry, metallurgy, fertilizer production, etc.

2.3 Degradation of EDTA and NTA

Degradation of EDTA and NTA in natural conditions proceeds due to growth of specific bacteria from the subclass of *Procteobacteria.* Several bacteria strains such as the gram-negative strain BNC1 were also found to be able to degrade EDTA. It was found that M(II)-EDTA complexes with the stability constants below 10^{12} such as Ba(II), Mg(II), Ca(II) and Mn(II) were degraded whereas chelates with higher stability constants such as Fe(III), Co(II), Cd(II), Pb(II), Ni(II) or Cu(II) (Table 1) were not metabolized (Nörtemann, 2005). The system of EDTA transport is based on its extracellular evolution with metal ions found in the solution and intracellular absorption and precipitation of metals combined with EDTA to form complexes (Witschel, 1999). It should be noticed that probably only the complexes of stability constants $<10^{14}$ can be transported inside the cell. Contrary to EDTA, transport of NTA inside the bacterium cell has not been well recognized yet. It is assumed that it gets inside the cell through the active transport. As established the metabolic intermediates of EDTA biodegradation include ethylenediaminetriacetic acid (ED3A), N,N'-ethylenediaminediacetic acid (N,N'-EDDA), ethylenediaminemonoacetic acid (EDMA), ethylenediamine (ED) and glyoxylate. The metabolic intermediates of NTA biodegradation are iminodiacetic acid (IDA), glycine, and glyoxylate (Yuan & VanBriesen, 2008). EDTA and DTPA are also reported to be photodegradable in the form of Fe(III) complexes (Cokesa, et al. 2004a). The process is pH dependent. It was found that it is faster in acidic conditions.

M(II)	EDTA	NTA	DTPA	NTMP
Al(III)	16.5	11.4	18.6	—
Ba(II)	7.9	4.8	8.7	—
Ca(II)	10.7	6.4	10.8	5.75
Cd(II)	16.5	9.8	—	6.4
Co(II)	16.5	10.4	18.8	7.5
Cu(II)	18.8	12.9	21.2	10.7
Fe(II)	14.3	8.3	16.2	—
Fe(III)	25.1	15.9	28.0	—
Hg(II)	21.5	—	26.4	—
Mg(II)	8.8	5.5	9.3	4.3
Mn(II)	13.9	7.5	15.2	—
Ni(II)	18.4	11.5	20.1	7.1
Pb(II)	18.0	11.4	18.8	—
Zn(II)	16.5	10.6	18.2	8.2

Table 1. Comparison of the stability constants of M(II)-L=1:1 complexes with EDTA, NTA, DTPA and NTMP (Martell & Smith, 1982). (— data not available)

2.4 Application of APCAs

Aminopolycarboxylic acids are used as components or process chemicals in a wide variety of applications, however, according to Fig.1 only pulp and paper, cleaning, chemical processing, agriculture and water treatment constitute 80% of their consumption.

APCAs are used for:

- **pulp and paper production** - stabilization of ozone and hydrogen peroxide action on pulp by complexing with metal ions (especially Fe, Cu and Mn that catalyse their decomposition), prevention from brightness reversion and protection of bleach potency;
- **household and industrial cleaning** - removal of hard water scale, soap film and inorganic scales, for example to improve the bottle cleaning in the beverage industry; *detergents, soaps processing* - prevention from precipitation of calcium and magnesium salts (deliming action) and their deposition on clothes, prevention soaps from becoming rancid, intensification of the adhesion of dirty surface and the cohesion of dirt particles to each other due to complexing metal ions (to better soil removal during laundering), to prevent from decomposition of bleaching agents such as sodium perborate, inhibition of colour changes, stabilization of hydrogen peroxide in liquid detergents for special requirements; to enhance the antibacterial effect, *cosmetics industry* - stabilization of creams, lotions and emulsions, reduction of allergic reactions to nickel and chromium, hair preparations, shampoos and almost every type of personal care formulation, blue colour of the $[Cu(edta)]^{2-}$ complex is used in many shampoos;
- **agrochemicals** - full utilization of micronutrients by plants (complexing agents are used to complex the microelements, which are then gradually released to the relevant ionic

form, the micronutrient chelates protect metal cations against reduction in soil and facilitate their uptake by leaves and roots);

- **water treatment** – to scale control; control water hardness and scale-forming calcium and magnesium ions;
- **photographic industry** - prevention from precipitation of calcium and magnesium salts onto the photosensitive layer, $[NH_4Fe(edta)]$ complex is used as an oxidizing agent for Ag in bleach baths;
- **textile industry** - to remove trace metal impurities in all phases of textile processing, particularly the scouring, dyeing and color stripping stages;
- **consumer products** – *food industry* by complexing metal ions to prevent from oxidation processes leading to colour changes especially in the case of canned food, for example vegetables; *pharmaceuticals* - stabilize formulations, antioxidants and anticoagulants added to stored blood in blood banks ($[K_2(H_2edta)]$ to prevent clotting; $[NaFe(edta)]$ and $[Na_2(H_2edta)]$ added to typical iron fortification compounds in cereals increasing the adsorption of iron in adult humans, Gd(III) complex with DTPA is used as a contrast agent in diagnosis by nuclear magnetic resonance imaging;
- **metalworking** – for surface preparation, metal cleaning, metal plating and in metalworking fluids;
- **Others**

  nuclear industry - formation of water soluble complexes with radionuclides in the decontamination of reactors; *adhesives productions* – production of latex-based coating compounds; *dispersants* – prevention from scale build up in industrial water handling processes which results in reduced water flow though pipes, reduced heat transfer in boilers and condensers, causes pump failures; *dispersants in paints and coatings production* – to disperse solid pigments such as titanium dioxide, sunscreen pigments during the manufacture of chemical dyes; *fuel gas cleaning and oil production* – to stabilize Na_2SO_3 solution used for adsorption of SO_2; components of cleaners for platforms and scale inhibitors especially for platforms which are not too deep; in drilling, production, and recovery of oil; *biodiesel production* - it was found that the addition of APCAs to biodisel

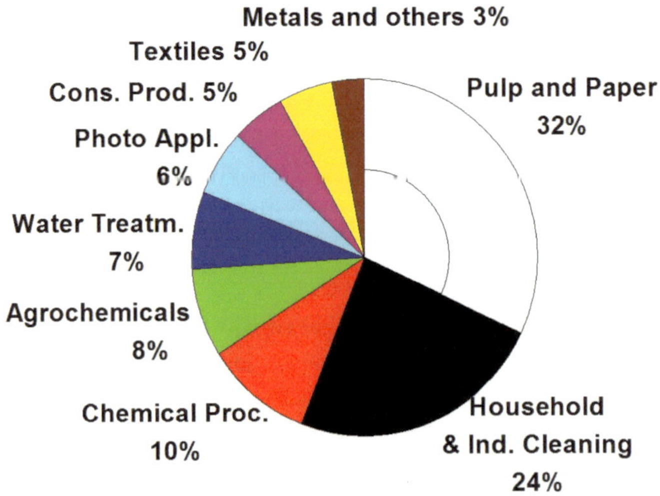

Fig. 1. The percentage contribution of different applications of APCAs.

reduces the humidity and kinematic viscosity; *polymerization* - for suspension, emulsion, and solution polymers, both in polymerization reactions and for finished polymer stabilization, *electroplating* - to control the level of Cu(II) ions and improve the quality of the Cu layer formed in copper plating, whereas in nickel and gold plating - to prevent from co-precipitation of other metals; *antiscalants* to prevent from scale formation. On the other hand, aminopolycarboxylates which act as dispersants are ableto keep the scale particles suspended in the bulk fluid by imparting a negative charge to the particles, *soil remediation and chelant-enhanced phytoremediation* - chelates have been shown to significantly increase metal concentrations in soil solution, the enhancement of plant uptake varies greatly, depending on the specific metal, chelate, plant combination and on soil conditions.

2.5 Heavy metal complexes with EDTA and NTA

EDTA includes six donor atoms and acts as a hexadentate ligand whereas NTA includes four donor atoms so it is a tetradentate ligand. EDTA and NTA form the complexes with metal ions of the molar ratio M(II):L=1:1. The complexes possess the octahedral structure. In the case of EDTA octahedral coordination is possible only with small size cations. With large cations limitations in the EDTA structure do not allow for such ideal structure and the complexing metal ion can be accessible for other ligands such as water molecules. X-ray investigations show that the structures of most metal complexes are different from octahedral ones and the cations coordination numbers are often larger than six. On the other hand, in some complexes such as with Cu(II) or Ni(II), EDTA does not make full use of donor possibilities leaving one carboxyl group uncoordinated. To octahedral system is completed by a water molecule.

3. Characterization of chelating agents of a new generation and their complexes with heavy metal ions

Despite obvious advantages, the conventional complexing agents, due to some undesired features such as their persistence or slow transformation in the environment, remobilization of toxic metal ions mainly from sediments and soils as well as radionuclides from radioactive waste and their implication in eutrophication of natural water systems, are of great concern, therefore their replacement and the use of chelating agents with improved biodegradability is necessary (Reinecke, et al. 2000). It should be stressed that most of the APCAs (such as EDTA - ethylenediaminetetraacetic acid, IDA - iminodiacetic acid, DTPA - diethylenetriaminepentaacetic acid) are resistant to conventional biological and physicochemical methods of waste water treatment and purification of drinking water. During the last few years complexing agents of a new generation have appeared on market as a solution to this problem. Their biodegradability is an important focus because of the renewed attention towards environmental protection issues. They are also proposed as a response to the changes in terms of legislation i.e. BREF documents, (EU EDTA risk reduction strategy documents, 2004, IPPC Directive 96/61/EC, BAT - Best Available Techniques Reference documents or OSPAR documents). For example, in the case of Austrian waste water emission legislations for the pulp and paper industry, the use of chelating agents with biodegradation levels below 70% or 80% is forbidden using the 28-day EN ISO (International Standards Organization) 7827 test. Nowadays, DTPA and EDTA are widely used in modern, total chlorine-free washing (TCF) and bleaching steps of pulps using hydrogen peroxide for complexing metals such as Mn(II) and Fe(III). As a result of their practical non-degradability, the attempts were made to find the alternative products (Matzinger, et al. 2007). New developed ligands to be used in practice should form strong complexes with the minimal content of nitrogen. The presented below biodegradable ligands of a new generation contain a basic nitrogen atom or two atoms (in the case of EDDS) with an electron pair capable of interacting with metal ions and acidic carboxylic groups capable of coordinating metal ions through the oxygen. The basic properties of the **IDS** (**N-(1,2-dicarboxyethyl)-D,L-aspartic acid** also known as **iminodisuccinic acid**), **DS** (**polyaspartic acid**), **EDDS** (**ethylenediamine-N,N'-disuccinic acid**), **GLDA** (**N,N-bis(carboxymethyl)glutamic acid**) and **MGDA (methylglycinediacetic acid)** and their complexes with heavy metal ions are described in detail in the next paragraphs.

3.1 IDS

Iminodisuccinic acid (N-1,2-dicarboxyethyl)-D,L-aspartate acid, (CAS No. 144538-83-0) also denoted as IDS has been commercialized since 1998 by the Lanxess (Germany, formerly Bayer AG) to the name Baypure CX 100. Its production is based on the reaction of maleic anhydride with ammonia and sodium hydroxide (Fig.2):

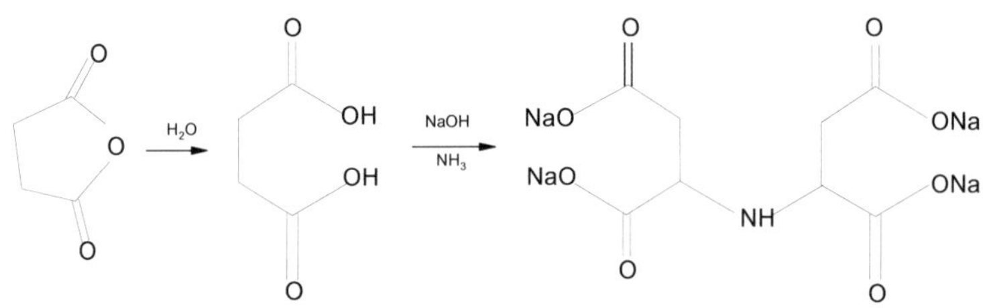

Fig. 2. Scheme of IDS production.

The process is extremely environment-friendly and does not generate any off-gases or effluents contaminating the environment. The sodium salts of iminodisuccinic acid (>32%), aspartic acid (<7%), fumaric acid (<3.5%), hydroxosuccinic acid (<0.9%), maleic acid (<0.9%) are the main ingredients of Baypure CX 100. It was also found that there are four possible stereoisomers of IDS: [R,R], [S,S], [R,S] and [S,R]. However, the two meso forms [R,S] and [S,R] are identical. The isomeric mixture of IDS consists of 25% [S,S], 25% [R,R] and 50% [R,S] forms (Fig. 3).

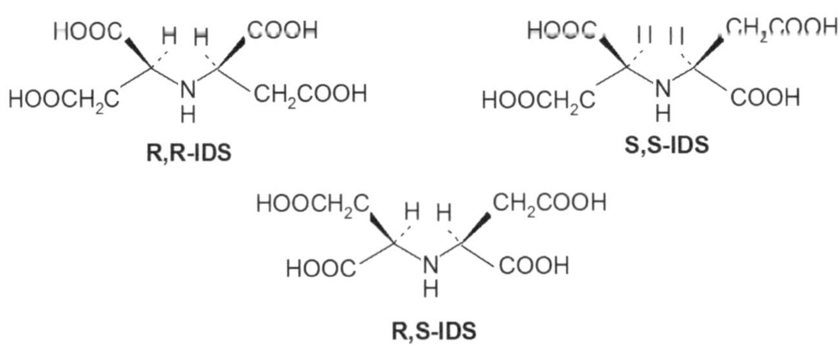

Fig. 3. The structural isomers of IDS.

IDS exhibits extremely rapid biodegradation, which equals approx. 80% after just 7 days. It is characterized by excellent calcium binding properties, stability in a wide pH, good complexation of heavy metal ions and low environmental impact due to low toxicity and good biodegradability according to the OECD (Organization of Economic Cooperation and Development) tests (OECD 301E i.e. modified OECD-screening test: > 78%, OECD 302 B i.e. Zahn-Wellens test: >89%). IDS undergoes lyased dependent degradation. The end product of transformation of [S,S]-IDS and [R,S]-IDS isomers by *Ralstonia sp. SLRS7* is fumaric acid. However, in the case of the [S,S]-IDS and [R,S]-IDS isomers as metabolites L-aspartic acid and D-aspartic acid were found. As further stages fumaric acid and ammonia are supposed (Cokesa, et al. 2004a). As for IDS the following pKs values were found: $pK_1=2.8$, $pK_2=3.6$, $pK_3=4.7$ and $pK_4=10.2$ the reaction of complexes formation with M(II) ions can be summarized as:

$$M^{2+} + H_n ids^{n-4} \rightleftarrows [M(H_n ids)]^{n-2}, \text{ where } n=0,1,2,3 \tag{1}$$

The comparison of the stability constants of metal complexes with IDS in the M(II):IDS=1:1 system is presented in Table 3 (Hyvönen & Aksela, 2010, Vasilev, et al. 1996, Vasilev, et al. 1998).

M(II)	IDS	DS	EDDS	GLDA	MGDA
Ag(I)	3.9	—	—	—	—
Al(III)	14.1	—	—	—	—
Ba(II)	2.1	—	3.0	3.5	**4.9**
Ca(II)	5.2	2.7	4.6	5.2	**7.0**
Cd(II)	**8.4**	**1.7**	**16.4**	**9.1**	—
Co(II)	10.5	—	13.6	10.0	—
Cr(III)	9.6	7.5	—	—	—
Cu(II)	**13.1**	**4.8**	**18.4**	**13.1**	13.9
Fe(II)	8.2	10.0	—	8.7	**8.1**
Fe(III)	15.2	18.5	22.0	11.7	**16.5**
Hg(II)	14.9	2.8	—	14.3	—
Mg(II)	6.1	2.0	6.0	6.1	**5.8**
Mn(II)	7.7	2.1	—	7.6	**8.4**
Ni(II)	12.2	—	16.7	10.9	**12.0**
Zn(II)	**10.8**	**2.2**	**13.4**	**10.0**	10.9
Pb(II)	**11.0**	—	**12.7**	**10.5**	—

Table 3. Comparison of the stability constants of M(II)-L=1:1 complexes with IDS, DS, EDDS, GLDA and MGDA (— data not available).

In the paper (Hyvönen, et al. 2003) it was found that in the dilute solutions of IDS more than 90% of metal ions are bound over a wide pH range. The percentage distribution of copper(II) complexes with IDS as a function of pH was presented in (Kołodyńska, 2009b).

For Cu(II) ions the effective complexation region was at the pH values from 3 to 12, for Zn(II) from 5 to 11 and for Mn(II) from 7 to 11. According to (Cokesa, et al. 2004b) some metal complexes with IDS satisfy the criteria of ready biodegradability. In this group Ca(II)-IDS and Fe(II)-IDS were found. Mn(II)-IDS and Cu(II)-IDS complexes revealed only 55% and 40% biodegradation after 28 days, respectively.
In perspective IDS can be used in oxidative bleaching of cotton with H_2O_2, as a component of detergents - very good results were achieved for spinach and tea removal by detergents containing Baypure CX 100, for cleaning of membranes blocked by diesel oil, silicon fluid, floors and glass. Baypure CX 100 brings good results in cutlery cleaning (Brochure of Baypure CX 100). A new proposal for the use of IDS is its application for coating seeds to enhance the adherence of trace metals and nutrient salts (Aksela, et al. 2004). IDS can be also applied for production of fertilizers. In Poland the leader in the production of micronutrient fertilizers based on IDS is the Production-Consulting Firm 'ADOB'. It has been producing iminodisuccinic acid since 2005. They also proposed a new Fe(III) chelates with N,N'-bis(2-hydroxy-5-methylbenzyl)ethylenediamine-N,N'-diacetic acid (HJB) and N,N'-di(2-hydroxybenzyl)ethylenediamine-N,N'-diacetic acid (HBED) (Stegient-Nowicka & Michałek, 2010).

3.2 DS

Polyaspartic acid (CAS No. 181828-06-8) is accessible on a commercial scale as Baypure DS 100 (denoted as DS), the commercial product of Lanxess (Germany). However, the Donlar Corporation were the first to develop an economical way to produce polyaspartates in high yield and with little or no waste products from L-aspartic acid, a natural amino acid. The way of the synthesis is presented in Fig. 4 (Wheeler & Koskan, 1993). In the first reaction, polysuccinimide (poly(anhydroaspartate)) is synthesized by dry thermal polycondensation of powdered aspartic acid. Subsequently, the polyimide rings are hydrolyzed with stoichiometric quantities of base to form poly(aspartate). Typically, the resulting polyamide contains a racemic mixture of D- and L-aspartic acid and is a copolymer in which the amide bonds are formed from either the α- or *β*-carboxyl groups (Mosig, et al. 2000).

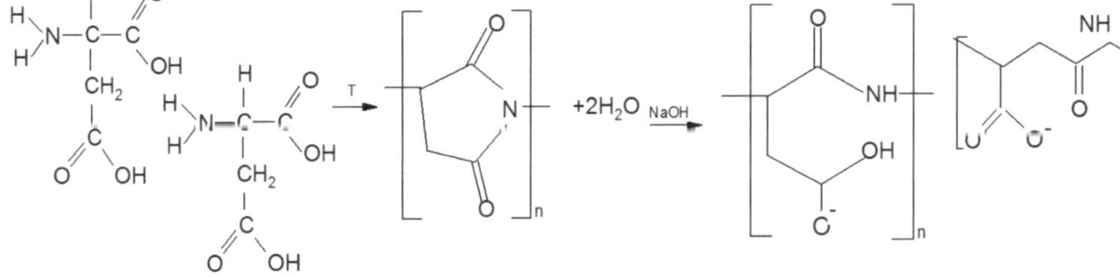

Fig. 4. Scheme of DS production.

Baypure DS 100 is characterized by excellent calcium(II) tolerance, corrosion inhibition properties, stability in a wide pH range as well as low environmental impact by low toxicity and good biodegradability (OECD 301E: > 60%, OECD 302 B: >75%). It is also an effective chelating agent of a new generation which can be applied for heavy metal ions removal (Freeman, et al. 1996). As follows from the structure, it consists of polymerized α- and β-aspartyl residues, each containing carboxylic functional groups that can combine with metal ions to form M(II)-DS complexes. In paper (Wu & Grant, 2002) it was found that aspartyl residues undergo protonation, deprotonation and complexation in solution. For sodium polyaspartate with a molecular weight of 10 000, 72 repeated units of aspartyl residues can be recognized as 18 independent hypothetical polyaspartic acid molecules (denoted as H_4ds) for which the pKs of DS are equal to: $pK_1=2.27$, $pK_2=3.60$, $pK_3=4.09$, $pK_4=5.17$, At pH 1 about 100 % of DS occurs as H_4ds, at pH 3 about 70 % occurs as H_3ds^-, at pH 4 about 48 % occurs as H_2ds^{2-}, at pH 4,8 about 65 % occurs as Hds^{3-} and at pH above 7 100% occurs as ds^{4-} (Burns, et al. 2003). The formation of metal complexes with DS can be summarized as:

$$M^{2+} + H_n ds^{n-4} \rightleftarrows [M(H_n ds)]^{n-2}, \text{ where } n=0,1,2,3 \tag{2}$$

The stability constants of M(II)-DS=1:1 complexes are presented in Table 3 (Brochure of Baypure DS 100). It should be also stressed that the theoretical capacity of polyaspartic acid for heavy metal ions is about 7.2 meq/g, whereas EDTA is only 6.0 meq/g. Although polyaspartic acid has also a quite high capacity, it shows poor efficiency at a low ion concentration (Sun, et al. 2005, Kołodyńska, et al. 2008a, 2008b).

Polyaspartates are used as corrosion and scale inhibitors, dispersing agents, waste water additives, superabsorbers, and also as agricultural polymers. It should be mentioned about polyaspartic resins obtained by modification of polyaspartic acid. They are characterized by high water absorbency and used as soil amendments, manufacturing of diapers, sanitary napkins, medicals, cosmetics, fabrics, metal absorbent materials and etc. (Zhao, et al. 2005). Among their other important properties super-absorption, fluid retention capacity and stability can be included (Zhao, et al. 2006). For example, AmiSorb® is the Donlar's brand name polymer used to enhance the growth and yield of corn, wheat, cotton, soybean and other crops. It contains polyaspartates which cause that mineral nutritions are more available and efficiently sorbed by plants. Recently it was proven that the overall yield change from the addition of AmiSorb® was equal to +1.75 bu/a for corn and +1.07 bu/a for wheat (Goos & Cattanach, 1996).

In June 2001 Lanxess was awarded with the Presidential Green Chemistry Award by the US Environmental Protection Agency (US EPA) for the best product in the field of industrial environment protection promotion. The product awarded was the Baypure CX 100. In 2002 Baypure DS 100 was given a prestigious award by the German Industry Federation (BDI). The Baypure based Danish product Groen Linie Maskinvask Color/Ultra became the first laundry detergent in the world to receive the 'EU Flower' ecolabel.

3.3 EDDS

Ethylenediamine-N,N'-disuccinic acid (H_4edds, EDDS) (CAS No. 144538-83-0) is a structural isomer of EDTA. On a commercial scale this complexing agent is produced, among others, by Innospec Inc., UK (formerly Octel) as Enviomet™. It forms four isomers: S,S- (25%), R,R- (25%), R,S- (50%), S,R- (50%). The S,S-isomer of EDDS, based on the naturally occurring amino acid i.e. L-aspartic acid, is readily biodegradable. According to the OECD 83% of S,S-EDDS convert to CO_2 within 20 days (Jaworska, 1999). However, the others are partly or completely non-biodegradable (Tandy, et al. 2006a). The chemical structure of EDDS isomers is presented in Fig.5.

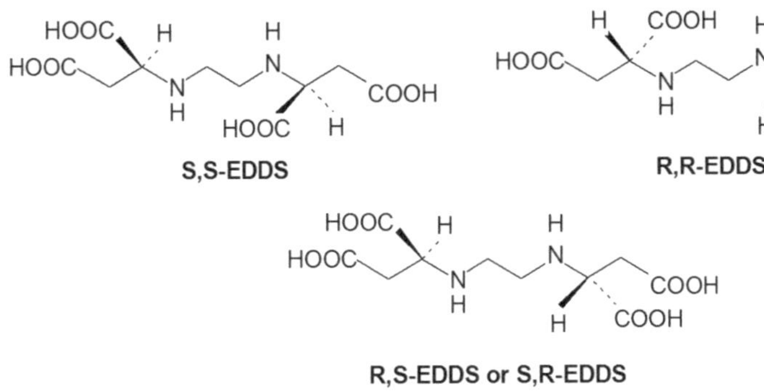

Fig. 5. The structural isomers of EDDS.

The S,S-EDDS isomer is produced by the actinomycete strain *Amycolatopsis orientalis* (Zwicker, et al. 1997). It can be also obtained in the reaction between 1,2-dibromoethane and L-aspartic acid (Fig.6).

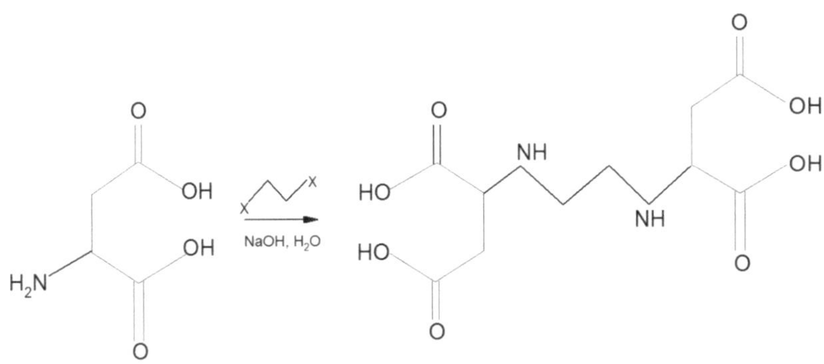

Fig. 6. Scheme of EDDS production.

The isomeric mixture of EDDS can be also synthesized from etyhylenediamine and maleic anhydride. On the contrary to the other biodegradable complexing agents like NTA, gluconic acid or citric acid it is characterized by high capacity for complexing cations as well as low toxicity (to fish and daphnia $EC_{50} > 1000$ mg/L).
As follows from the speciation diagram of EDDS at pH > 10 the predominant form of H_4edds is $edds^{4-}$ presented in 62% whereas the $Hedds^{3-}$ form is present in 38%(Crouch, et al. 2001). The appropriate ligand protonation constants pKs are as follows: 3.03, 3.92, 7.01, and 9.70 (Loonen, et al. 1999). They are very close to the data for EDTA. The process of heavy metal ions complexation by EDDS is characterized by the formation of stable 1:1 metal to ligand complexes as the major species according to the reaction:

$$M^{2+} + H_n edds^{n-4} \rightleftarrows [M(H_n edds)]^{n-2}, \text{ where } n=0,1,2,3 \qquad (3)$$

Metal ions in the formed complexes were bound to two nitrogen atoms and one oxygen atom of each of the four carboxylate groups. For example, for Cu(II) ions there was assumed formation of the following complexes: $[Cu(H_2edds)]$, $[Cu(Hedds)]^-$, $[Cu(edds)]^{2-}$, $[Cu(OH)(edds)]^{3-}$, $[Cu(H_2edds)_2]^{2-}$, $[Cu(H_2edds)(Hedds)]^{3-}$, $[Cu(Hedds)_2]^{4-}$, $[Cu(Hedds)edds]^{5-}$ and $[Cu_2edds]$ whereas for Zn(II) they are as follows: $[Zn(H_2edds)]$,

$[Zn(Hedds)]^-$, $[Zn(edds)]^{2-}$, $[Zn(OH)(edds)]^{3-}$ (Knepper , et al, 2005). It was shown that under some conditions of pH EDDS is more efficient than EDTA as far as chelation of metal ions is concerned. For example, in the case of Cs(I), Mg(II), Sr(II) they are not well complexed by EDDS at acidic and neutral pH and for them EDTA appears to be superior to EDDS (Jones & Williams, 2002), for Cr(III), Mn(II) their total amount is almost equal whereas in the case of Fe(III), Co(II), Ni(II) the complexation occurs in an identical manner. The obtained complexes are mostly not biodegradable (Loonen, et al. 1999; Vandevivere, et al. 2001). For example, Pb(II) and Zn(II) complexes with EDDS (Kos & Lěstan, 2003; Neal & Roshe, 1973) have practically the same stability constants but the Pb(II)-EDDS complex is much more readily biodegradable than Zn(II) one (Orama, et al. 2002). For M(II) complexes with EDDS the stability constants are presented in Table 3 (Vandevivere, et al. 2001) and compared to IDS, DS, GLDA and MGDA.

According to the literature data EDDS showed good chelation properties towards heavy metal ions. As mentioned earlier, it is also characterized by better extraction efficiency in the *ex situ* washing tests and phytoextraction process of heavy metals from contaminated soils than EDTA (Tandy, et al. 2004; Tandy, et al. 2006b). It can be also used as the H_2O_2 stablizer in the pulp and paper industry.

In 2002 EDDS production achieved 10,000 tons. Nowadays EDDS is a chelating reagent available in Europe on a large scale. The demand for ethylenediaminedisuccinic acid has been growing at the rate of 15 % per year. Also Innospec, the producer of EDDS was awarded by RoSPA (Royal Society for the Prevention of Accidents). Moreover, the Enviomet™ was the winner of the UK Green Chemical Technology Award in 2003 as a chemical product and then awarded by CRYSTAL Faraday Partnership (Institute of Chemical Engineers, Royal Society of Chemistry, Chemical Industry Association) (Brochure of Enviomet™).

3.4 GLDA

A novel readily biodegradable chelating agent, tetrasodium of N,N-bis(carboxymethyl) glutamic acid (also denoted as Dissolvine GL-38 or GLDA, CAS No. 51981-21-6) was also introduced on the commercial scale by AkzoNobel Functional Chemicals (The Netherlands). Its production process is based on the flavour enhancer monosodium glutamate (MSG) from the fermentation of readily available corn sugars and according to the following reaction (Fig.7) (Hauthal, 2009):

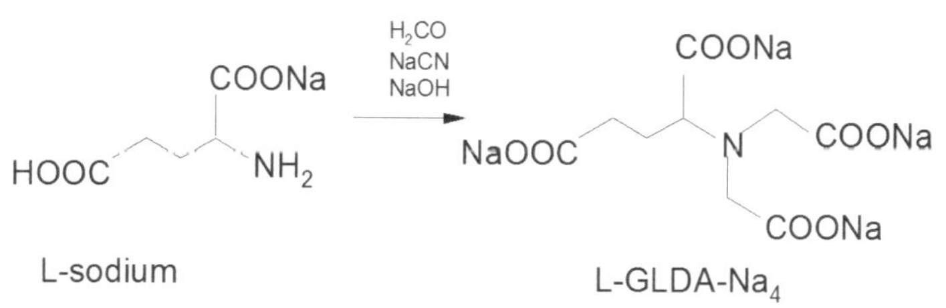

Fig. 7. Scheme of GLDA production.

In contrast to the manufacture of EDTA whose carbon content is fossil based, the carbon source of GLDA is primarily biobased. Therefore, GLDA is the only chelating agent with 'green' carbon atoms. The biodegradation of GLDA is initiated by mono-oxygenases catalysing the removal of carboxymethyl groups. According to the Swedish Society for Nature Conservation GLDA is 86% based on natural, raw materials. It also possesses good solubility at both low and high pH. Greater than 60% of L-GLDA degrades within 28 days. Dissolvine GL-38 consists only of the L-form. This is significant, because the D-form is not biodegradable. The thermal stability of GLDA is surprisingly high. When tested at temperatures above 573 K it showed no significant decomposition. This property has been used to develop water treatment systems for operating boilers to reduce the effect of hard water. Tests have also shown up to 10 times higher solubility of GLDA in 25% NaOH sodium hydroxide solution compared to EDTA and NTA. It is also characterized by excellent solubility at low pH (Seetz, 2007; Seetz & Stanitzek, 2008).

GLDA possesses complexing properties comparable to EDTA and NTA. In the case of GLDA 1:1 metal to ligand complexes are formed according to the reaction:

$$M^{2+} + H_n glda^{n-4} \rightleftarrows [M(H_n glda)]^{n-2}, \text{ where } n=0,1,2,3 \quad (4)$$

The stability constants for this type of complexes are presented in Table 3 (Brochure of Dissolvine GL-38). Therefore, it can be an alternative to EDTA and NTA. The benefit of GLDA in such formulations is its broad effective pH range for the chelation of calcium and magnesium ions equal to 4-12.

GLDA is directly used in all ordinary industrial cleaning applications. It has further potential uses in the production of micronutrient fertilizers (Borowiec & Hoffmann, 2005; Borowiec, et al. 2007).

3.5 MGDA

Methylglycinediacetic acid (MGDA) also known as glycine-N,N-diacetic acid is manufactured by BASF the Chemical Company as a brand name Trilon® M. MGDA and its derivatives are prepared by reacting glycine with formaldehyde and alkali metal cyanide in the aqueous alkaline medium (Schneider, et al. 1999).

The BASF started the new production line in 2008, which can supply over 120,000 tons of this complexing agent after 2010.

MGDA is readily biodegradable >68% and does not require adapted bacteria for decomposition, but instead is degraded under the standard conditions. Consequently, the EU does not require any safety or hazard labelling for MGDA. In the case of MGDA high stability throughout the entire pH range and even at elevated temperatures is found (Table 3). This sets Trilon M apart from the chemicals like EDDS or IDS which show much lower stability. Furthermore, the metal complexing power of MGDA is stronger than that of citrates. The complexes formed by it are characterized by high stability over a wide range of both pH and temperature. In the case of MGDA the reaction of complexes formation can be as follows:

$$M^{2+} + H_n mgda^{n-3} \rightleftarrows [M(H_n mgda)]^{n-2}, \text{ where } n=0,1,2 \quad (5)$$

MGDA can be applied as an excellent ingredient of products used for lime scale removing and for CIP (cleaning in place) cleaning processes (due to its dissolving power for fatty acids containing stains) (Brochure of Trilon M®; Jachuła, et al. 2011).

Owing to its properties it can be a perfect substitute for phosphates (STPP) in detergents and cleaners as well as complexing agent used in phytoremediation (Cao, et al. 2007, Jachuła, et al. 2011).

The properties of the above-described chelating agents are summarised in Table 3.

Name	Schematic structure	Physicochemical characteristic
IDS N-(1,2-dicarboxyethylene)-D,L-asparagine acid	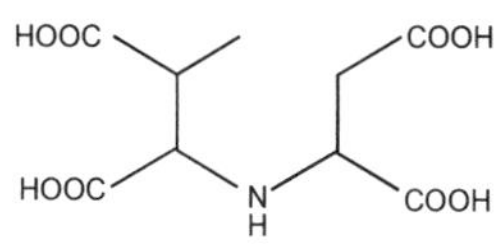	**molecular weight:** 337.1 **appearance:** colourless to light yellow **pH:** 10.3-11.4 **density:** 1.32-1.35 g/mL **solubility in H_2O:** in any ratio **biodegradability:** > 80%
DS polyaspartic acid	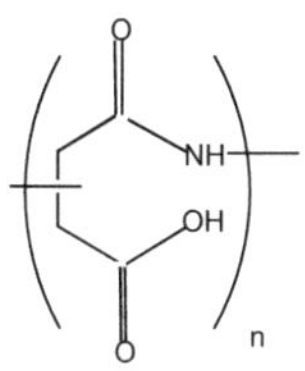	**molecular weight:** $[C_4H_4NO_3Na]_x$ **appearance:** clear reddish **pH:** 9.5-10.5 **density:** 1.3 g/mL **solubility in H_2O:** in any ratio **biodegradability:** > 60%
EDDS ethylenediamine-disuccinic acid	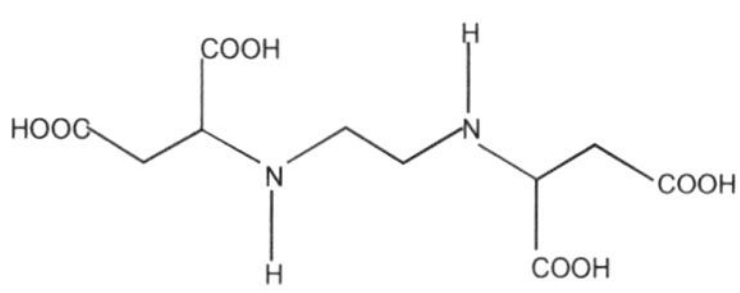	**molecular weight:** 358.1 **appearance:** colourless to yellowish **pH:** 9.2 **density:** 1.26 g/mL **solubility in H_2O:** in any ratio **biodegradability:** > 60%
GLDA N,N-bis(carboxylmethyl)-L-glutamic acid	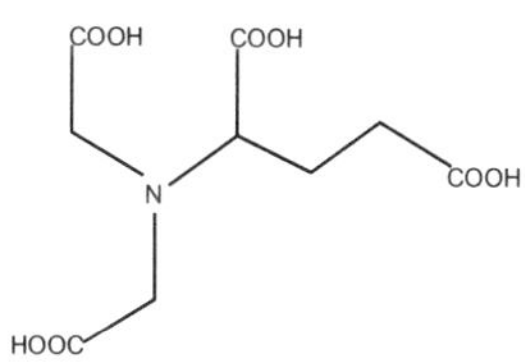	**molecular weight:** 351.1 **appearance:** colourless to yellowish **pH:** 13.5 **density:** 1.38 g/mL **solubility in H_2O:** in any ratio **biodegradability:** > 83%
MGDA methylglycine-diacetic acid	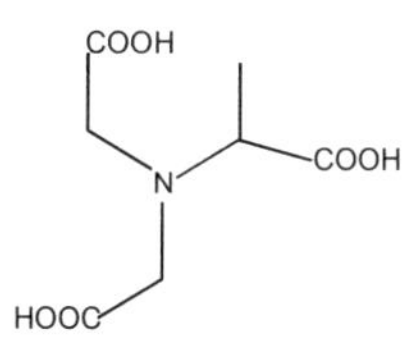	**molecular weight:** 271.0 **appearance:** clear yellowish **pH:** 11.0 **density:** 1.31 g/mL **solubility in H_2O:** in any ratio **biodegradability:** > 68%

Table 3. Physicochemical properties of IDS, DS, EDDS, GLDA and MGDA.

4. Determination of aminopolycarboxylic acids

It should be mentioned that speciation determination of the above presented chelating agents can be done by the gas chromatographic method using the international standard ISO 16588:2002. According to this norm the EDTA, NTA, DTPA, β-ADA (β-alaninediacetic acid) and 1,3-PDTA (1,3-propylenedinitrilotetraacetic acid) as well as MGDA can be determined in drinking, ground, surface and waste water samples. In waste waters and especially in pulp and paper wastes EDTA and DTPA can be determined by the liquid chromatography method in the form of Fe(III) complexes. The high performance liquid chromatography and electrophoresis methods can be used to determine Fe(III)-IDS complexes in cosmetic products (Sillanpää, et al. 1995; Katata, et al. 2006). The rapid UV-Vis method of determination of IDS was described in (Reinecke, et al. 2000). The samples containing IDS should be contacted with an appropriate amount of $CuSO_4$ in order to form blue $[Cu(ids)]^{2-}$ complexes. Their absorption was measured at 305 nm and 710 nm. The found absorption coefficients were equal to 327.7 and 67.5 for IDS concentrations in the range from 2 mM to 16 mM. It was also stated that the absorption of the $[Cu(ids)]^{2-}$ complexes was almost constant at pH values between 4 and 9. The modification of this method was also proposed and pH of the samples was adjusted at 1.3 by phosphoric acid to avoid $Cu(OH)_2$ precipitation. The obtained detection limit was 0.075 mM. The IDS can be also analysed by the ion chromatography (IC) method with isocratic elution (methanol in formate buffor) on the Purospher RP18 column. The method allows separation of [R,S]-IDS from [R,R]- and [S,S]-IDS isomers. In the paper by (Knepper , et al. 2005) the determination of EDTA and DTPA in aqueous solutions by the IC-MS method using the A Supp5 anion exchange column was described. The advantages of the method are: simple sample preparation, analysis without derivatization and possibility of determination of chelating agents in complex matrices (with high salt content). Quantification to a concentration down to the level of 1 μg/L was achieved by applying the ^{13}C marked internal standards.

5. Sorption of heavy metal ions in the presence of IDS, EDDS and GLDA

5.1 Experimental

In the presented paper the results of the sorption of heavy metal ions such as, Cd(II), Pb(II), Cu(II), Zn(II) as well as Cr(VI) and As(V) in the presence of the above-mentioned complexing agents on commercially available anion exchangers and chelating ion exchangers are presented. It provides overview of the information obtained using batch and column studies and descriptions of various sorption mechanisms. The obtained sorption results were fitted using the Langmuir, Freundlich, Dubinin-Radushkevich (D-R) or Temkin sorption isotherm models, whereas the kinetic data were analysed using the sorption kinetic Lagergren, Ho and McKay, Elovich, Weber and Morris (W-B) as well as Boyd models. It should be stressed that the choice of a particular chelating agent is dependent on several factors, including the pH and temperature range of the system in which they are used, which metal ions are to be controlled and the overall economy of the system. Therefore, the studies were carried out considering the effects of some important parameters such as phase contact time, initial concentration, pH and temperature. The aim of the presented studies was to compare sorption of heavy metal ions in the presence of the biodegradable chelating agents IDS, EDDS and GLDA. After comparison the most suitable treatment ways of waste waters containing heavy metal ions and the complexing agents of a new generation will be discussed. The literature lacks the data concerning sorption of heavy metal ions in the presence of biodegradable complexing agents of a new generation on ion exchangers.

In the investigations the following strongly basic anion exchangers were used: Amberlite IRA 900, Amberlite IRA 910, Amberjet 4200, Amberjet 4400, Amberjet 4600 and Purolite A520 E. In the chapter they were also denoted as: IRA 900, IRA 910, 4200, 4400, 4600 and A520E. For comparison the anion exchangers of Lewatite group: Lewatit MonoPlus M 500, Lewatit MonoPlus M 600 and Lewatit MonoPlus MP 500 as well as the chelating ion exchangers: Dowex M 4195 and Diphonix® were also used (denoted as M 500, M 600, MP 500, M 4195 and Diphonix, respectively). The physicochemical properties of ion exchangers used are summarized in Table 4.

Anion exchanger	Amberlite IRA 900	Amberlite IRA 910	Amberjet 4200	Amberjet 4400	Amberjet 4600	Purolite A520E
Manufacturer	Rohm and Haas, France	Rohm and Haas, France	Rohm and Haas, France	Rohm and Haas, France	Rohm and Haas, France	Purolite International, Ltd., UK
Functional groups	$-N^+(CH_3)_3$	$-N^+(CH_3)_2CH_2CH_2OH$	$-N^+(CH_3)_3$	$-N^+(CH_3)_3$	$-N^+(CH_3)_2CH_2CH_2OH$	$-N^+(CH_3)_3$
Matrix	PS-DVB	PS-DVB	PS-DVB	PS-DVB	PS-DVB	PS-DVB
Structure	macroporous	macroporous	gel	gel	gel	macroporous
Capacity [eq/L]	1.0	1.0	1.3	1.4	1.25	0.9
Bead size [μm]	650-800	530-800	600-800	530-630	600-800	300-1200
Zingg classification	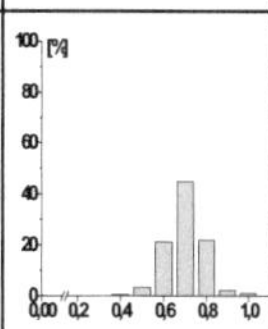	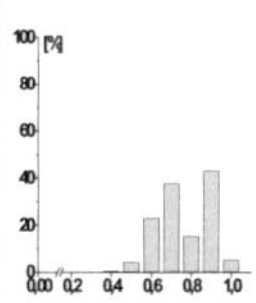	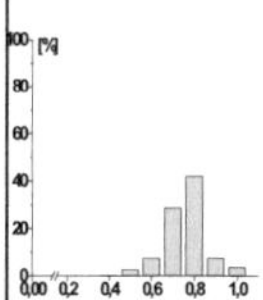	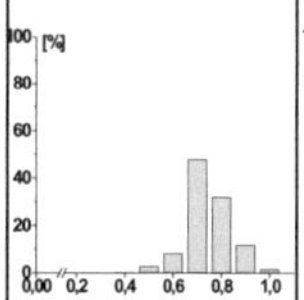	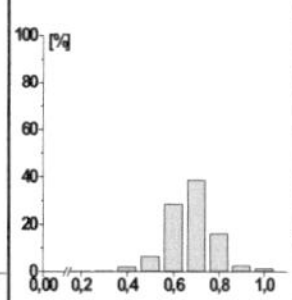	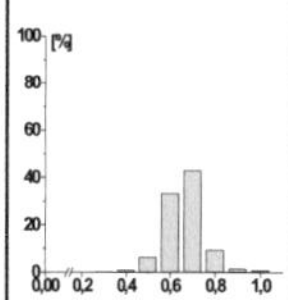
pH range	0-14	0-14	0-14	0-14	0-14	0-14
Max.temp.range [K]	350	350	350	350	350	350
BET surface area [m²/g]	36.00	39.25	5.27	1.96	1.47	23.52
Aver. pore diameter [nm]	15.45	16.02	9.03	6.14	17.53	16.37

Table 4. Physicochemical properties of used anion exchangers.

The detailed characteristics of Lewatit MonoPlus M 500, Lewatit MonoPlus M 600 and Lewatit MonoPlus MP 500 as well as Dowex M4195 and Diphonix were presented in (Kołodyńska, 2009a, 2010b, 2011b). The metals chosen for the evaluation were Cu(II), Zn(II), Pb(II), Cd(II) as well as Cr(VI) and As(V) due to their occurrence and abundance in waste water streams. The sorption tests included: evaluation of the rate and extent of complexes sorption on ion exchangers (%S), single vs. multi-component sorption, pH and temperature effects, interfering ions influence as well as exposition to UV light. The rate of metal complexes sorption was measured under the batch and the column conditions.

The objectives of the isotherm studies were to describe and quantify the partition of metals between the liquid phase and the ion exchanger phases with varying concentrations, and to determine the equilibrium capacity of the ion exchangers for the metal complexes. In order to compare the surface morphologies of the studied ion exchangers before and after the sorption process AFM (Atomic Force Microscopy) scans were made. The infrared spectroscopy was used to identify the changes after the sorption process. The details for the experimental procedures were presented in (Kołodyńska, et al. 2009; Kołodyńska, 2009ab; 2010a-c; 2011a-d).

5.2 Sorption of heavy metal complexes with IDS, EDDS and GLDA – the batch studies

5.2.1 pH effect

The chelating ability of IDS, EDDS and GLDA depends on the pH value, however, in the case of the use of anion exchangers as potential materials for heavy metal ions removal only complexes with more negative charge will be sorbed effectively. As has already been shown, IDS, EDDS and GLDA form 1:1 complexes with M(II). The effective regions for forming this type of complexes are above 4. As sorption of anionic complexes onto the selected ion exchangers depends on the acidity of the medium, pH of the metal complexes solution must be rigorously controlled. Therefore, to determine the optimum working conditions for sorption of the $[ML]^{2-}$ and $[ML]^{-}$ complexes by Lewatit MonoPlus M 500, Lewatit MonoPlus M 600, Lewatit MonoPlus MP 500, Amberlite IRA 900, Amberlite IRA 910, Amberjet 4200, Amberjet 4400, Amberjet 4600 and Purolite A520E, the effect of acidity was studied. The exemplary results for the Cu(II), Zn(II), Cd(II) and Pb(II) complexes with IDS and EDDS sorption on Amberlite IRA 900 (anion exchanger of type 1) and Amberlite IRA 910 (anion exchanger of type 2) are presented in Fig.8a-d.

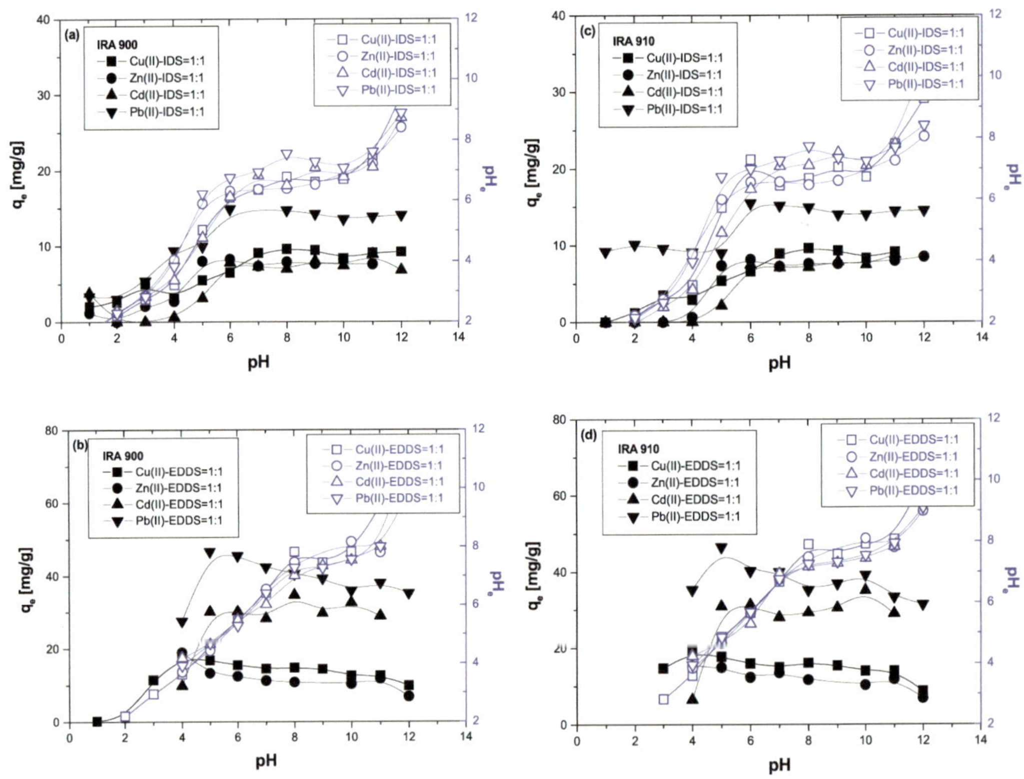

Fig. 8. a-d Effect of pH on the sorption capacities of Cu(II), Zn(II), Cd(II) and Pb(II) complexes with IDS and EDDS on the Amberlite IRA 900 (a, b) and Amberlite IRA 910 (c, d) (c_0 1×10^{-3} mol/L for IDS and c_0 2×10^{-3} mol/L for EDDS complexes, shaking time 180 min, shaking speed 180 rpm, room temperature).

It was found that the sorption capacity increased in the pH range from 4.0 to 9.0 for IDS and GLDA (not presented) complexes and achieved the maximum values at pH >6.0. For EDDS

complexes the sorption effectiveness slightly decreases with the increasing pH and the optimal sorption conditions were at pH >4.0. The pHs of the metal complex solutions were also measured during the sorption process. For the initial pH values from 4 to 12 the decrease in the equilibrium pH was observed due to the release of the Cl^- ions from the resin. Therefore, for the studied strongly basic anion exchangers from the Lewatite, Amberjet and Amberlite groups the following reactions of the process can be written:
the anion exchangers of type 1:

$$2R{-}N^+(CH_3)_3Cl^- + [ML]^{2-} \rightleftarrows [R{-}N^+(CH_3)_3]_2[ML]^{2-} + 2Cl^- \quad (6)$$

$$R{-}N^+(CH_3)_3Cl^- + [MHL]^- \rightleftarrows [R{-}N^+(CH_3)_3][MHL]^- + Cl^- \quad (7)$$

the anion exchangers of type 2:

$$2R{-}N^+(CH_3)_2CH_2CH_2OHCl^- + [ML]^{2-} \rightleftarrows [R{-}N^+(CH_3)_2CH_2CH_2OH]_2[ML]^{2-} + 2Cl^- \quad (8)$$

$$R{-}N^+(CH_3)_2CH_2CH_2OHCl^- + [MHL]^- \rightleftarrows [R{-}N^+(CH_3)_2CH_2CH_2OH][MHL]^- + Cl^- \quad (9)$$

where R is the resin matrix.
The highest values of the pH decrease were found for the initial pH equal to 11-12 and for the IDS complexes. For these values the decrease of the complexes removal was also observed. In the system containing, for example Cu(II) ions without the presence of complexing agents, their precipitation in the form of $Cu(OH)_2$ should be observed at pH > 6.5. However, in the presence of complexing agents the hydrolysis accompanied by precipitation of metal hydroxides does not occur. It is obviously connected with the stability constants of formed complexes (Table 3), which are the lowest for IDS. For Lewatit MonoPlus M 500, Lewatit MonoPlus M 600, Lewatit MonoPlus MP 500, Amberlite IRA 900 and Amberlite IRA 910 the obtained results were almost analogous to those for Amberjet 4200, Amberjet 4400, Amberjet 4600 and Purolite A520E anion exchangers and presented in (Kołodyńska, et al. 2009, Kołodyńska, 2011d). However, in the case of the chelating ion exchangers Dowex M 4195 and Diphonix a slight decrease of the sorption effectiveness with the increasing pH values was observed. Additionally, it was also found that at lower initial concentrations the amount of the sorbed complexes decreases while the pH increases. At higher concentrations, the sorbed amount does not increase (up to 2×10^{-2} mol/L) due to the low concentration of free functional groups near the saturation stage. On the other hand, the increase of the concentration of the solution is accompanied by further pH decreasing. Since Cl^- ions are less competitive with respect to the metal complexes with IDS, EDDS and GLDA in the ion exchange process, the ion exchange increases or stays constant in spite of pH lowering.

5.2.2 Effect of interfering ions

Besides the effect of pH, the sorption of Cu(II), Zn(II), Cd(II) and Pb(II) complexes with IDS, GLDA and EDDS is strongly affected by the interfering ions presence. In the case of the system without interfering ions, strongly basic anion exchangers have high affinity for anionic complexes of $[ML]^{2-}$ or $[MHL]^-$ types. At the initial concentration of 1×10^{-3} M, almost 100 % of the sorption percentage (%S) of Cu(II), Zn(II), Cd(II) and Pb(II) in the presence of IDS was observed for the studied anion exchangers. However, only 55-60% removal was found in the presence of SO_4^{2-} ions, while sorption in the presence of Cl^- ions was reduced minimally which confirms the results obtained for pH changes. In the presence of NO_3^- interfering ions S%

changes in the range of 81-92% depending on the anion exchanger used. The described results for the IDS complexes presented in Fig.9, show the general trend which is consistent with those obtained for other complexing agents. The %S values are also in agreement with the affinity of the anion exchangers for anions (Helfferich, 1962; Minczewski et al. 1982):
the anion exchangers of type 1:

$$ClO_4^- > I^- > HSO_4^- > NO_3^- > Br^- > NO_2^- > Cl^- > HCO_3^- > CH_3COO^- > OH^- > F^-,$$

the anion exchangers of type 2:

$$ClO_4^- > I^- > HSO_4^- > NO_3^- > Br^- > NO_2^- > Cl^- > HCO_3^- > OH^- > CH_3COO^- > F^-.$$

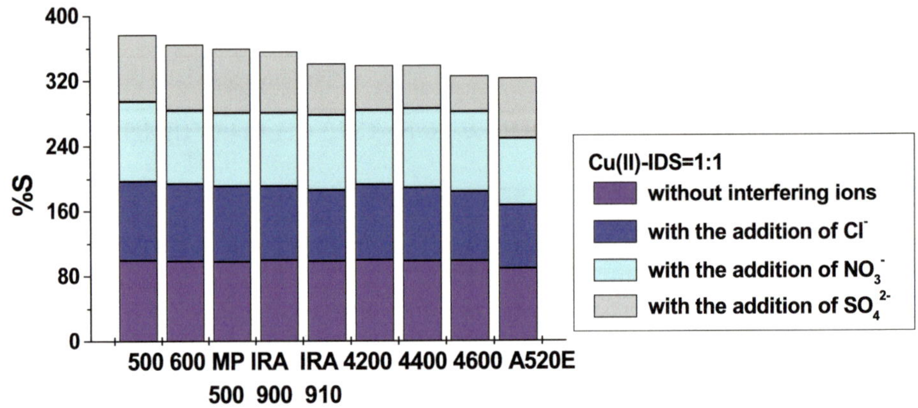

Fig. 9. Effect of interfering ions on the sorption percentage (%S) of Cu(II) complexes with IDS on the Lewatit MonoPlus M 500, Lewatit MonoPlus M 600, Lewatit MonoPlus MP 500, Amberlite IRA 900, Amberlite IRA 910, Amberjet 4200, Amberjet 4400, Amberjet 4600 and Purolite A520E (c_0 1×10^{-3} mol/L for IDS complexes, shaking time 180 min, shaking speed 180 rpm, room temperature).

The multi-component studies of sorption of metal complexes on the selected anion exchangers are also very important for assessing the degree of interferences in waste waters treatment. In the case of the binary (Cu(II)-Zn(II)) and quaternary (Cu(II)-Zn(II)-Cd(II)-Pb(II)) sorption studies of Cu(II), Zn(II), Cd(II) and Pb(II) complexes with IDS, EDDS and GLDA, their co-sorption induces a decrease in equilibrium sorption capacity. However, the percentage of decrease depends on co-metal ion present in the system. Other authors also presented such competitive sorption in the multi-component systems. This can be connected with the structure of the sorbed complexes and the ease with which they are formed. It should be stated that the ionic radius of Pb(II) equal to 1.12Å is greater than that for Cd(II) 0.97Å, Zn(II) 0.83Å and Cu(II) 0.72Å. Pb(II) has also the maximum hydrated radius and molecular weight in comparison to Cd(II), Zn(II) and Cu(II). Therefore, its connection with such ligands as EDDS or DS may be difficult taking into account the spatial reasons. This is evidenced, among others by large values of the maximum sorption capacity of the chelating ion exchanger Diphonix towards Pb(II) ions (357 mg/g) in the presence of EDDS compare to those obtained for Amberjet 4200 (216 mg/g), Amberjet 4400 (159 mg/g), Amberjet 4600 (164 mg/g) and much higher to the Lewatit MonoPlus M 500 (128 mg/g), Lewatit MonoPlus M 600 (139 mg/g) and Lewatit MonoPlus MP 500 (120 mg/g).

5.2.3 Anion exchanger properties - particle size, cross-linking, type of functional groups

The size of the polymer droplets formed during polymerisation and thus the size of the beads depend on the polymerisation technique, the suspension medium and the monomer concentration. Several producers now offer ion exchange resins with a very uniform size distribution. The example of this type of resins are Amberjets (Rohm and Haas) or Lewatit MonoPlus manufactured by Lanxess. Standard ion exchange resins have the mean particle sizes of 500-800 μm. In the presented studies the used anion exchangers possess the following bead sizes: 650-800 μm - Amberlite IRA 900, 530-800 μm - Amberlite IRA 910, 600-800 μm - Amberjet 4200, 530-630 μm - Amberjet 4400, 600-800 μm - Amberjet 4600 and 300-1200 μm - Purolite A520E. The scans of the Amberjet 4200 and Amberjet 4400 anion exchangers with the uniform size are presented in Fig.10a-b.

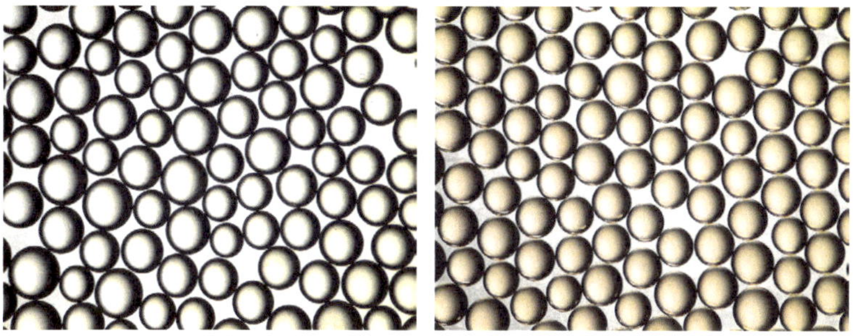

(a) (b)

Fig. 10. The scans of the Amberjet 4200 (a) and Amberjet 4400 (b) anion exchangers (http://dardel.info/IX/).

Using AWK 3D Analyser (Kamika Instruments, Poland) for determining size and shape of particles based on the Zingg classification the granulometric analysis was performed (Zingg, 1935). The Zingg classification is based on the proportions between three dimensions of ion exchanger beads a - the longest, b - mean and c - the shortest. However, the following assumptions should be made: (i) beads with the proportions $b/a > 0.67$ and $c/b > 0.67$ are similar to spheres, (ii) for $b/a = c/b = 1$ beads are spheres (iii) for $b/a > 0.67$ and $c/b < 0.67$ beads are similar to discs, (iv) for $b/a < 0.67$ and $c/b > 0.67$ beads are similar to cylinders, and (v) for $b/a < 0.67$ and $c/b < 0.67$ beads are similar to blades. Volume fractions of beads for $b/a = 1$ were shown in the figures in Table 4 (in these figures x axis presents c/b). As follows from the obtained results the bead size of the ion exchangers used has an approximately the Gaussian distribution. It was found that with the increase of beads dimensions, the volume fractions of disc-similar beads decrease and the beads are more spherical. For example, for Amberlite IRA 900 it is 45% at c/b=0.6-0.7; for Amberlite IRA 910 38%, for Amberjet 4200 36,5% and 40% at c/b=0.7-0.8; for Amberjet 4400 48% and 32% at c/b=0.7-0.8; for Amberjet 4600 38,8% and 28.6 at c/b=0.5-0.6; for Purolite A520E 47% and 27.5% at c/b=0.5-0.6. Moreover, for chelating ion exchangers these values are as follows: for Dowex M4195 43% at c/b=0.6-0.7 and 33% at c/b=0.5-0.6; for Diphonix 44% at c/b=0.6-0.7. The ideal spheres do not occur.

The particle size of the ion exchanger material and its uniformity are the most important parameters influencing on hydraulics and kinetics of the ion exchange process especially in the column mode technique. It is also obvious that the flow rate decreases with the decreasing particle size, however, smaller particles have larger outer surface, but cause larger head loss in the column processes. In the investigations the influence of size and shape of the used anion exchangers on sorption effectiveness was compared based on the results of sorption of Cu(II), Zn(II), Cd(II) and Pb(II) complexes with IDS. To this aim, Amberjet 4200 and Amberjet 4400 strongly basic anion exchangers of type 1 were used. The obtained results are presented in Fig. 11a-b and for the complexes with GLDA in Fig.12.

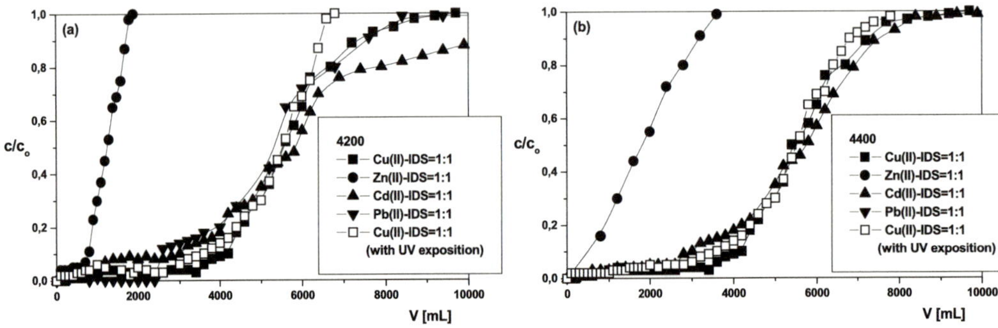

Fig. 11. a-b Breakthrough curves of Cu(II), Zn(II), Cd(II) and Pb(II) complexes with IDS on the Amberjet 4200 (a) and Amberjet 4400 (b) anion exchangers (c_0 1×10^{-3} mol/L for IDS complexes, bed volume 10 mL, flow rate 0.6 mL/min)

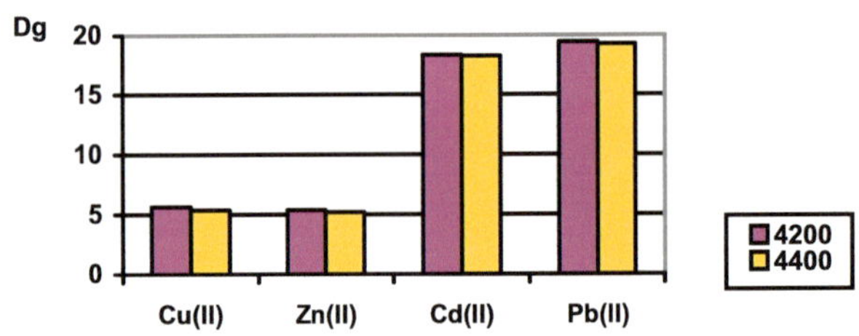

Fig. 12. Comparison of the mass distribution coefficients (D_g) for sorption of Cu(II), Zn(II), Cd(II) and Pb(II) complexes with GLDA on the Amberjet 4200 and Amberjet 4400 anion exchangers.

It was found that the anion exchanger with a large size of beads and moreover, high percentage of spherical beads is characterized by better sorption properties toward the GLDA complexes with Cu(II), Zn(II), Cd(II) and Pb(II). Both anion exchangers exhibit higher affinity for the Cu(II) complexes with IDS than for the corresponding Zn(II) one (Fig.11a-b). These breakthrough curves indicate also that their affinity for the Cu(II) complexes is higher in the case of Amberjet 4200 (c/c_o=0,5; V= about 5560 mL) than Amberjet 4400 (c/c_o=0,5; V= about 5400 mL). The affinities for the complexes with GLDA are arranged in a similar order which is confirmed by the mass distribution coefficients (D_g) (Fig.12). The obtained breakthrough curves have typical S-shape. As follows from Fig.11a-b the UV exposition does not have a significant effect on the decomposition of the complexes in the resin phase.

In the waste water treatment polystyrene ion exchangers usually have a DVB (divinylbenzene) concentration of ca. 8 %. Ion exchangers with a higher degree of cross-linking (10-12%) are used to obtain water of very high purity, and with a lower degree of cross-linking (5-7%) for softening. In the presented studies the anion exchangers have probably the same %DVB (this parameter is not given by the manufacturers) and therefore it is difficult to estimate.

Among the factors, which strongly affect selectivity of anion exchangers are type of functional groups, active site spacing and hydrophobic or hydrophilic nature of the resin matrix. The resins with the quaternary ammonium groups $-N^{+}(CH_3)_3$ are strongly basic (Lewatit MonoPlus M 500, Lewatit MonoPlus MP 500, Amberlite IRA 900, Amberjet 4200, Amberjet 4400, Purolite A520E). These are also known as type 1. The strongly basic anion exchangers possessing $-N^{+}(CH_3)_2CH_2CH_2OH$ groups (Lewatit MonoPlus M 600, Amberlite IRA 910, Amberjet 4600,) are denoted as type 2. The behaviour of the above-mentioned groups depends mainly on the acidity of the solution. The anion exchangers of type 1 are more strongly basic compared to type 2 and have the greatest affinity for weak acids. However, the efficiency of their regeneration to the hydroxide form is lower. Taking this parameter into account the effect of this type of functional groups was compared for Cu(II), Zn(II), Cd(II) and Pb(II) complexes(II) with IDS, EDDS and GLDA. As expected, under these conditions the most effective results were obtained for the anion exchangers of type 1. For example, the amount of Cu(II), Zn(II), Cd(II) and Pb(II) complexes with IDS sorbed on Amberlite IRA 900 was 4.32 mg/g; 3.25 mg/g; 18.51 mg/g and 24.35 mg/g whereas only 4.26 mg/g; 3.68 mg/g; 17.96 mg/g and 21.35 mg/g on Amberlite IRA 910. The analogous results were obtained in the case of EDDS and GLDA complexes. Comparing the strongly basic anion exchangers supplied by different manufacturers, it was found that only in some cases Amberlite anion exchangers are slightly less effective than Amberjet anion exchangers (4.32 mg/g and 5.17 mg/g for Amberlite IRA 900 and Amberjet 4200 in the case of Cu(II)-IDS sorption, respectively). However, the strongly basic anion exchangers from Lewatit group are characterized by the highest sorption capacities for these complexes. For example, for Cu(II)-EDDS complexes at the initial concentration $2x10^{-3}$ M the obtained sorption capacities for Lewatit MonoPlus M 500, Lewatit MonoPlus M 600, Lewatit MonoPlus MP 500 and Amberjet 4200, Amberjet 4400, Amberjet 4600 are as follows: 12.66 mg/g; 12.58 mg/g; 12.72 mg/g and 12.14 mg/g; 11.85 mg/g; 11.43 mg/g, respectively. On the other hand, taking into account the matrix of the used anion exchangers (gel or macroporous) such as for Lewatit MonoPlus M 500 and Lewatit MonoPlus MP 500, the anion exchangers of macroporous type are slightly more effective than the gel ones.

5.3 Kinetic evaluation

From the industrial point of view, where a solution flows through the ion exchanger the equilibrium is not necessarily reached and the results are influenced by kinetic consideration. Therefore, the experimental data for the M(II)-L=1:1 complexes sorption on the anion exchangers were fitted with the pseudo first-order (PF-order), pseudo second-order (PS-order) models proposed by Lagergren and Ho and McKay according (Blanchard, et al. 1984; Lagergren, 1898) as well as the intraparticle Webber and Morris (W-B) model (Weber & Morris, 1963). The kinetic parameters can be obtained by both linear and non-linear regression analysis. Linear regression is frequently used to determine the best fitting kinetic model, and the method of least squares is used for finding the parameters of the kinetic models. In the case of PS-order kinetic constant, q_e and k_2, they were calculated from the following plots: t/qt vs. *t* (PS-order of type 1), $1/q_t$ vs. 1/t (PS-order of type 2), qt vs. qt/t (PS-order of type 3), qt/t vs. qt (PS-order of type 4), respectively. However, as was stated taking into account the sum of the squares of errors (SSE), except for the PS-order equation of type 1, no other model provided a better fit to the experimental kinetic data. In these cases the SSE values were for the PS-order model lower than for the PF-order one. Moreover, in the case of the intraparticle kinetic model of Weber and Morris, the linear part of the plot of q_t vs. $t^{1/2}$ does not pass through the origin.

On the plots three different regions: before the equilibrium, the intraparticle diffusion and the stage to the final equilibrium can be distinguished. Therefore, it could be stated that the intraparticle diffusion is not the only rate controlling step. Similar results were obtained applying the Boyd model (Boyd, et al. 1947; Reichenberg, 1953). Based on it, two rate limiting steps for ion exchange kinetics – intradiffusion of counter-ions within the ion exchanger and interdiffusion of counter-ions in the adherent films can be distinguished. In this case the calculated diffusion rate constants B_t, which are related to the internal diffusion process plotted vs. t, do not give the lines which pass through the origin. This indicates that the rate limiting process is connected with film diffusion. It is well known that the film diffusion is a rate controlling step for the value of film diffusion coefficient D_f in the range 10^{-6}–10^{-8} cm²/s. On the other hand, the pore diffusion is rate controlling, the pore diffusion coefficient D_p is in the range 10^{-11}–10^{-13} cm²/s. For sorption processes of Cu(II), Zn(II), Cd(II) and Pb(II) complexes with IDS, EDDS and GLDA, the values of film D_f and pore D_p diffusion coefficients were found to be in the range 10^{-6}–10^{-7} cm²/s and 10^{-8}–10^{-9} cm²/s, respectively which indicates the film diffusion has some influence in the rate determining step. In order to establish affinity of anion exchangers for the above-mentioned complexes, the q_2 values for the PS-order model of type 1 were compared and presented in Fig.13a-c.

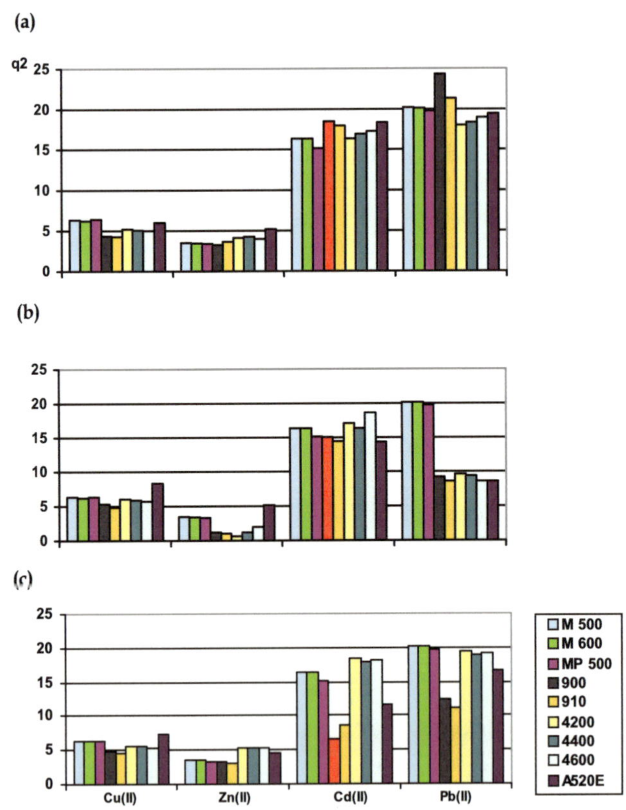

Fig. 13. a-c Comparison of sorption capacities q_2 (mg/g) of Cu(II), Zn(II), Cd(II) and Pb(II) complexes with IDS (a), EDDS (b) and GLDA (c) on the studied anion exchangers (c_0 1×10^{-3} mol/L for IDS and GLDA complexes as well as c_0 2×10^{-3} mol/L for EDDS complexes, shaking speed 180 rpm, shaking time 1-60 min, room temperature).

Kinetic sorption experiments were also carried out with the increased complexes concentrations from 1×10^{-3} mol/L to 2×10^{-2} mol/L, with the ion exchanger amount of 10 g/L, time 120 min, pH without adjustment, at room temperature. With an increase of metal complexes concentrations a continuous increase in the amount adsorbed per unit mass of ion exchanger was observed till the equilibrium was achieved. The amount of Cu(II)-IDS complexes sorbed on Dowex M4195 and Diphonix at different initial concentrations is shown in Fig.14a-b.

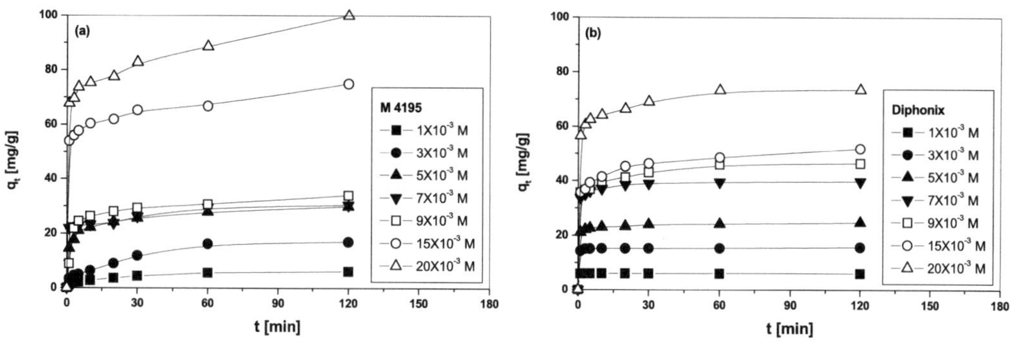

Fig. 14. a-b Effect of the phase contact time on the sorption capacities of Cu(II)-IDS complexes at different initial concentrations on the Dowex M 4195 (a) and Diphonix (b) (c_0 1×10^{-3} mol/L-2×10^{-2} mol/L, shaking speed 180 rpm, shaking time 1-120 min, room temperature).

The sorption kinetics data of these complexes at different initial concentrations fitted to different kinetic models are presented in Fig. 15a-f. As follows from the obtained results for the most suitable kinetic model, the rate k_2 values decrease with the increasing initial concentrations, while h increases.

In the case of the chelating resin Dowex M 4195 possessing the bis(2-pyridylmethyl)amine (BPMA) functional groups, depending on the pH value, nitrogen atoms are capable of coordinating metal ions. Based on the pK_a values of bis(2-pyridylmethyl)amine (pK_1=0.5, pK_2=2.2, pK_3=3.4), it can be stated that at low pH values three nitrogen atoms would be protonated, while in the middle range of pH only one:

$$R–N(BPMA)_2 + H^+ \rightleftarrows R–NH^+(BPMA)_2 \quad (10)$$

Therefore, in the presence of complexing agents decomposition of the complexes should be taken into account:

$$[ML]^{2-} \rightleftarrows M^{2+} + L^{4-} \quad (11)$$

$$[MHL]^{-} \rightleftarrows M^{2+} + HL^{3-} \quad (12)$$

$$[ML]^{-} \rightleftarrows M^{2+} + L^{3-} \quad (13)$$

$$R–NH^+(BPMA)_2 + M^{2+} \rightleftarrows [R–N(BPMA)_2]_2 \rightarrow M^{2+} + H^+ \quad (14)$$

The ionic interaction mechanism between the protonated amines and the anionic complexes of the $[ML]^{2-}$, $[MHL]^{-}$ and $[ML]^{-}$ is also possible. The appropriate reaction can be as follows:

$$2R\text{–}NH^{+}(BPMA)_2\,Cl^{-} + [ML]^{2-} \rightleftarrows [R\text{–}NH^{+}(BPMA)_2]_2[ML]^{2-} + 2Cl^{-} \quad (15)$$

$$R\text{–}NH^{+}(BPMA)_2\,Cl^{-} + [MHL]^{-} \rightleftarrows [R\text{–}NH^{+}(BPMA)_2][MHL]^{-} + Cl^{-} \quad (16)$$

or

$$R\text{–}NH^{+}(BPMA)_2\,Cl^{-} + [ML]^{-} \rightleftarrows [R\text{–}NH^{+}(BPMA)_2][ML]^{-} + Cl^{-} \quad (17)$$

The analogous mechanism of sorption is in the case of Diphonix chelating ion exchanger:

$$R\text{–}(PO_3H_2)_2^{2-} + M^{2+} \rightleftarrows R\text{–}(PO_3H_2)^{2-}\rightarrow M^{2+} \quad (18)$$

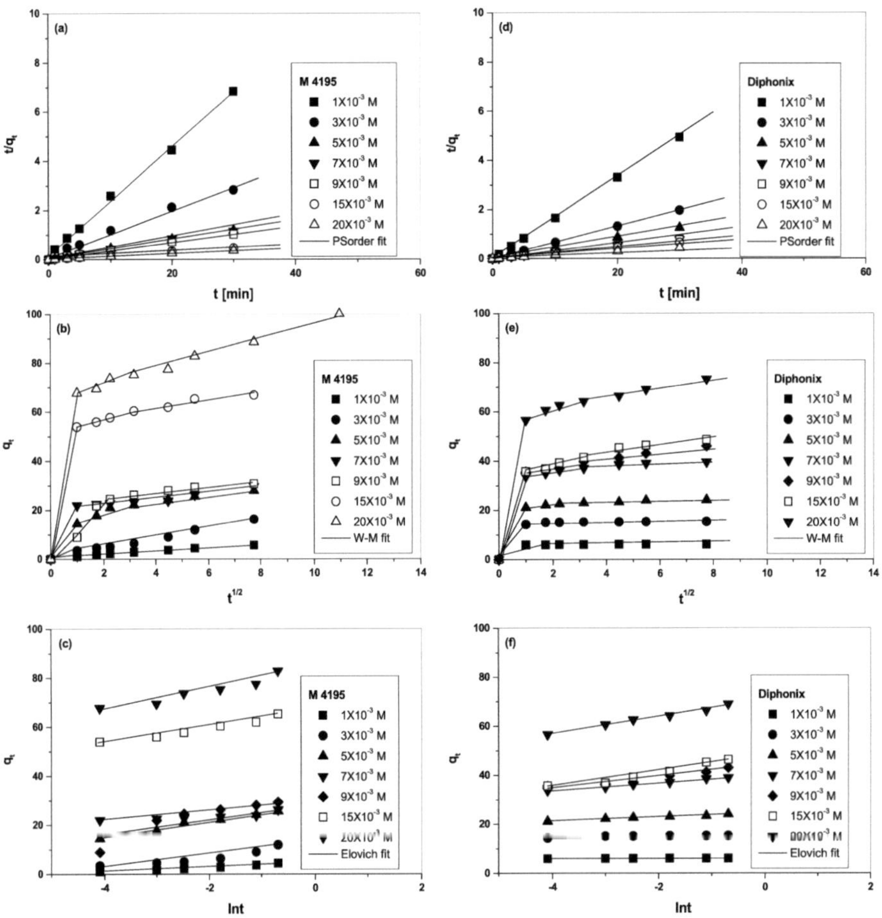

Fig. 15. a-f The comparison of the PS-order, Weber and Morris as well as Elovich kinetic models for the sorption of Cu(II)-IDS complexes at different initial concentrations on the Dowex M 4195 (a-c) and Diphonix (d-f) (c_0 1×10^{-3} mol/L-2×10^{-2} mol/L, shaking speed 180 rpm, room temperature).

As for sorption of As(V) ions, the obtained results reveal that rapid sorption of As(V) on strongly basic anion exchangers, similarly to the Cr(VI) sorption, is attributed to the surface sites initially available and is very large compare to the concentration of sorbed ions. However, the effect of the complexing agent added is neglectable (the obtained results are not presented).

5.4 Adsorption isotherm

To investigate the sorption capacity, various concentrations of Cu(II), Zn(II), Cd(II) and Pb(II) complexes with IDS, EDDS and GLDA were contacted with the studied ion exchangers. For interpretation the Langmuir, Freundlich, Dubinin-Radushkevich (D-R) and Temkin sorption isotherm models were used. The sorption data were drawn as the plots of c_e/q_e vs. c_e (Langmuir model); log q_e vs. log c_e (Freundlich model); ln q_e vs. ε^2 (D-R model) and q_e vs. ln c_e (Temkin model). The exemplary results for the Cu(II)-IDS, EDDS or GLDA systems for Amberjet 4200 and Lewatit MonoPlus M 600 anion exchangers are presented in Fig. 16a-c and e-f, respectively.

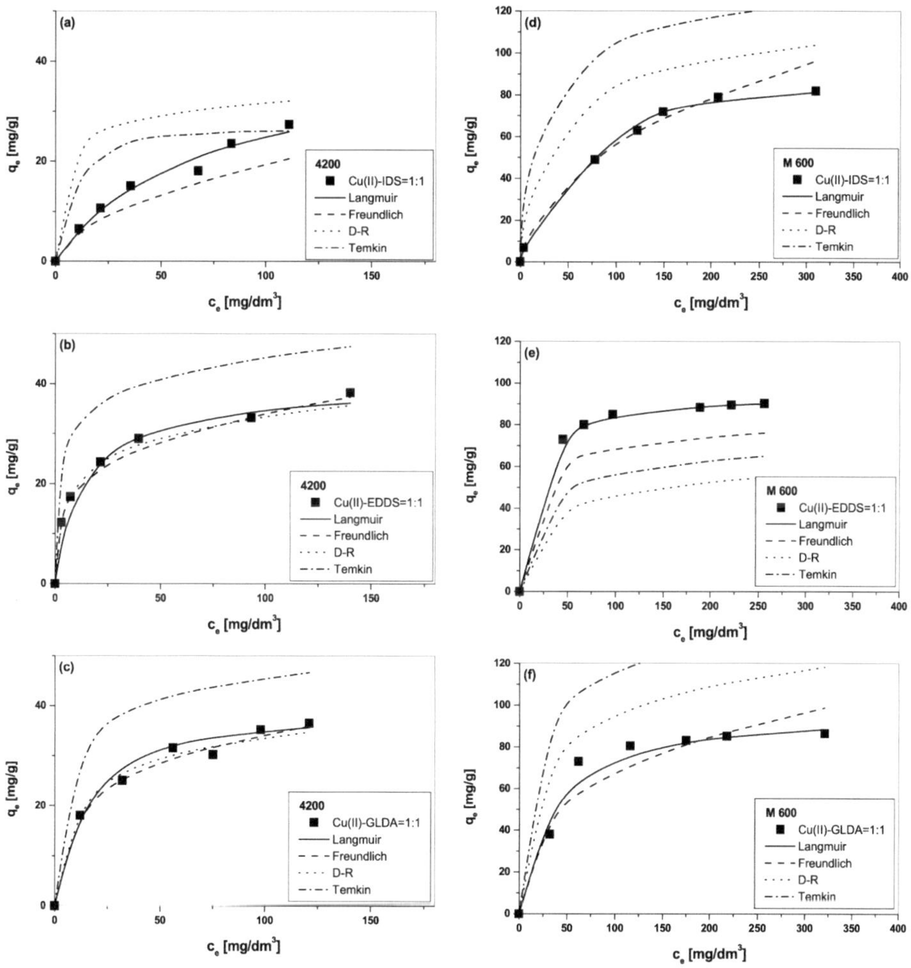

Fig. 16. a-f The isotherm of Cu(II)-IDS, EDDS and GLDA complexes sorption on Amberjet 4200 (a-c) and Lewatit MonoPlus M 600 (d-f) fitted by the Langmuir, Freundlich, D-R and Temkin models (c_0 1×10^{-3} M–2×10^{-2} M, shaking speed 180 rpm, shaking time 120 min, room temperature).

As follows from the obtained results the Langmuir equation better fit the experimental data than the Freundlich, D-R or Temkin equations. However, to the removal of the heavy metal ions in the presence of IDS, EDDS or GLDA under industrial condition fitting of the obtained results by D-R equation should be also taken into account.

5.5 Effect of temperature

Anion exchangers are temperature sensitive. When heated, the Hofmann degradation occurs transforming quaternary ammonium functional groups into tertiary amines or even destroing the functional groups completely. The strongly basic anion exchangers of type 1 such as Lewatit MonoPlus M 500, Lewatit MonoPlus MP 500, Amberlite IRA 900, Amberjet 4200, Amberjet 4400 or Purolite A520E are the most stable. However, for the strongly basic anion exchangers of type 2 such as Lewatit MonoPlus M 600, Amberlite IRA 910 and Amberjet 4600 the degradation undergoes more readily. Therefore, in the presented paper the effect of the temperature was studied in the range from 293 K to 333 K. The evaluated mean free energy value (ΔG^o) indicated that the sorption of Cu(II), Zn(II), Cd(II) and Pb(II) complexes with IDS, EDDS and GLDA onto Lewatit MonoPlus M 500, Lewatit MonoPlus M 600, Lewatit MonoPlus MP 500, Amberlite IRA 900, Amberlite IRA 910, Amberjet 4200, Amberjet 4400, Amberjet 4600 and Purolite A520E proceeded by anion exchange. The calculated thermodynamic parameters, ΔG^o, ΔH^o and ΔS^o showed that the sorption was favourable, spontaneous and endothermic under the examined conditions.

6. Conclusions

Heavy metal complexes of Cu(II), Zn(II), Cd(II) and Pb(II) with IDS, EDDS and GLDA were removed from aqueous solution using the commercially available strongly basic anion exchangers Lewatit MonoPlus M 500, Lewatit MonoPlus M 600, Lewatit MonoPlus MP 500, Amberlite IRA 900, Amberlite IRA 910, Amberjet 4200, Amberjet 4400, Amberjet 4600 and Purolite A520E as well as chelating ion exchangers Dowex M 4195 and Diphonix. The influence of phase contact time (1-180 min), anion exchanger dosage (0.25-1.0g), solution pH (1-12), initial concentration (1×10^{-3} mol/L–2×10^{-2} mol/L), temperature (293-333K) and interfering ions (Cl^-, NO_3^-, SO_4^{2-}) in the single and multi-component systems was studied by the batch method. Additionally, the obtained results were compared with those for the column studies. The amounts of complexes sorbed at equilibrium using the strongly basic anion exchangers were differentiated and the following affinity series were found:
in the M(II)-IDS=1:1 system:

Cu(II):MP500>M500>M600>M4195>IRA910>IRA900>A520E≈4200>4400>Diphonix≈4600,

Zn(II):IRA900>IRA910>M500>M600>MP500>M4195>4200≈4400≈4600>A520E>Diphonix,

Cd(II):M500>MP500>M600>IRA900>Diphonix>IRA910>M4195>4200>4400>4600>A520E,

Pb(II):Diphonix>M600>M500>MP500>M4195>4200>IRA910≈4400>IRA900≈4600>A520E;

in the M(II)-EDDS=1:1 system:

Cu(II):M4195≈M500>M600>MP500>A520E>Diphonix>IRA900>IRA910>4200>4600>4400,

Zn(II):Diphonix>M4195>MP500>M500>M600>A520E>IRA900>IRA910>4400>4600≈4200,

Cd(II):Diphonix>MP500>4200>M500>M600>4400>IRA900>IRA910>4600>M4195>A520E,

Pb(II):Diphonix>4200>A520E>4600>4400>M600>M500>MP500>IRA900>IRA910>M4195;

in the M(II)-GLDA=1:1 system:

Cu(II):MP500>M500>M600>Diphonix>IRA900>IRA910>M4195>4200>A520E>4400>4600,

Zn(II):M600≈MP500≈M500>Diphonix>4200>M4195≈4400≈IRA910≈4600>IRA900>A520E,

Cd(II):M500>M600>Diphonix>4200>4400>4600>A520E>MP500>IRA900>IRA910>M4195,

Pb(II):M500≈M600>MP500>Diphonix>4200≈4400≈M4195≈4600≈A520E>IRA900≈IRA910.

Comparing the types of the used anion exchangers, the strongly basic ones of type 1 were found to offer much greater capacities for the sorbed complexes than the anion exchangers of type 2. Also anion exchangers of macroporous type are slightly better than gel ones. For them the anion exchange process occurs. However, for Dowex M 4195 and Diphonix the decomposition of the M(II)-IDS, EDDS and GLDA complexes should be taken into account.
It was also found that the equilibrium sorption capacity increased when the temperature of solutions increased from 293 to 333 K. The Langmuir, Freundlich, Temkin and Dubinin-Radushevish (D-R) models of adsorption were used for analysis of the experimental data. The Langmuir isotherm model was found to be the most appropriate. The kinetic data were fitted using the PF-order, PS-order, W-M as well as Boyd equations. The experimental data were well described by the PS-order kinetic model.
The anion exchangers can be regenerated very efficiently using 1 M HCl. They can be used in multiple cycles of operation without loss in capacity.

7. Acknowledgments

Author is thankful to Prof. Z. Hubicki for his valuable suggestions.

8. References

Aksela, R.; Peltonen, J. & Weckman, A. WO 2004110961, 2004.

Anderson, R.L.; Bishop, E.B. & Campbell, R.L. (1985). A review of the environmental and mammalian toxicology of nitrilotriacetic acid, *Critical Reviews in Toxicology*, Vol. 15, pp. 1-102.

Blanchard, G.; Maunaye, M. & Martin, G. (1984). Removal of heavy metals from waters by means of natural zeolites, *Water Research*, Vol. 18, pp. 1501-1507.

Borowiec, M. & Hoffmann, J. (2005). New Biodegradable chelating compounds for micronutricient fertilizers. In: *Development in Production and use of new agrochemicals* (eds.: Górecki, H.; Dobrzański, Z. & Kafarski, P.) Czech-Pol Trade, Prague, Brussels, pp. 38-41.

Borowiec, M.; Polańska, P. & Hoffmann, J. (2007). Biodegradability of the compounds introduced with microelement fertilizers into the environment, *Polish Journal of Chemical Technology*, Vol. 9, pp. 38-41.

Boyd, G.E.; Adamson, A.W. & Myers, L.S. Jr. (1947). The exchange adsorption of ions from aqueous solutions by organic zeolites. II. Kinetics, *Journal of American Chemical Society*, Vol. 69, pp. 2836-2848.

Brochure of Baypure CX 100. Baypure CX 100 a new environmentally friendly alternative to conventional complexing agents - product guide of Lanxess, http://www.baypure.com/bp/en/

Brochure of Baypure DS 100. Polyaspartic acid sodium salt. A new biodegradable dispersant - product guide of Lanxess, http://www.baypure.com/bp/en/

Brochure of Enviomet™. Biodegradable chelat Enviomet TM - product guide of Innospec Inc., 2009.

Brochure of Dissolvine GL-38. Master the element - product guide of Akzo Nobel Functional Chemicals, 2007.

Brochure of Trilon M®, Trilon M® liquid chelating agent - product guide of BASF Co., 2008.

Burns, K.; Wu, Y.T. & Grant, Ch. (2003). Mechanisms of calcite dissolution using environmentally benign polyaspartic acid: a rotating disk study, *Langmuir*, Vol. 19, pp. 5669-5679.

Cao A.; Carucci, A.; Lai, T.; La Colla, P. & Tamburini, E. (2007). Effect of biodegradable chelating agents on heavy metals phytoextraction with *Mirabilis jalapa* and on its associated bacteria, *European Journal of Soil Biology*, Vol. 43, pp. 200-206.

Crouch, A.M.; Khotseng, L.E.;. Polhuis, M. & Williams, D.R. (2001). Comparative study of cyclic voltammetry with potentiometric analysis for determining formation constants for polyaminocarboxylate–metal ion complexes, *Analytica Chimica Acta*, Vol. 48, pp. 231–237.

Cokesa, Ž.; Lakner, S.; Knackmuss, H.J. & Rieger, P.G. (2004a). A stereoselective carbon-nitrogen lyase from Ralstonia sp. SLR7 cleaves two of three isomers of iminodisuccinate, *Biodegratadion*, Vol. 15, pp. 229-239.

Cokesa, Ž.; Knackmuss, H.J. & Rieger, P.G. (2004b). Biodegradation of all stereoisomers of the EDTA substitute iminodisuccinate by *Agrobacterium tumefaciens* BY6 requires an epimerase and a stereoselective C-N Lyase, *Applied and Environmental Microbiology*, Vol. 70, pp. 3941-3947.

Freeman, M.B.; Paik, Y.H.; Wilczynski, R.; Wolks, S.K. & Yocom, K.M. (1996). Biodegradability of polycarboxylates: structure-activity studies. In: *Hydrogels and biodegradable polymers for bioapplications*, (eds.: Ottenbrite, R.M.; Huang, S.J. & Park, K.) American Chemical Society, Washington, DC, chapter 10.

Goos, J. R. & Cattanach, A. (1996). Effect of "Amisorb" on spring wheat and sugarbeet. *Sugarbeet Recearch and Extention Reports*, Vol. 27, pp.143-145.

Helfferich, F. Ion exchange, McGraw-Hill, New York 1962.

Hauthal, H.G. (2009). Suitanable detergents and cleaners. Progress in ingredients, nanoparticles, analysis, environment, *Tenside Surfactant Detergents*, Vol. 46, pp. 53-62.

Heintz, W. (1862) Uber dem amoniaktypus angehorige suren. *Analitical Chemistry and Pharmacy*, Vol. 122, pp. 257-294.

Hyvönen, H.; Orama, M.; Saarinen, H. & Aksela, R. (2003) Studies on biodegradable chelating ligands: complexation of iminodisuccinic acid (ISA) with Cu(II), Zn(II), Mn(II) and Fe(III) ions in aqueous solution, *Green Chemistry*, Vol. 5, pp. 410-414.

Hyvönen, H. & Aksela, R. (2010). Complexation of 3-hydroxy-2,2'-iminodisuccinic acid (HIDS) with Mg^{2+}, Ca^{2+}, Mn^{2+}, Fe^{3+}, Fe^{2+}, Co^{2+}, Ni^{2+}, Cu^{2+}, and Zn^{2+} ions in aqueous solution, *Journal of Coordination Chemistry*, Vol. 63, pp.2013-2025.

ISO 16588:2002, Water quality - Determination of six complexing agents - gas chromatographic method. International Organization for Standardization, Geneva, 2002.

Jachuła, J.; Kołodyńska, D.; Hubicki, Z. (2011). Sorption of Cu(II) and Ni(II) ions in presence of novel chelating agent methylglycinediacetic acid by microporous ion exchangers and sorbents from aqueous solutions, *Central European Journal of Chemistry*, Vol. 9, pp. 52-65.

Jaworska, J.S.; Schowaneck, D. & Feitzel, T.C.J. (1999). Environmental risk assessment for trisodium [S, S]-ethylene diaminedisuccinate, a biodegradable chelator used in detergent applications, *Chemosphere*, Vol. 38, pp. 3597-3625.

Jones, P.W. & Williams, D.R. (2002). Chemical speciation simulation used to assess the efficiency of environment-friendly EDTA alternatives for use in the pulp and paper industry, *Inorganica Chimica Acta*, Vol. 339, pp. 41-50.

Katata, L.; Nagaraju, V. & Crouch, A.M. (2006). Determination of ethylenediaminetetraacetic acid, ethylenediaminedisuccinic acid and iminodisuccinic acid in cosmetic products by capillary electrophoresis and high performance liquid chromatography, *Analytica Chimica Acta*, Vol. 579, pp. 177-184.

Knepper, T.P. (2003). Synthetic Chelating Agents and Compounds Exhibiting Complexing Properties in the Aquatic Environment. *Trends in Analytical Chemistry*, Vol. 22, pp. 708-724.

Knepper, T.P. ; Werner, A. & Bogenschütz, G. (2005). Determination of synthetic chelating agents in surface and waste water by ion chromatography–mass spectrometry, *Journal of Chromatography A*, Vol. 1085, pp. 240-246.

Kołodyńska, D.; Hubicki, Z. & Gęca, M. (2008a). Application of a new generation complexing agent in removal of heavy metal ions from aqueous solutions, *Industrial & Engineering Chemistry Research*, Vol. 47, pp. 3192-3199.

Kołodyńska, D.; Hubicki, Z. & Gęca, M. (2008b). Polyaspartic acid as a new complexing agent in removal of heavy metal ions on polystyrene anion exchangers, *Industrial & Engineering Chemistry Research*, Vol. 47, pp. 6221-6227.

Kołodyńska, D.; Hubicka, H. & Hubicki, Z. (2009). Studies of application of monodisperse anion exchangers in sorption of heavy metal complexes with IDS, *Desalination*, Vol. 239, pp. 216-228.

Kołodyńska, D. (2009a). Polyacrylate anion exchangers in sorption of heavy metal ions with the biodegradable complexing agent, *Chemical Engineering Journal*, Vol. 150, pp. 280-288.

Kołodyńska, D. (2009b). Iminodisuccinic acid as a new complexing agent for removal of heavy metal ions from industrial effluents, *Chemical Engineering Journal*, Vol. 152, pp. 277-288.

Kołodyńska, D. (2010a). Biodegradable complexing agents as an alternative to chelators in sorption of heavy metal ions, *Desalination and Water Treatment. Science and Engineering*, Vol. 16, pp. 146-155.

Kołodyńska, D. (2010b). Diphonix Resin® in sorption of heavy metal ions in the presence of biodegradable complexing agents of a new generation, *Chemical Engineering Journal*, Vol. 159, pp. 27-36

Kołodyńska, D. (2010c). The effect of the novel complexing agent in removal of heavy metal ions from waters and wastewaters, *Chemical Engineering Journal*, Vol. 165, pp. 835-845.

Kołodyńska, D. (2011a). Cu(II), Zn(II), Co(II) and Pb(II) removal in the presence of the complexing agent of a new generation, *Desalination*, Vol. 267, pp. 175-183.

Kołodyńska, D. (2011b). The effects of the treatment conditions on metal ions removal in the presence complexing agents of a new generation, *Desalination*, doi: 10.1016/j.desal.2010.06.053

Kołodyńska, D. (2011c). Application of strongly basic anion exchangers for removal of heavy metals in the presence of green chelating agent, *Chemical Engineering Journal*, doi:10.1016/j.cej.2011.01.073.

Kołodyńska, D. (2011d). Green complexing agent - EDDS in removal of heavy metal ions on strongly basic anion exchangers, *Desalination*, in press.

Kos, B. & Lěstan, D. (2003). Influence of a biodegradable ([S,S]-EDDS) and nondegradable (EDTA) chelate and hydrogel modified soil water sorption capacity on Pb phytoextraction and leaching, *Plant and Soil*, Vol. 253, pp. 403-411.

Lagergren, S. (1898). Zur Theorie der sogenannten adsorption gelöster. Stoffe, Kungliga Svenska Vetenskapsakademiens, Handlingar , Vol. 24, 1-39.

Loonen, H.; Lindgren, F.; Hansen, B.; Karcher, W.; Niemelä, J.; Hiromatsu, K.; Takatsuki, M.; Peijnenburg, W.; Rorije, E. & Struijs, J. (1999). Prediction of biodegradability from chemical structure: Modeling of ready biodegradation test data, *Environmental Toxicology and Chemistry*, Vol. 18, pp. 1763-1768.

Martell A.E. & Smith R.M. (1982). *Critical Stability Constants*, Vol. 5: First Supplement, Plenum Press, New York.

Matzinger, I.; Brunthaler, J.; Fürhapper, C.; Pollak, U. & Aschacher, G. (2007). Biodegradation of new complexing agents in compliance with Austrian waste water emission legislation for the pulp & paper industry. In: *Complexing Agents between Science, Industry, Authorities and Users.* (eds.: Nowack, B. & Giger, W.) Monte Verità, Ascona, Switzerland, March 11-16, 2007, pp. 19.

Minczewski, J.; Chwastowska, J. & Dybczyński. R. *Separation and preconcentration methods in inorganic trace analysis.* John Wiley & Sons, New York 1982.

Mosig, J.; Gooding, C.H. & Wheeler, A.P. (2000). Kinetic and thermal characterization of the hydrolysis of polysuccinimide, *Industrial and Engineering Chemistry Research*, Vol. 36, pp. 2163-2170.

Neal, J.A. & Roshe, N.J. (1973). Further studies of the coordination of othylonediaminedisuccinic acid, an isomer of ethylenediaminetetraacetic acid, *Inorganic Chemistry*, Vol. 12, pp. 1226-1232.

Nowack, B. (1998). The behaviour of phosphonates in wastewater treatment plants of Switzerland, *Water Research*, Vol. 32, pp. 1271-1279.

Nowack, B. (2003). Environmental chemistry of phosphonates, *Water Research*, Vol. 37, pp. 2533-2546.

Nowack, B. (2007). Chelating agents - overview and historical perspective, In: *Complexing Agents between Science, Industry, Authorities and Users.* (eds.: Nowack, B. & Giger, W.) Monte Verità, Ascona, Switzerland, March 11-16, 2007, pp. 25.

Nörtemann, B. (2005). Biodegradation of chelating agents: EDTA, DTPA, PDTA, NTA, and EDDS. In: *Biogeochemistry of Chelating Agents*, (eds.: Nowak, B. & VanBriesen, J.M.) ACS Symposium Series, Vol. 910, pp. 150-170.

Orama, M.; Hyvönen, H.; Saarinen, H. & Aksela, R. (2002). Complexation of [*S,S*] and mixed stereoisomers of *N,N'*-ethylenediaminedisuccinic acid (EDDS) with Fe(III), Cu(II), Zn(II) and Mn(II) ions in aqueous solution, *Dalton Transactions*, pp. 4644-648.

Reichenberg, D. (1953). Properties of ion-exchange resins in relation to their structure. III. Kinetics of exchange, *Journal of American Chemical Society*, Vol. 75, pp. 589-597.

Reinecke, F.; Groth, T.; Heise, K.P.; Joentgen, W.; Müller, N. & Steinbüchel, A. (2000) Isolation and characterization of an *Achromobacter xylosoxidans* strain B3 and other bacteria capable to degrade the synthetic chelating agent iminodisuccinate, *FEMS Microbiology Letters*, Vol. 188, pp. 41-46.

Schneider, J.; Potthoff-Karl, B.; Kud, A.; Baur, R.; Oftring, A. & Greindl, T. (1999). Use of glycine-N,N-diacetic acid derivatives as biodegradable complexing agents for alkaline earth metal ions and heavy metal ions, United States Patent 6,008,176.

Seetz, J. (2007). Bound by biodegradability, *Soap Perfumery and Cosmetics*, Vo. 4, pp. 75-76.

Seetz, J. & Stanitzek, T. (2008). GLDA: The new green chelating agent for detergents and cosmetics. *SEPAWA Congress and European Detergents Conference Proceedings*, 15-17 October 2008, pp. 17-22.

Sillanpää, M.; Kokkonen, R.; Sihvonen, M. (1995). Determination of EDTA and DTPA as their Fe(III) complexes in pulp and paper mill process and waste waters by liquid chromatography, *Analytica Chimica Acta*, Vol. 303, pp. 187-192.

Stegient-Nowicka, J. & Michałek, B. (2010). Fe(III)HJB and Fe(III) HBED chelates as efficient microelement fertilizers, *Przemysł Chemiczny*, Vol. 89, pp. 1520-1522 (in Polish).

Sun, B.; Mi, Z.T.; Ouyang, J.; An, G. & Song, X.K. (2005). Pb^{2+} binding by polyaspartyl polymers and their application to Pb^{2+} removal from glycyrrhizin, *Journal of Applied Polymer Science*, Vol. 97, pp. 2215-2220.

Tandy, S; Bossart, K.; Mueller, R.; Ritschel, J.; Hauser, L.; Schulin, R. & Nowack, B. (2004). Extraction of heavy metals from soils using biodegradable chelating agents, *Environmental Science and Technology*, Vol. 38, pp. 937-944.

Tandy, S.; Ammann, A.; Schulin, R. & Nowack, B. (2006a). Biodegradation and speciation of residual SS-ethylenediaminedisuccinic acid (EDDS) in soil solution left after soil washing, *Environmental Pollution*, Vol. 142, pp. 191-199.

Tandy, S.; Schulin, R. & Nowack, B. (2006b). Uptake of metals during chelant-assisted phytoextraction with EDDS related to the solubilized metal concentration, *Environmental Science and Technology*, Vol. 40, pp. 2753-2758.

Tucker, M.D.; Barton, L.L.; Thomson, B.M.; Wagener, B.M. & Aragon. A. (1999). Treatment of waste containing EDTA by chemical oxidation, *Waste Management*, Vol. 19, pp. 477-482.

Vasilev, V.P.; Katrovtseva, A.V.; Gorelov, I.P. & Tukumova, N.V. (1996). Stability of complexes of Ni(II) with iminodisuccinic acid, *Journal of Inorganic Chemistry*, Vol. 41, pp. 1320-1323 (in Russian).

Vasilev, V.P.; Katrovtseva, A.V.; Bychkova, S.A. & Tukumova, N.V. (1998) Stability of complexes of Co(II) and Cu(II) with iminodisuccinic acid, *Journal of Inorganic Chemistry*, Vol. 43, pp. 808-809 (in Russian).

Vandevivere, P.C.; Saveryn, H.; Verstraete, W.; Feijtel, T.C.J. & Schowanek, D.R. (2001). Biodegradation of metal-[S,S]-EDDS complexes, *Environmental Science and Technology*, Vol. 35, pp. 1765-1770.

Weber W.J. Jr. & Morris, J.C. (1963). Kinetics of adsorption on carbon from solution, *Journal of the Sanitary Engineering Division*, Vol. 89, pp. 31-59.

Wheeler, A.P. & Koskan, L.P. (1993). Large scale thermally synthesized polyaspartate as a substitute in polymer applications, *Materials Research Society Symposium Proceedings*, Vol. 292, pp. 277-283.

Witschel, M.; Egli, T.; Zehnder, A.J.B.; Wehrli, E. & Spyeher, M. (1999). Transport of EDTA into cells of the EDTA degradingstrain DSM 9103, *Microbiology*, Vol. 154, pp. 973-983.

Wu, Y.T. & Grant, Ch. (2002). Effect of chelation chemistry of sodium polyaspartate on the dissolution of calcite, *Langmuir*, Vol. 18, pp. 6813-6820.

Yuan, Z.; VanBriesen J.M. (2008). Bacterial growth yields on EDTA, NTA, and their biodegradation intermediates, *Biodegratadion*, Vol. 19, pp. 41-52.

Zhao, Y.; Su, H.; Fang, L. & Tan, T. (2005). Superabsorbent hydrogels from poly(aspartic acid) with salt-,temperature- and pH-responsiveness properties, *Polymer*, Vol. 46, pp. 5368-5376.

Zhao, Y.; Kang, J. & Tan, T. (2006). Salt-, pH- and temperature-responsive semi-interpenetrating polymer network hydrogel based on poly(aspartic acid) and poly(acrylic acid), *Polymer*, Vol. 47, pp. 7702-7710.

Zingg, T., (1935). Beitrag zur schotteranalyse. *Schweizerische Mineralogische und Petrologische Mitteilungen*, Vol. 15, pp. 39-140

Zwicker, N.; Theobald, U.; Zähner, H. & Fiedler, H.P. (1997) Optimization of fermentation conditions for the production of ethylenediaminedisuccinic acid by *Amycolatopsis orientalis*, *Journal of Industrial Microbiology and Biotechnology*, Vol. 19, pp. 280-285.

http://dardel.info/IX/

18

Operating Experience of Desalination Unit Coupled to Primary Coolant System of Cirus

R.C. Sharma and Rakesh Ranjan
Research Reactor Services Division and Reactor Operations Division,
Bhabha Atomic Research Centre, Mumbai,
India

1. Introduction

CIRUS is a 40 MWth, heavy water moderated, light water cooled, natural uranium fuelled research reactor located at Mumbai (Bombay), India. Seawater, after removing heat from primary coolant, is released back into sea in an open loop. The reactor had been in operation since 1960 with an average availability factor of about 70% for about three decades. During early nineties reactor exhibited signs of ageing. Systematic ageing studies were undertaken to identify refurbishment requirements with the aim of extending the life of reactor by 10–15 years. Refurbishment plan was finalized and reactor was shut down for refurbishment during October 1997. During refurbishment, a 30 Ton/day capacity Desalination unit had been integrated with the primary cooling water system of CIRUS towards demonstration of nuclear reactor waste heat utilization for desalination. The desalination unit, developed and designed by BARC, works on low temperature vacuum evaporation process. It was refurbished before shifting it to CIRUS.

A task force evaluated the various available heat sources at CIRUS and worked out the scheme for integrating primary coolant circuit of CIRUS with the desalination unit. The design, procurement, erection, installation and commissioning of the unit was done in-house. During post-refurbishment operation of the reactor, the unit has been operated at its rated capacity and its performance has been satisfactory. The product water has been utilized to meet demineralized water inventory requirement of primary coolant system.

2. Low Temperature Vacuum Evaporation (LTVE) desalination unit

Please refer fig 1. The 30 t/d LTVE desalination unit is a self-contained unit suitable for installation on a stationary structure or a floating platform. Water jet ejectors are used to create vacuum and pump out concentrated seawater from the evaporator. Scale formation is practically eliminated because of the low boiling temperature and brine density. This unit is eco-friendly, since it does not require chemical pre-treatment of the feed seawater. Apart from the electric energy required for the pump, no other power or fuel is necessary, except for waste heat at around 65˚C to heat seawater (feed water) to at least 41˚C.

2.1 Selection of heat source

Hot outlet seawater (SW) and primary cooling water (PCW) were considered as heat source for the evaporator - condenser section of the Desalination unit. Minimum temperature and flow of outlet seawater from the reactor to be used as direct source of heat was estimated as 50°C and 2500 lpm respectively. Maximum temperature of outlet seawater from CIRUS is 49°C at 40 MW operation (at 29°C inlet temperature). At lower power levels, the temperature will still be lower. Hence use of outlet seawater as heat source was not feasible. The outlet hot water as feed water itself was found unsuitable due to various operational and radiological considerations.

Since the primary coolant system equipment / piping handles radioactive de-mineralized water, they need to be housed within the reactor building and can't be taken out for direct coupling to the desalination unit installed outside the reactor building from radiological considerations. This necessitated an intermediate circuit as a heat carrier between the primary coolant circuit and the evaporator of the Desalination unit. The need for an intermediate circuit was also necessitated as a barrier between active primary coolant circuit and inactive desalination unit in the event of any leakage from the primary coolant circuit. The intermediate circuit coolant temperature (minimum) and flow, to be effective heat carrier, was envisaged as 53°C and 1600 lpm respectively. To achieve this temperature, minimum temperature of primary coolant at the inlet of intermediate circuit heat exchanger was estimated as 58°C. During normal operation of CIRUS at rated power of 40 MW, primary coolant temperature is 80 °C while at 20 MW operation, it is about 65 °C. Thus adequate margins were available for operation of the unit even when reactor is operated at 50% of its rated capacity. Adequate space was available in reactor annulus building for installation of intermediate circuit heat exchanger and associated piping and desalination unit just outside the building. Hence primary coolant circuit of CIRUS was found suitable for integration of the desalination unit. Coupling of the primary coolant circuit pipelines with the intermediate circuit heat exchanger was easy as suitable isolation / pipe tapping already existed.

Fig. 1. Plate type H.Ex to transfer heat from primary coolant to intermediate circuit.

2.2 Integration of desalination unit with cirus

A scheme was worked out to transfer heat from primary coolant circuit of reactor to desalination unit through an intermediate (DM water) circuit. An intermediate heat exchanger with associated piping and pump was installed to transport heat to desalination unit. Description of the desalination unit is given below:

About 1 MW of heat energy is transferred from the primary coolant to the desalination unit through an intermediate DM water circuit. For this, 1500 lpm flow is tapped off from the upstream of PCW/ SW heat exchangers. The hot coolant after transferring heat to intermediate circuit in a heat exchanger joins back to primary circuit at the downstream of PCW/ SW heat exchangers.

Intermediate (DM water) circuit works as a heat carrier to the condenser-evaporator unit. Recirculation flow through the circuit is maintained at about 1600 lpm. During normal operation, pressure of the intermediate circuit in the heat exchanger remains more than the primary coolant side to prevent ingress of active primary coolant water into the intermediate circuit. About 15 lpm flow is maintained through a de-oxygenator and ion exchanger to control water chemistry.

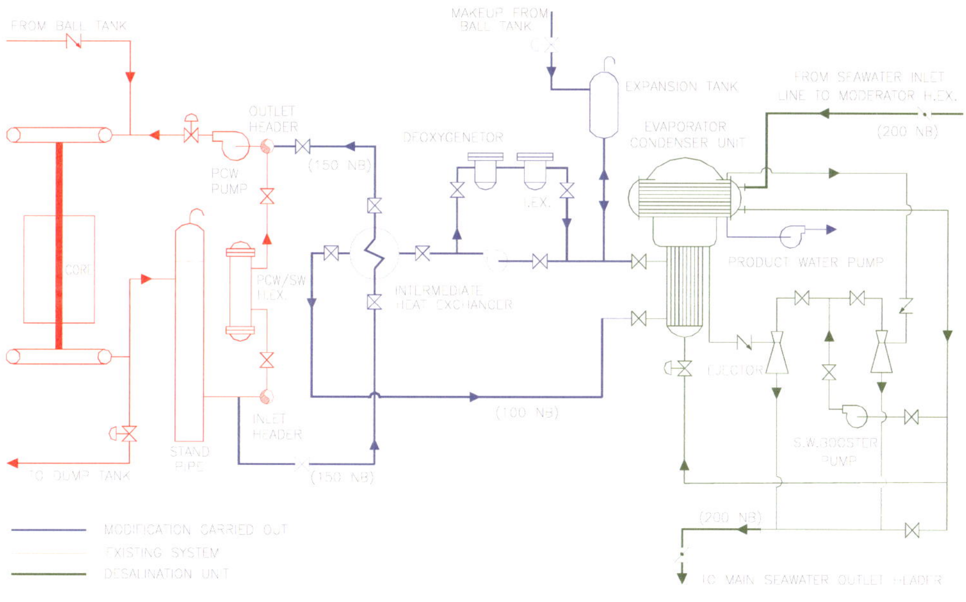

Fig. 2. Schematic of integration of Desalination unit with Cirus

Seawater at 2000 lpm is tapped from main seawater inlet line and is supplied to condenser of the condenser-evaporator unit after passing through a strainer. From the condenser outlet, 65 lpm seawater is supplied as feed water to the evaporator and 1100 lpm through booster pump to the vacuum ejectors as motive force. Normal vacuum created by ejectors is higher than 700 mm of Hg. The outlet from these units together with the balance flow from condenser outlet joins the main seawater outlet header. The desalination unit is located at about 1.5 m above ground level to maintain atmospheric pressure at ejector outlets and also to ensure flow of seawater from ejector outlet to main seawater outlet line under gravity.

Product water circuit works in an open loop. One conductivity meter is provided on product water pump discharge line. The pump transfers product water from condenser to

Dump Tank # 1(underground concrete tank), if specific conductivity of water is below preset value (20 μS/cm). If specific conductivity is high, air operated control valve at pump discharge line diverts product water to storm drain. From Dump Tank # 1, water is transferred to Dump Tank # 3 (underground concrete tank) through mixed bed resin cartridge for augmenting the make up requirement of primary cooling water system.

Instrumentation for monitoring various process parameters like flow, temperature, pressure and level has been provided with alarm at appropriate level. All alarms are hooked up to a local panel and a common "off-normal" alarm is provided in control room.

Fig. 3. Evaporator section of Desalination unit and Booster pump

Fig. 4. Desalination unit located outside Cirus annulus area

3. Operation of desalination unit

Operation of desalination unit was commenced after raising reactor power to NP-60%. Effect of operating the unit on primary coolant circuit of the reactor was assessed and found matching well with design stipulations. All operating parameters of the unit were found as intended. Subsequently reactor power was raised to its rated capacity in steps and all operating parameters were found normal at the corresponding power level. The unit operation at its rated capacity was satisfactory.

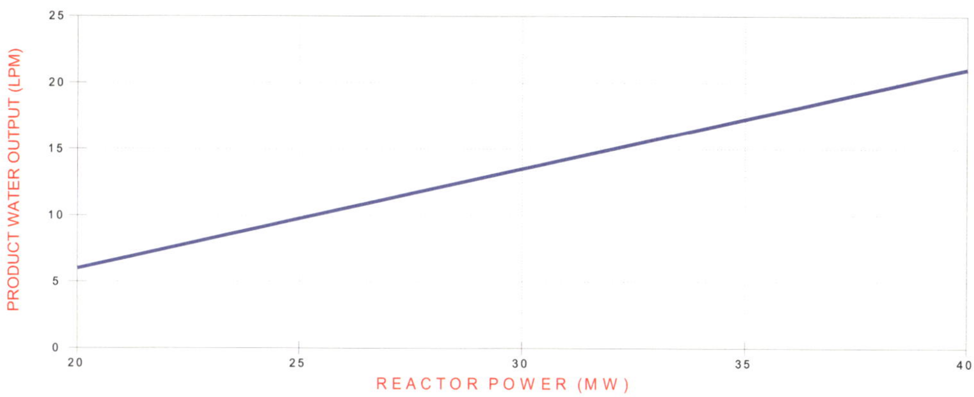

REACTOR POWER VS. PRODUCT WATER OUTPUT

3.1 Product water quality

Typical quality of product water generated from desalination unit was as below:

pH	K (μS/cm)	TDS ppm	Na ppm	K ppm	Ca ppm	Mg ppm	Fe ppm	Cu ppm	Ni ppm	SiO_2 ppm	Cl^- ppm
5.8	15-18	9.8	1.4	<0.1	0.1	0.2	0.59	0.31	0.17	<0.2	3.9

Table 1.

To generate de-mineralized water for primary coolant inventory, service water supplied by City Municipal Corporation, is passed through ion exchangers. The specific conductivity of service water varies between 100 – 160 μS/cm while that of product water generated by desalination unit could be achieved as low as 15 μS/cm. Hence use of product water as feed stock for DM water has resulted into reduction in demand for operation of ion exchangers. Consequently consumption of regenerants, generation of liquid waste for regeneration of ion exchangers and Person-Seivert consumption has also come down.

3.2 Installation of demister

Towards improving the quality of product water, demister pads (made of knitted wire mesh of layered configuration of SS 316 having wire diameter of 0.3 mm, 50 mm thick and provided with SS 316 support grid at the bottom) were installed in the evaporator section of the unit. With this, product water specific conductivity came down from 10-15 μS/cm to 3-4 μS/cm. As pads were not knitted as per the size of evaporator, but fabricated from the rectangular pieces available in the market, bypassing of entrainments could not be eliminated completely. Also available pads were of 50 mm thick only against the designed thickness of 75 mm. Procurement of pads of required thickness is on hand and its knitting at site as per the actual shape will result into achieving the desired quality of DM water of 2 μS/cm.

Fig. 5. Demister Pads

With the installation of demister pads, initial start up time of the unit (time for the product water specific conductivity to come down less than 20 uS/cm) has come down to less than 1.0 hour from 3-4 hours earlier. It has further resulted into reduction in i) demand for operation of ion exchangers, ii) generation of liquid waste for regeneration of ion exchangers, iii) consumption of regenerants and iv) Person-Seivert consumption.

4. Experience with the operation of desalination unit

The successful integration of low temperature vacuum evaporation unit with CIRUS has demonstrated the technology for use of low temperature waste heat for generation of desalinated potable water. The unit has been operating successfully at rated capacity. The operation of the unit for over five years has provided valuable experience related to behavior of the unit during normal as well as off-normal conditions which will serve as feedback for further improvement. Some of the important observations are described below:

- High leakage through gland packing of intermediate circuit pump and product water pump used to be a constant problem during the initial days of the unit operation. Apart from loss of water from system and spillage in area, the pumps used to lose suction resulting into shut down of the unit. Both the pumps have been replaced with the pumps having mechanical seals. The pump availability and hence the unit availability has significantly improved.
- Sea water booster pumps also used to develop frequent leak from gland packing. The pumps used to lose suction resulting into shut down of the unit. Installation of mechanical seal did not help as suspended matter in sea water used to erode the sealing surface of the seals. External cooling with service water has been provided and pump availability has significantly improved.
- Specific conductivity meter of product water was fouling frequently due to water stagnancy as it was located at one dead end of the pipe. The conductivity meter cell has

been shifted to appropriate location with minor modification in piping lay out such that continuous water flow takes place through its housing.

- Condenser & Evaporator level - Tiny air bubbles were observed in the product water level gauge of condenser if the level was below 250 mm. To maintain more than 250 mm level, frequent adjustment of product water flow was needed by throttling product water pump discharge valve. Hence one ½" line with a valve was installed in parallel to product water discharge valve (1" size) for better control of product water flow and condenser water level in desired range (250 – 350 mm) could be achieved.
- On a typical occasion when unit was started, specific conductance of product water was not coming down below 1000 μS/cm although all other operating parameters of the unit showed normal values except quick level build up in the condenser. Detailed inspection revealed that a dished shape separator located between evaporator and condenser section which returns back the brine carried over with vapors to the evaporator and hence controls the quality of the product water; had got detached and fallen down. The separator was held in position by a M.S. hanger attached to the condenser. The hanger failed after two years of service due to corrosion and hence the separator fell down over the evaporator. The separator was fixed back in its position with hangers made of SS 316. An additional hanger was also provided and normal operation of the unit was resumed.
- The evaporator-condenser section of desalination unit employs two ejectors. Ejector #1 is primarily used for creation of vacuum while ejector #2 sucks high salinity water from the evaporator and discharges it to the common seawater outlet line to avoid scale formation. It also helps in maintaining water level in the evaporator. Frequent failures of these ceramic lined MS ejectors due to erosion-corrosion was solved by replacing them with SS 316 made ejectors. This improved availability of the unit.
- The suction line of product water pump started leaking frequently due to corrosion as the line remains in contact with high salinity water till the required quality of product water is achieved. The MS line was replaced with that of SS 316.
- Installation of demister improved the quality of product water to near DM water quality.

5. Conclusion

Integration of desalination unit with CIRUS and its successful operation has demonstrated the utilization of low temperature waste heat of a nuclear reactor in generation of potable water (or near DM water quality) from seawater. The quality of product water is suitable for generation of demineralised water after polishing through ion exchanger. The valuable experience gained in integrating the unit with an old reactor, commissioning, operation at rated capacity and modifications carried out to improve availability; will be of immense help in installation of such units at the places having potential for utilization of low temperature waste heat.

6. References

[1] Design manual of Desalination unit at CIRUS
[2] Commissioning of desalination unit integrated with CIRUS: An internal report of Reactor Group, BARC by Rakesh Ranjan, Surinder Kumar and Alok Srivastava.

[3] Low temperature vacuum evaporation process for seawater desalination using waste heat from PHWRs by Shri P.K. Tiwari, R.K. Verma, B.M. Misra and H.K. Sadhukhan; IAEA -SM -347/34 proceeding of a symposium Taejon, Republic of Korea, 26-30 May, 1997.
[4] Fresh water generation on board a floating platform: Floating nuclear plants for seawater desalination, IAEA TECHDOC -940, IAEA VIENNA, (1997) 65-70.
[5] Introduction to nuclear desalination - IAEA guide book, Technical series 400, Vienna(2000)

19

Nanofiltration Used for Desalination and Concentration in the Manufacture of Liquid Dyes Production

Petr Mikulášek and Jiří Cuhorka
University of Pardubice
Czech Republic

1. Introduction

Reactive dye is a class of highly coloured organic substances, primarily used for tinting textiles. The dyes contain a reactive group, either a haloheterocycle or an activated double bond, which, when applied to a fibre in an alkaline dye bath, forms a chemical bond between the molecule of dye and that of the fibre. The reactive dye therefore becomes a part of the fibre and is much less likely to be removed by washing than other dyestuffs that adhere through adsorption. Reactive dyeing, the most important method for the coloration of cellulosic fibres, currently represents about 20-30% of the total market share for dyes, because they are mainly used to dye cotton which accounts for about half of the world's fibre consumption.

Generally, reactive dyes are produced by chemical synthesis. Salt, small molecular weight intermediates and residua compounds are produced in the synthesis process. These salt and residual impurities must be removed before the reactive dyes are dried for sale as powder to meet product quality requirement. Conventionally, the reactive dye is precipitated from an aqueous solution using salt. The slurry is passed through a filter press, and the reactive dye is retained by a filter press. The purity of the final reactive dye product in conventional process is low, having a salt content around 30%. Furthermore, the conventional process is carried out in various batches, which makes the entire process highly labor intensive and causes inconsistency in the production quality.

In dye manufacture, like most other processes, there is a continual search for production methods that will improve product yield and reduce manufacturing costs. Dye desalting and purification, the process by which impurities are removed to improve the quality of the product, is currently one of the biggest applications for nanofiltration (NF) technology. Dye manufacturers are now actively pursuing the desalting of the finished dye prior to spray drying because it not only improves product quality, but makes spray drying more efficient because the granulation of the dye takes place without the production dust. NF is proving to be an ideal method for this salt removal (Yu et al., 2001; He et al., 2007).

Nanofiltration is the most recently developed pressure-driven membrane separation process and has properties that lie between those of ultrafiltration (UF) and reverse osmosis (RO). The nominal molecular weight cut-off (MWCO) of NF membranes is in the range 200-1000 Da. Separation may be due to solution diffusion, sieving effects, Donnan and dielectric effects. The rejection is low for salts with mono-valent anion and non-ionized organics with a molecular weight below 150 Da, but is high for salts with di- and multi-valent anions and organics with a molecular weight above 300 Da. Thus, NF can be used for the simultaneous removal of sodium chloride (salt) and the concentration of aqueous dye solutions (Mulder, 2000; Oatley et al., 2005).

Diafiltration is the process of washing dissolved species through the membrane, which is to improve the recovery of the material in permeate, or to enhance the purity of the retained stream. Typical applications can be found in the recovery of biochemical products from their fermentation broths. Furthermore, diafiltration can be found in removal of free hydrogel present in external solution to purify of a semi-solid liposome (SSL), purification of polymer nanoparticles, enhancing the protein lactose ratio in whey protein products, separating sugars or dye from NaCl solution (desalting), and many other fields. According to the property of the solute and the selectivity of membrane, diafiltration can be used in the process of MF, UF or NF (Al-Zoubi et al., 2007; Bowen & Mohammad, 1998; Foley, 2006; Kovács et al., 2009a; Wang et al., 2008; Weselowska et al., 2004).

The aim of this study is also devoted to the mathematical modelling of nanofiltration and description of discontinuous diafiltration by periodically adding solvent at constant pressure difference.

The proposed mathematical model connects together the design equations and model of permeation through the membrane. The transport through the membrane depends on the different approaches. Firstly the membrane is regarded to a dense layer and in this case transport is based on solution-diffusion model (Cséfalvay et al., 2009; Das et al., 2008). Second approach is regarded membrane to porous medium. Models with this approach are based mainly on extended Nernst-Planck equation. Through this approach, a system containing any number of n ions can be described using set of (3n + 2) equations. In this approach, it is assumed that the flux of every ion through the membrane is induced by pressure, concentration and electrical potentials. These models describe the transport of ions in terms of an effective pore radius r_p (m), an effective membrane thickness/porosity ratio $\Delta x/A_k$ (m) and an effective membrane charge density X_d (mol/m^{-3}). Such a model requires many experiments for determination of these structural parameters. These models are hard to solve (Bowen & Mohammad, 1998; Hussain et al., 2008; Kovács et al., 2009a; Oatley et al., 2005). The last approach is based on irreversible thermodynamics. These models assume the membrane as "black box" and have been applied in predicting transport through NF membranes for binary systems (Kedem-Katchalsky, Spiegler-Kedem models). Perry and Linder extended the Spiegler and Kedem model to describe the salt rejection in the presence of organic ion. This model describes transport of ion through membrane in terms of salt permeability P_s, reflection coefficient σ (Al-Zoubi et al., 2007; Kovács et al., 2009a; Kovács et al., 2009b; Koyuncu & Topacik, 2002; Schirg & Widmer, 1992). In our work is solution-diffusion model used. The solution-diffusion model can be replaced by more theoretical model in future.

The process of desalination of aqueous dye-salt solutions by polymeric nanofiltration membranes using commercially available modules was studied. The influence of dye and salt concentration on the salt rejection and pressure applied on the flux as well as comparison of individual NF membranes for desalting purposes is presented. Extended Spiegler-Kedem model including Donnan exclusion mechanism and the term of concentration dependence of salt permeability was used for prediction of salt rejection. The great interest is also devoted to the mathematical modelling of nanofiltration and description of discontinuous diafiltration by periodically adding solvent at constant pressure difference. This study illustrates that nanofiltration is a promising process for dye desalting and concentrating.

2.1 Mathematical modelling

Mathematical model connects together balance equations and solution-diffusion model, which is extended by dependence of salt permeability on the salt concentration in feed and Donnan equilibrium.

The balances for the concentration mode can be written as:

Solvent mass balance:

$$\frac{d(V_F \rho_F)}{d\tau} = -JA^* \rho_P \tag{1}$$

Mass balances of dye and salt:

$$\frac{d(V_F c_{D,F})}{d\tau} = -JA^* c_{D,P} \tag{2}$$

$$\frac{d(V_F c_{S,F})}{d\tau} = -JA^* c_{S,P} \tag{3}$$

Eq. (1) is possible to write in the form:

$$\frac{dV_F}{d\tau} = -J_V A^* \tag{4}$$

Mass balances of dye and salt are formally same and we can solve them together. Subscripts i represent dye and salt. Eq. (2) (or (3)), may be re-written as:

$$\frac{d(V_F c_{i,F})}{d\tau} = -J_V A^* (1 - R_i) c_{i,F} \tag{5}$$

where R_i is real rejection.

In the concentration mode, the volume and the concentration in feed depends on the time. Expanded differential equation with using the product rule can be written as:

$$V_F \frac{dc_{i,F}}{d\tau} + c_{i,F} \frac{dV_F}{d\tau} = -J_V A^* (1 - R_i) c_{i,F} \tag{6}$$

Substituting Eq. (4) into Eq. (6) leads to:

$$V_F \frac{dc_{i,F}}{d\tau} = J_V A^* R_i c_{i,F} \tag{7}$$

Dividing Eq. (7) by Eq. (4) leads to:

$$V_F \frac{dc_{i,F}}{dV_F} = -R_i c_{i,F} \tag{8}$$

If we assume constant rejection and permeate flux (for small change of volume in feed tank, or better of yield - permeate volume divided by feed volume, it is achieved) or the average values integrations of Eq.(8) and Eq.(4) with the boundary conditions (V_{F0} to V_F) resulting in Eq.(9) and Eq.(10):

$$c_{i,F} = c_{i,F0}\left(\frac{V_{F0}}{V_F}\right)^{R_i} \tag{9}$$

$$\tau = \frac{V_{R0} - V_R}{J_V A^*} \tag{10}$$

On the base of Eq. (9) and Eq. (10) we can obtain the concentration in feed tank and the time for separation of certain permeate volume in concentration mode, respectively. Next process is diluting. Pure solvent (water) is used as diluant. Salt concentration in feed tank after this operation (c_S') is:

$$c_S' = c_{i,F0}\left(\frac{V_F}{V_{F0}}\right) \tag{11}$$

This concentration (c_S') is now equal to the salt concentration in feed tank ($c_{S,F0}$) for the next concentration mode in the second diafiltration step.
For solving of these equations we need to know dependence of rejection and permeate flux on salt concentration in feed.
The basic equations for rejection can be written as:

$$J_S = B(c_F - c_P) \tag{12}$$

$$c_P = \frac{J_S}{J_V} \tag{13}$$

This model can be extended by the dependence of salt permeability on salt concentration in the feed (Koyuncu & Topacik, 2002). To avoid some in conveniences with units, here c* is introduced and chosen to be 1 g/l.

$$B = \alpha.\left(\frac{c_{S,F}}{c^*}\right)^{\beta} \tag{14}$$

Assuming equilibrium on the membrane - solution interface we can obtain (approximating activities with concentrations) (Koyuncu & Topacik, 2002):

$$c_{S,W} = c_{S,F}\cdot\left(1 + \frac{v_D \cdot c_{D,F} \cdot M_S}{c_{S,F} \cdot M_D}\right) \tag{15}$$

In the presence of Donnan exclusion forces, induced by the impermeable organic ions, the salt transport across the membrane proceeds as if the membrane were exposed to a salt solution having concentration $c_{S,W}$ instead $c_{S,F}$. Thus the value of $c_{S,W}$ and not that of $c_{S,F}$ determines the driving force for the salt passage.

Then the expression for salt passage in the presence of retained organic ion can be written as:

$$J_S = \alpha\left(\frac{c_{S,F}}{c^*}\right)^{\beta}\left(c_{S,W} - c_{S,P}\right) \tag{16}$$

and then salt concentration in permeate can be expressed as

$$c_{S,P} = \frac{\alpha \dfrac{c_{S,F}^{\beta+1}}{c^{*\beta}}\left(1 + \dfrac{v_D . c_{D,F} . M_S}{c_{S,F} . M_D}\right)}{J_V + \alpha c_{S,F}^{\beta} c^{*-\beta}} \tag{17}$$

For the permeate flux these equations can be used:

$$J_V = A\left(\Delta P - \Delta\pi_s - \delta\right) \tag{18}$$

Eq. (18) is the osmotic pressure model. This model is used in similar form by many authors (Al-Zoubi et al., 2007; Cséfalvay et al., 2009; Das et al., 2008; Kovács et al., 2009a; Kovács et al., 2009b; Koyuncu & Topacik, 2002; Mulder, 2000). Parameter A (water permeability) can be concentration or viscous depended (Cséfalvay et al., 2009; Kovács et al., 2009a). For our model we assume this parameter as constant. Coefficient δ represents the effect of dye on flux. This means mainly osmotic pressure of dye. If this parameter represents only osmotic pressure of dye, then it is constant too (constant dye concentration).

The osmotic pressure gradient for salt is related to the difference of the concentration Δc by the van´t Hoff law:

$$\Delta\pi_S = \frac{\nu R^* T}{M}\Delta c_S \tag{19}$$

where

c is concentration,
A^* - membrane area,
A - water permeability,
B - salt permeability,
J - flux,
R - rejection,
R^* - universal gas constant,
M - relative molecular mass,
Δ - coefficient for dye solution,
σ - reflection coefficient
ν - valence (for NaCl is v = 2 and for dye v = 3).
α - coefficient for salt permeability,
B - coefficient for concentration dependence of salt permeability

subscripts

S - salt
D - dye
V - water

F - feed
P - permeate
R - retentate
W - membrane interface (wall)
0 - beginning of the concentration mode

2.2 Characterization of membranes

Before diafiltration experiments characterizations of commercial membranes are carried out. For these characterizations pure water and water solutions of salt are used. From experiments with pure water model parameter A (water permeability) can be estimated. This parameter is slope of the curve (straight line) $J = f(\Delta P)$ (see Eq. (18) and $\Delta\pi = \delta = 0$ because no salt and dye are used). In our model we assume water permeability as constant. However, an increase in concentration can cause significant changes in viscosity and a consequent modification of the water permeability. According to resistance model ($A = 1/(\mu R_M)$) the dependence of water permeability on viscosity can be expressed as:

$$A_\mu = \frac{A}{\mu_{REL}} \tag{20}$$

where A is the water permeability respect to pure water and μ_{REL} is the relative viscosity of feed solution to pure water (Kovács et al., 2009a).
In case of diafiltration fouling or gel layer effects can occur and then parameter A is depended on dye/salt concentration ratio (in resistance model is added next resistance $A=1/\mu_{REL}(R_M+R_F)$, where R_M and R_F are membrane and fouling resistance (Cséfalvay et al., 2009).
Similar experiments are made with salt solutions. Four salt concentrations (1, 5, 10 and 35 g/l) are used. From these experiments can be obtained parameter B (salt permeability) and then α and β (plotting B versus c_F). Values obtained from these experiments are not used direct but are used as first approximation values for best fit parameters (see Table 3). From results (salt rejection and flux) the suitable membranes for desalination were chosen, Desal 5DK, NF 70, NF 270 and TR 60. Membranes NF 90 and Esna 1 had higher rejection (see Fig. 3). For desalination membrane with small rejection of salt are suitable.

2.3 Comparison of membranes

For comparison of membranes, three factors were used.
The first factor is separation factor of diafiltration, S:

$$S = \frac{\dfrac{c_D}{c_D^0}}{\dfrac{c_S}{c_S^0}} = \frac{c_D c_S^0}{c_D^0 c_S} \tag{21}$$

where c^0_D, c^0_S are concentrations of dye and salt at the beginning of experiment, c_D, c_S are concentrations of dye and salt in the end of experiment.
The separation factor, *S*, represents how well the dye will be desalinated. With higher separation factor the dye desalination is better. But it is also clear that with bigger separation factor the loss of the dye will be bigger because real membranes have not 100% rejection of dye.

The dye loss factor, **Z**, can be defined as the rates of amount of the dye in permeate to amount of the dye at the beginning of experiment:

$$Z = \frac{V.\left(c_D^0 - c_D\right)}{V.c_D^0} = 1 - \frac{c_D}{c_D^0} \tag{22}$$

The third parameter is time of diafiltration needed to reach certain separation factor, **S**. The total time of diafiltration with n steps, τ, can be expressed (constant permeate flux in each concentration mode) as:

$$\tau_{total} = \sum_{i=1}^{n} \frac{\Delta V}{Q} = \sum_{i=1}^{n} \frac{V_{F0} - V_F}{A.J} \tag{23}$$

where Q is flow of permeate.

3. Methods

3.1 Membranes

Six NF membranes were chosen for this study. Properties of membranes used are given in Table 1.

Indication	Type	Producer	MWCO [Da]	Material	Module
Desal 5DK	Desal 5DK	GEW & PT	200	polyamide	spiral-wound
Esna 1	Esna 1	Hydranautics	100-300	polyamide	spiral-wound
NF 90	NF 90	Dow	90	polyamide	spiral-wound
NF 270	NF 270	Dow	270	polyamide	spiral-wound
NF 70	CSM NE 2540-70	Saehan	250	polyamide	spiral-wound
TR 60	TR 60 - 2540	Toray	400	polyamide	spiral-wound

Table 1. Properties of the membranes used

3.2 Materials

Dye was obtained from VÚOS a.s. Pardubice, Czech Republic. The commercial name is Reactive Orange 35, and a molecular weight is 748.2 Da in free acid form (three acidic groups) or 817.2 Da as the sodium salt. Fig. 1 shows structural formula of the free acid form.

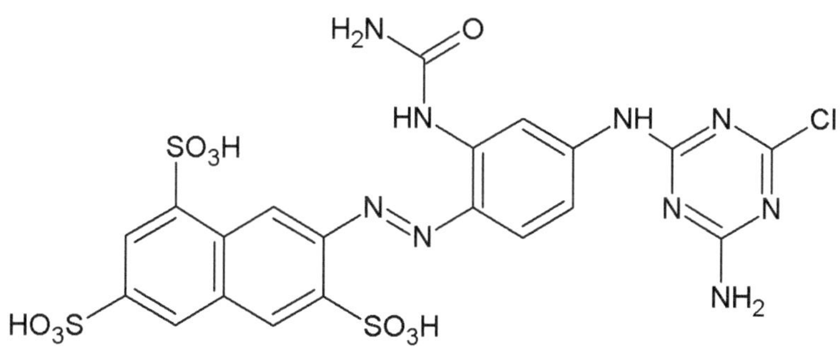

Fig. 1. Structural formula of dye (free acid)

NaCl and $MgSO_4$ used for all experiments were analytical grade. The demineralised water with the conductivity between 4-15 μS/cm was used in this study.

3.3 Experimental system

Experiments were carried out on system depicted schematically on Fig. 2. Feed (F) was pumped by pump (3) (Wanner Engineering, Inc., type Hydracell G13) from feed vessel (2) to membrane module (1). Pressure was set by valve (4) placed behind membrane module. Permeate (P) and retentate (R) were brought back to feed vessel. Pressure was measured by manometer (5). Temperature was detected by thermometer (6). Stable temperature was maintained by cooling system (7).

3.4 Analytical methods

Dye concentrations were analysed using a spectrophotometer (SPECOL 11). NaCl and $MgSO_4$ concentrations were calculated from conductivity measurements using a conductivity meter (Cond 340i). Permeate and retentate salt concentrations during diafiltration experiments were analysed using potentiometric titration.

3.5 Separation procedure

The system was operated in the full recirculation mode while both retentate and permeate were continuously recirculated to the feed tank except sampling and concentration mode of diafiltration. By changing applied pressure (from 5 to 30 bar) and concentration of salt (1, 5, 10 and 35 g/l) in characterization of membranes both the retentate and permeate were returned back to the feed tank for 0.5 h or 10 min, respectively to reach a steady state before sampling. Before first concentration mode in diafiltraton experiments and after each diluting mode the total recirculation was used 1h and minimally 5 min, respectively. The permeate flux was measured by weighing of certain permeate volume and using a stopwatch.

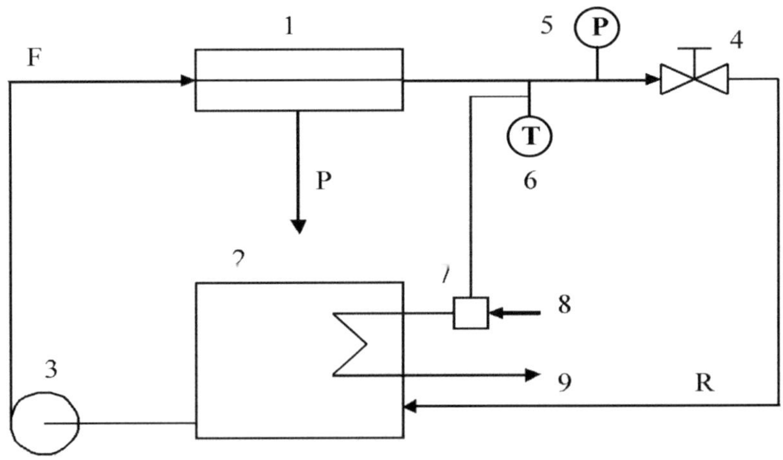

Fig. 2. Schematic diagram of the experimental set-up used: 1 membrane module, 2 feed vessel, 3 high pressure pump, 4 back pressure valve, 5 manometer, 6 temperature controller, 7 cooling system, 8 cooling water input, 9 cooling water output, F feed, P permeate, R retentate (concentrate)

4. Results and discussion

4.1 Pure water flux

Water permeability is one of the basic characteristic of NF membranes. The pure water permeability of the six membranes was determined by measuring the de-ionized water flux at different operating pressures. According to Darcy´s law, the permeate flux is directly proportional to the pressure difference across the membrane. The slope of this line corresponds to the water permeability (A).

Membrane	A [l/m^2.h.bar]
Desal 5DK	3.365
Esna 1	4.824
NF 90	5.845
NF 270	6.801
NF 70	2.650
TR 60	3.952

Table 2. Water permeability of membranes used

4.2 Rejection of NaCl

Basic membrane characteristics are the dependence of the permeate flux and the salt rejection on other operation parameters, i.e. the applied pressure difference and the salt concentration in feed.

The permeate flux increases with increasing pressure and decreases as the feed concentration of salt increases. For the lowest concentration of salt (1 g/l), the values of permeate flux were similar to the values of clean water. The lower values of permeate flux were obtained with the increasing salt concentrations in feed (increasing osmotic pressure). For membrane NF 90 fluxes were not measured at the highest salt concentration for pressure smaller than 25 bar, because the osmotic pressure was too high. Opposite problem was with membrane NF 270 at the smallest salt concentration in feed. The permeate flux was too high and pump was not able to deliver necessary volumetric flow of retentate (600 l/h) for constant conditions at all experiments.

The observed rejection increases as the pressure difference increases, and decreases with the increasing salt concentration in feed for all tested membranes. However, the minimal values were obtained during experiments with membrane NF 270. Low values of the salt rejection and higher values of the permeate flux are suitable for desalting. Fig. 3 shows the comparison of tested membranes for the lowest (1 g/l) and the highest salt concentrations in feed (35 g/l), respectively.

4.3 Diafiltration

The concentration of macrosolutes by batch NF is frequently accompanied by a diafiltration step to remove microsolutes such as salts. Batch diafiltration with periodically adding solvent at 20 bars and constant retentate flow 600 l/h was provided. Aqueous dye solutions with dye concentrations 100, 50 and 10 g/l and salt concentration between 20-23 g/l were desalted at 23°C. Volume of the pure solvent added in every dilute mode was 4l (the same volume of permeate was remove before in concentration mode). Total feed volume in tank was 52l. For every membrane and every concentration of the dye in feed fifty diafiltration steps were made. One point in Figs. 4-7 is one diafiltration step before concentration mode.

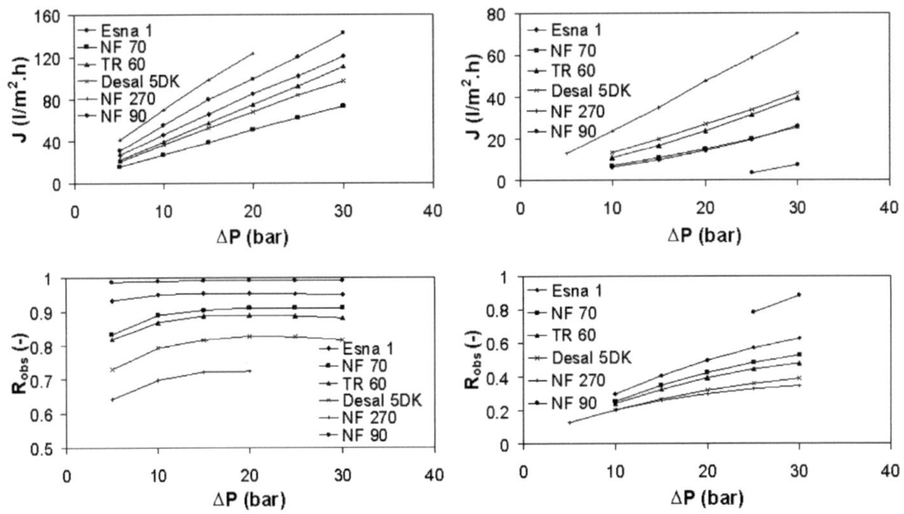

Fig. 3. Permeate flux and rejection as a function of pressure for the lowest salt concentration (1 g/l) - left figures and the highest salt concentration (35 g/l) - right figures

Four membranes only - Desal 5DK, NF 70, NF 270 and TR 60- were used for diafiltration experiments. Diafiltration with membrane NF 270 was provided only with dye concentrations 100 and 50 g/l and with membrane TR 60 only at the highest dye concentration, which is the best for desalination. For the reason of low values of permeate flux, membrane NF 90 and Esna 1 were not used for those experiments.

Dependences of rejection on salt concentration in feed are given in Fig. 4 for Desal 5DK, NF 70, NF 270 and TR 60, respectively. Membranes are compared at dye concentration 100 g/l. The lowest values of rejection (max. 0.29) were obtained for membrane Desal 5DK. The membrane NF 70 had the highest values.

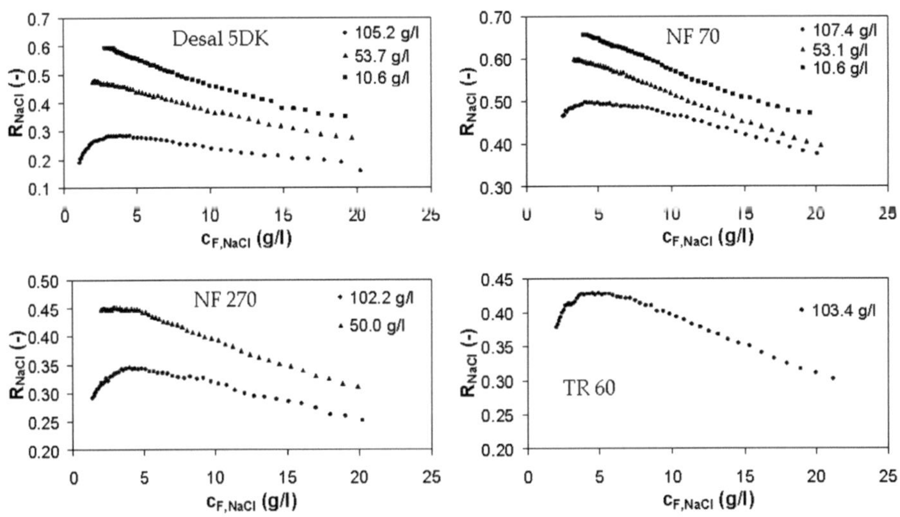

Fig. 4. Salt rejection as a function of salt concentration in feed

Dependences of flux on salt concentrations are shown in Fig. 5. The highest values of flux (70.3 $l/m^2.h$) were obtained in experiments with membrane NF 270. The permeate flux decreased while salt concentration increased. This is due to the effect of osmotic pressure along with the concentration polarization. Due to the concentration polarization phenomenon, the osmotic pressure of the aqueous solution adjacent to the membrane active layer is higher than the corresponding value of the feed solution. As a result, the osmotic pressure would increase dramatically while the salt concentration increased.
In Fig. 6 dependences of salt concentrations on time of diafiltration are shown.

Membrane	A [$l/m^2.h.bar$]	α [$l/m^2.h$]	β [-]	δ [bar]
Desal 5DK	3.365	5.758	0.620	7.503
NF 70	2.650	4.839	0.381	4.664
NF 270	6.801	10.350	0.529	8.561
TR 60	3.952	6.693	0.474	5.888

Table 3. Model parameters.

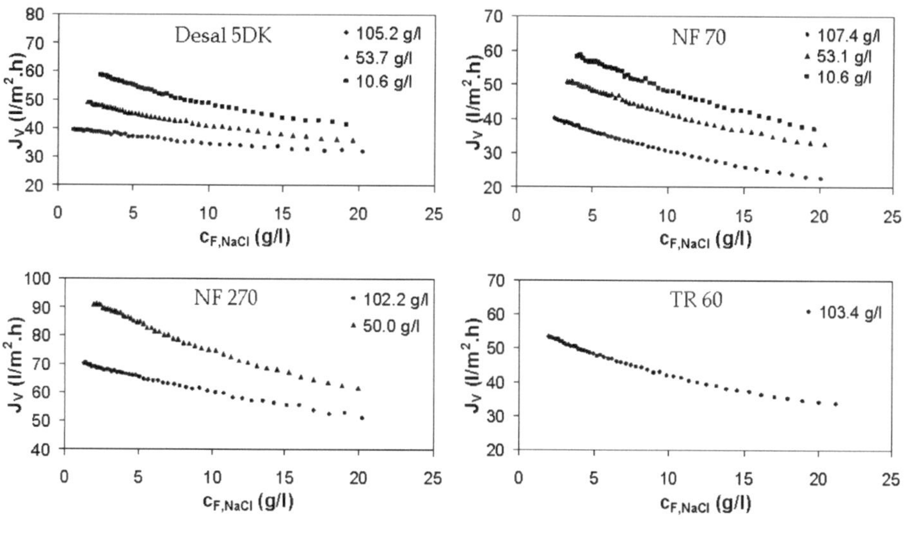

Fig. 5. Permeate flux as a function of salt concentration in feed

The comparisons of experimental and model data for the highest dye concentration (100 g/l) are shown in Fig. 7. Salt concentrations are calculated using Eq. (9) and Eq. (11). Rejection needed for these equations is calculated on basis of Eq. (17). Best fit parameters for proposed model are given in Table 3.
From Table 3 can be shown that δ is not only osmotic pressure (if we assume water permeability as constant), because the values of δ are different. From these results we can assume, the highest effect of dye on flux is for membrane NF 270. This membrane is the most fouled from these membranes. It is appropriate assumed the change in water permeability (A) in case of desalination of dyes.

In our experiments, the decrease of permeate flux was mainly caused by the effect of concentration polarization and the increase of the viscosity of dye solution. The dye formed a boundary layer over the membrane surface (concentration polarization) and consequently, increased the resistance against the water flux through the membrane. At the same time, the viscosity of solution increased with higher concentration.

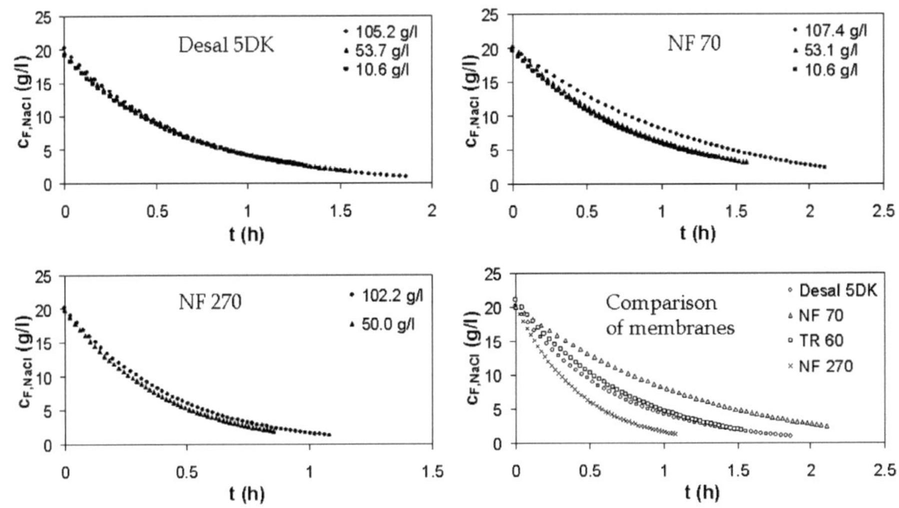

Fig. 6. Salt concentration as a function of diafiltration time for membranes Desal 5DK, NF 70, NF 270 and comparison of tested membranes at dye concentration 100 g/l

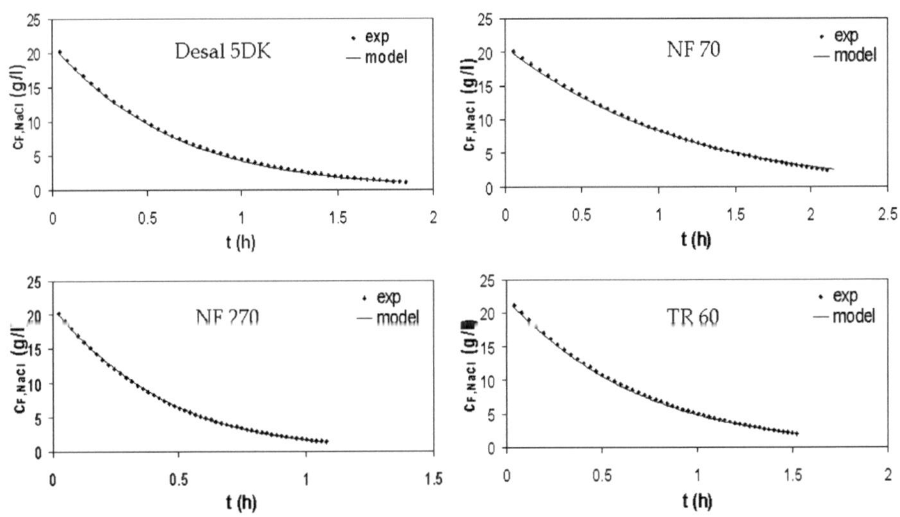

Fig. 7. Comparison of experimental and model data for the highest dye concentration

From Fig. 7 can be shown that the experimental results in permeate fit the model very well. Due to considerably low salt concentrations in permeate, concentration polarization was minimized. The diafiltration process benefits to obtain pure salt product and this can be predicted by a mathematic model on the basis of description of discontinuous diafiltration by periodically adding solvent at constant pressure difference.

DESAL 5DK				
$c_{F,NaCl,Z}$	(g/l)	22.00	19.62	19.21
$c_{F,NaCl,K}$	(g/l)	0.97	1.88	2.70
$c_{F,dye,Z}$	(g/l)	105.17	53.75	10.61
$c_{F, dye,K}$	(g/l)	105.15	53.74	10.60
τ_{total}	(hod)	1.86	1.55	1.32
S	(-)	22.71	10.44	7.10
Z	(%)	0.02	0.01	0.07
NF 70				
$c_{F,NaCl,Z}$	(g/l)	20.08	20.33	19.64
$c_{F,NaCl,K}$	(g/l)	2.40	3.21	3.86
$c_{F, dye,Z}$	(g/l)	107.42	53.05	10.56
$c_{F, dye,K}$	(g/l)	107.38	53.04	10.56
τ_{total}	(hod)	2.11	1.57	1.37
S	(-)	8.37	6.34	5.09
Z	(%)	0.03	0.03	0.05
TR 60				
$c_{F,NaCl,Z}$	(g/l)	21.14	20.56	19.72
$c_{F,NaCl,K}$	(g/l)	1.88	2.22	3.16
$c_{F, dye,Z}$	(g/l)	103.41	53.65	10.71
$c_{F, dye,K}$	(g/l)	102.74	53.16	10.52
τ_{total}	(hod)	1.52	1.25	0.98
S	(-)	11.17	8.98	6.25
Z	(%)	0.65	0.48	0.18
NF 270				
$c_{F,NaCl,Z}$	(g/l)	22.00	19.92	20.64
$c_{F,NaCl,K}$	(g/l)	1.28	1.90	2.68
$c_{F, dye,Z}$	(g/l)	102.15	49.98	10.52
$c_{F, dye,K}$	(g/l)	102.12	49.97	10.53
τ_{total}	(hod)	1.08	0.86	0.67
S	(-)	17.19	10.50	6.89
Z	(%)	0.03	0.04	0.04

Table 4. Total time of diafiltration, τ_{total}, the separation factor, **S**, and the loss of dye, **Z**. (subscript $_{Z, K}$ are start and end of diafiltration, respectively)

From Table 4 is clearly shown, the total time of diafiltration τ_{total} decrease with decreasing dye concentration. The shortest time had membrane NF 270. Time for the highest dye concentration is not two times higher than with medium dye concentration for all tested membranes (the time/amount of dye desalted ratio is the smaller for higher concentration of dye). Separation factor decreases with decreasing concentration of dye and it is the second reason why the highest dye concentration was used as the best mode for desalination. The best separation factor had membrane Desal 5DK (very similar values, except the highest dye concentration, had membrane NF 270). The loss of dye is almost same for membrane Desal 5DK, NF 70 and NF 270 at all concentrations of dye. Only for membrane TR 60 are obtained higher loss of dye.

5. Conclusion

The separation performance of dye, salt and dye solution with six different nanofiltration membranes were investigated, followed by the study of the optimum of diafiltration and concentration process of dye solution.

Asymmetric and negatively charged polyamide thin-film composite membranes of near similar molecular weight cut-off were characterized for key physical and surface properties, and employed to perform the laboratory-scale experiments to investigate the impacts of membranes properties on reactive dye removal from dye/salt mixtures through NF process. It was found that properties of the NF membrane play an important role in dye removal rate, stable permeate flux and their change behaviour with operational conditions.

The electrostatic repulsive interaction between dye and membrane surface promotes the dye removal and decreases concentration polarization and dye adsorption on the membrane surface. But, the action will be weakened as the dye concentration or salt concentration increased.

It was seen that salt concentration decreases the dye retention of the membrane. The domination of the steric (size exclusion) effect over the Donnan (charge) effect is the main reason for this. At very high salt concentration, the surface charge of the membrane is neutralized, whereby the separation mechanism depends only on the molecular size of the dye. However, as there is a loss of membrane selectivity due to neutralization of charge the separation, there is a diffusion of dye in to the permeate side of the membrane due to accumulation of the dye layer over the surface.

From the results presented above it is clear that the best concentration of the dye in feed for desalination of reactive dye by batch diafiltration is 100 g/l. In this case the salt rejection reaches minimal value due to Donnan potential which strengthens the flow of salt through the membrane.

The best membrane for desalination is NF 270 which has smaller dye loss factor and the shortest time of diafiltration. Very suitable membrane is also Desal 5DK, which has the best separation factor and dye loss factor, but this membrane has longer time of diafiltration (see Table 4). For desalination qualitative description it is convenient to use the proposed model.

Reactive dye desalination and purification processes generate a large volume of effluent but with low quantities of pollutants. The process is one of the high water-consumption processes in the dye production industry that results in high wastage of water which becomes discharge effluent. Membrane processes can be applied to treat the wastewater and recycle it back into the manufacturing process. The cleaner process can not only improve the rejection efficiencies and flux recovery, but can also recycle permeate from the desalting and purification process.

6. Acknowledgement

This project was financially supported by Ministry of Education, Youth and Sports of the Czech Republic, Project MSM0021627502.

7. References

Al-Zoubi, H.; Hilal, N.; Darwish, N.A. & Mohammed, A.W. (2007). Rejection and modelling of sulphate and potassium salts by nanofiltration membranes: neural network and Spiegler-Kedem model. *Desalination*, Vol.206, No.1-3, (February 2007), pp. 42-60, ISSN 0011-9164

Bowen, W.R. & Mohammad, A.W. (1998). Diafiltration by nanofiltration: Prediction and optimization. *AIChE Journal*, Vol.44, No.8, (August 1998), pp. 1799-1812, ISSN 0001-1541

Caéfalvay, E.; Pauer, V. & Mizsey, P. (2009). Recovery of copper from process waters by nanofiltration and reverse osmosis. *Desalination*, Vol.240, No.1-3, (May 2009), pp. 132-146, ISSN 0011-9164

Das, Ch.; Dasgupta, S. & De, S. (2008). Steady-state modeling for membrane separation of pretreated soaking effluent under cross flow mode. *Environmental Progress*, Vol.27, No.3, (October 2008), pp. 346-352, ISSN 0278-4491

Foley, G. (2006). Water usage in variable volume diafiltration: comparison with ultrafiltration and constant volume diafiltration. *Desalination*, Vol.196, No.1-3, (September 2006), pp. 160-163, ISSN 0011-9164

He, Y.; Li, G.M.; Zhao, J.F. & Su, H.X. (2007). Membrane technology: Reactive dyes and cleaner production. *Filtration & Separation*, Vol.44, No.4, (May 2007), pp. 22,24, ISSN 0015-1882

Hussain, A.A.; Nataraj, S.K.; Abashar, M.E.E.; Al-Mutaz, I.S. & Aminabhavi T.M. (2008). Prediction of physical properties of nanofiltration membranes using experiment and theoretical models. *Journal of Membrane Science*, Vol.310, No.1-2, (March 2008), pp. 321-336, ISSN 0376-7388

Kovács, Z.; Discacciati, M. & Samhaber, W. (2009a). Modelling of batch and semi-batch membrane filtration processes. *Journal of Membrane Science*, Vol.327, No.1-2, (February 2009), pp. 164-173, ISSN 0376-7388

Kovács, Z.; Discacciati, M. & Samhaber, W. (2009b). Modelling of amino acid nanofiltration by irreversible thermodynamics. *Journal of Membrane Science*, Vol. 332, No.1-2, (April 2009), pp. 38-49, ISSN 0376-7388

Koyuncu, I. & Topacik, D. (2002). Effect of organic ion on the separation of salts by nanofiltration membranes. *Journal of Membrane Science*, Vol.195, No.2, (January 2002), pp. 247-263, ISSN 0376-7388

Mulder M. (2000). *Basic Principles of Membrane Technology*, 2nd Ed., Kluwer Academic Publishers, ISBN 0-7923-4248-8, Dordrecht, The Netherlands

Oatley, D. L.; Cassey, B.; Jones, P. & Bowen, W. R. (2005). Modelling the performance of membrane nanofiltration – recovery of high-value product from a process waste stream. *Chemical Engineering Science*, Vol.60, No.7, (April 2005), pp. 1953-1964, ISSN 0009-2509

Schirg, P. & Widmer, F. (1992). Characterisation of nanofiltration membranes for the separation of aqueous dye-salt solution. *Desalination*, Vol.89, No.1, (November 1992), pp. 89-107, ISSN 0011-9164

Wang, L.; Yang, G.; Xing, W. & Xu, N. (2008). Mathematic model of yield for diafiltration. *Separation and Purification Technology*, Vol.59, No.2, (February 2008), pp. 206-213, ISSN 1383-5866

Weselowska, K.; Koter, S. & Bodzek, M. (2004). Modelling of nanofiltration in softening water. *Desalination*, Vol.162, No.1-3, (March 2004), pp. 137-151, ISSN 0011-9164

Yu, S.; Gao, C.; Su, H. & Liu, M. (2001). Nanofiltration used for desalination and concentration in dye production. *Desalination*, Vol.140, No.1, (October 2001), pp. 97-100, ISSN 0011-9164

20

New Type Filtration, Ion-Exchange, Sorption Small Multi Process Water Conditioning Device Used as a Multi Cell Water Deionizer

Angel Zvezdov and Dilyana Zvezdova
Prof. Dr. Asen Zlatarov University of Bourgas, Department of Water treatment Technology, Bourgas, Bulgaria

1. Introduction

Sand filtration, ion-exchange and sorption methods are often used in the design of small water treatment devices. Such designs are suitable for use in the technological water conditioning plants before small ultra-pure water point-of-use systems, typical laboratory application, a feed to autoclaves, and environmental cabinets. Present work is an attempt to create a simple and easily used water treatment device for the cost effective production of general laboratory grade water (Catalog ELGA Co.,2008; Catalog Purity Ltd., 2007). This trend utilizes a combination of technologies including membrane processes, ion-exchange, adsorption, sanitization etc. (Adjarov G.,1998). That sets our device apart is the unique process which provides a constant supply of high purity laboratory water and is more cost-effective and convenient for the user. Usually, small water treatment deionization systems include ion-exchange cartridges, the ion-exchange capacity which is exhausted during the working period has to be regularly regenerated for continued use. Some producers (Catalog ELGA Co.,2008; Catalog Purity Ltd., 2007; Catalog CHROMSERVISE s.r.o.-Chehia, 2003) accept as a more cost-effective and convenient solution the replacement of ion-exchange cartridges with ready regenerated cartridges by specialized work departments. Purchasing ready made replacement cartridges supplied by specialized dealers, requires the user to be a subscriber to this department. Other producers design ion-exchange plants with regeneration systems (reagent vessels, pumps, valves, tubes etc.). It enables local regeneration procedures to be completed by the user (Plehov G.,1997 ; Zvezdov A., et.al).The decision on which kind of design should be purchased to equip the laboratory depends on many considerations: distances and transport prices; manual service expenses; operative procedures and their complication; personal qualification etc. On the basis of the above mentioned items it should be concluded that there is a necessity to research and enhance the activity for creation of new designs for small water deionization plants. The researchers and designers are looking for a simple, cheaper laboratory water treatment system. These research works are typically important in hard water areas where ion-exchange cartridges are replaced frequently. It is important as a cost–effective choice for labs that require higher volumes of water, as well. The purpose is to create and evaluate a multi cellular small water treatment ion-exchange and sorption device with a simple, cost-effective and user friendly design. The design should allow the regeneration of exhausted ion-exchange and sorption beds to be made locally by the user. From a scientific point of view it is important to clarify the influence of the physical (pore) structure of ion-exchange resins and adsorbents used. The ability to be effectively regenerated and washed to remove the organic humus substances retained into their pore structure during the working functional period of deionized water production.

2. Experimental

As an example for its different functionality, the multi process small water treatment filtra-tion device has been designed to provide uncompromising deionized water quality while still being affordably priced. The device consists of three separated vertical cells which could be loaded by different filtration and sorption materials. It has only three valves for treated water and regeneration flows regulation (Fig.1). In the researched case, the three cells filtra-tion body was loaded by ion-exchange resins and active carbon sorption material. The unit is evaluated as a device to produce deionized and sorption purified water. It produces se-conddary grade purified water in accordance with standards (BDS ISO 3696, 1993) from a potable source in a wide range of volumes for laboratory and small industrial applications. The device created can handle most laboratory water needs, from reagent preparation and glassware washing to sensitive instrumentation analyses. The ion-exchange and sorption methods have been researched are incorporated in an innovative design of a three cellular one body small water deionizer. It could be localized on a mobile fundament frame and can be moved easy if it is necessary. A flow type conductance apparatus is included to the de-sign for a continuous monitoring and control of the deionized water quality. The device could operate from a range of tap water pressure between 0,2 to 0,6 MPa. The flow rate of the deionized water produced was up to 200 dm^3/h. The continuous water monitoring was achieved by means of a flow conductance control apparatus which is situated on the front panel.

The regeneration of the exhausted ion-exchange and sorption beds could be accomplished on the basis of gravitation phenomenon (by means of reagent containers localized to a suitable height), locally by the user. For greater convenience there is a choice of bench, under bench and wall mounting options to best suit the layout of the laboratory or site. A major technological contribution was possibility to realize five processes into one filtration body with just three valves for flow rate controlling included in the construction (Fig.1). During the water deionizing experiment, we loaded the three chambers as follows:

- layer of weak acid cation-exchange resin (Dowex MWC-2) $H\text{-}R_w$ (hygrogen form) and a layer of strong acid cation-exchange resin (Dowex MSC-1) $H\text{-}R_{str}$ (hygrogen form) in the chamber between walls 1 and 2;
- layer of weak base anion-exchange resin (Dowex MWA-1) $R_w\text{-}OH$ (hydroxyl form) and a layer of strong base anion-exchange resins (alternative: Dowex MSA-1, Dowex MSA-2, Amberlite IRA-938) $R_{str}\text{-}OH$ (hydroxyl form) in the chamber between walls (2) and (3). This chamber was loaded with polyreticular (Dobrevsky Iv.& A. Zvezdov, 1999) true porous (Kunin R., 1968) ion-exchange resins to be able to reduce humus natural organic impurities from the water treated (Einsatz und Schaltung Leverkusen,Bayer-AG, 1970);
- layer of active carbon or bioadsorbent hitosan sorption material in the central chamber.

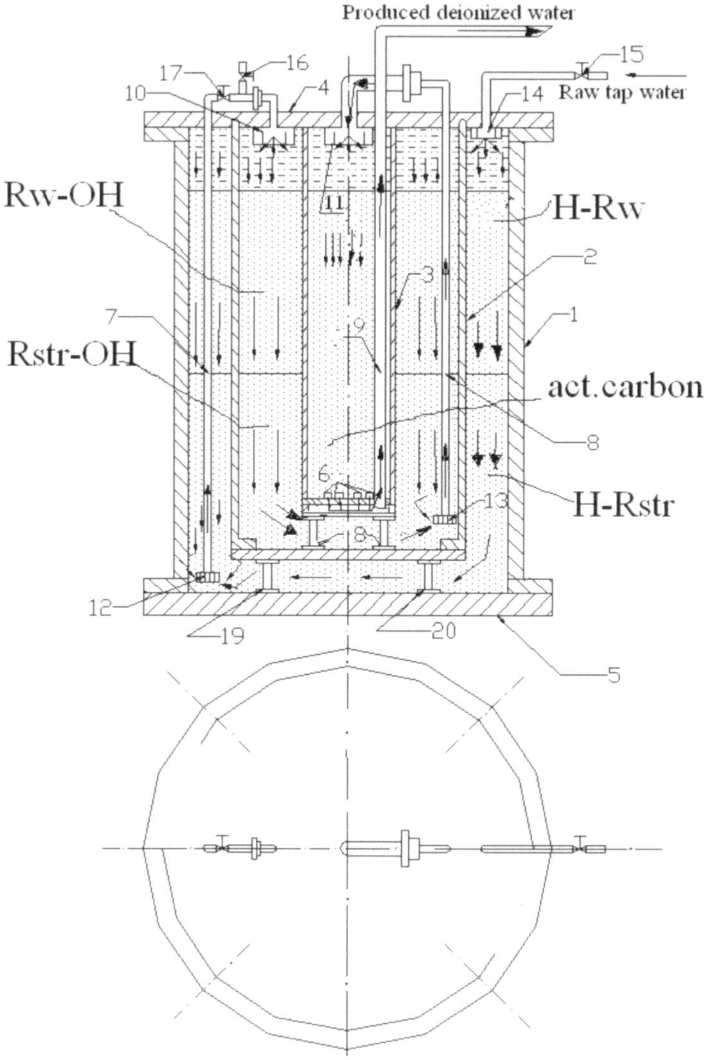

Fig. 1. Design of Multi process ion-exchange and sorption small water conditioning device: 1-outside body; 2,3- inside bodies; 4- upper cover; 5- bottom; 6,10÷14-scatter nozzles; 15÷17 – valves; 18÷20–bearings; H-R_w-weak acid cation-exchanger (hydrogen ion form); H-R_{str}–strong acid cation-exchanger hydrogen ion form); R_w-OH-weak base anion-exchanger (hydroxyl ion form); R_{str} -OH - strong base anion-exchanger (hydroxyl ion form

One very important role was weak base anion-exchange resin (Dowex MWA-1). The importance was due to the sorption-desorption ability of this resin. Actually it possesses an ability to realize an adsorption process of natural humus insoluble substances on its surface area during the working period, purifying the treated water flow. Then it was able to clean its porous structure and surface area, respectively from this organic natural substances during the period of reagent regeneration process by sodium hydroxide. This resin did play an important protection role to the strong base anion exchange resin against an irreversible poisoning of the ion-exchange functionality (Dobrevsky Iv.,1991).

In order to clarify the effect of physical structure of the filtration materials loaded in the device on the technological behaiviour of water treatment device, their physical structure was investigated by means of electron scanning microscope and mercury intrusion apparatus („POROSIMETER-1500"CarloErba-Strumentazione-Italy). The physical structures of adsorption active carbon materials are shown in (fig.2-5). Pore size distribution curves of ion-exchange resins used are shown in Fig. 6, 7.
The electron scanning microscope photographs for active carbon physical structure show uniformity of pore sizes. Actually it is an identity pore channel physical structure. Average size of both types active carbon pores is approx. 13-14 µm. It is a suitable pore sizes for a sorption process to remove the organic colloidal natural impurities that possess sizes between 0,10-0,001 µm (Alekin.O.A., 1970).

Fig. 2. Electron scanning microphotography of active carbon "AP-4-60" **(magnified 10⁴times)**

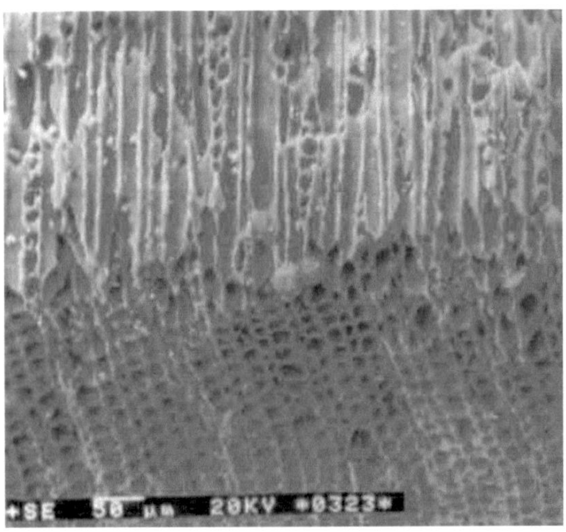

Fig. 3. Electron scanning microphotography of active carbon AC-Luise **(magnified 10⁴times)**

(a) (b)

Fig. 4. Electron scanning microphotography of strong base anion exchangers: a) Amberlite IRA-938; b) Dowex MSA-1 (**magnified 10^4times**)

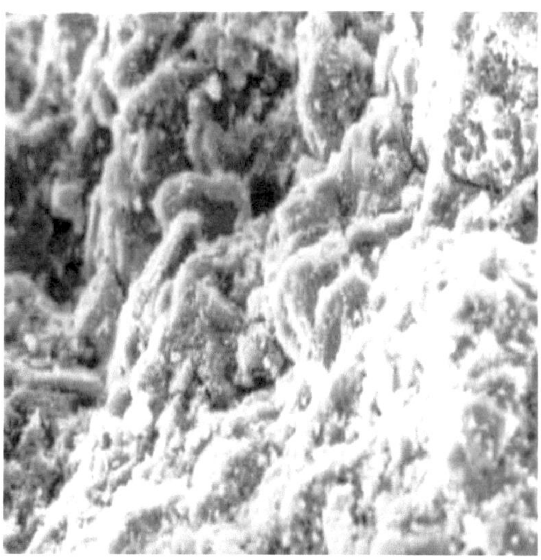

Fig. 5. Electron scanning microphotography of weak base anion-exchange Dowex MWA-1 (**magnified 10^4times**)

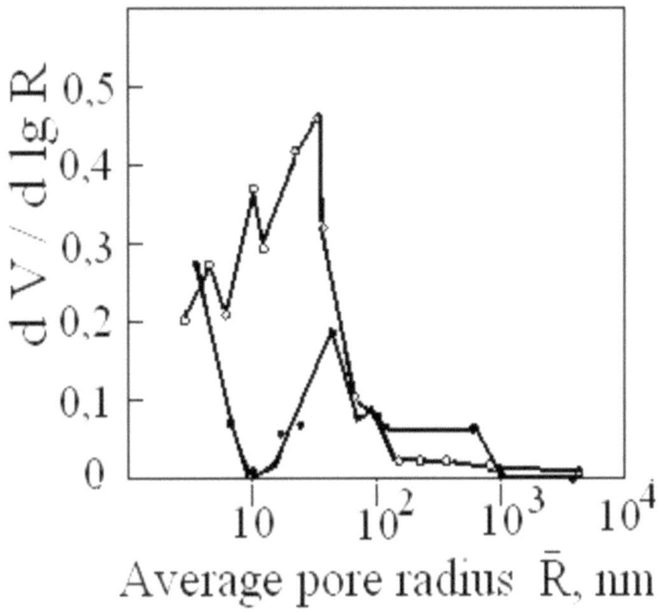

Fig. 6. Differential pore size distribution curves of weak base ion-exchange resins: **o-o**-Dowex MWA-1 (weak base anion-exchanger); •-•-Dowex MWC-2 (weak acid cation-exchanger)

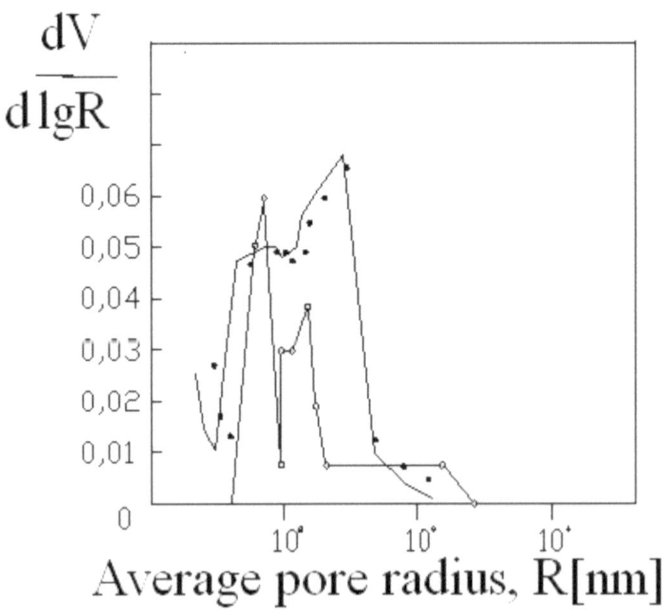

Fig. 7. Differential pore size distribution curves of strong base anion exchangers resins: **o-o**:Dowex MSA-1; •-•:Dowex MSA-2

Pore size distribution curves of ion-ex-change resins used, show the average pore size of predominated pores as follows:

- strong base anion exchanger Amberlite IRA-938: $R_{average}$=7500 nm
- strong base anion exchanger Dowex MSA-1: $R_{average}$=6-13 nm
- strong base anion exchanger Dowex MSA-2: $R_{average}$=5-45 nm
- weak base anion exchanger Dowex MWA-1: $R_{average}$=5-50 nm;
- weak acid cation exchanger Dowex MWC-2: $R_{average}$=5-60 nm;

The physical pore structure properties of adsorption and ion–exchange filtration materials loaded in the small water treatment device researched explain the significant reduction of permanganate demand value of raw natural water treated (from P = 3,2 mg O_2 / dm^3 to the values shown in Table 3).

The multi process ion-exchange and adsorption small water conditioning device functioned as follows:

2.1 Working functional period

The raw tap water (Table 1,2) was fed through the valve (15). It was processed with the following:

- layer of weak acid cation-exchange (H-R_w) ;
- layer of strong acid cation-exchange resin (H-R_{str}); a weak base anion-exchange resin (R_w-OH) layer;
- layer of strong base anion-exchange resin (R_{str}-OH) layer;
- layer of active carbon layer (central cell). Actually in the central (last) chamber was realized a sorption process, but in the surrounding besides chambers before the sorption process were realized four ion-exchange processes as follows:
- weak acid cation-exchange process and strong acid cation-exchange process in the first two layers [ion-exchange chamber between walls (1) and (2)];
- weak base anion-exchange and strong base anion-exchange processes in the second two layers [ion-exchange chamber between walls (2) and (3)].

In the first cation-exchange chamber was realized a cation-exchange removal process and by this process H-ions from cation-exchanger`s polymer matrix were changed with all cations included in the treated row tap water. All the mineral salts dissolved in the water treated were transformed to acids. Then the acids contents water flow goes in the second anion exchange two layers chamber between walls (2) and (3). This chamber was loaded by weak base and strong base anion-exchangers in OH-ion form (R-OH). By the contact with anion-exchange resins loaded there, all these acids were transformed to equivalent new water molecules. Such a way the salty row tap water became to be desalinated. Then the water flow was going trough the central adsorption chamber (3) and it contacts the active carbon to be purified from organic halogen, natural humus organic impurities and partially from heavy metals compounds. Finally the purified water flow product went out by a special situated (Zvezdov A., 1999) suitable pipe to be used by a user.

2.2 Scarify backwash functional period

Functioning as a scarify backwash water flow, the feeding row tap water did enter through the tube (9), carbon bed, tube (8), nozzles (13), (10), valve (17), nozzles (12),(14), valve (15) and the flow was going out to the drainage waste water communication of the building for approx. of 10 minutes.

2.3 Reagent regeneration periods

2.3.1 Acid reagent regeneration period

Above the water treatment device was situated reagent tank connected by reinforced hose to valve (15). During the first regeneration step it was filled with 10 % HCl solution. On the basis of gravitation phenomenon, the opened valves (15) and (17) did allow the acid regeneration solution to pass through valve (15), nozzles (14,12,10,13,11,6) contacting the weak acid cation-exchange, strong acid cation-exchange, weak base anion-exchange, strong base anion-exchange resins and active carbon beds. Such a way the acid solution did transform (regenerate) to hydrogen ion-form of the functional ion-exchange groups into the polymer matrix of weak and strong acid cation-exchange resins ($H\text{-}R_w; H\text{-}R_{str}$) (Dobrevsky Iv.,1991). In parallel it exhausted the functional ion-exchange groups into the polymer matrix of weak and strong base anion exchange resins. Passing through the active carbon adsorbent bed, the acid solution did remove from its porous sorption structure the impurities which are soluble at pH< 7 (it is called acid activation process).

2.3.2 Washing period for $H\text{-}R_W$ and $H\text{-}R_{STR}$ filtration beds

A. Slow washing process

A feeding row tap water was going in through the same flow way (opened valves 15, and 17) as the reagent regeneration period, but slowly with approximately same linear velocity as the reagent regeneration period (10 minutes).

B. Fast washing process

A feeding row tap water was going in through the same flow way (opened valves (15) and (17) as the reagent regeneration period. It was faster than during the slow washing process (for 30 minutes with approximately same linear velocity run like during the working functional period).

2.3.3 Alkaline reagent regeneration period

After the washing processes, the situated above the water treatment device reagent tank was connected by reinforced hose to opened valve (16) (but valve 17 was closed). To perform the second regeneration step reagent tank was filled with 5 % NaOH solution. The opened valves (16) and produced water output did allow the alkaline regeneration solution to pass trough nozzles (10, 13, 11, 6) to contact the weak base anion-exchange, strong base anion-exchange, and active carbon beds. Such a way the alkaline regeneration solution did transform (regenerate) to hydroxyl ion form the functional ion-exchange groups into the polymer matrix of weak and strong base anion exchange resins ($R_w\text{-}OH; R_{str}\text{-}OH$) Passing through the active carbon adsorbent bed, the alkaline regeneration solution did remove from its porous sorption structure the impurities which are soluble at pH > 7 (it is called alkaline activation process).

2.3.4 Washing period for R_w-OH and R_{str}-OH filtration beds

A. Slow washing process

The feeding row tap water was connected by reinforced hose to valve (16). It went through the valve (16) nozzles (10,13,11) and (6) to wash the weak base anion-exchange, strong base anion-exchange, and active carbon beds (for 20 minutes with same flow way and flow rate as the reagent regeneration period).

B. Final washing process

The feeding row tap water went through the valve (15), weak acid resin H-R $_w$ layer; strong acid cation-exchange resin H-R_{str} layer; weak base anion-exchange resin R_w-OH; strong base anion-exchange resin R_{str}–OH; active carbon bed (same flow way and approximately same velocity as working functional period, but the flow was dumped to the building drainage system until the flow conductance apparatus indicates the necessary water electric conductivity (after approx. 30÷40 minutes).Then the working functional period producing deioinized water began again.

To achieve a better quality of deionization water processes the module could be doubled (fig. 10.).

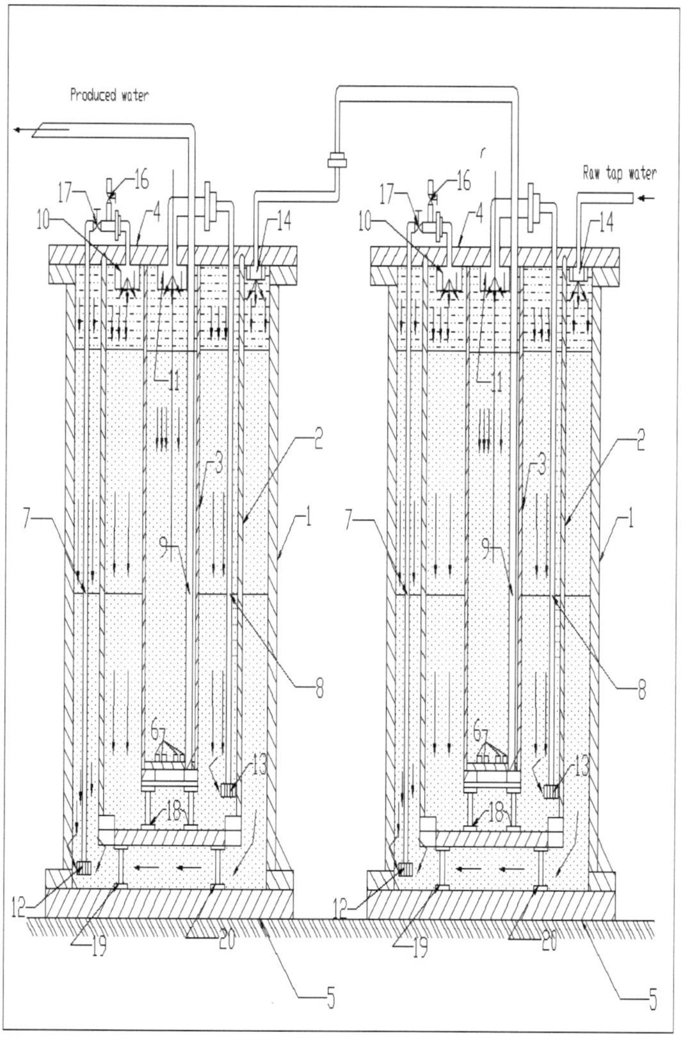

Fig. 10. Duplex module deionization bench scale tap water treatment device for laboratory grade water production according BDS ISO 3696: 1-outside body; 2,3- inside body; 4- upper cover; 5- bottom; 6,10 -14 - scatter nozzles; 15-17 – valves; 18-20 – bearings

The water treatment device had not automation. It was reasonable because the quantity of deionized water produced between two regeneration procedures was aprox. 2000 dm^3 (depending on the size of filter body and row tap water salt content, the productivity could be increased or decreased). This quantity was consumed by the laboratory approximately during 150÷200 days. It means that for laboratory use, two times a year (maximum) was necessary to change the flow directions using regeneration solutions. The regeneration procedures could be easy done by means of two manually turned valves, without necessity of electronic automation. A broad range of optional accessories were also available to further complement deionization device: remote display; full range of water storage reservoirs with an availability for water produced level governor etc. (if it was defined like necessity).

3. Results and discussion

A series of pilot experiments have been carried out The R-H and R-OH ion-exchange beds were loaded by polyreticular [10] ion-exchange resins to be able to reduce humus natural organic impurities from the water treated (Kunin R., 1968; Dobrevsky Iv.,1991). The specific electrical conductivity of the water produced was measured by a flow conductivity meter apparatus.Two working filter cycles were performed:

- The first working cycle was carried out and 2300 dm^3 deionized water was produced. It is ended when the conductivity meter showed $\chi \geq 5$ µS/cm. For the initial 1900 dm^3 the specific electrical conductivity was 1 to 2 µS /cm and then slowly increased up to 5 µS / cm.
- The second working cycle was with the duplex module device (Fig.10). The specific elec-trical conductivity achieved of the water produced 0,27 to 0,38 µS/cm. About 250 dm^3 water was produced with this value of specific electrical conductivity.
- The analytical measurements of the properties of the deionized water obtained at a flow rate Q = 150÷200 dm^3/h have been carried out in accordance with the parameters required by BDS ISO 3696 provides pure water to ISO 3696 and BS and ASTM D1193-99. The results obtained are presented in Table 3.

The results in Table 3 show that the properties of the water produced by the small water deionization device comply with the requirements for water with primary or secondary degree of purity (BDS ISO 3696,1993; ASTM D 1193-99,1999). Therefore, it is concluded that the product can be used for usual laboratory needs, as well as for more high quality laboratory needs, instead of the expensive water prepared by electrical distillers (Petkov K., et al.,2007; Petkov K.& A. Zvezdov, 2007).

An example for ultra pure water production by small local water treatment plant used for high quality urea solution AUS 32 defined by DIN 700070 is shown in next text. On the base of design and results discussed above a small local water treatment plant for high quality urea solution AUS 32 defined by DIN 700070 is created. The technological parameters of this plant are researched during the ultra pure water production used as a solvent for urea solution AUS 32. A very salty under-ground water was treated to get a deionized ultra pure water with specific electric conductivity $\chi \leq 1$ µS/cm. It was used as a solvent during the automotive grade urea solution (AUS 32) production. The product obtained was according to quality requirements of DIN 700070. The role and use of deionized ultra pure water addressed to quality assurance guidance document (QAGD) is discussed.

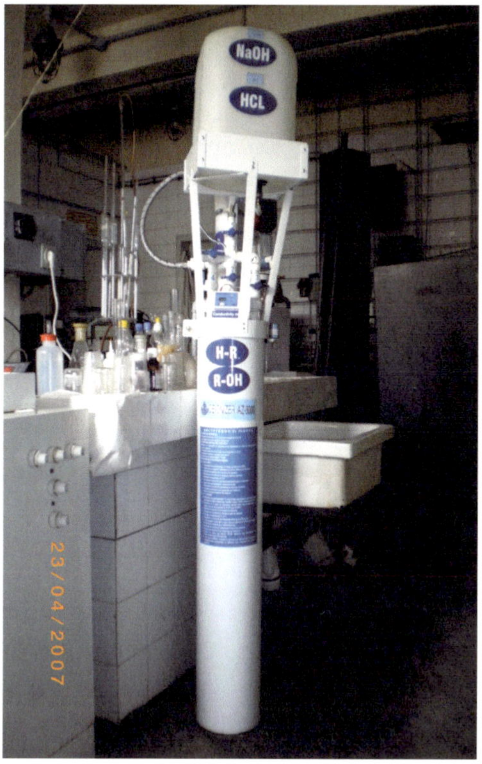

Fig. 8. Photography of a pilot one body multi stage multi cell water deionization filtration device

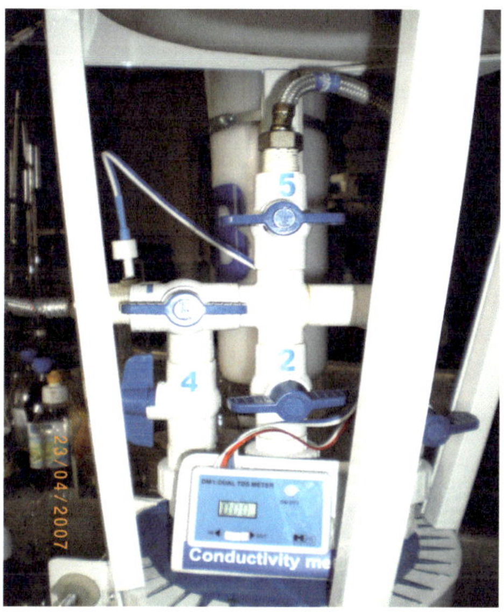

Fig. 9. Photography of a flow distributing and conductivity measuring part of the pilot one body multi stage multi cell water deionization filtration device

After evaluation different concepts to meet the Euro IV (10/2006) and Euro V (10/2009) standards for heavy–duty vehicles, the automotive industry has decided to use the Selective catalytic reduction (SCR)-technology with AUS 32 (Quality assurance Guidance on AUS 32, 2005) as a reducing agent for most of these vehicles. The SCR-technology has demonstrated the capability to reduce nitrogen oxides emissions from heavy–duty tracks in test cycles and in on the real road operation. The combination of fuel-economy-optimized engines with low particulates and SCR is able to meet the Euro IV and Euro V regulation. The mentioned above technology with AUS 32 as a reducing agent has already been applied successfully to stationary applications, waste incinerators, power plants and to mobile Diesel engines in application such as ships and locomotives.
The basic chemical reaction (Quality assurance Guidance on AUS 32, 2005) of the urea SCR-process is as follows:
AUS 32 is injected into the hot exhaust gas and is hydrolysed above apprrox.180 ^{o}C forming ammonia (NH_3) according the following reaction equation:

$$(NH_2)_2CO + H_2O \rightarrow 2NH_3 + CO_2 \quad (1)$$

At the homogeneous extruded, base metal SCR catalyst the following reaction takes place to convert nitrogen oxides with NH_3 from the hydrolysis reaction to nitrogen and water (very positive ecological waste products).

$$4\,NH_3 + NO + O_2 \rightarrow 4\,N_2 + 6\,H_2O \quad (2)$$

$$8\,NH_3 + 6\,NO_2 \rightarrow 7\,N_2 + 12\,H_2O \quad (3)$$

On the base of the design discussed above, a small local water treatment technology plant was researched for this purpose. This paper describes the technological scheme, its compo-nents and recommended production parameters in producing a ultra pure deionized water to get 32.5% aqueous urea solution AUS 32 DIN 70070 grade, to preserve the quality of the solution from production via storage, handling and distribution to the enduser. A water as a solvent plays very important role to be obtained necessary quality of the product and additionally the water properties are important also for washing procedures of product handling, storage reservoirs, distribution pipe system etc. A water takes a great deal to environmental protection aspects also. Its purity deals all participants in the distribution chain, producers of AUS 32, logistics partners, public filling and truck stations. They should draw their attention to common guidelines concerning producing, handling in such a way that the intended high quality of AUS 32 will be maintained from the point of production to the point of enduse.
The water treatment plant is designed as a combination of different water technology methods (percolation; ion-exchange softening (Type DMEaWZD); reverse osmosis; ion-exchange deionization) (Fig.11).The technological scheme is created on the base of raw water properties (Table1,2). It is a suitable decision, because of relatively high salinity of this underground water.
Because of very serious probability to obtain a dangerous scaling phenomenon on the reverse osmosis membrane surfaces, a first module on the raw water treated line is a softening module (DMEaWZD). It is an automatically controlled duplex type ion-exchange softening plant (Fig.11). After that, the water flow is treated by reverse osmosis module and then is finally conditioned by a filtration trough multi process mixed-functioning module (H-cation exchange / OH–anion exchange; adsorption processes). This module has a design (Zvezdov A.,1997) shown on (Fig. 1, 8,13)

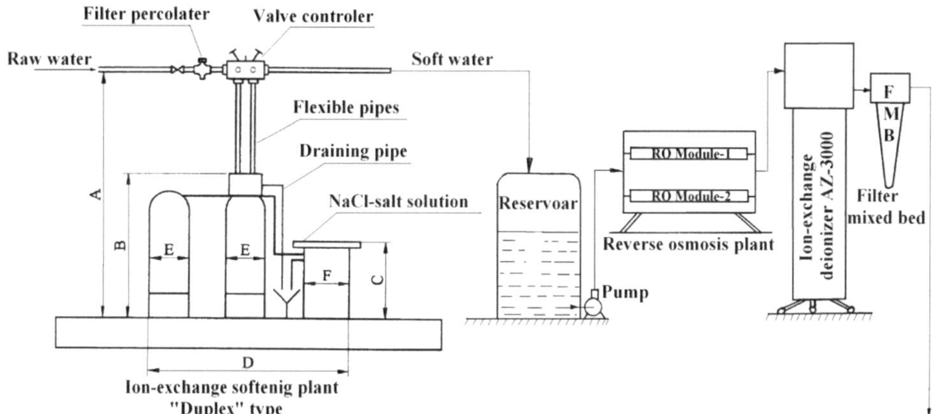

Type WZD	A Conec-tion height	B Assem-bly height	C Salt rese-rvoar height	D Assem-bly width	E Filter body diamet.	F Salt reze-rvoar di-meter	Assem-bly depth	Wei-ght	Capa city
	mm	mm	mm	mm	mm	mm	mm	kg	m^3/h
0100150060	1432	1082	677	1300	184	487	500	130	1

Fig. 11. A technological scheme for ultra pure deionized water production used as a solvent for 32.5% aqueous urea solution AUS 32 DIN 70070 grade

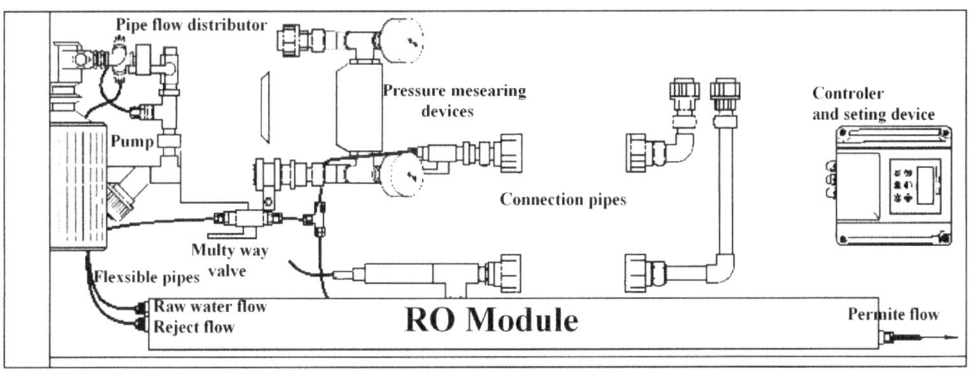

Fig. 12. Reverse osmosis module included into the technological scheme for ultra pure deionized water production

Such designs are suitable for use in the technological water conditioning plants after membrane water treatment modules as a point-of-use final systems (typical finally used module for a deep deionized water-product). Present investigation is an attempt to include a simple and easily used ion-exchange water treatment device in the technological plant for cost effective production of aqueous urea solution AUS 32 DIN 70070 grade 32 (Quality assurance Guidance on AUS 32, 2005). This trend utilizes a combination of technologies including membrane processes, ion-exchange, adsorption, etc. That sets our device for the unique process which provides a constant supply of high purity deionized water and is more cost-effective and convenient for the user.

The properties of raw water used for a production of solvent in AUS 32 production processes are presented in Tables 1 and 2. The schemes for water treatment plants are shown on Fig.11 and 12 with all main technological components. The water treatment plant includes the created new type small ion-exchange water deionizer (Zvezdov A.,1997; (Zvezdov A.,1999) (fig.1, 13).The regular reagent-regeneration processes of ion-exchange beds into main filtration deionizer body are necessary. The reagent flows pass through the filter under gravitation (see Fig.1).The presence of an operator, control or monitoring are not necessary during this process. The device is designed and constructed so as the reagent hydrostatic pressure from the regeneration reservoir drives the two regenerating solutions through the ion-exchange beds at linear velocities suitable for the kinetic of regeneration processes and the flows are drained after their mixing in a neutralizing collection module (it was explaned above). Technological project calculations were carried out and then a technological scheme for ultra pure deionized water production was realized. A pilot experimental water treatment plant was made and series of pilot experiments were carried out in two alternative ways : with one or with two FMB modules. The deionized water produced was used as a solvent for 32.5% aqueous urea solution AUS 32 DIN 70070 grade. The R-H and R-OH ion-exchange modules and FMB modules were loaded with polyreticular ion-exchange resins to reduce humus and natural organic impurities from the treated water. The specific electrical conductivity of the water produced was measured by a flow type conductance measuring apparatus equipped with two sensors. When the capacity of the device is exhausted, its functionality as a deionizer is recovered by treatment of the ion-exchange beds with reagent regeneration solutions as follows (see Fig.1,Fig 8 and Fig.13):

- a solution of 5÷8% hydrochloric acid is passed through the exhausted ion-exchange R-H beds;
- a solution of 5÷6% sodium hydroxide is passed through the exhausted ion-exchange R-OH beds;
- the FMB module is filled with a mixture of a strong acid cation-exchanger and a strong base anion-exchanger(type MB), preliminarily regenerated into H^+ and OH^- ion forms, respectively *;
- a possible alternative is to use the central cell of main iltration body as a sorption module charged with active carbon (Fig.1) in order to achieve the requested value of permanganate demand of the water obtained required by BDS ISO 3696 and DIN 70070 (there is such option instead of additional adapted separate module after mixed bed (FMB) ion-exchange module).

When the periods of washing fillers from the residual reagents and waste regeneration products are completed, the ion-exchange device switches to working period (according to technological requirements - when the target specific electrical conductivity measured by the flow conductivity meter is reached). The results obtained are presented in Table 3.

The results in Table 3 show that the properties of the water produced by the small water deionization device comply with the requirements for water with primary or secondary degree of purity (BDS ISO 3696,1993;ASTM 1193, 1999; DIN 70070) Therefore, it was concluded that the product can be used as a solvent of urea to produce AUS 32; for usual laboratory needs, as well as for more precise laboratory analyses both for laboratory and small industrial purposes.

Fig. 13. Photography of a multi process ion-exchange and sorption small water conditioning device (Zvezdov A.,1997; Zvezdov A.,1999; Zvezdov A., et al.2006)

№	**Parameter**	**Dimention**	**Stand. method**	**Results obtained**	**Decree 9/2001**
	Gustatory properties				
1	*Smell*	-	*BDS 8451-77*	absent	*acceptable*
2	*Appearance*	-	*BDS 8451-77*	*Visually clear*	*acceptable*
3	*Transparency*	-	*ISO 7026603*	*38 cm*	*Acceptable*
	Physical and physic-chemical properties				
4	*Taste*	-	*BDS 8451-77*	*no bad taste.*	*acceptable*
5	*COLOUR*		*BDS 8451-77*	*acceptable*	*acceptable*
6	*Temperature*			$20{,}4^{O}C$	
7	*pH*	*pH units*	*BDS 3423-81*	*7,6*	*6,5÷9,5*
8	*Turbidity*	mg/dm^3	*ISO70266-03*	*2,4*	*acceptable*
9	*Plug. abbility*	-	-	*K= 0,12(weakly)*	-
10	*Perm. demand*	$mg\ O_2/dm^3$	*BDS 3413-77*	*3,2*	*≤5,0*
11	*Dry rest*	g/dm^3	*BDS171408-77*	*0,76*	*≤0,030*

Table 1. Data on gustatory properties, physical and physic-chemical properties of raw water dealing technological creation of water treatment plant

№	Parameter:	Dimention	Stand. method	Results obtained	Decree 9/2001
			Cation content:		
1	*Total hardnes*	*mg-eq/dm³ (°H)*	*BDS ISO 6059-02*	*8,87 (24,84)*	*≤ 12 mg-eq/dm³*
2	*Carbonate hardnes*	*mg-eq/dm³ (°H)*	*BDS ISO 6059-02*	*4,8 (13,44)*	-
3	*Noncarbonate hardnes*	*mg-eq/dm³ (°H)*	*BDS ISO 6059-02*	*4,07 (11,4)*	-
4	Ca^{2+}	*mg-eq/dm³ (mg/dm³)*	*BDS ISO 6058-02*	*6,83 (136,87)*	*≤ 150 mg/dm³*
5	Mg^{2+}	*mg-eq/dm³ (mg/dm³)*	*BDS ISO 6059-02*	*2,04 (24,83)*	*≤ 80 mg/dm³*
6	$Na^{+}+K^{+}$ *cations*	*mg-eq/dm³ (mg/dm³*	-	*1,98 (45,53)*	-
7	*Total Fe*	*µg/dm³*	*ISO 6332-02*	*0,11)*	*200 µg/dm³*
8	Mn^{2+}	*µg/dm³*	*ISO 6333-02*	*UMAL*[1]	*≤ 50 mg/dm³*
9	Cu^{2+}	*mg/dm³*	*BDS 3770-80*	*UMAL*[1]	*≤ 2,0 mg/dm³*
10	NH_4^{+}	*mg/dm³*		*UMAL*[1]	*≤ 0,5 mg/dm³*
11	*TCC* [2]	*mg-eq/dm³*	-	*10,85*	-
			Anion content:		
12	HCO_3^{-}	*mg-eq/dm³ mg/dm³*	*M3-85*	*4,8 (292,9)*	-
13	CO_3^{2-}	*mg/dm³*	-	*UMAL*[1]	-
14	SO_4^{2-}	*mg-eq/dm³ mg/dm³*	*BDS 3588-77*	*1,79 (85,97)*	*≤ 250 mg/dm³*
15	Cl^{-}	*mg-eq/dm³ mg/dm³*	*BDS 3414-80*	*3,21 (113,89)*	*≤ 250 mg/dm³*
16	NO_3^{-}	*mg-eq/dm³ mg/dm³*	*BDS 3758-85*	*0,049 (3,04)*	*≤ 50 mg/dm³*
17	NO_3^{-}	*mg-eq/dm³ mg/dm³*	*BDS EN 26777-97*	*UMAL*[1]	*≤ 0,5 mg/dm³*
18	$HSiO_3^{-}$ *as* SiO_3^{2-}	*mg-eq/dm³ mg/dm³*	-	*0,78 (23)*	-
19	F^{-}	*mg-eq/dm³ mg/dm³*	-	*UMAL*[1]	*≤ 1,5*
20	PO_4^{2-}	*mg-eq/dm³ mg/dm³*	*BDS EN ISO 6878-04*	*UMAL*[1]	*≤ 0,5*
	TAC [3]	*mg-eq/dm³*	-	*10,779*	-
	TDS [4]	*mg/dm³*	-	*731,32*	-
4	*Spec. el. cond.*	*µS/ cm*	*BDS 27888-93*	*898*	*≤ 2000*

[1] UMAL–under minimal analytical level; [2] TCC-total cation content; [3] TAC-total anion content; [4] TDS-total dissolved salts;

Table 2. Data on chemical properties and salt content of raw water

Parameters	**Purity**			**Purity**									
	By BDS ISO 3696			With one FMB modules, bed vol.					With two FMB modules, bed vol.				
	I	*II*		*I*	*II*		*I*	*II*		*I*	*II*		*I*
pH at 25°C	Not incl.	Not incl.	pH at 25°C	Not incl.	Not incl.	pH at 25°C	Not incl.	Not incl.	pH at 25°C	Not incl.	Not incl.	pH 25°C	Not incl.
Spec.electric conductivity, μs/cm³	0,1	1	5	0,53	0,49	0,51	0,6	0,46	0,12	0,09	0,10	0,08	0,09
Permanganate oxygen demand,[1] mgO_2/dm^3	Not incl.	0,08	0,4	0,08	0,07	0,07	0,06	0,07	0,06	0,06	0,07	0,06	0,06
Absorption at 254 nm cuvete - 1 cm	0,001	0,01	Not incl..	0,01	0,01	0,01	0,01	0,01	0,01	0,01	0,01	0,01	0,01
Dry residue at 110°C, mg/dm^3	Not incl.	1	2	0,9	0,9	0,9	0,9	0,9	0,8	0,8	0,9	0,7	0,8
SiO_2 μg/dm³	10	20	Not incl.	14	16	15	16	12	8	9	17	19	11

[1] Permanganate demand of the water produced was measured by samples taken from the flow after the final adsorption module charged with active carbon

Table 3. Data on standard parameters and purity of deonized water produced dealing technological functionality of water treatment plant created

4. Conclusion

A water treatment device is designed to realize different processes in one filtration body. It is a multistage multiprocess filter. It can be used for different processes (filtration, ion-exchange, active carbon sorption, biopolymer-chitosan sorption, etc.) Experimentally, the device is researched for a deionized water production using ion-exchange resins and active carbon sorption methods in one body.

The device`s features and price are minimized, but the technological behaiviour of the device is very good. It produces secondary and third grade deionized water in accordance with BDS ISO 3696;ASTM D 1193-99]; DIN 70070 from a potable source for laboratory and other small industrial applications.

A small local water treatment plant for high quality urea solution AUS 32 defined by DIN 700070 is also created. The technological parameters of this plant are researched during the ultra pure water production used as a solvent for urea solution AUS 32. The water treatment plant is designed as a combination of different water technology methods (percolation; ion-exchange softening; reverse osmosis; ion-exchange deionization.

The properties of the deionized water produced are monitored and controlled by two built-in flow sensors providing signals to conductometric digital gauge measuring the specific electrical conductivity of the product after the main ion-exchange module or after the mixed bed (FMB) module, depending on the user`s wish.
The cost of the device is much chipper than the market available deionizers, but the operating expenses are much lower and the device allows easy control of the processes, technological maintenance and repair, as well as much longer technological life.

5. References

Adjarov G. (1998) *Device for ultra pure water,* Sofia (private announ-cement)

Alekin O. A., *The base of hydrochemistry,* Ed.by"Hydrometeoizdat", Peterburg .(1970)

ASTM D 1193-99 (1999) *Annual Book of ASTM Standards,* Vol. 11

BDS ISO 3696 "Water for LaboratoryAnalyses"(1993)"*Standardization*"Publisher, Sofia

Catalogue of Prospect materials US ELGA Corp. (2008)

Catalogue of Prospect materials Purity Ltd.(2007)

Catalogue of Prospect materials CHROMSERVIS s.r.o., 1300 Praha 3, Chehia, (2003)

Dobrevsky Iv., A. Zvezdov (1999) *Proceed. of 23 th Microsymposium Selective Pollimeric Sorbents,* Prague, p.M18-1-5.Chehoslovakia

Dobrevsky Iv., *Water treatment technology – Part 1,* Ed. by "Technika" Sofia (1991)

Kunin R.(1968) *Desalination,* vol. 4, p.38

Makroporose Anioneaustauscher Einsatz (1970) *Auswahl und Schaltung,* Leverkusen, Bayer-AG, Best. № 1 ,1411

Petkov, K.T.,A.Zvezdov, St.Pavlova, D.Zvezdova (2007) *Journ. of Env. Protection and Ecology,* 8, Part 1, p.43-49

Petkov K.T., A. Zvezdov, D. Zvezdova (2007) *Journ.of Env. Protection and Ecology,* 8, Part1, p.188-196

Plehov G.,1997) *Method for purifi-cation of water and device for its implementation,* Pat.Appl.95100104/02, Russia, MPK C 1251/14, 5.1995, Bull. 10

Quality assurance Guidance Document on AUS 32 according to DIN 70070, p.121, (2005)

Zvezdov A., D. Zvezdova, Ecologically and economically expedent small water treatment device 3978, *Proceed. of 10-th International Symposium Ecology 2001,* p.444-49

Zvezdov A., D. Zvezdova, Kr. Petkov (2006) Deionization of water for small factories or laboratories by four-stage ion-exchange device, *Desalination,* 189, p. 243-247

Zvezdov A., (1997), *Pat. BG 67 Y 1 B 01 D 24 /, 12 C02 F 1/42*

Zvezdov A.,(1999), *Pat. BG 234 Y 1, C 02 F 1/42*